AF791492

THERMAL STRESS ANALYSIS OF COMPOSITE BEAMS, PLATES AND SHELLS

THERMAL STRESS ANALYSIS OF COMPOSITE BEAMS, PLATES AND SHELLS

Computational Modelling and Applications

Erasmo Carrera
Politecnico di Torino,
Department of Mechanical and Aerospace Engineering,
Italy

Fiorenzo A. Fazzolari
University of Cambridge,
Department of Engineering,
United Kingdom

Maria Cinefra
Politecnico di Torino,
Department of Mechanical and Aerospace Engineering,
Italy

ELSEVIER

AMSTERDAM • BOSTON • HEIDELBERG • LONDON
NEW YORK • OXFORD • PARIS • SAN DIEGO
SAN FRANCISCO • SINGAPORE • SYDNEY • TOKYO

Academic Press is an imprint of Elsevier

ACADEMIC PRESS

Academic Press is an imprint of Elsevier
125 London Wall, London EC2Y 5AS, United Kingdom
525 B Street, Suite 1800, San Diego, CA 92101-4495, United States
50 Hampshire Street, 5th Floor, Cambridge, MA 02139, United States
The Boulevard, Langford Lane, Kidlington, Oxford OX5 1GB, United Kingdom

Notices

Knowledge and best practice in this field are constantly changing. As new research and experience broaden our understanding, changes in research methods, professional practices, or medical treatment may become necessary.

Practitioners and researchers must always rely on their own experience and knowledge in evaluating and using any information, methods, compounds, or experiments described herein. In using such information or methods they should be mindful of their own safety and the safety of others, including parties for whom they have a professional responsibility.

To the fullest extent of the law, neither the Publisher nor the authors, contributors, or editors, assume any liability for any injury and/or damage to persons or property as a matter of products liability, negligence or otherwise, or from any use or operation of any methods, products, instructions, or ideas contained in the material herein.

Library of Congress Cataloging-in-Publication Data
A catalog record for this book is available from the Library of Congress

British Library Cataloguing-in-Publication Data
A catalogue record for this book is available from the British Library

ISBN: 978-0-12-849892-7

For information on all Academic Press publications
visit our website at https://www.elsevier.com

Publisher: Joe Hayton
Acquisition Editor: Brian Guerin
Editorial Project Manager: Carrie Bolger
Production Project Manager: Mohanapriyan Rajendran
Designer: Matthew Limbert

Typeset by VTeX

CONTENTS

ABOUT THE AUTHORS

Erasmo Carrera

After earning two degrees (Aeronautics, 1986, and Aerospace Engineering, 1988) at the Politecnico di Torino, Erasmo Carrera received his PhD in Aerospace Engineering in 1991, jointly at the Politecnico di Milano, Politecnico di Torino, and Università di Pisa. He began working as a Researcher in the Department of Aeronautics and Space Engineering at the Politecnico di Torino in 1992 where he held courses on Missiles and Aerospace Structures Design, Plates and Shells, and the Finite Element Method, and where he has been Professor of Aerospace Structures and Aeroelasticity since 2000. He has visited the Institut für Statik und Dynamik, Universität Stuttgart twice, the first time as a PhD student (6 months in 1991) and then as Visiting Scientist under a GKKS Grant (18 months from 1995). In the summer of 1996, he was Visiting Professor at the ESM Department of Virginia Tech. He was also Visiting Professor for two months at SUPMECA, Paris, in 2004. His main research topics are: composite materials, FEM, plates and shells, postbuckling and stability, smart structures, thermal stress, aeroelasticity, multibody dynamics, non-classical lifting systems and multifield problems. Professor Carrera has made significant contributions to these topics. In particular, he proposed the Carrera Unified Formulation to develop hierarchical beam/plate/shell theories and finite elements for multilayered structure analysis as well as the generalization of classical and advanced variational methods for multifield problems. He has been responsible for various research contracts with the EU and national and international agencies/industries. Presently, he is Full Professor and Deputy Director of his department. He is the author of more than 500 articles, many of which have been published in international journals. He serves as a referee for many journals, such as Journal of Applied Mechanics, AIAA Journal, Journal of Sound and Vibration, International Journal of Solids and Structures, International Journal of Numerical Methods in Engineering, and as contributing editor for Mechanics of Advanced Materials and Structures. He has also served on the Editorial Boards of many international conferences.

Fiorenzo A. Fazzolari

Fiorenzo A. Fazzolari received his BSc and MSc in Aerospace Engineering from the Polytechnic of Turin in 2007 and 2010, respectively. In his MSc thesis, which led to two journal articles, he was principally involved in the development of spectral methods for static and dynamic analysis of multilayered composite plates and shells. He spent six months at San Diego State University in partial fulfilment of his dissertation work. In January 2011 he joined the structural dynamics and aeroelasticity group at City University London as a PhD candidate to work on the doctoral programme, *Dynamic Stiffness Modelling of Composite Plate and Shell Assemblies*, funded by the American Air Force Research Laboratory (AFRL). At the end of the same year he was awarded with the title of MPhil (Master of Philosophy). He successfully completed his PhD in 2013. In March 2014 he was appointed as a Research Fellow at the University of Southampton to work within the Computational and Engineering Design (CED) research group on the *ReBioStent* project (Reinforced Bioresorbable Biomaterials for Therapeutic Drug Eluting Stents) funded by the European 7*th* Framework Programme for Research and Technological Development. His work was manly focused on the advanced modelling and design of stent prototypes for cardiovascular applications and on the nonlinear analysis of composite biomaterials. Since February 2016 he has been working in the Dynamics and Vibration Research Group (DVRG) at the University of Cambridge as a Research Associate. He is now involved in the *Engineering Nonlinearity* project funded by the Engineering and Physical Sciences Research Council (EPSRC). His research is primarily related to the theoretical and computational modelling of complex nonlinear dynamical systems. His main research interests include: dynamic stiffness formulation for structural elements, advanced modelling of metallic and laminated composite structures: beams, plates and shells, structural dynamics, coupled and uncoupled thermoelasticity, thermal stress analysis of metallic and composite structures, multifield analysis for multilayered structures, development of unconventional variational principles, nonlinear vibration and Statistical Energy Analysis (SEA). He has authored more than 60 publications, including international journal and conference papers, reports, books and book chapters. He serves as a referee for more than 30 international journals, such as Composite Structures, European Journal of Mechanics - A/Solids, Composites Part B: Engineering, Journal of Sound and Vibration and Journal of Thermal Stresses, amongst others. He is Associate Editor of *Shock and Vibration* (ISSN: 1070-9622) and *Mathematical Problems in Engineering* (ISSN: 1024-123X).

Maria Cinefra

Since July 2015, Maria Cinefra is Associate Professor at the Department of Mechanics and Aerospace Engineering of Politecnico di Torino. After earning two degrees (Bachelor, March 2007, and Master, December 2008) at the Politecnico di Torino, she was enrolled in a Ph.D. (from January 2009 to April 2012) under the supervision of Prof. Carrera at Politecnico di Torino and a foreign co-advisor, Prof. Olivier Polit at the University of Paris Ouest Nanterre. Her research project, related to the "Thermo-mechanical design of multi-layered plates and shell embedding FGM layers", was funded by the Fonds National de la Recherche of Luxembourg and it was performed in collaboration with the CRP Henri Tudor of Esch (Lux). She was awarded for the best Ph.D paper (Ian Marshall's Award) at the 16th International Conference on Composite Structures (28–30 June 2011, Porto, Portugal). She was involved in a collaboration with the Department of Mathematics of Università di Pavia in order to develop an advanced shell finite element based on the Unified Formulation for the analysis of structures made of composite, functionally graded and piezo materials. Contemporaneously, she collaborated with Professor Ferreira, editor of the Journal "Composite Structures", about the meshless method "Radial Basis Functions" combined with the Unified Formulation for the analysis of advanced structures. Since 2010, she holds teaching activity at the Politecnico di Torino in different courses at Bachelor and Master levels: "Fundamentals of Structural Mechanics", "Aeronautic Legislation, human factors and safety", "Non-linear analysis of structures", "Structures for spatial vehicles" and "Aeroelasticity". She held also a Ph.D. course at Università del Salento with the title "1D and 2D models for the analysis of advanced material structures with multi-field properties through analytical and numerical methods". Her research topics cover: composite materials, shell finite elements, FE analysis, meshless methods, smart structures, functionally graded materials, thermal stress analysis, multifield interaction, panel flutter, advanced kinematic theories for plates and shells, mixed variational methods, local-global methods, failure analysis of laminated structures, non-linear problems. M. Cinefra is author and co-author of about 60 papers on the above topics, most of which have been published in first rate international journals, as well as a recent book published by J Wiley & Sons with the title "Finite Element Analysis of Structures through Unified Formulation". Cinefra's papers have had about 1000 citations with h-index=17 (data taken from Scopus). She was invited to hold a plenary talk in the international conference ICCS18 (Lisbon, June 2015) about "FE Shell Formulations for Layered Composite Structures". She made a contribution as reviewer to about ten international peer-reviewed journals and was on the Editorial Board of some international conferences. M. Cinefra was co-organizer of the ICMNMMCS (Torino, June 2012, chaired by Prof. Carrera and Prof. Ferreira) and of the ECCOMAS SMART 13 conference (Torino, June 2013, chaired by Prof. Carrera). Professor Cinefra is a member of the MUL2 group at the Politecnico di Torino. The MUL2 group is considered one of the most active research teams at the Politecnico; it has acquired a significant international reputation in the field of multilayered structures subjected to multifield loadings.

PREFACE

Since the first article on thermoelasticity proposed by J.M.C. Duhamel [1], where the derivation of the governing equations for the coupling between the temperature field and the continuum elastic deformation was proposed for various boundary value problems, a significant amount of contributions have been given in this field, some remarkable books can be found in Refs. [2–8] amongst many others that have been proposed all along the book. Nowadays, thermoelasticity is an established topic with a consistent theoretical development. However, the authors deemed useful to write this book in order to elucidate and shed light on some more subtle aspects that rise when dealing with thermal stresses of laminated composite and functionally graded material beam, plate and shell structures.

The main purpose of this book is to propose a comprehensive study of thermal stresses related to the most advanced composite structures, while providing the fundamentals of thermoelasticity and its basic governing equations. This approach has been deliberately applied by the authors in order to make the book self-contained without the need from readers to consult other sources and references.

The book has been subdivided in three main parts. In the first part the fundamentals of thermoelasticity have been briefly recalled, various sample problems ranging from 1-D to 3-D thermoelasticity and for both steady-state and transient conditions have been provided. The second part has been entirely devoted to the derivation of the weak- and strong-form of the governing equations for the pure mechanical case, and for both uncoupled and coupled thermoelasticity. In particular, the governing equations have been obtained by using advanced variational principles combined with refined beam, plate and shell theories generated by using the Unified Formulation (UF) proposed by the first author and significantly enhanced by the second. In the same part, a significant development of both exact and approximate solution techniques has been provided.

Finally, the third and last part of the book has been completely focused on the application of the theoretical models, previously derived, on some opportunely selected thermal stress problems. Attention has been mainly given at the importance of using refined quasi-3D beam, plate and shell formulations when coping with advanced thermal structures. In this respect, the solutions of complex thermoelastic problems for both metallic and composite structures have been proposed.

The book is intended for a wide range of readers who are directly or indirectly interested in the subject matter, including graduate and PhD students, research fellows/associates as well as professors in various universities around the word but also scientists and engineers involved in government and industrial research activities.

REFERENCES

1. Duhamel JMC. Second mémoire sur les phénomènes thermomécaniques. J Éc Polythech 1837;15(25):1–57.
2. Nowacki W. Thermoelasticity. First edition. New York: Pergamon Press; 1962.
3. Nowinski JL. Theory of thermoelasticity with applications. First edition. Netherlands: Springer; 2011 [reprinting].
4. Parkus H. Thermoelasticity. Second edition. New York (USA): Springer-Verlag; 1976.
5. Hetnarski R, Eslami RM. Thermal stresses – advanced theory and applications. First edition. Springer; 2008.
6. Thorton EA. Thermal structures for aerospace applications. First edition. Reston (Virginia, USA): AIAA Education Series; 1996.
7. Boyle BA, Weiner JH. Theory of thermal stresses. First edition. New York (USA): John Wiley & Sons; 1960.
8. Carlson DE. Linear thermoelasticity. In: Encyclopedia of physics, mechanics of solids. Berlin: Springer; 1972. p. 297–345.

INTRODUCTION

THERMAL STRUCTURES AND THEIR APPLICATIONS

From a theoretical standpoint the first studies of the thermoelasticity and thermal stresses were proposed by J.M.C. Duhamel in a paper published in 1837 [1]. Following Duhamel's paper many other contributions were given in the nineteenth century [2–7] and the first decades of the twentieth century [8–13] in order to both derive the complete set of the governing thermoelasticity equations and provide a number of solutions in the case of specific problems, more references can be found in Hetnarski and Eslami [14]. Several other papers of noteworthy interest were published before the World War II some of them can be found in Ref. [15]. In the late 1940s increased the need to understand the structures behaviour when subjected to thermal loads. The World War II acted as a catalyst in this respect. Indeed, the majority of the new produced technologies required a significant amount of research in the thermal stresses field. In the aftermath of the World War II, by means of the Bell X-1 program, was proved the capability to fly faster than the speed of sound of manned aircraft. Further enhancement of the Bell X-1 led to the X-1B. Thereafter, the aerospace research moved towards higher and higher supersonic speeds, an example of high supersonic speed aeroplane is the SR-71 blackbird shown in Fig. 1, it was long-range strategic reconnaissance aircraft, able to reach speeds higher than Mach 3. Therefore, phenomena like aerothermal heating became of primary importance, leading researchers and designers to introduce the concept of thermal barrier coating (TBC) for aerospace structures.

Turbomachinery blade is a further aerospace application in which the use of the TBC is mandatory. This is due to the harsh environment condition in which the temperature can be extremely high. Figure 2 shows a typical turbine blade with TBC, it must be noted the significant temperature reduction provided by the TBC. Coatings are used throughout the gas turbine to improve and enhance surface properties of bulk materials that do not have the combination of all the required characteristics demanded by the operating conditions. The turbine blade coatings enable the turbine materials to withstand the very high combustion chamber exit temperatures. For instance, the turbine blades present in the *Trent Engine* produced by Rolls-Royce are made of a single crystal of nickel super alloy which has a melting point of 1350 °C. Each high pressure (HP) turbine blade has to

Figure 1 Lockheed SR-71.

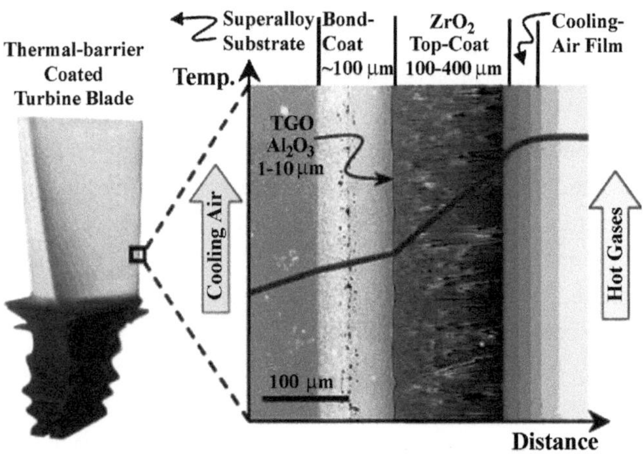

Figure 2 Temperature distribution through a cross-section of a turbine blade with thermal barrier coating.

work in temperatures up to 1600 °C while spending 10000 rpm, which is the reason why they are coated in a ceramic TBC to protect them from the hot gas temperature. In particular, the TBC on HP turbine blades provides a good example of a complex multi-step coating process. A bond coat is applied by electroplating or thermal spray, followed by heat treatment and then finally, the application of a zirconia-based ceramic coating by EB-PVD or thermal spray.

The combustion chamber represents an other jet engine component in which the temperatures are the hottest in the engine. Fuel is burned in the combustion chamber at temperatures of over 2000 °C yet they are still expected to last the many hours of operation between overhauls. This requires substantial technology in the form of wall materials (both metallic and ceramic), coatings and manufacturing technology. In particular, the TBC is sprayed on the wall until create a thickness commonly of the order of 200–300 μm.

An advanced Thermal Protection System (TPS) is also required during the re-entry procedure of space vehicles. The Orbiter's external surface reached extreme temperatures up to 1648 °C (3000 °F). The TPS was designed to provide a smooth, aerodynamic surface while protecting the underlying metal structure from excessive temperature. The loads endured by the system included launch acoustics, aerodynamic loading and associated structural deflections, and on-orbit temperature variations as well as natural environments such as salt fog, wind, and rain. In addition, the TPS had to resist pyrotechnic shock loads as the Orbiter separated from the External Tank (ET). The TPS consisted of various materials applied externally to the outer structural skin of the Orbiter to passively maintain the skin within acceptable temperatures, primarily during the re-entry phase of the mission. During this phase, the TPS materials protected the Orbiter's outer skin from exceeding temperatures of 176 °C (350 °F). In addition, they were reusable for 100 missions with refurbishment and maintenance. These materials performed in temperatures that ranged from −156 °C (−250 °F) in the cold soak of space to re-entry temperatures that reached nearly 1648 °C (3000 °F). The TPS also withstood the forces induced by deflections of the Orbiter airframe as it responded to various external environments. At the vehicle surface, a boundary layer developed and designed to be laminar-smooth and non-turbulent fluid flow. However, small gaps and discontinuities on the vehicle surface could cause the flow to transition from laminar to turbulent, thus increasing the overall heating. Therefore, tight fabrication and assembly tolerances were required of the TPS to prevent a transition to turbulent flow early in the flight when heating was at its highest. Requirements for the TPS extended beyond the nominal trajectories. For abort scenarios, the systems had to continue to perform in drastically different environments. These scenarios included: Return-to-Launch Site; Abort Once Around; Transatlantic Abort Landing; and others. Many of

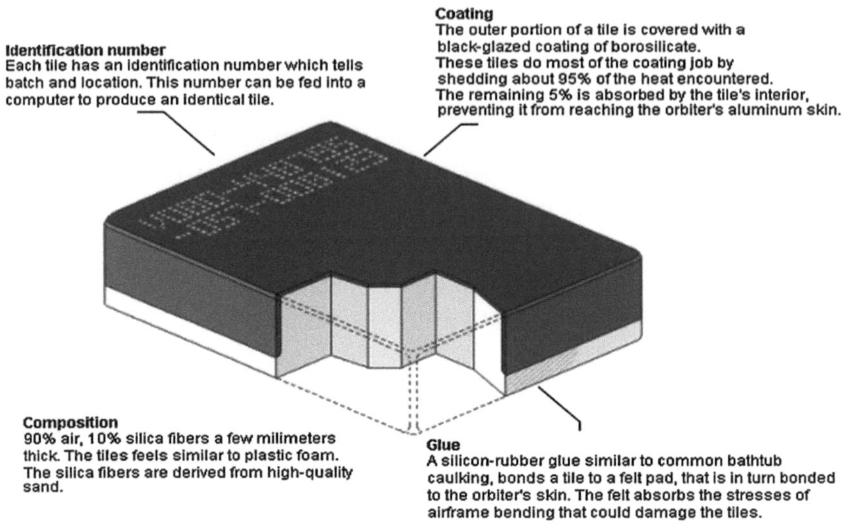

Figure 3 Temperature distribution through a cross-section of a turbine blade with thermal barrier coating.

these abort scenarios increased heat load to the vehicle and pushed the capabilities of the materials to their limits. Several types of TPS materials were used on the Orbiter. These materials included tiles, advanced flexible reusable surface insulation, reinforced carbon-carbon, and flexible reusable surface insulation. All of these materials used high-emissivity coatings to ensure the maximum rejection of incoming convective heat through radiative heat transfer. Selection was based on the temperature on the vehicle. In areas in which temperatures fell below approximately 1260°C (2300°F), NASA used rigid silica tiles or fibrous insulation. At temperatures above that point, the agency used reinforced carbon-carbon. An example of tiles used in the TPS are the high-temperature reusable surface insulation tiles shown in Fig. 3 made of black borosilicate glass coating that had an emittance value greater than 0.8 and covered areas of the vehicle in which temperatures reached up to 1260°C (2300°F).

Civil engineering structures are subjected to long-term thermal effects induced by solar radiation and ambient air temperature. The structural components may receive heat energy from the direct radiation, diffuse radiation, and reflected radiation caused by seasonal and daily temperature changes. The time varying thermal loadings in the structure may induce structural deformations and thermal stresses which, due to their indeterminacy, may cause the damage of the components or even the collapse of the entire structure. Therefore, the thermal loading and thermal effects have attracted great attention across the world for the past decades. The thermal effects on civil engineering structures have been investigated to simulate the temperature distribution and predict the structural responses by using solar radiation model and heat transfer analysis with the aid of finite element method (FEM).

Elevated thermal stresses are the main responsible to the limited service life of microelectronics and photonic structures, packages and systems. Indeed, thermal stresses and strains can result in the functional failures, such as, in the loss of the electrical or optical performance of a component or device. For instance, transistor junction failure can occur due to the elevated thermal stress in the integrated circuit, if the heat produced by the chip cannot readily escape. Optical performance failure occurs, when the lateral thermally induced displacement in the gap between two light guides exceeds the allowable limit. In laser packages, such displacements can be due also to thermal stress relaxation in a laser weld. The ability to understand the sources of thermal stresses and displacements, predict their distribution and the maximum values, and possibly minimize them is of clear practical importance.

ADVANCED STRUCTURAL THEORIES IN THE MODELLING OF THERMAL STRESS PROBLEMS

As already mentioned, the primary aim of the this book is to provide new insights in the thermal stress analysis of laminated composite and FGM beam, plate and shell structures. To introduce the discussion that will be broadly enriched in the following chapters, let us consider a temperature distribution $T(x, y, z)$ in the plate domain V (structural element volume). The temperature $T\left(x, y, z\right)$ when acting on a structural element is a field load and by exploiting the separation of variable technique can be written in the following form

$$T\left(x, y, z\right) = T_p\left(z\right)\; T_\Omega\left(x, y\right) \tag{1}$$

where $T_p\left(z\right)$ represents the temperature profile across the plate thickness coordinate z while $T_\Omega\left(x, y\right)$ is the temperature distribution over the reference surface domain Ω. In order to show which are the implications, in the modelling of thermal stress problems, of the different forms of the temperature profile $T_p\left(z\right)$, some considerations and examples are needed.

The use of classical structural theories based on a displacement approach such as those developed for beam structures by da Vinci-Euler-Bernoulli [16–18] and Timoshenko [19]

$$\underbrace{\begin{aligned} u_z &= u_{z_0} + x\,\frac{\partial u_{z_0}}{\partial x} \\ u_x &= u_{x_0} \end{aligned}}_{\text{da Vinci-Euler-Bernoulli beam theory}} \qquad \underbrace{\begin{aligned} u_z &= u_{z_0} + x\,u_{z1} \\ u_x &= u_{x_0} \end{aligned}}_{\text{Timoshenko beam theory}} \tag{2}$$

and the analogous models for plate structures provided by Kirchhoff-Love [20,21] and Reissner-Mindlin [22,23]

$$\underbrace{\begin{aligned} u_x &= u_{x_0} + z\,\frac{\partial u_{z_0}}{\partial x} \\ u_y &= u_{y_0} + z\,\frac{\partial u_{z_0}}{\partial x} \\ u_z &= u_{z_0} \end{aligned}}_{\text{Kirchhoff-Love plate theory}} \qquad \underbrace{\begin{aligned} u_x &= u_{x_0} + x\,u_{x1} \\ u_y &= u_{y_0} + x\,u_{y1} \\ u_z &= u_{z_0} \end{aligned}}_{\text{Reissner-Mindlin plate theory}} \tag{3}$$

have been successful for decades in the case of mechanical problems. Nonetheless, the above defined kinematic models for beams and plates, although widely used in the analysis of thermal structures, are not suitable to deal with thermal stress problems. The reason of that lies within the axiomatic assumptions made to describe the kinematics of the structural element under investigation. More specifically, as will become more clear from what follows, the main drawback is in the assumption of constant distribution of the transverse displacement u_z in the z direction of the plate. To elucidate this issue the following examples, which take into account various temperature distributions through-the-thickness plate direction, are proposed.

- Case I. $T_p(z)$ has constant distribution in the thickness direction
  The analyzed plate is considered made up of isotropic materials and heated by a through-the-thickness constant distribution

$$T_p(z) = T_0 \tag{4}$$

The associated in-plane and transverse normal thermal strains are

$$\varepsilon^{\theta}_{xx} = \varepsilon^{\theta}_{yy} = \varepsilon^{\theta}_{zz} = \alpha\,T_0 \tag{5}$$

where α is the coefficient of thermal expansion and subscript T underlines that the written strains represent the contribution coming from the temperature loadings. In the most general case, in fact, strains are also due to geometrical (constraints) and/or mechanical (forces) boundary conditions. The following statement can be made:

- The order of magnitude of transverse thermal strains $\varepsilon^{\theta}_{zz}$ is the same as that of the in-plane $\varepsilon^{\theta}_{xx}$ and $\varepsilon^{\theta}_{yy}$ ones.

The field of transverse displacement u_z in the direction of the plate thickness can be calculated from

$$\varepsilon^{\theta}_{zz} = u^{\theta}_{z,z} \tag{6}$$

where the comma denotes partial derivatives. Upon integration in z one obtains

$$u^{\theta}_{z} = u_{z_0} + z\alpha\, T_0 \tag{7}$$

The following statement can be made:

- Plate theories with at least a linear transverse displacement field in the z direction are required to capture the transverse normal strain due to the constant distribution of temperature across the thickness. CLT and FSDT are inadequate to this purpose.

The in-plane displacement could be constant, but in order to capture the bending due to the constraints and/or to other loadings, the linear form is usually required for u_x and u_y.

- Case II. $T_p\,(z)$ is a linear distribution in the thickness direction
  Let us consider a linear temperature profile across the plate thickness,

$$T_p\,(z) = T_0 + z\, T_1 \tag{8}$$

According to what has been done for Case I, one obtains:

$$\varepsilon^{\theta}_{xx} = \varepsilon^{\theta}_{yy} = \varepsilon^{\theta}_{zz} = \alpha\,(T_0 + z\, T_1) \tag{9}$$

So that

$$u^{\theta}_{z} = u_{z_0} + z\alpha\, T_0 + z^2\alpha\, T_1 \tag{10}$$

The following consideration can be drawn:

- Plate theories with at least a quadratic transverse displacement field in the z direction are required to capture transverse normal strain due to the linear distribution of temperature across the thickness.

The in-plane displacements u_x and u_y are at least required to be linear in z. In the most general case, the form of $T_p(z)$ is a result of the solution of a heat conduction problem. This calculated $T_p(z)$ turns out to be a nonlinear polynomial of z, often of a transcendental form. Moreover, in the case of a thick multilayered structure, $T_p(z)$ would require a layer-wise description.

- Plate theories with higher order (or layer-wise) displacement fields are required to capture a temperature profile obtained from the solution of a heat conduction problem.

It is concluded that the use of advanced structural theories is a mandatory requirement in the modelling of thermal stress problems, above all for those featured by nonlinear temperature distributions. Only advanced structural theories are able to model the temperature according to its nature of *field load*.

CLASSIFICATION OF THERMOELASTIC PROBLEMS

The thermoelasticity represents an extension of the classical theory of elasticity in which deformation and stresses are generated not only by the application of mechanical forces, but also by temperature variation. The introduction of the temperature field in the analysis of the continuous deformable body complicates significantly the set of the differential governing equations. Complication arises due to the nature itself of the phenomenon. In particular, the mechanical and thermal aspects are coupled, and inseparable. When a continuum is subjected to a deformation this will result in a change in its temperature.

However, from a practical standpoint it is generally possible to discount the coupling and to evaluate the temperature and deformation fields, separately.

In this respect, in the classical thermoelasticity, more generally it is possible to identify and distinguish several classes of thermoelastic problems:

- **Coupled** problems which are based on the most general and comprehensive set of governing differential equations (GDEs) which couple the thermal and mechanical aspects of the phenomenon under consideration. In particular, coupled problems of thermoelasticity take into account the time rate of change of the first invariant of the strain tensor in the first law of the thermodynamics causing the dependence between the temperature and strain fields, and thus creating the coupling between elastic and thermal fields. Coupled problems can also be generated under adiabatic conditions.

- **Coupled quasi-static** problems in which the change of temperature proceed slowly and the inertia terms in the GDEs can be discarded.
- **Uncoupled** problems in which the term of mechanical origin in the heat conduction equation is neglected and the temperature field is no more coupled with the elastic one.
- **Uncoupled quasi-static** problems in which along with the elimination of the mechanical term in the heat conduction equation also the inertia term is neglected. Despite the fact that the real nature of a thermoelastic process is dynamic due to the involvement of heat flow, however the uncoupled quasi-static approach is the most common.
- **Stationary** problems in which the loads are applied slowly and the time derivatives in the GDEs are suppressed. This approach automatically uncouples the temperature field from elastic one.

The mathematical description of some of the above listed coupled and uncoupled thermoelastic formulations is given in chapter 3, where all the thermoelastic formulations have been derived with a variational approach and the relative GDEs and boundary conditions (BCs) have been written in a unified and compact manner.

In some specific thermal applications the time rate of change of thermal boundary conditions on a structure, or the time rate of change of thermal sources, is comparable with the structural vibration characteristics, in this case thermal stress waves are produced. Under these circumstances the solution of the problem for the stresses and temperature fields must be evaluated through the coupled equations of thermoelasticity. Generally the classical coupled dynamic thermoelasticity is based on the paradox of infinite propagation speed of thermal signals or thermoelastic waves, which leads to a parabolic (diffusion-type) equation in the heat conduction analysis of solids. Although the infinite propagation speed of thermoelastic waves is not exact for most solids, it usually occurs with such high velocity that the approximation leads to very accurate results above all in aerospace applications. However, on the other hand, this result is physically unacceptable and inconsistent with the heat transfer mechanisms. This drawback has been successfully overcome by enhancing the mathematical models and allowing them to account for the *finite speed of the thermoelastic waves/disturbances* [24]. The GDEs opportunely modified are now waves (hyperbolic-type) equations. The proposed hyperbolic-type heat conduction or wave-type heat propagation is usually referred to as *second sound*, being the first sound the usual sound (pressure waves). These new improved theories which admit finite speed for the thermoelastic waves have been formulated either by incorporating a heat flux within the classical Fourier's law or by including the temperature-rate among the constitutive variables. The resulting theories are referred to as coupled thermoelasticity with second sound effect or generalized theory of coupled thermoelasticity. Some of the most used theories are those of Lord-Shulman [25], Green-Lisday [26] and Hetnarski [27] amongst others.

BOOK'S CONTENT

This volume covers applications of thermal stress problems from classical to advanced topics on the subject matter. Some chapters of the present book are suitable to be used as teaching material in postgraduate courses. An overview of the book's content is given in the following.

Chapter 1 is devoted to the introduction of fundamental concepts. In particular, the stress tensor and the strain tensor along with the compatibility and equilibrium equations have been described. Moreover, the conservation laws have been proposed in a concise manner. Finally, the governing equations of the 3D and 2D thermoelasticity have been derived.

Chapter 2 is dedicated to the presentation and solution of sample problems in the classical theory of thermoelasticity. More specifically, in the first part of the chapter classical thermoelastic problems involving bars, beams, plates, cylinders, spheres and disks have been proposed. In the second part of the chapter basic and complex 1D, 2D and 3D steady-state and transient heat conduction problems have been addressed.

Chapter 3 presents classical and unconventional variational principles for pure mechanical problems and for both coupled and uncoupled thermoelasticity. Most notably, the Principle of Virtual Displacements (PVD) and Reissner's Mixed Variational Theorem (RMVT) have been significantly used.

Chapter 4 deals with the fundamentals of mechanics of beam, plate and shell structures. The asymptotic and axiomatic approaches in structural mechanics have been introduced. The most known structural models have been thoroughly reviewed. Various examples have been shown.

Chapter 5 copes with the development of advanced beam, plate and shell theories with hierarchical capabilities generated by using the Unified Formulation (UF). In the first part of the chapter the UF notation has been introduced. The concept of fundamental nucleus has been proposed. Advanced 1D, 2D and 3D Finite Element (FE) formulations based on refined UF models have been presented. Several examples involving the derivation of the FE matrices have been shown. In the second part of the chapter more advanced theories, such as those that involve 1×1 UF nuclei and those based on the axiomatic/asymptotic approach have been described. Finally, the chapter ends with the construction of the best theory diagram.

Chapter 6 provides the fundamental equations of the anisotropic elasticity. Various constitutive laws, such as those of isotropic, orthotropic, anisotropic and functionally graded materials have been proposed. In the latter case, temperature-dependent materials for different types of FGMs have also been presented. Moreover, the constitutive law for the RMVT case have been derived.

Chapter 7 proposes the development of advanced computational methods developed within the framework of the UF for beam, plate and shell structures. In the framework of meshless methods three advanced Hierarchical Ritz, Galerkin and Generalized Galerkin Formulations (HRF, HGF, HGGF) have been proposed. The governing differential equations have been derived. Weak- and strong-form of the equations have also been presented for the RMVT case and for coupled and uncoupled thermoelastic formulations.

Chapter 8 studies the influence of through-the-thickness temperature profile on the thermoelastic response of multilayered composite and sandwich thick and thin plates. Results have been proposed by using various PVD and RMVT-based plate theories.

Chapter 9 investigates the static response of uncoupled thermoelastic problems of metallic and composite plate and shell structures. The temperature profile through-the-thickness direction is both assumed considering a linear variation and calculated by solving Fourier's heat conduction equation. Various advanced plate and shell theories have been assessed against three dimensional elasticity results.

Chapter 10 deals with the study of free vibration of uncoupled thermoelastic problems. The effect of thermal gradients on the modal characteristics of the plate structures has been thoroughly investigated. Assessment of advanced plate models has been carried out. Results have been compared to those present in literature and/or obtained by using commercial FE software.

Chapter 11 copes with the static and dynamics response of coupled thermoelastic problems. The accuracy of refined FEs based on the UF has been tested for various static and dynamic thermoelastic problems of plate structures.

Chapter 12 examines the thermal buckling phenomenon of various laminated composite and sandwich structures. Refined plate theories have been used in the investigations. Advanced Ritz (RM), Galerkin (GM) and Generalized Galerkin (GGM) methods have been employed in the analysis.

Chapter 13 is utterly dedicated to an accurate thermoelastic stability and thermoelastic vibration analysis of FGM isotropic and sandwich plate structures by using hierarchical plate models. The analyses have been carried out by considering the effect of various nonlinear temperature gradients both on the critical temperatures and on the modal characteristics of the investigated structures.

Chapter 14 studies the thermal effect on the panel flutter phenomena. The classical panel flutter for flat and curved panels is investigated. Thermomechanical load interactions have been taken into account and the aero-thermo-elastic stability diagrams for various panel configurations, and different temperatures and dynamic pressure have been obtained.

REFERENCES

1. Duhamel JMC. Second mémoire sur les phénomènes thermomécaniques. J Éc Polythech 1837;15(25):1–57.
2. Navier CLMH. Mémoire sur les lois de l'équilibre et du mouvement des corps solides élastiques. Mém Acad Sci, Paris 1827;VII:375–93.
3. Fourier JBJ. Théorie analytique de le chaleur. Paris: Firmin Didot; 1822.
4. Neumann F. Vorlesung über die Theorie des Elasticität der festen Körper und des Lichtäthes. Leipzig: Teubner; 1855.
5. Almansi E. Use of the stress functions in thermoelasticity. Mem R Accad Sci Torino (2) 1897;47.
6. Borchardt CW. Untersuchungen über die Elastizität fester isotropen Körper unter Berücksichtigung der Wärme. Mber Acad Wiss, Berlin, vol. 9. 1873.
7. Hopkinson J. Thermal stresses in a sphere, whose temperature is a function of *r* only. Mess Math 1879;8:168.
8. Tedone O. Allgemeine Theoreme der matematishen Elastizitätslehre (Integrationstheorie). In: Encyklopädie der matematishen Wissenschaften, Vol 4, Part D. 1906. p. 54–124, 125–214 [second article written with A. Timpe].
9. Voigt W. Lehrbuch der Kristallphysik. Berlin: Teubner; 1910.
10. Leon A. On thermal stresses. Bautechnik 1904;26:968.
11. Timoshenko S. Bending and buckling of bi-metallic strips. J Opt Soc Am 1925;11:233.
12. Muschelishvili N. Sur l'équilibre des corpes élastiques soumis l'action de le chaleur. Bull Univ Tiflis 1923;3.
13. Signorini A. Sulle Deformazioni Termoelastiche Finite. In: Proc. 3rd Int. Congr. Appl. Mech., Vol 2. 1930. p. 80–9.

14. Hetnarski R, Eslami RM. Thermal stresses – advanced theory and applications. First edition. Springer; 2008.
15. Thorton EA. Thermal structures for aerospace applications. First edition. Reston (Virginia, USA): AIAA Educaton Series; 1996.
16. Reti L. The unknown Leonardo. New York: McGraw-Hill; 1974.
17. Ballarini R. The da Vinci–Euler–Bernoulli beam theory? Mech Eng Mag 2003. http://www.memagazine.org/.
18. Euler L. De Curvis Elasticis, Methodus Inveniendi Lineas Curvas Maximi Minimive Proprietate Gaudentes, Sive Solutio Problematis Isoperimetrici Lattissimo Sensu Accepti, Ser. 1, vol. 24. Bousquet (Geneva): Opera Omnia; 1744.
19. Timoshenko SP. On the correction for shear of the differential equation for transverse vibrations of prismatic bars. Philos Mag Ser 1921;6:742–6.
20. Kirchhoff GR. Überdas Gleichgewicht und die Bewegung einer elastischen Sheibe. J Reine Angew Math 1850;40:51–88.
21. Love AEH. The small free vibration and deformation of a thin elastic shell. Philos Trans R Soc Lond A 1888;179:491–546.
22. Reissner E. The effect of transverse shear deformation on the bending of elastic plates. J Appl Mech 1945;67:A67–77.
23. Mindlin RD. Influence of rotatory inertia and shear on flexural motions of isotropic, elastic plates. J Appl Mech 1951;18:31–8.
24. Ignaczak J, Starzewski MO. Thermoelasticity with finite wave speeds. New York: Oxford University Press; 2010.
25. Lord HW, Shulman Y. A generalized dynamical theory of thermoelasticity. J Mech Phys Solids 1967;15:229–309.
26. Green AE, Lindsay KA. Thermoelasticity. J Elast 1972;2:1–7.
27. Hetnarski RB, Ignaczak J. Soliton-like waves in a low-temperature nonlinear thermoelastic solid. Int J Eng Sci 1996;34:1767–87.

PART 1

Thermoelasticity

CHAPTER 1

Fundamentals of thermoelasticity

1.1 STRESS TENSOR

In a continuum deformable body in the original state of equilibrium related to the undeformed configuration the resultant of the forces acting in a portion of the body is zero. When a deformation occurs some internal forces arise to return the body in equilibrium, these forces are referred to as *stresses*. The stress state is characterized by nine stress components at a point within the continuum. In an orthogonal Cartesian coordinate system (x, y, z) the stress tensor is generally represented by σ_{ij} with $i, j = x, y, z$. The stress tensor is symmetric and the six independent stress components can be divided into normal stresses σ_{xx}, σ_{yy}, σ_{zz} and shear stresses σ_{xy}, σ_{xz}, σ_{yz}. The latter are even commonly defined as τ_{xy}, τ_{xz}, τ_{yz}. The knowledge of the stress tensor at a given point permits the evaluation of the *traction vector* by simply using the Cauchy's formula

$$t_i^n = \sigma_{ij}\, n_j \tag{1.1}$$

where n_j are the cosine directors. The magnitude of the traction vector $\mathbf{t}^n$ depends upon the orientation of the local coordinate system. In the particular case in which the direction of the traction vector and the normal vector $\mathbf{n}$ coincides, the plane is referred to as *principal plane*. The reference system is featured by principal axes and as the traction vector has no projection on this plane the shear stresses are zero and extreme values for normal stresses are produced. The computation of the principal stresses is carried out by solving the characteristic equation,

$$-\sigma^3 - I_1\, \sigma^2 - I_2\, \sigma + I_3 = 0 \tag{1.2}$$

The roots of the characteristic equation are always real due to the symmetry of the stress tensor. The coefficients of the characteristic equation, I_1, I_2 and I_3, are the first, second, and third stress invariant, respectively, have always the same value regardless of the coordinate system's orientation and take the following form

$$\begin{aligned} I_1 &= \sigma_{xx} + \sigma_{yy} + \sigma_{zz} \\ I_2 &= \sigma_{xx}\,\sigma_{yy} + \sigma_{yy}\,\sigma_{zz} + \sigma_{xx}\,\sigma_{zz} - \sigma_{xx}^2 - \sigma_{yy}^2 - \sigma_{zz}^2 \\ I_3 &= \det\left(\sigma_{ij}\right) \end{aligned} \tag{1.3}$$

Thermal Stress Analysis of Composite Beams, Plates and Shells.
DOI: http://dx.doi.org/10.1016/B978-0-12-849892-7.00002-9

The first and third invariant are the trace and determinant respectively, of the stress tensor. The stress invariants can also be expressed in terms of the principal stresses,

$$\begin{aligned} I_1 &= \sigma_{11} + \sigma_{22} + \sigma_{33} \\ I_2 &= \sigma_{11}\,\sigma_{22} + \sigma_{22}\,\sigma_{33} + \sigma_{11}\,\sigma_{33} \\ I_3 &= \sigma_{11}\,\sigma_{22}\,\sigma_{33} \end{aligned} \tag{1.4}$$

In the particular case of creep and plasticity analysis, it is useful to define an alternative form of the stress tensor generally called *deviatoric stress tensor* which assumes the following form

$$s_{ij} = \sigma_{ij} - \sigma_m\,\delta_{ij} \tag{1.5}$$

where σ_m is the *mean normal stress tensor* or alternatively defined *hydrostatic stress tensor*

$$\sigma_m = \frac{\sigma_{xx} + \sigma_{yy} + \sigma_{zz}}{3} = \frac{1}{3}\sigma_{kk} = \frac{1}{3} I_1 \tag{1.6}$$

and δ_{ij} is the *Kronecker delta* or *Kronecker symbol* defined by

$$\delta_{ij} = \begin{cases} 1 & \text{if } i = j \\ 0 & \text{if } i \neq j \end{cases} \tag{1.7}$$

In Eq. (1.6) the Einstein summation convention is used and the terms are summed for $k = 1, 2, 3$. The principal deviatoric stresses are obtained by solving the characteristic equation

$$s^3 + J_2\,s - J_3 = 0 \tag{1.8}$$

where

$$\begin{aligned} J_1 &= s_{11} + s_{22} + s_{33} = 0 \\ J_2 &= \frac{1}{2} s_{ij}\,s_{ji} \\ J_3 &= \frac{1}{3} s_{ij}\,s_{jk}\,s_{ki} \end{aligned} \tag{1.9}$$

are the three invariants of the deviatoric stress tensor. The first invariant J_1 is zero so that the square term in Eq. (1.8) does not appear. The second invariant J_2 is important in the plasticity and creep analysis of structures and is directly involved in the Huber-von Mises-Hencky yield criterion as it is proportional to the distortion strain energy. Indeed the criterion states that yielding of the material at a point occurs when the stress state reaches the level to cause J_2. In particular, if the principal stresses fall on the yield surface described by J_2, plastic deformation occurs.

1.2 DISPLACEMENT AND STRAIN TENSOR

When subjected to surface forces, body forces or temperature changes a continuum deformable body undergoes to deformation, i.e., it changes its geometry in shape and volume. Similarly to the stress state, the deformation state at a point is characterized by the nine strain components of the strain tensor ε_{ij}. Being the latter symmetric, only six components are independent, the three normal strains along the Cartesian axes, ε_{xx}, ε_{yy}, ε_{zz}, which represent a simple extension or compression, and the shear strains ε_{xy}, ε_{xz}, ε_{yz}. In some cases it is convenient the introduction of the engineering shear strains $\gamma_{xy} = 2\,\varepsilon_{xy}$, $\gamma_{xz} = 2\,\varepsilon_{xz}$, $\gamma_{yz} = 2\,\varepsilon_{yz}$, which individuate the distortion of the angle between any two directions.

During the deformation each point experiences a displacement which can be given in terms of Cartesian components u_x, u_y, u_z. These three displacement components are related to the strains by virtue of the *geometrical relationships*. In the case of finite strain theory, these relations are identified by the *Green-Lagrange strain tensor*

$$\varepsilon_{ij} = \frac{1}{2}\left(\partial_j u_i + \partial_i u_j + \partial_i u_k\, \partial_j u_k\right) \tag{1.10}$$

where $i, j, k = x, y, z$ and the summation convention on the repeated indexes is used. By writing out the strain components the following equations are given

$$\begin{aligned}
\varepsilon_{xx} &= \frac{\partial u_x}{\partial x} + \frac{1}{2}\left[\left(\frac{\partial u_x}{\partial x}\right)^2 + \left(\frac{\partial u_y}{\partial x}\right)^2 + \left(\frac{\partial u_z}{\partial x}\right)^2\right] \\
\varepsilon_{yy} &= \frac{\partial u_y}{\partial y} + \frac{1}{2}\left[\left(\frac{\partial u_x}{\partial y}\right)^2 + \left(\frac{\partial u_y}{\partial y}\right)^2 + \left(\frac{\partial u_z}{\partial y}\right)^2\right] \\
\varepsilon_{zz} &= \frac{\partial u_z}{\partial z} + \frac{1}{2}\left[\left(\frac{\partial u_x}{\partial z}\right)^2 + \left(\frac{\partial u_y}{\partial z}\right)^2 + \left(\frac{\partial u_z}{\partial z}\right)^2\right] \\
\varepsilon_{xy} &= \frac{1}{2}\left(\frac{\partial u_y}{\partial x} + \frac{\partial u_x}{\partial y}\right) + \frac{1}{2}\left[\frac{\partial u_x}{\partial x}\frac{\partial u_x}{\partial y} + \frac{\partial u_y}{\partial x}\frac{\partial u_y}{\partial y} + \frac{\partial u_z}{\partial x}\frac{\partial u_z}{\partial y}\right] \\
\varepsilon_{xz} &= \frac{1}{2}\left(\frac{\partial u_z}{\partial x} + \frac{\partial u_x}{\partial z}\right) + \frac{1}{2}\left[\frac{\partial u_x}{\partial x}\frac{\partial u_x}{\partial z} + \frac{\partial u_y}{\partial x}\frac{\partial u_y}{\partial z} + \frac{\partial u_z}{\partial x}\frac{\partial u_z}{\partial z}\right] \\
\varepsilon_{yz} &= \frac{1}{2}\left(\frac{\partial u_z}{\partial y} + \frac{\partial u_y}{\partial z}\right) + \frac{1}{2}\left[\frac{\partial u_x}{\partial y}\frac{\partial u_x}{\partial z} + \frac{\partial u_y}{\partial y}\frac{\partial u_y}{\partial z} + \frac{\partial u_z}{\partial y}\frac{\partial u_z}{\partial z}\right]
\end{aligned} \tag{1.11}$$

In many practical applications the displacement gradient $\partial_i u_k$ is small and being the third term in Eq. (1.10) of the second order of smallness then can be neglected. The strain-displacement relations are reduced to a linear form which is referred to as *Cauchy strain tensor*

$$\varepsilon_{ij} = \frac{1}{2}\left(\partial_j u_i + \partial_i u_j\right) \tag{1.12}$$

The six strain-displacement relations for the infinitesimal theory of elasticity reduce to

$$\begin{aligned}
&\varepsilon_{xx} = \frac{\partial u_x}{\partial x} && \varepsilon_{yy} = \frac{\partial u_y}{\partial y} && \varepsilon_{zz} = \frac{\partial u_z}{\partial z} \\
&\varepsilon_{xy} = \frac{1}{2}\left(\frac{\partial u_y}{\partial x} + \frac{\partial u_x}{\partial y}\right) && \varepsilon_{xz} = \frac{1}{2}\left(\frac{\partial u_z}{\partial x} + \frac{\partial u_x}{\partial z}\right) && \varepsilon_{yz} = \frac{1}{2}\left(\frac{\partial u_z}{\partial y} + \frac{\partial u_y}{\partial z}\right)
\end{aligned} \tag{1.13}$$

For the Cauchy stain tensor the engineering shear strains assume the following forms

$$\gamma_{xy} = \left(\frac{\partial u_y}{\partial x} + \frac{\partial u_x}{\partial y}\right) \quad \gamma_{xz} = \left(\frac{\partial u_z}{\partial x} + \frac{\partial u_x}{\partial z}\right) \quad \gamma_{yz} = \left(\frac{\partial u_z}{\partial y} + \frac{\partial u_y}{\partial z}\right) \tag{1.14}$$

As any other symmetric tensor, the strain tensor ε_{ij} can be diagonalised at any point within the continuum. This means that, at any point exists a particular orientation of the coordinate axes called principal axes which can be found produces principal strains. The principal strains are normal to the principal planes and the shear strain components are zero. The strain tensor has many properties similar to the stress tensor and mathematical procedure involved in the computation of the principal strains and strain invariants coincide.

When the six independent strain components are known at a given point within the continuum, the displacement components can be obtained by an integration procedure of the strain-displacement relations. Nevertheless, for a prescribed value of the strain components, the strain tensor represents a system of six differential equations for the computation of three displacement components, then the system is overdetermined. This drawback can be overcome by invoking the compatibility equations

$$\varepsilon_{ij,kl} + \varepsilon_{kl,ij} - \varepsilon_{ik,jl} - \varepsilon_{jl,ik} = 0 \tag{1.15}$$

Equation (1.15) represents the indicial form of a system of 81 equations. Nonetheless, some simplification occurs because of the symmetry in the indices *ij* and *kl* and because some of them are identically satisfied. At the end only 6 of the 81 equations are essential and their explicit form in terms of orthogonal Cartesian coordinates is

$$
\begin{aligned}
&\frac{\partial^2 \varepsilon_{xx}}{\partial y^2} + \frac{\partial^2 \varepsilon_{yy}}{\partial x^2} = 2\frac{\partial^2 \varepsilon_{xy}}{\partial x \partial y} \\
&\frac{\partial^2 \varepsilon_{yy}}{\partial z^2} + \frac{\partial^2 \varepsilon_{zz}}{\partial y^2} = 2\frac{\partial^2 \varepsilon_{yz}}{\partial y \partial z} \\
&\frac{\partial^2 \varepsilon_{zz}}{\partial x^2} + \frac{\partial^2 \varepsilon_{xx}}{\partial z^2} = 2\frac{\partial^2 \varepsilon_{xz}}{\partial x \partial z} \\
&\frac{\partial^2 \varepsilon_{xx}}{\partial y \partial z} = \frac{\partial}{\partial x}\left(\frac{\partial \varepsilon_{xz}}{\partial y} + \frac{\partial \varepsilon_{xy}}{\partial z} - \frac{\partial \varepsilon_{yz}}{\partial x}\right) \\
&\frac{\partial^2 \varepsilon_{yy}}{\partial x \partial z} = \frac{\partial}{\partial y}\left(\frac{\partial \varepsilon_{xy}}{\partial z} + \frac{\partial \varepsilon_{yz}}{\partial x} - \frac{\partial \varepsilon_{xz}}{\partial y}\right) \\
&\frac{\partial^2 \varepsilon_{zz}}{\partial x \partial y} = \frac{\partial}{\partial z}\left(\frac{\partial \varepsilon_{xz}}{\partial y} + \frac{\partial \varepsilon_{yz}}{\partial x} - \frac{\partial \varepsilon_{xy}}{\partial z}\right)
\end{aligned}
\tag{1.16}
$$

The compatibility equations were first derived by Saint-Venant in 1860. The proof which assures that the fulfilment of the compatibility equations is a necessary and sufficient conditions for the continuity and single-valuedness of the displacement field is due to E. Cesàro [1–4].

1.3 CONSERVATION LAWS

The conservation of mass, momentum and energy is locally expressed by virtue of governing differential equations. These governing equations, even known as conservation laws, are derived from the integral forms of the equations of balance, which represent fundamental postulates of the continuum mechanics. In the derivation of what follows certain integral transformation formulas are needed, and in particular, the Divergence theorem (Gauss's theorem) will be extensively used. In the present discussion the density is assumed to be constant, and the conservation of mass is identically satisfied.

1.3.1 Conservation of linear and angular momenta

The principle of conservation of linear momentum, even known as Newton's second law of motion is derived in its integral form, considering an elastic body subjected to arbitrary traction surface unit $\mathbf{t}^n$ and a body force $\mathbf{b}$. At instant t of time, the elastic body occupies the volume V and is bounded by the surface Ω. The resulting body force acting on the body is given by

$$\mathcal{F}_i = \int_{\Omega} t_i^n \, \mathrm{d}\Omega + \int_V \rho\, b_i \, \mathrm{d}V \tag{1.17}$$

At this stage recalling the Cauchy's formula in Eq. (1.1) and using the Divergence Theorem, the surface integral of the traction forces is transformed to a volume integral as follows

$$\int_{\Omega} t_i^n \, \mathrm{d}\Omega = \int_{\Omega} \sigma_{ij}\, n_j \, \mathrm{d}\Omega = \int_V \sigma_{ij,j} \, \mathrm{d}V \tag{1.18}$$

Thereby, Eq. (1.17) can be written as

$$\mathcal{F}_i = \int_V \left(\sigma_{ij,j} + \rho\, b_i\right) \mathrm{d}V \tag{1.19}$$

The definition of the linear momentum is given in the following

$$\mathcal{P} = \int_V \rho\, \dot{u}_i \, \mathrm{d}V \tag{1.20}$$

The linear momentum conservation expressed by the postulate is

$$\mathcal{F}_i = \dot{\mathcal{P}}_i \tag{1.21}$$

Substituting Eqs. (1.19) and (1.20) into Eq. (1.21), the linear momentum conservation takes the form

$$\int_V \left(\sigma_{ij,j} + \rho\, b_i\right) \mathrm{d}V = \int_V \rho\, \ddot{u}_i \, \mathrm{d}V \tag{1.22}$$

Equation (1.22) has to be satisfied for any arbitrary volume, hence at each point we have

$$\sigma_{ij,j} + \rho\, b_i = \rho\, \ddot{u}_i \tag{1.23}$$

In the rectangular Cartesian coordinates the equations of motion in expanded form are

$$\begin{aligned}
\frac{\partial \sigma_{xx}}{\partial x} + \frac{\partial \sigma_{xy}}{\partial y} + \frac{\partial \sigma_{xz}}{\partial z} + \rho\, b_x = \rho\, \ddot{u}_x \\
\frac{\partial \sigma_{yx}}{\partial x} + \frac{\partial \sigma_{yy}}{\partial y} + \frac{\partial \sigma_{yz}}{\partial z} + \rho\, b_y = \rho\, \ddot{u}_y \\
\frac{\partial \sigma_{zx}}{\partial x} + \frac{\partial \sigma_{zy}}{\partial y} + \frac{\partial \sigma_{zz}}{\partial z} + \rho\, b_z = \rho\, \ddot{u}_z
\end{aligned} \tag{1.24}$$

The solution of the equations of motions is found upon specification of boundary conditions. These latter are required for the evaluation of the integration constants and are usually expressed in terms of the known tractions on the boundary, which come from Eq. (1.1). Indeed, when the traction forces are known, the Cauchy's formula can be used to evaluate the stresses on the boundary, as follows

$$\begin{aligned}\hat{t}_x &= \hat{\sigma}_{xx}\, n_x + \hat{\sigma}_{xy}\, n_y + \hat{\sigma}_{xz}\, n_z \\ \hat{t}_y &= \hat{\sigma}_{yx}\, n_x + \hat{\sigma}_{yy}\, n_y + \hat{\sigma}_{yz}\, n_z \\ \hat{t}_z &= \hat{\sigma}_{zx}\, n_x + \hat{\sigma}_{zy}\, n_y + \hat{\sigma}_{zz}\, n_z\end{aligned} \tag{1.25}$$

where $\hat{\sigma}_{ij}$ are the unknown stresses at the boundary, n_x, n_y and n_z are the cosine directions of the vector normal at the boundary and $\hat{t}_x$, $\hat{t}_y$ and $\hat{t}_z$ are the known components of the traction force on the boundary.

The equilibrium conditions further require that the resultant moment of the surface and volume forces vanishes with respect to an arbitrary point. Thus, given x_i as a component of the position vector at the point of application of forces, the moment of momentum of the body with respect to the origin of the coordinate system is

$$\mathcal{H}_i = \int_V e_{ijk}\, x_j\, \rho\, \dot{u}_k \mathrm{d}V \tag{1.26}$$

where e_{ijk} is the Levi-Civita symbol following defined

$$e_{ijk} = \begin{cases} 1 & \text{if } (i,j,k) \text{ is an even permutation of } (1,2,3) \\ 0 & \text{if } (i,j,k) \text{ otherwise} \\ -1 & \text{if } (i,j,k) \text{ is an odd permutation of } (1,2,3) \end{cases} \tag{1.27}$$

The moment of the traction forces t_i^n and the body forces b_i with respect to the origin is given by

$$\mathcal{M}_i = \int_\Omega e_{ijk}\, x_j\, t_k^n \mathrm{d}\Omega + \int_V e_{ijk}\, x_j\, b_k \mathrm{d}V \tag{1.28}$$

Then, by introducing the Cauchy's formula and by using the Divergence theorem, to transform the surface integral to a volume integral, Eq. (1.28) can be rewritten as

$$\mathcal{M}_i = \int_V e_{ijk} \left[x_j \left(\sigma_{lk,l} + b_k \right) + \delta_{lj}\, \sigma_{lk} \right] \mathrm{d}V \tag{1.29}$$

where the identity $x_{j,l} = \delta_{jl}$ has been used, δ_{jl} is Kronecker delta defined in Eq. (1.7). The principle of conservation of angular momentum arise from the equilibrium condition, which requires that *the time rate of change of the total moment of momentum for a continuum is equal to the vector sum of the moments of external forces acting on the continuum.* Then,

$$\mathcal{M}_i = \dot{\mathcal{H}}_i \tag{1.30}$$

or equivalently

$$\int_V e_{ijk}\left[x_j\left(\sigma_{lk,l}+b_k-\rho\,\ddot{u}_k\right)+\delta_{ij}\sigma_{lk}\right]\mathrm{d}V=0 \tag{1.31}$$

Referring to the equations of motion, Eq. (1.31) becomes

$$\int_V e_{ijk}\sigma_{lk}\mathrm{d}V=0 \tag{1.32}$$

If no couples act on the body then, Eq. (1.32) leads to the symmetry of the stress tensor, $\sigma_{lk}=\sigma_{kl}$ or in matrix form

$$\boldsymbol{\sigma}=\begin{bmatrix}\sigma_{xx} & \tau_{xy} & \tau_{xz}\\ & \sigma_{yy} & \tau_{yz}\\ \text{Sym} & & \sigma_{zz}\end{bmatrix} \tag{1.33}$$

1.3.2 Conservation of energy

The principle of the conservation of energy for an elastic continuum represents a balance equation of all energy forms possible arising from the thermal and the mechanical behaviour. Thus, it is assumed that

$$\dot{\mathcal{E}}_I+\dot{\mathcal{E}}_G=\dot{\mathcal{E}}_S+\dot{\mathcal{E}}_O \tag{1.34}$$

where $\dot{\mathcal{E}}_I$ is the rate of energy entering in the control volume (i.e., either a fixed region in space through which material flows or a fixed material body that is in motion), such as the energy transfered across the surface by heat transfer; $\dot{\mathcal{E}}_G$ is the rate of energy generated within the control volume, for instance by means of sources within the continuum; $\dot{\mathcal{E}}_S$ is the rate of energy stored within the control volume, for example the energy stored in the material by the material's thermal capacity; $\dot{\mathcal{E}}_O$ is the rate of energy leaving the control volume.

Equation (1.34) is now applied to an elastic continuum. No energy is entering within the control volume, so that

$$\dot{\mathcal{E}}_I=0 \tag{1.35}$$

The energy generated within the control volume by the volumetric heating rate Q gives a contribution over the volume equal to

$$\dot{\mathcal{E}}_G=\int_V Q\mathrm{d}V \tag{1.36}$$

In considering the energy stored within the elastic continuum, it is useful to point out that we are assuming a partial coupling, that denotes a situation in which a change in the thermal energy state generates a change in the mechanical state, but the converse is not true. Full coupling instead denotes the general situation in which thermal and mechanical states are coupled, so that changes in the thermal state affect the mechanical one and vice-versa. Thus, for partial coupling, we assume that the internal energy $u = u(T)$, which states that the internal energy depends upon the temperature of the material and the material's intrinsic capability to store internal energy by its thermal capacity. Then, for a change of temperature ΔT at a point, it is possible to write

$$\Delta u = \rho\, c(T)\, \Delta T \tag{1.37}$$

where $c(T)$ is the material specific heat. Equation (1.37) in rate form can be written

$$\frac{\partial u}{\partial t} = \rho\, c(T)\, \frac{\partial T}{\partial t} \tag{1.38}$$

The rate of energy stored for the entire control volume can, therefore, be expressed as

$$\dot{\mathcal{E}}_S = \int_V \rho\, c(T)\, \frac{\partial T}{\partial t}\, \mathrm{d}V \tag{1.39}$$

The final contribution is given by the energy leaving the control volume, which can be obtained by integration of the normal component of the heat flux vector over the surface Ω. The normal component of the heat-flux by performing the dot product of the heat flux with a unit vector normal to the surface. Then, integrating over the total surface, it is possible to write

$$\dot{\mathcal{E}}_O = \int_\Omega h_i\, n_i\, \mathrm{d}\Omega \tag{1.40}$$

After applying the Divergence theorem at Eq. (1.40) and substituting Eqs. (1.35), (1.36), (1.39) and (1.40) into Eq. (1.34), the final result is

$$\dot{\mathcal{E}}_O = \int_V \left[\rho\, c(T)\, \frac{\partial T}{\partial t} + \frac{\partial h_i}{\partial x_i} - Q\right] \mathrm{d}V \tag{1.41}$$

and, because the volume V is arbitrary the integrand must vanish. Thus, the conservation energy equation for an elastic continuum, assuming partial coupling, is

$$\rho\, c(T)\, \frac{\partial T}{\partial t} + \frac{\partial h_i}{\partial x_i} - Q = 0 \tag{1.42}$$

Equation (1.42) can also be obtained for full coupling state, and in this case takes the following form

$$\rho \frac{\partial u}{\partial t} + \frac{\partial h_i}{\partial x_i} - \sigma_{ij} \frac{\partial \varepsilon_{ij}}{\partial t} - Q = 0 \tag{1.43}$$

As can be observed in Eq. (1.43), the internal energy is now function of both temperature and strain state, $u = u(\varepsilon_{ij}, T)$. This equation can be applied in some special cases such as thermal shock phenomena in which extremely rapid deformation occurs and a significant amount of mechanical energy is converted in thermal energy. Fortunately, however, in many cases of practical interest Eq. (1.42) can be comfortably and effectively used instead of Eq. (1.43). This is due to the fact that the heat produced within the body by the small deformation may be neglected. This means that the energy equation is independent of the structure's state of stress, strain and deformation.

1.3.3 Constitutive equations of linear thermoelasticity

In the theory of linear thermoelasticity, usually referred to as *thermoelasticity of small deformation and of small and slow temperature changes*, the components of the strain tensor are linear functions of the components of the stress tensor and the components of the strain tensor due to thermal change (see [5–8]). The mechanical strain tensor ε_{ij}^M is produced by the stress tensor and it is linearly proportional at the latter as

$$\varepsilon_{ij}^M = \frac{1}{2G}\left(\sigma_{ij} - \frac{\nu}{1+\nu}\sigma_{kk}\,\delta_{ij}\right) \tag{1.44}$$

where G is the shear modulus and ν is the Poisson's ratio. Equation (1.44) is referred to as constitutive law of linear elasticity, or Hooke's law.

Consider $T = T_0 + \Theta$ as the temperature of an elastic continuum cube element subjected to a uniform temperature increase Θ from the reference temperature T_0. Then, the thermal strain tensor ε_{ij}^T of the element due to the uniform temperature change is

$$\varepsilon_{ij}^T = \alpha\,\Theta\,\delta_{ij} \tag{1.45}$$

where α is the coefficient of the thermal expansion. It should be borne in mind that the temperature rise is considered uniform so that the elastic continuum will simply increase its volume without any change of shear angle. This is due to the fact that uniform temperature rise does not produce stresses.

Adding the mechanical and thermal strain contributions, the total strain tensor is

$$\varepsilon_{ij} = \frac{1}{2\,G}\left(\sigma_{ij} - \frac{\nu}{1+\nu}\,\sigma_{kk}\,\delta_{ij}\right) + \alpha\,\Theta\,\delta_{ij} \tag{1.46}$$

Equation (1.46) is known as constitutive law of linear thermoelasticity. Solving this equation in terms of stress tensor σ_{ij} leads to

$$\sigma_{ij} = 2\,G\left[\varepsilon_{ij} + \frac{\nu}{1-2\,\nu}\left(\varepsilon_{kk} - \frac{1+\nu}{\nu}\,\alpha\,\Theta\right)\delta_{ij}\right] \tag{1.47}$$

At this stage is useful to introduce the Lamé constants, which can be written in terms of the traditional engineering constants as

$$\lambda = \frac{\nu\,E}{(1+\nu)\,(1-2\,\nu)}; \qquad \mu = G = \frac{E}{2\,(1+\nu)} \tag{1.48}$$

The stress and strain relationships in Eqs. (1.46) and (1.47) can be conveniently written in terms of Lamé constants as follows, respectively

$$\varepsilon_{ij} = \frac{1}{2\,\mu}\left(\sigma_{ij} - \frac{\lambda}{3\,\lambda + 2\,\mu}\,\sigma_{kk}\,\delta_{ij}\right) + \alpha\,\Theta\,\delta_{ij} \tag{1.49}$$

and

$$\sigma_{ij} = 2\,\mu\,\varepsilon_{ij} + [\lambda\,\varepsilon_{kk} - \alpha\,(3\,\lambda + 2\,\mu)\,\Theta]\,\delta_{ij} \tag{1.50}$$

Equation (1.50) is referred to as the Duhamel-Neumann law. It is interesting to note that from one of the two Eqs. (1.49) and (1.50), it is possible to derive the relation between the first invariant of the stress tensor I_1 defined in Eq. (1.3) and the first invariant of the strain tensor given by

$$e = \varepsilon_{xx} + \varepsilon_{yy} + \varepsilon_{zz} \tag{1.51}$$

This relation can be written as

$$e = \frac{1-2\,\nu}{1+\nu}\,\frac{I_1}{2\,G} + 3\,\alpha\,\Theta \tag{1.52}$$

If the elastic continuum made of isotropic material is considered in a state of uniform temperature, then the stress and strain invariants are directly related between them by means of the bulk modulus K as

$$e = \frac{I_1}{3\,K}; \qquad K = \frac{2\,G(1+\nu)}{3\,(1-2\,\nu)} \tag{1.53}$$

In the case of uniform hydrostatic pressure of intensity p

$$\sigma_{xx} = \sigma_{yy} = \sigma_{zz} = -p; \qquad \sigma_{xy} = \sigma_{xz} = \sigma_{yz} = 0 \tag{1.54}$$

For this special case $I_1 = -3p$ and remembering that the first invariant of the strain tensor e, for the infinitesimal displacement field, represents the change of volume ΔV per unit volume V called *cubic dilatation*, then Eq. (1.53) takes the following form

$$\frac{\Delta V}{V} = -\frac{p}{K} \tag{1.55}$$

Equation (1.55) provides useful information. In particular, it is possible to understand as the bulk modulus K measures the substance's resistance to uniform compression. Indeed, for incompressible materials its value approaches infinity. The latter case allows to define the upper bound of Poisson's ratio value, that is, $\nu = 1/2$ (see [2–4] for further details).

In linear thermoelasticity the stress and strain relationships in Eq. (1.49) express terms of Young's modulus, Poisson's ratio and thermal expansion coefficient, and in an unabridged form can be written as

$$\begin{aligned}
\varepsilon_{xx} &= \frac{1}{E}\left[\sigma_{xx} - \nu\left(\sigma_{yy} + \sigma_{zz}\right)\right] + \alpha\,\Theta \\
\varepsilon_{yy} &= \frac{1}{E}\left[\sigma_{yy} - \nu\left(\sigma_{zz} + \sigma_{xx}\right)\right] + \alpha\,\Theta \\
\varepsilon_{zz} &= \frac{1}{E}\left[\sigma_{zz} - \nu\left(\sigma_{xx} + \sigma_{yy}\right)\right] + \alpha\,\Theta \\
\gamma_{xy} &= 2\,\varepsilon_{xy} = \frac{\sigma_{xy}}{G} \\
\gamma_{yz} &= 2\,\varepsilon_{yz} = \frac{\sigma_{yz}}{G} \\
\gamma_{zx} &= 2\,\varepsilon_{zx} = \frac{\sigma_{zx}}{G}
\end{aligned} \tag{1.56}$$

For the sake of completeness it is useful at this stage to introduce some important relationships among the isotropic elastic constants

$$\begin{aligned}
\lambda &= \frac{2\,G\nu}{1-2\,\nu} = \frac{G(E-2\,G)}{3\,G-E} = K - \frac{2}{3}\,G = \frac{E\nu}{(1+\nu)\,(1-2\,\nu)} = \frac{3\,K\,\nu}{1+\nu} \\
&= \frac{3\,K\,(3\,K-E)}{9\,K-E} \\
\mu &= G = \frac{E}{2\,(1+\nu)} = \frac{\lambda\,(1-2\,\nu)}{(2\,\nu)} = \frac{2}{3}\,(K-\lambda) = \frac{3\,K\,(1-2\,\nu)}{2\,(1+\nu)} \\
&= \frac{3\,KE}{9\,G-E} \\
\nu &= \frac{E}{2\,G} - 1 = \frac{\lambda}{2\,(\lambda+G)} = \frac{\lambda}{(3\,K-\lambda)} = \frac{3\,K-2\,G}{2\,(3\,K+G)} = \frac{3\,K-E}{6\,K}
\end{aligned}$$

$$
\begin{aligned}
E &= \frac{\lambda(1+\nu)(1-2\nu)}{\nu} = \frac{G(3\lambda+2G)}{\lambda+G} = \frac{9K(K-\lambda)}{3K\quad\lambda} = 2G(1+\nu) \\
&= \frac{9KG}{3K+G} = 3K(1-2\nu) \\
K &= \frac{E}{3(1-2\nu)} = \lambda + \frac{2}{3}G = \frac{\lambda(1+\nu)}{3\nu} = \frac{2G(1+\nu)}{3(1-2\nu)} = \frac{GE}{3(3G-E)}
\end{aligned}
\tag{1.57}
$$

Of interest are also the following identities

$$
\begin{aligned}
&\frac{G}{\lambda - G} = 1 - 2\nu; \qquad \frac{\lambda}{\lambda - 2G} = \frac{\nu}{1-\nu} \\
&\frac{\lambda + 2G}{E} = \frac{1-\nu}{(1+\nu)(1-2\nu)}; \qquad \frac{E}{1-\nu^2} = \frac{4G(\lambda+G)}{\lambda+2G}
\end{aligned}
\tag{1.58}
$$

1.4 THREE-DIMENSIONAL THERMOELASTICITY

Three-dimensional thermoelastic problems of an elastic continuum are defined by the equations given in Section 1.3. The solution of these equations can be obtained by imposing appropriate boundary conditions. In the development of what follows the following assumptions are taken into account:

1. Small displacement gradients and use of the classical linear strain-displacement relations given by the Cauchy strain tensor in Eq. (1.12).
2. Deformations are assumed to be small, and the governing differential equations are referred to the undeformed configuration of the solid and are provided in Eq. (1.24).
3. The strain and stress tensor are symmetric.
4. Partial thermoelastic coupling, which means that conservation of energy does not depend on deformations, and temperature may be determined independently from displacements and stresses.
5. The temperature is evaluated separately by solving the uncoupled conservation of energy law given in Eq. (1.42).
6. The material is linear and elastic, and the generalized Hooke's law for homogeneous isotropic materials is employed as constitutive equations.
7. The material elastic properties, including Young's modulus, Poisson's ratio and the coefficient of thermal expansion, are constant, so that the material properties are independent of the temperature.

The implication of these assumptions is that it is possible to formulate thermoelastic problems as linear boundary/initial-value problems that have unique solutions. The basic unknowns in a classical three-dimensional thermoelasticity problem may vary with position as denoted by Cartesian coordinates x, y, z and with the time t. Globally there are 15 unknowns: 3 displacement components, 6 strain components and 6 stress components. The problem is well formulated because there are 15 equations which allow the evaluation of the unknowns, these equations are: 6 strain-displacement equations, 3 equations of motions and 6 constitutive equations. In practice these equations are not solved simultaneously but are reduced to smaller set by combining them algebraically. This approach leads to the traditional two formulations: 1) Displacement formulation and 2) Stress formulation.

1.4.1 Displacement formulation

In the displacement formulation the 15 equations are combined algebraically into 3 equations in which the unknowns are the three displacements components u_x, u_y and u_z. The equations of motion in terms of stresses were derived in Section 1.2, see Eq. (1.24)

$$\sigma_{ij,j} + \rho\, b_i = \rho\, \ddot{u}_i \tag{1.59}$$

The stresses can be expressed in terms of strains and then in terms of displacements. Substituting for the strain tensor ε_{ij} in terms of the displacements u_i, Eq. (1.50) gives

$$\sigma_{ij} = \mu\, \left(u_{i,j} + u_{j,i}\right) + \left[\lambda\, u_{k,k} - \alpha\,(3\,\lambda + 2\,\mu)\,\Theta\right]\delta_{ij} \tag{1.60}$$

Taking the partial derivative of Eq. (1.60) and substituting into the equilibrium equations (1.59) yields

$$\mu\, u_{i,kk} + (\lambda + \mu)\, u_{k,ki} - \alpha\,(3\,\lambda + 2\,\mu)\,\Theta_{,i} + \rho\, b_i = \rho\, \ddot{u}_i \tag{1.61}$$

Equation (1.61) is referred to as *Cauchy-Navier equation.* It is expressed in terms of the displacement components along the three coordinate axes, and in an unabridged form it gives

$$\begin{aligned}
&\mu\,\nabla^2 u_x - \alpha\,(3\,\lambda + 2\,\mu)\,\frac{\partial\,\Theta}{\partial\,x} + (\lambda + \mu)\,\frac{\partial\,e}{\partial\,x} + \rho\,b_x = \rho\,\ddot{u}_x\\
&\mu\,\nabla^2 u_y - \alpha\,(3\,\lambda + 2\,\mu)\,\frac{\partial\,\Theta}{\partial\,y} + (\lambda + \mu)\,\frac{\partial\,e}{\partial\,y} + \rho\,b_y = \rho\,\ddot{u}_y\\
&\mu\,\nabla^2 u_z - \alpha\,(3\,\lambda + 2\,\mu)\,\frac{\partial\,\Theta}{\partial\,z} + (\lambda + \mu)\,\frac{\partial\,e}{\partial\,z} + \rho\,b_z = \rho\,\ddot{u}_z
\end{aligned} \tag{1.62}$$

where e is the first invariant of the strain tensor (see Eq. (1.51)). The boundary conditions must be satisfied on the surface boundary of the body. If the traction components on the boundary are given as t_i^n, then by using Cauchy's formula

$$t_i^n = \sigma_{ij}\, n_j \tag{1.63}$$

Since the problem has been formulated in terms of displacements, the prescribed traction on the boundary can be related to the displacement components by combining Eq. (1.63) and Eq. (1.60) as follows

$$t_i^n = \left\{\mu\,\left(u_{i,j} + u_{j,i}\right) + \left[\lambda\, u_{k,k} - \alpha\,(3\,\lambda + 2\,\mu)\,\Theta\right]\delta_{ij}\right\} n_j \tag{1.64}$$

and in unabridged form

$$\begin{aligned}
t_x^n &= \lambda\, e\, n_x + \mu\left(n_x\frac{\partial u_x}{\partial x} + n_y\frac{\partial u_x}{\partial y} + n_z\frac{\partial u_x}{\partial z}\right) + \mu\left(n_x\frac{\partial u_x}{\partial x} + n_y\frac{\partial u_y}{\partial x} + n_z\frac{\partial u_z}{\partial x}\right)\\
&\quad - \frac{E\,\alpha\,\Theta}{1-2\nu}\,n_x\\
t_y^n &= \lambda\, e\, n_y + \mu\left(n_x\frac{\partial u_y}{\partial x} + n_y\frac{\partial u_y}{\partial y} + n_z\frac{\partial u_y}{\partial z}\right) + \mu\left(n_x\frac{\partial u_x}{\partial y} + n_y\frac{\partial u_y}{\partial y} + n_z\frac{\partial u_z}{\partial y}\right)\\
&\quad - \frac{E\,\alpha\,\Theta}{1-2\nu}\,n_y\\
t_z^n &= \lambda\, e\, n_z + \mu\left(n_x\frac{\partial u_z}{\partial x} + n_y\frac{\partial u_z}{\partial y} + n_z\frac{\partial u_z}{\partial z}\right) + \mu\left(n_x\frac{\partial u_x}{\partial z} + n_y\frac{\partial u_y}{\partial z} + n_z\frac{\partial u_z}{\partial z}\right)\\
&\quad - \frac{E\,\alpha\,\Theta}{1-2\nu}\,n_z
\end{aligned} \tag{1.65}$$

where n_x, n_y and n_z are the cosine directions of the unit outer normal vector to the boundary. The traction boundary conditions along with the thermal boundary conditions will fully define the displacement and temperature fields. It should be noted that the solution of Cauchy-Navier equation (1.61) simultaneously satisfied compatibility condition and the constitutive law and, therefore, is an acceptable solution of a problem of thermoelasticity.

1.4.2 Stress formulation

In a considerable amount of thermoelasticity problems the boundary conditions are given in terms of stresses. Then, in this case it is more convenient to derive the differential governing equations of thermoelasticity in terms of the stress tensor, which represent a generalization of Beltrami and Michell equations, to the case of thermal stresses. Introducing the stain tensor ε_{ij} given in Eq. (1.12) into the compatibility equations (1.16), and exploiting the use of the equilibrium equations presented in Eq. (1.59), for stationary and quasi-static thermal problems, the following system of six equations is obtained

$$\sigma_{ij,kk} + \frac{1}{1+\nu} I_{1,ij} + \frac{\alpha E}{1+\nu}\left(T_{,ij} + \frac{1+\nu}{1-\nu} T_{,kk}\,\delta_{ij}\right) \tag{1.66}$$

The above equation can be contracted with respect to the indices i and j in the following form

$$(1-\nu)\, I_{1,kk} + 2\alpha\, E\, T_{,kk} = 0 \tag{1.67}$$

In the particular case of a stationary temperature field and no heat source acting inside the body, Eq. (1.66) is significantly simplified. Since in this case $T_{,kk} = 0$ according to Eq. (1.67), $I_{1,kk} = 0$. Thus

$$\sigma_{ij,kk} + \frac{1}{1+\nu}\left(I_{1,ij} + \alpha\, E\, T\right)_{,ij} = 0 \tag{1.68}$$

The set of Eqs. (1.66) can be derived by making use only of the displacement equations and the stress strain relationship. It is convenient to this aim starting from the Lamé equation

$$\mu\, u_{i,kk} + (\lambda + \mu)\, u_{k,ki} = \alpha\,(3\,\lambda + 2\,\mu)\, T_{,i} \tag{1.69}$$

Taking into account the relation

$$2\,\varepsilon_{ij} = u_{i,j} + u_{j,i} \tag{1.70}$$

Equation (1.69) becomes

$$\mu\,\varepsilon_{ij,kk} + (\lambda + \mu)\,\varepsilon_{kk,ij} = \alpha\,(3\,\lambda + 2\,\mu)\, T_{,ij} \tag{1.71}$$

Introducing in the equation the Duhamel-Neumann relations

$$2\,\mu\,\varepsilon_{ij,kk} = \sigma_{ij} - \lambda\,\frac{\delta_{ij}}{3\,\lambda + 2\,\mu}\sigma_{kk} + 2\,\mu\,\alpha\, T\,\delta_{ij} \tag{1.72}$$

the following equation is obtained

$$\sigma_{ij,kk} + \frac{2\,(\lambda+\mu)}{3\,\lambda + 2\,\mu}\sigma_{kk,ij} - \frac{\lambda\,\delta_{ij}}{3\,\lambda + 2\,\mu}\sigma_{ss,kk} + 2\,\mu\,\alpha\left(T_{,ij} + \delta_{ij}\,T_{,kk}\right) = 0 \tag{1.73}$$

Contraction of Eq. (1.73), with respect to the indices i and j, leads to

$$\sigma_{ss,kk} + \frac{4\mu 3\,\lambda + 2\,\mu}{\lambda + 2\mu} T_{,kk} = 0 \tag{1.74}$$

Finally, by using Eqs. (1.74) and (1.73)

$$\sigma_{ij,kk} + \frac{2\,(\lambda+\mu)}{3\,\lambda + 2\,\mu}\sigma_{kk,ij} + 2\,\mu\,\alpha\left(T_{,ij} + \frac{3\,\lambda + 2\,\mu}{\lambda + 2\,\mu}\,\delta_{ij}\,T_{,kk}\right) = 0 \tag{1.75}$$

and in an unabridged form

$$
\begin{aligned}
&\nabla^2\sigma_{xx} + \frac{2(\lambda+\mu)}{3\lambda+2\mu}\frac{\partial^2 I_1}{\partial x^2} + 2\mu\alpha\left(\frac{\partial^2 T}{\partial x^2} + \frac{3\lambda+2\mu}{\lambda+2\mu}\nabla^2 T\right) = 0\\
&\nabla^2\sigma_{yy} + \frac{2(\lambda+\mu)}{3\lambda+2\mu}\frac{\partial^2 I_1}{\partial y^2} + 2\mu\alpha\left(\frac{\partial^2 T}{\partial y^2} + \frac{3\lambda+2\mu}{\lambda+2\mu}\nabla^2 T\right) = 0\\
&\nabla^2\sigma_{zz} + \frac{2(\lambda+\mu)}{3\lambda+2\mu}\frac{\partial^2 I_1}{\partial z^2} + 2\mu\alpha\left(\frac{\partial^2 T}{\partial z^2} + \frac{3\lambda+2\mu}{\lambda+2\mu}\nabla^2 T\right) = 0\\
&\nabla^2\sigma_{xz} + \frac{2(\lambda+\mu)}{3\lambda+2\mu}\frac{\partial^2 I_1}{\partial x\partial z} + 2\mu\alpha\frac{\partial^2 T}{\partial x\partial z} = 0\\
&\nabla^2\sigma_{yx} + \frac{2(\lambda+\mu)}{3\lambda+2\mu}\frac{\partial^2 I_1}{\partial y\partial x} + 2\mu\alpha\frac{\partial^2 T}{\partial y\partial x} = 0\\
&\nabla^2\sigma_{yz} + \frac{2(\lambda+\mu)}{3\lambda+2\mu}\frac{\partial^2 I_1}{\partial y\partial z} + 2\mu\alpha\frac{\partial^2 T}{\partial y\partial z} = 0
\end{aligned}
\tag{1.76}
$$

1.5 TWO-DIMENSIONAL THERMOELASTICITY

The displacement and the stress formulations of the preceding section state the mathematical formulations for the thermoelasticity of a three-dimensional elastic solid. Often, however, the solution of these general equations is difficult to carry out, even by exploiting the use of advanced computational methods substantiated by a huge computing power. For many problems of practical interest, some useful assumptions can be opportunely introduced to reduce drastically the complication of the problem. The simplifications lead to the well-known *plane stress* and *plane strain* states.

1.5.1 Plane stress

The plane stress state simplification is adopted to thin plate that are subjected to temperature distributions that do not vary through the thickness of the plate. For thin-walled structures this assumption is realistic, indeed, in this case the temperature gradients through-the-thickness are small and can be generally neglected without loss of accuracy in the formulation. According to this formulation the two-dimensional stresses are function of two variables x and y, and the transverse stresses are zero, i.e.,

$$
\begin{aligned}
&\sigma_{xx} = \sigma_{xx}(x, y); \quad \sigma_{yy} = \sigma_{yy}(x, y); \quad \sigma_{xy} = \sigma_{xy}(x, y)\\
&\sigma_{zz} = \sigma_{xz} = \sigma_{yz} = 0
\end{aligned}
\tag{1.77}
$$

The Hooke's law for this case is reduced to

$$\begin{aligned}\varepsilon_{xx} &= \frac{1}{E}\left(\sigma_{xx} - \nu\,\sigma_{yy}\right) + \alpha\,\Theta \\ \varepsilon_{yy} &= \frac{1}{E}\left(\sigma_{yy} - \nu\,\sigma_{xx}\right) + \alpha\,\Theta \\ \varepsilon_{xy} &= \frac{\gamma_{xy}}{2} = \frac{1}{2\,G}\sigma_{xy}\end{aligned} \tag{1.78}$$

Solving Eq. (1.78) in terms of stresses leads to

$$\begin{aligned}\sigma_{xx} &= \frac{E}{1-\nu^2}\left[\left(\varepsilon_{xx} + \nu\,\varepsilon_{yy}\right) - \alpha\,(1+\nu)\,\Theta\right] \\ \sigma_{yy} &= \frac{E}{1-\nu^2}\left[\left(\varepsilon_{yy} + \nu\,\varepsilon_{xx}\right) - \alpha\,(1+\nu)\,\Theta\right] \\ \sigma_{xy} &= \frac{E}{1+\nu}\varepsilon_{xy}\end{aligned} \tag{1.79}$$

The equations of motion in the plane stress case are reduced to

$$\begin{aligned}\frac{\partial\,\sigma_{xx}}{\partial\,x} + \frac{\partial\,\sigma_{xy}}{\partial\,y} + \rho\,b_x &= \rho\,\frac{\partial^2 u_x}{\partial t^2} \\ \frac{\partial\,\sigma_{yx}}{\partial\,x} + \frac{\partial\,\sigma_{yy}}{\partial\,y} + \rho\,b_y &= \rho\,\frac{\partial^2 u_x}{\partial t^2}\end{aligned} \tag{1.80}$$

The compatibility equations (1.16) are all identically satisfied except

$$\frac{\partial^2\,\varepsilon_{xx}}{\partial\,y^2} + \frac{\partial^2\,\varepsilon_{yy}}{\partial\,x^2} = 2\,\frac{\partial^2\,\varepsilon_{xy}}{\partial\,x\,\partial\,y} \tag{1.81}$$

Substituting the strain-displacements relations (1.13) into Eqs. (1.79) and introducing them in the equilibrium equations (1.80), a new set of equilibrium equations in terms of displacement components is obtained

$$\begin{aligned}G\left(\frac{\partial^2 u_x}{\partial x^2} + \frac{\partial^2 u_x}{\partial y^2}\right) + G\frac{1+\nu}{1-\nu}\frac{\partial}{\partial x}\left(\frac{\partial u_x}{\partial x} + \frac{\partial u_x}{\partial y}\right) - \frac{E\alpha}{1-\nu}\frac{\partial\Theta}{\partial x} + b_x &= \rho\,\frac{\partial^2 u_x}{\partial t^2} \\ G\left(\frac{\partial^2 u_y}{\partial x^2} + \frac{\partial^2 u_y}{\partial y^2}\right) + G\frac{1+\nu}{1-\nu}\frac{\partial}{\partial x}\left(\frac{\partial u_y}{\partial x} + \frac{\partial u_y}{\partial y}\right) - \frac{E\alpha}{1-\nu}\frac{\partial\Theta}{\partial y} + b_y &= \rho\,\frac{\partial^2 u_y}{\partial t^2}\end{aligned} \tag{1.82}$$

Substituting the strains given in Eq. (1.78) into the compatibility equation (1.81), gives

$$\frac{\partial^2}{\partial y^2}\left(\sigma_{xx} - \nu\,\sigma_{yy}\right) + \frac{\partial^2}{\partial x^2}\left(\sigma_{yy} - \nu\,\sigma_{xx}\right) + E\,\alpha\,\nabla^2\Theta = 2\,(1+\nu)\,\frac{\partial^2\sigma_{xy}}{\partial x\partial y} \tag{1.83}$$

Differentiating the first of Eqs. (1.80) with respect to x and the second with respect to y and adding, yields

$$\frac{\partial^2 \sigma_{xx}}{\partial x^2} + \frac{\partial^2 \sigma_{yy}}{\partial y^2} + \frac{\partial b_x}{\partial x} + \frac{\partial b_y}{\partial y} - \rho\left[\frac{\partial}{\partial x}\left(\frac{\partial^2 u_x}{\partial t^2}\right) + \frac{\partial}{\partial y}\left(\frac{\partial^2 u_y}{\partial t^2}\right)\right] = -2\frac{\sigma_{xy}}{\partial x \partial y} \tag{1.84}$$

Eliminating σ_{xy} between Eqs. (1.83) and Eqs. (1.84) leads to

$$\begin{gathered}\left(\frac{\partial^2}{\partial x^2} + \frac{\partial^2}{\partial y^2}\right)(\sigma_{xx} + \sigma_{yy}) + E\alpha\,\nabla^2\Theta = -(1+\nu) \\ \left\{\frac{\partial b_x}{\partial x} + \frac{\partial b_y}{\partial y} - \rho\left[\frac{\partial}{\partial x}\left(\frac{\partial^2 u_x}{\partial t^2}\right) + \frac{\partial}{\partial y}\left(\frac{\partial^2 u_y}{\partial t^2}\right)\right]\right\}\end{gathered} \tag{1.85}$$

Either Eqs. (1.82) or (1.85) are the governing equations of two-dimensional simple plane stress problem. The solution for the stresses must satisfy the following boundary conditions

$$\begin{aligned} t_x^n &= n_x\,\sigma_{xx} + n_y\,\sigma_{xy} \\ t_y^n &= n_x\,\sigma_{yx} + n_y\,\sigma_{yy} \\ t_z^n &= 0 \end{aligned} \tag{1.86}$$

where n_x and n_y are the cosine directions of the unit outer normal vector $\mathbf{n}$.

1.5.2 Plane strain

In the case of simple plane strain conditions the displacement components u_x and u_y are assumed to be independent of the z-coordinate and $u_z = 0$, then

$$u_x = u_x(x, y)\,; \quad u_y = u_y(x, y)\,; \quad u_z = 0 \tag{1.87}$$

The strain-displacements relations in this case are

$$\begin{aligned} \varepsilon_{xx} &= \varepsilon_{xx}(x, y) = \frac{\partial u_x}{\partial x} \\ \varepsilon_{yy} &= \varepsilon_{yy}(x, y) = \frac{\partial u_{yx}}{\partial x} \\ \varepsilon_{xy} &= \varepsilon_{xy}(x, y) = \frac{1}{2}\left(\frac{\partial u_y}{\partial x} + \frac{\partial u_x}{\partial y}\right) \\ \varepsilon_{zz} &= \varepsilon_{xz} = \varepsilon_{yz} = 0 \end{aligned} \tag{1.88}$$

The imposition of the condition $\varepsilon_{zz} = 0$ in the Hooke's law gives

$$\sigma_{zz} = \nu(\sigma_{xx} + \sigma_{yy}) - E\alpha\,\Theta \tag{1.89}$$

The introduction of Eq. (1.89) in the stress-strain relations leads to

$$
\begin{aligned}
\varepsilon_{xx} &= \frac{1-\nu^2}{E}\left(\sigma_{xx} - \frac{\nu}{1-\nu}\sigma_{yy}\right) + \alpha\,(1+\nu)\,\Theta \\
\varepsilon_{yy} &= \frac{1-\nu^2}{E}\left(\sigma_{yy} - \frac{\nu}{1-\nu}\sigma_{xx}\right) + \alpha\,(1+\nu)\,\Theta \\
\varepsilon_{xy} &= \frac{1}{2\,G}\sigma_{xy}
\end{aligned}
\tag{1.90}
$$

The equations of motion and the compatibility equations remain the same as given by Eqs. (1.80) and (1.81). The equations of motion in terms of displacement components are obtained using similar procedure, and for the plane strain state are

$$
\begin{aligned}
& G\left(\frac{\partial^2 u_x}{\partial x^2} + \frac{\partial^2 u_x}{\partial y^2}\right) + G\frac{1}{1-2\nu}\frac{\partial}{\partial x}\left(\frac{\partial u_x}{\partial x} + \frac{\partial u_y}{\partial y}\right) \\
& \quad - \frac{E\alpha}{1-2\nu}\frac{\partial \Theta}{\partial x} + b_x = \rho\,\frac{\partial^2 u_x}{\partial t^2} \\
& G\left(\frac{\partial^2 u_y}{\partial x^2} + \frac{\partial^2 u_y}{\partial y^2}\right) + G\frac{1}{1-2\nu}\frac{\partial}{\partial y}\left(\frac{\partial u_x}{\partial x} + \frac{\partial u_y}{\partial y}\right) \\
& \quad - \frac{E\alpha}{1-2\nu}\frac{\partial \Theta}{\partial y} + b_y = \rho\,\frac{\partial^2 u_y}{\partial t^2}
\end{aligned}
\tag{1.91}
$$

The compatibility equation in terms of stresses for the plane strain is given as

$$
\begin{aligned}
& \left(\frac{\partial^2}{\partial x^2} + \frac{\partial^2}{\partial y^2}\right)\left(\sigma_{xx} + \sigma_{yy}\right) + \frac{E\alpha}{1-\nu}\nabla^2\Theta = -\frac{1}{1-\nu} \\
& \left\{\frac{\partial b_x}{\partial x} + \frac{\partial b_y}{\partial y} - \rho\left[\frac{\partial}{\partial x}\left(\frac{\partial^2 u_x}{\partial t^2}\right) + \frac{\partial}{\partial y}\left(\frac{\partial^2 u_y}{\partial t^2}\right)\right]\right\}
\end{aligned}
\tag{1.92}
$$

REFERENCES

1. Cesàro E. Rendiconto dell'Accademia delle scienze fisiche e matematiche. Società reale di Napoli; 1906.
2. Malvern LE. Introduction to the mechanics of a continuum medium. First edition. USA: Prentice-Hall; 1969.
3. Sokolnikoff IS. Introduction to the mechanics of a continuum medium. Second edition. USA: McGraw Hill; 1956.
4. Fung YC. Foundations of solid mechanics. First edition. New Jersey (USA): Prentice-Hall; 1965.
5. Parkus H. Thermoelasticity. Second edition. New York (USA): Springer-Verlag; 1976.

6. Hetnarski R, Eslami RM. Thermal stresses – advanced theory and applications. First edition. Springer; 2008.
7. Nowacki W. Thermoelasticity. First edition. New York: Pergamon Press; 1962.
8. Thorton EA. Thermal structures for aerospace applications. First edition. Reston (Virginia, USA): AIAA Education Series; 1996.

CHAPTER 2

Solution of sample problems in classical thermoelasticity

2.1 SAMPLE PROBLEMS IN THERMOELASTICITY

The classical mathematical theory of three-dimensional and two-dimensional thermoelasticity, concisely proposed in Chapter 1, despite its noteworthy usefulness, has a restricted and limited field of application. This drawback has been overcome over the past decades with the introduction of the theory of thermal structures. In spite of the drastic simplification introduced by the latter, it has, however, allowed to cope successfully with many engineering problems of practical interest.

In the present section a comprehensive analysis of thermal structures, such as bars, beams, plates, cylinders, spheres and disks, based on the aforementioned approximations, is proposed. Various sample problems are presented in the three main classical coordinate systems, that is, the rectangular Cartesian coordinates, the cylindrical coordinates, and the spherical coordinates.

Several other examples of thermal structures problems have been proposed by Thorton [1], Boyle [2], Hetnarski [3,4], Gatewood [5] and Jhons [6] amongst others.

2.1.1 Bars

Bars are usually defined as long and slender member that are subjected to a 1D state of normal stresses σ_{zz}. The other stresses are negligible, because lower in terms of order of magnitude. In the derivation of the governing differential equations of thermal bar structures the following hypotheses are taken into account:

1. The stress σ_{zz} is considered uniform over the cross section, and it may be function of spatial and time coordinate, z and t, respectively.
2. The cross-section area A of the bar may vary along the z direction, i.e., $A = A(z)$.
3. The stress σ_{zz} may be tensile or compressive.
4. Constraints are such that transverse bending is precluded.

Thermal Stress Analysis of Composite Beams, Plates and Shells.
DOI: http://dx.doi.org/10.1016/B978-0-12-849892-7.00003-0

5. The bar is in a thermal environment of known temperature $T = T(z, t)$.
6. The material follows the Hooke's law.
7. The only allowed displacement component is the $u_z(z, t)$, which may vary in space z and time t but not over the cross-section.
8. The density ρ of the material is assumed to be constant.

The governing differential equation is following derived by using the classical Newtonian mechanics. To this purpose it is useful to consider a segment of bar subjected to the stress σ_{zz} and the body force F_z. By using the second Newton's law

$$\sum F_z = m\, a_z \tag{2.1}$$

the equilibrium leads to

$$\begin{aligned} &-\sigma_{zz} A + (\sigma_{zz} + \Delta\sigma_{zz})(A + \Delta A) + \left(F_z + \frac{1}{2}\Delta F_z\right)\left(A + \frac{1}{2}\Delta A\right)\Delta z \\ &= \rho\, A\, \Delta z \frac{\partial^2 u_z}{\partial z^2} \end{aligned} \tag{2.2}$$

dividing by Δz Eq. (2.2) becomes

$$\frac{\Delta(\sigma_{zz} A)}{\Delta z} + F_z A + \frac{1}{2}\Delta(F_z A) + \frac{1}{4}\Delta F_z\, \Delta A = \rho\, A \frac{\partial^2 u_z}{\partial z^2} \tag{2.3}$$

The limit as $\Delta z \to 0$ yields

$$\frac{\partial}{\partial z}(\sigma_{zz} A) + F_z A = \rho\, A \frac{\partial^2 u_z}{\partial z^2} \tag{2.4}$$

In this case the Hooke's law gives

$$\sigma_{zz} = E\, \varepsilon_{zz} + \alpha\, \Theta \tag{2.5}$$

where $\varepsilon_{zz} = \frac{\partial u_z}{\partial z}$ is the only non-zero deformation in the bar and $\Theta = T - T_0$, with T_0 being the reference temperature. Equation (2.4) can, therefore, be written as

$$\frac{\partial}{\partial z}\left(EA \frac{\partial u_z}{\partial z}\right) + P_z(z, t) - P_z^T(z, t) = \rho\, A \frac{\partial^2 u_z}{\partial z^2} \tag{2.6}$$

where P_z and P_z^T represent mechanical and thermal forces per unit length, respectively. The second order equation (2.6), can be solved, once defined one boundary condition at each end of the bar. It is possible to specify either the displacement or the stress. Thus, for a bar of length L,

$$\begin{aligned} u_z(0, t) = u_{z_1}(t) \quad &\text{or} \quad \sigma_{zz}(0, t) = \sigma_{zz_1}(t) \\ u_z(L, t) = u_{z_2}(t) \quad &\text{or} \quad \sigma_{zz}(L, t) = \sigma_{zz_2}(t) \end{aligned} \tag{2.7}$$

In addition to the above boundary conditions, other two initial conditions must be imposed as follows

$$u_z(z,0) = u_{z_0}(z) \quad \text{or} \quad \frac{\partial u_z}{\partial t}(z,0) = \dot{u}_{z_0}(z) \tag{2.8}$$

Equations (2.6), (2.7) and (2.8) represent a classical linear boundary/initial value problem for determination of the axial displacement $u_z(z,t)$.

Problem 1

A hollow cylinder with a bar of the same length l and the same centerline are subjected to different temperature changes ΔT_i with $i = 1, 2$, respectively. The hollow cylinder and the bar are connected to two rigid plates. Calculate the thermal stresses σ_i produced in both the hollow cylinder and the bar, and the elongations λ_i.

Solution

The elongation λ_i due to both the free thermal elongation and the thermal stresses are

$$\lambda_1 = \alpha_1 \Delta T_1 l + \frac{\sigma_1}{E_1} l, \quad \lambda_2 = \alpha_2 \Delta T_2 l + \frac{\sigma_2}{E_2} l \tag{2.9}$$

where A_i, E_i and α_i indicate cross-sectional area, Young's modulus and the coefficient of linear thermal expansion of the i-th material, respectively. Since the final length of both the cylinder and the bar is the same, the following relation holds

$$l + \alpha_1 \Delta T_1 l + \frac{\sigma_1}{E_1} l = l + \alpha_2 \Delta T_2 l + \frac{\sigma_2}{E_2} l \tag{2.10}$$

The equilibrium of the inertial forces is described by

$$\sigma_1 A_1 + \sigma_2 A_2 = 0 \tag{2.11}$$

Solving Eqs. (2.10) and (2.11) gives the stresses

$$\lambda_1 = \lambda_2 = \frac{\alpha_1 \Delta T_1 E_1 A_1 + \alpha_2 \Delta T_2 E_2 A_2}{E_1 A_1 + E_2 A_2} \tag{2.12}$$

Problem 2

A bar of mild steel of cross-sectional area A_s is placed between two parallel bars of copper of cross sectional area A_c. When the three bars of same length l are bounded together and are subjected to a temperature change ΔT_s in the bar of mild steel and ΔT_c in the bar of copper, calculate the thermal stresses in each bar.

Solution

The final lengths of mild steel and two copper bars are the same

$$l+\alpha_s\,\Delta T_s\,l+\frac{\sigma_s\,l}{E_s}=l+\alpha_c\,\Delta T_c\,l+\frac{\sigma_c\,l}{E_c} \tag{2.13}$$

Moreover, the equilibrium condition of the internal forces gives

$$\sigma_s\,A_s+2\,\sigma_c\,A_c=0 \tag{2.14}$$

From Eqs. (2.13) and (2.14) we get

$$\sigma_s=\frac{E_s\,(\alpha_c\,\Delta T_c-\alpha_s\,\Delta T_s)}{1+\frac{A_s\,E_s}{2\,A_c\,E_c}},\qquad \sigma_c=-\frac{\sigma_s\,A_s}{2\,A_c} \tag{2.15}$$

2.1.2 Beams

We consider the thermal stresses in beams under da Vinci-Euler-Bernoulli hypothesis. The plane which is perpendicular to the neutral axis before deformation remains plane and perpendicular after deformation. The neutral axis passes through the centroid of the cross section of the beam which is defined by

$$\int_A y\mathrm{d}A=0 \tag{2.16}$$

where dA denotes a small element area of the cross section at a distance y from the neutral plane. The thermal stress is given as

$$\sigma_x=-\alpha\,E\,\Delta T+E\varepsilon_0+E\frac{y}{k} \tag{2.17}$$

where k denotes the radius of curvature at the neutral plane and ε_0 is the axial strain at the neutral plane. When the beam is subjected to an axial force N and a mechanical bending load M^M, the axial strain ε_0 and the curvature $1/k$ at the neutral plane $y=0$ are

$$\begin{aligned}\varepsilon_0&=\frac{1}{EA}\left[N+\int_A\alpha\,E\,\Delta T(y)\,\mathrm{d}A\right]\\ \frac{1}{k}&=\frac{1}{EI}\left[M^M+\int_A\alpha\,E\,\Delta T(y)\,y\,\mathrm{d}A\right]\end{aligned} \tag{2.18}$$

where I is the moment of inertia of the cross section which is defined by

$$I=\int_A y^2\mathrm{d}A \tag{2.19}$$

The thermal stress is

$$\sigma_x(y) = -\alpha E \Delta T(y) + \frac{1}{A}\left[N + \int_A \alpha E \Delta T(y)\, \mathrm{d}A\right] + \frac{y}{I}\left[M^M + \int_A \alpha E \Delta T(y)\, y \mathrm{d}A\right] \tag{2.20}$$

The thermal stress in the beam with free boundary condition is

$$\sigma_x(y) = -\alpha E \Delta T(y) + \frac{1}{A}\left[\int_A \alpha E \Delta T(y)\, \mathrm{d}A\right] + \frac{y}{I}\left[\int_A \alpha E \Delta T(y)\, y \mathrm{d}A\right] \tag{2.21}$$

The thermal stress in the beam with rectangular cross section with width b and the height h is

$$\sigma_x(y) = -\alpha E \Delta T(y) + \int_{-h/2}^{h/2} \alpha E \Delta T(y)\, \mathrm{d}y + \frac{12y}{h^3}\int_{-h/2}^{h/2} \alpha E \Delta T(y)\, y \mathrm{d}y \tag{2.22}$$

Next, we consider the thermal stress in the beam subjected to an arbitrary temperature change $\Delta T(x, y, z)$. The thermal stress σ_x is

$$\sigma_x(y) = -\alpha E \Delta T(x, y, z) + E\varepsilon_0 + E\frac{y}{k_y} + E\frac{y}{k_z} \tag{2.23}$$

where k_y and k_z denote the radii of curvature in the y and z directions, respectively, at the centroid of the cross section. When the external forces and moments act on the beam, the condition of the equilibrium of both forces and moments are

$$N = \int_A \sigma_x \mathrm{d}A, \quad M^{M_y} = \int_A \sigma_x z \mathrm{d}A, \quad M^{M_z} = \int_A \sigma_x y \mathrm{d}A \tag{2.24}$$

where N is the axial forces and M^{M_y} and M^{M_z} denote the mechanical bending moments with respect to y and z axes, respectively. The axial strain ε_0 and the curvatures $1/k_y$ and $1/k_z$ at $y = z = 0$, from Eq. (2.18), are

$$\begin{aligned}
\varepsilon_0 &= \frac{P}{EA} \\
\frac{1}{k_y} &= \frac{I_y M_z - I_{yz} M_y}{E\left(I_y I_z - I_{yz}^2\right)} \\
\frac{1}{k_z} &= \frac{I_z M_y - I_{yz} M_z}{E\left(I_y I_z - I_{yz}^2\right)}
\end{aligned} \tag{2.25}$$

where I_y and I_z are the moments of inertia of the cross section about the y and z axes, respectively, and I_{yz} is the product of inertia about these axes, they are defined as

$$I_y = \int_A z^2 \, \mathrm{d}A, \quad I_z = \int_A y^2 \, \mathrm{d}A, \quad I_{yz} = \int_A y\, z \, \mathrm{d}A \tag{2.26}$$

and

$$P = N + P^T, \quad M_y = M^{M_y} + M^{T_y}, \quad M_z = M^{M_z} + M^{T_z} \tag{2.27}$$

where

$$P^T = \int_A \alpha\, E\, \Delta T(x, y, z)\, \mathrm{d}A \tag{2.28}$$

and

$$M^{T_y} = \int_A \alpha\, E\, \Delta T(x, y, z)\, z\, \mathrm{d}A, \quad M^{T_z} = \int_A \alpha\, E\, \Delta T(x, y, z)\, y\, \mathrm{d}A \tag{2.29}$$

in which P^T represents the thermally induced force, and M^{T_y} and M^{T_z} the thermally induced moments about the y and z axes, respectively, P the total force, and M^y and M^z the total moments due to both mechanical and thermal loadings. The thermal stress due to both thermal and mechanical loadings is given from Eq. (2.23) by

$$\sigma_x(x, y, z) = -\alpha\, E\, \Delta T + \frac{P}{A} + \frac{I_y M_z - I_{yz} M_y}{E\left(I_y I_z - I_{yz}^2\right)} y + \frac{I_z M_y - I_{yz} M_z}{E\left(I_y I_z - I_{yz}^2\right)} z \tag{2.30}$$

When the cross section of the beam is symmetric about the y axis Eq. (2.30) is reduced to

$$\sigma_x(x, y, z) = -\alpha\, E\, \Delta T + \frac{P}{A} + \frac{M_z}{I_z} y + \frac{M_y}{I_y} z + \tag{2.31}$$

This is because the product of inertia $I_{yz} = 0$. Next, we consider the shearing stress in a beam. The equilibrium of an element of the beam with arbitrary cross section, small length $\mathrm{d}x$ and width of the beam $b(y)$ is

$$-\int_{-b_1}^{b_1} (\sigma_{xy}\, \mathrm{d}x)\, \mathrm{d}z + \int_{-b_1}^{b_1} \left[\int_y^{e_1} \left(\sigma_x + \frac{\partial \sigma_x}{\partial x} \mathrm{d}x\right) \mathrm{d}y\right] \mathrm{d}z - \int_{-b_1}^{b_1} \int_y^{e_1} \sigma_x\, \mathrm{d}x\, \mathrm{d}z = 0 \tag{2.32}$$

where σ_{xy} is the shearing stress. The latter is then derived by using Eqs. (2.20) and (2.32) as

$$\begin{aligned} \sigma_{xy} = \int_y^{e_1} \frac{\partial}{\partial x} \Big(& -\alpha\, E\, \Delta T(x, y) + \frac{1}{A}\left[N + \int_A \alpha\, E\, \Delta T(x, y)\, \mathrm{d}A\right] \\ & + \frac{y}{I}\left[M_M + \int_A \alpha\, E\, \Delta T(x, y)\, y \mathrm{d}A\right]\Big) \end{aligned} \tag{2.33}$$

If the bending stress is independent of the coordinate x, the shearing stress does not occur.

Problem 1

Find the thermal stress in the beam subjected to a temperature rise ΔT, when the origin of the coordinate system does not coincide with the centroid of the section.

Solution

The variable y denotes the distance from the origin of the coordinate system. If ε_0 and k denote the axial strain and the radius of curvature at $y=0$, respectively, then the axial stress is

$$\sigma_{xx} = -\alpha E \Delta T + E\varepsilon_0 + E\frac{y}{k} \tag{2.34}$$

Since external forces do not act on the beam

$$\int_A \sigma_{xx}\, dA = 0, \quad \int_A \sigma_{xx}\, y\, dA = 0 \tag{2.35}$$

Substitution Eq. (2.34) into Eq. (2.35) leads to

$$\begin{aligned} E\varepsilon_0 A + E\frac{1}{k}\int_A y\, dA &= \int_A \alpha E \Delta T\, dA \\ E\varepsilon_0 \int_A y\, dA + E\frac{1}{k}\int_A y^2\, dA &= \int_A \alpha E \Delta T\, y\, dA \end{aligned} \tag{2.36}$$

The solutions of algebraic Eqs. (2.36) are

$$\begin{aligned} \varepsilon_0 &= \frac{1}{E(AI_2 - I_1^2)}\left(I_2\int_A \alpha E \Delta T\, dA - I_1\int_A \alpha E \Delta T\, y\, dA\right) \\ \frac{1}{k} &= \frac{1}{E(AI_2 - I_1^2)}\left(A\int_A \alpha E \Delta T\, y\, dA - I_1\int_A \alpha E \Delta T\, dA\right) \end{aligned} \tag{2.37}$$

where

$$I_1 = \int_A y\, dA, \quad I_2 = \int_A y^2\, dA \tag{2.38}$$

The stress σ_{xx} is

$$\begin{aligned} \sigma_{xx} = &-\alpha E \Delta T(y) \\ &+ \frac{1}{AI_2 - I_1^2}\left[I_2\int_A \alpha E \Delta T(y)\, dA - I_1\int_A \alpha E \Delta T(y)\, y\, dA\right] \\ &+ \frac{y}{AI_2 - I_1^2}\left[A\int_A \alpha E \Delta T(y)\, y\, dA - I_1\int_A \alpha E \Delta T(y)\, dA\right] \end{aligned} \tag{2.39}$$

Problem 2

We consider a statically determinant cantilever beam of symmetrical cross section. The beam experiences a temperature distribution that varies through the depth y of the beam, but it is independent of x and z as well as time; i.e., $T = T(y)$. We seek equations for the elongation, transverse bending deflections, and normal stress.

Solution

Because the temperature depends only on y, the thermal force and thermal moments are constant and are given by

$$N_T = E\alpha \int_A [T_y - T_{ref}]\, dA\,, \quad M_{yT} = 0\,, \quad M_{zT} = E\alpha \int_A [T_y - T_{ref}] y\, dA \tag{2.40}$$

The thermal bending moment M_{yT} is zero because the temperature is independent of z, and it is symmetrical about the y axis. The axial force N_x and bending moments M_y and M_z are zero at any cross section since the beam experiences no external forces. Because the cross-sectional area is symmetric about the y axis, the product of inertia $I_yz = 0$. From Eq. (6.13), we find that

$$\frac{\partial u_0}{\partial x} = \frac{N_T}{EA}\,, \quad \frac{\partial^2 v}{\partial x^2} = -\frac{M_{zT}}{EI_z}\,, \quad \frac{\partial^2 w}{\partial x^2} = 0 \tag{2.41}$$

The boundary conditions at the fixed end are

$$u_0(0) = 0\,, \quad v(0) = \frac{\partial v}{\partial x}(0) = 0\,, \quad w(0) = \frac{\partial w}{\partial x}(0) = 0 \tag{2.42}$$

Integrating the differential equations and imposing the boundary conditions yields

$$u_0(x) = \frac{N_T x}{EA}\,, \quad v(x) = -\frac{M_{zT} x^2}{2EI_z}\,, \quad w(x) = 0 \tag{2.43}$$

In this case, because of the symmetrical cross section and temperature distribution, the beam deforms in the $x - y$ plane. Unless we know the temperature distribution, we cannot tell the sense of the deflection. Intuitively, we expect that, if the top of the beam is hotter than the bottom, the top fibers of the beam will expand more than the bottom fibers, and the beam will deflect down. Thus, for this situation, we expect $M_{zT} > 0$ and $v < 0$.

The normal stress is given by Eq. (6.14) and, after substitutions, we find that

$$\sigma_x = \frac{N_T}{A} + \frac{M_{zT} y}{I_z} - E\alpha[T(y) - T_{ref}] \tag{2.44}$$

If the beam was subjected to mechanical loads, we could include their effects in the preceding analysis, or we could perform the analyses separately and superimpose the results because the problem is linear.

2.1.3 Plates

We consider the thermal stresses in plates under Kirchhoff-Love hypothesis. That is, the plane initially perpendicular to the neutral plane of the plate, remains plane and perpendicular in its deformed configuration. We consider a uniform change of temperature ΔT through-the-thickness plate direction, and the plate is subjected to the same bending along both x and y, the in-plane strain components ε_{xx} and ε_{yy} are

$$\varepsilon_{xx} = \varepsilon_{yy} = \varepsilon_0 + \frac{z}{k} \tag{2.45}$$

where z is the thickness coordinate, and ε_0 and k are the strain and radius of curvature at the neutral plane $z = 0$, respectively. When the plate is subjected to the in-plane force P per unit length and the bending moment M^M per unit length in the x and y directions, the thermal stress components in the x and y directions are given by

$$\begin{aligned}\sigma_{xx} = \sigma_{yy} = {} & \frac{P}{h} + \frac{12 M^M}{h^3} z + \frac{\alpha E}{1-\nu}\Big[- \Delta T(z) + \frac{1}{h}\int_{-h/2}^{h/2} \Delta T(z)\, \mathrm{d}z \\ & + \frac{12 z}{h^3}\int_{-h/2}^{h/2} \Delta T(z)\, z\, \mathrm{d}z\Big]\end{aligned} \tag{2.46}$$

In the case of pure thermal stress problem without external loadings, Eq. (2.46) reduces to

$$\sigma_{xx} = \sigma_{yy} = \frac{\alpha E}{1-\nu}\left[-\Delta T(z) + \frac{1}{h}\int_{-h/2}^{h/2} \Delta T(z)\, \mathrm{d}z + \frac{12 z}{h^3}\int_{-h/2}^{h/2} \Delta T(z)\, z\, \mathrm{d}z \right] \tag{2.47}$$

It is interesting to note that the thermal stresses given by Eq. (2.47) for the plate are $\frac{1}{(1-\nu)}$ times the values for the beam given by Eq. (2.22).

Referring to Kirchhoff-Love hypothesis the displacement components u, v and w in the x, y and z directions, respectively, are given as

$$u = u_0 - \frac{\partial w}{\partial x}, \quad v = v_0 - \frac{\partial w}{\partial y}, \quad w = w_0 \tag{2.48}$$

where u_0, v_0 and w_0 are the displacement components in the x, y and z directions at the reference plane $z = 0$.

The in-plane strain components are

$$\begin{aligned}
\varepsilon_{xx} &= \frac{\partial u}{\partial x} = \frac{\partial u_0}{\partial x} - z\,\frac{\partial^2 w}{\partial x^2} \\
\varepsilon_{yy} &= \frac{\partial v}{\partial y} = \frac{\partial v_0}{\partial y} - z\,\frac{\partial^2 w}{\partial y^2} \\
\varepsilon_{xy} &= \frac{1}{2}\left(\frac{\partial u}{\partial y} + \frac{\partial v}{\partial x}\right) = \frac{1}{2}\left(\frac{\partial u_0}{\partial y} + \frac{\partial v_0}{\partial x}\right) - z\,\frac{\partial^2 w}{\partial x \partial y}
\end{aligned} \tag{2.49}$$

Hooke's law is given as

$$\begin{aligned}
\varepsilon_{xx} &= \frac{1}{E}\left(\sigma_{xx} - \nu\,\sigma_{yy}\right) + \alpha\,\Delta T \\
\varepsilon_{yy} &= \frac{1}{E}\left(\sigma_{yy} - \nu\,\sigma_{xx}\right) + \alpha\,\Delta T \\
\varepsilon_{xy} &= \frac{1}{2}\frac{\sigma_{xy}}{G} = \frac{1+\nu}{E}\sigma_{xy}
\end{aligned} \tag{2.50}$$

The stress components are

$$\begin{aligned}
\sigma_{xx} &= \frac{E}{1-\nu^2}\left[\frac{\partial u_0}{\partial x} + \nu\frac{\partial v_0}{\partial y} - z\left(\frac{\partial^2 w_0}{\partial x^2} + \nu\,\frac{\partial^2 w_0}{\partial y^2}\right) - (1+\nu)\,\alpha\,\Delta T\right] \\
\sigma_{yy} &= \frac{E}{1-\nu^2}\left[\frac{\partial v_0}{\partial y} + \nu\frac{\partial u_0}{\partial x} - z\left(\frac{\partial^2 w_0}{\partial y^2} + \nu\,\frac{\partial^2 w_0}{\partial x^2}\right) - (1+\nu)\,\alpha\,\Delta T\right] \\
\sigma_{xy} &= \frac{E}{2\,(1+\nu)}\left(\frac{\partial u_0}{\partial y} + \frac{\partial v_0}{\partial x} - 2\,z\,\frac{\partial^2 w_0}{\partial x \partial y}\right)
\end{aligned} \tag{2.51}$$

Let us introduce the resultant forces N_{xx}, N_{yy} and N_{xy}, and the resultant moments M_{xx}, M_{yy} and M_{xy} per unit length of the plate

$$\begin{aligned}
N_{xx} &= \int_{-h/2}^{h/2} \sigma_{xx}\,\mathrm{d}z, \quad N_{yy} = \int_{-h/2}^{h/2} \sigma_{yy}\,\mathrm{d}z, \quad N_{xy} = \int_{-h/2}^{h/2} \sigma_{xy}\,\mathrm{d}z \\
M_{xx} &= \int_{-h/2}^{h/2} \sigma_{xx}\,z\,\mathrm{d}z, \quad M_{yy} = \int_{-h/2}^{h/2} \sigma_{yy}\,z\,\mathrm{d}z, \quad M_{xy} = \int_{-h/2}^{h/2} \sigma_{xy}\,z\,\mathrm{d}z
\end{aligned} \tag{2.52}$$

Moreover, we introduce N^T and M^T which are the so-called thermally induced resultant force and resultant moment

$$N^T = \alpha\,E\int_{-h/2}^{h/2} \Delta T\,\mathrm{d}z, \quad M^T = \int_{-h/2}^{h/2} \Delta T\,z\,\mathrm{d}z \tag{2.53}$$

The resultant forces and resultant moments can be expressed in terms of displacement components u_0, v_0 and w (being $w - w_0$) as

$$N_{xx} = \frac{E}{1-\nu^2}\left(\frac{\partial u_0}{\partial x} + \nu\frac{\partial v_0}{\partial y}\right) - \frac{1}{1-\nu}N^T$$
$$N_{yy} = \frac{E}{1-\nu^2}\left(\frac{\partial v_0}{\partial y} + \nu\frac{\partial u_0}{\partial x}\right) - \frac{1}{1-\nu}N^T \tag{2.54}$$
$$N_{xy} = \frac{E}{2(1+\nu)}\left(\frac{\partial u_0}{\partial y} + \frac{\partial v_0}{\partial x}\right)$$

and

$$M_{xx} = -D\left(\frac{\partial^2 w}{\partial x^2} + \nu\frac{\partial^2 w}{\partial y^2}\right) - \frac{1}{1-\nu}M^T$$
$$M_{yy} = -D\left(\frac{\partial^2 w}{\partial y^2} + \nu\frac{\partial^2 w}{\partial x^2}\right) - \frac{1}{1-\nu}M^T \tag{2.55}$$
$$M_{xy} = (1-\nu)\,D\frac{\partial^2 w}{\partial x \partial y}$$

where D is the bending rigidity of the plate defined by

$$D = \frac{E h^3}{12(1-\nu^2)} \tag{2.56}$$

The thermal stress components can be given in terms of resultant forces and resultant moments as

$$\sigma_{xx} = \frac{1}{h}N_{xx} + \frac{12z}{h^3}M_{xx} + \frac{1}{1-\nu}\left(\frac{1}{h}N^T + \frac{12z}{h^3}M^T - \alpha E \Delta T\right)$$
$$\sigma_{yy} = \frac{1}{h}N_{yy} + \frac{12z}{h^3}M_{yy} + \frac{1}{1-\nu}\left(\frac{1}{h}N^T + \frac{12z}{h^3}M^T - \alpha E \Delta T\right) \tag{2.57}$$
$$\sigma_{xy} = \frac{1}{h}N_{xy} - \frac{12z}{h^3}M_{xy}$$

The equilibrium equations in terms of forces in the in-plane directions x and y are

$$\frac{\partial N_{xx}}{\partial x} + \frac{\partial N_{xy}}{\partial y} = 0$$
$$\frac{\partial N_{xy}}{\partial x} + \frac{\partial N_{yy}}{\partial x} = 0 \tag{2.58}$$

By defining the in-plane force resultants in terms of a thermal stress function F as follows

$$N_{xx} = \frac{\partial^2 F}{\partial x^2}, \quad N_{yy} = \frac{\partial^2 F}{\partial x^2}, \quad N_{xy} = -\frac{\partial^2 F}{\partial x \partial y} \tag{2.59}$$

the equilibrium equations (2.58) are automatically satisfied. The governing equation in terms of F becomes

$$\nabla^2\nabla^2 F = -\nabla^2 N^T \tag{2.60}$$

Considering the equilibrium in the out-of-plane direction (z-axis), the resultants of twisting moment M_{yx} and shear forces Q_x and Q_y per unit length parallel to the x and y axis, respectively, are given as

$$Q_x = \int_{-h/2}^{h/2} \sigma_{xz}\,\mathrm{d}z, \quad Q_y = \int_{-h/2}^{h/2} \sigma_{yz}\,\mathrm{d}z, \quad M_{yx} = \int_{-h/2}^{h/2} \sigma_{yx}\, z\,\mathrm{d}z \tag{2.61}$$

Comparison between the definition of the M_{xy} given in Eq. (2.52) and that provided in Eq. (2.61) leads to

$$M_{yx} = -M_{xy} \tag{2.62}$$

The governing equation of deflection w for the thermal bending problem is

$$\nabla^2\nabla^2 w = -\frac{1}{(1-\nu)\,D}\nabla^2 M^T \tag{2.63}$$

The coordinate transformation of the moments and shearing forces between a Cartesian coordinate system (x, y) and an other Cartesian coordinate system (n, s), where n and s are the coordinate normal and tangential to the domain boundary, respectively, are

$$\begin{aligned}
M_n &= M_{xx}\cos^2(\alpha) + M_{yy}\sin^2(\alpha) - 2\,M_{xy}M_{xx}\sin(\alpha)\cos(\alpha)\\
M_s &= M_{xx}\sin^2(\alpha) + M_{yy}\cos^2(\alpha) + 2\,M_{xy}M_{xx}\sin(\alpha)\cos(\alpha)\\
M_{ns} &= \left(M_{xx} - + M_{yy}\right)\sin(\alpha)\cos(\alpha) + M_{xy}\left(\cos^2(\alpha) - \sin^2(\alpha)\right)\\
Q_n &= Q_x\cos(\alpha) + Q_y\sin(\alpha)\\
Q_s &= -Q_x\sin(\alpha) + Q_y\cos(\alpha)
\end{aligned} \tag{2.64}$$

where α is the angle between the x-axis and the n-axis. The moments M_n, M_s and M_{ns} and the shearing forces Q_n and Q_s, can also be expressed in terms of the transverse displacement w in the coordinate system (n, s) as

$$\begin{aligned}
M_n &= -D\left(\frac{\partial^2 w}{\partial n^2} + \nu\frac{\partial^2 w}{\partial s^2}\right) - \frac{1}{1-\nu}M^T\\
M_s &= -D\left(\frac{\partial^2 w}{\partial s^2} + \nu\frac{\partial^2 w}{\partial n^2}\right) - \frac{1}{1-\nu}M^T\\
M_n &= (1-\nu)\,D\frac{\partial^2 w}{\partial n\partial s}
\end{aligned} \tag{2.65}$$

$$Q_n = -\frac{\partial}{\partial n}\left(D\nabla^2 w + \frac{1}{1-\nu}M^T\right)$$
$$Q_s = -\frac{\partial}{\partial s}\left(D\nabla^2 w + \frac{1}{1-\nu}M^T\right)$$

Three kinds of boundary conditions for bending problems due to thermal loads are:

1. Fixed end

$$w = 0, \quad \frac{\partial w}{\partial n} = 0 \tag{2.66}$$

2. Simply supported edge

$$\begin{aligned} &w = 0, \quad M_n = 0 \\ &\text{or alternatively} \\ &w = 0, \quad \frac{\partial^2 w}{\partial n^2} + \nu\frac{\partial^2 w}{\partial s^2} = -\frac{1}{(1-\nu)D}M^T \end{aligned} \tag{2.67}$$

3. Free edge

$$\begin{aligned} &M_n = 0, \quad V_n = Q_n - \frac{M_{ns}}{\partial s} = 0 \\ &\text{or alternatively} \\ &\frac{\partial^2 w}{\partial n^2} + \nu\frac{\partial^2 w}{\partial s^2} = -\frac{1}{(1-\nu)D}M^T \\ &\frac{\partial}{\partial n}\left[\frac{\partial^2 w}{\partial n^2} + (2-\nu)\frac{\partial^2 w}{\partial s^2}\right] = -\frac{1}{(1-\nu)D}\frac{\partial M^T}{\partial n} \end{aligned} \tag{2.68}$$

Problem 1

We consider a rectangular plate of dimensions a and b with all edges simply supported. The plate experiences an arbitrary temperature distribution $T(x, y, z)$ and the transverse loading $q = 0$. We assume small deflections; hence, the problem is linear, and solutions due to temperature and the transverse loading may be developed separately and superimposed. Our objective is to determine the bending deflection $w(x, y)$; once w is known, bending moments and stresses may be derived directly.

Solution

The boundary-value problem consists of solving the equation

$$D\nabla^4 w + \nabla^2 M_T = 0 \tag{2.69}$$

subject to two boundary conditions on each of the four edges of the plate. Here, M_T is the thermal moment defined as

$$M_T = \frac{E\alpha}{1-\nu}\int_{-h/2}^{h/2} \Delta T z \, dz \tag{2.70}$$

where $\Delta T = T(x, y, z) - T_{ref}$. For an arbitrary temperature distribution, the thermal moment varies over the plate, i.e., $M_T = M_T(x, y)$. The boundary conditions are

$$\begin{aligned} w(0, y) &= 0 \quad w(x, 0) = 0 \\ w(a, y) &= 0 \quad w(x, b) = 0 \end{aligned} \tag{2.71}$$

and

$$\begin{aligned} M_x(0, y) &= 0 \quad M_y(x, 0) = 0 \\ M_x(a, y) &= 0 \quad M_y(x, b) = 0 \end{aligned} \tag{2.72}$$

Let us consider the moment boundary conditions in more detail. On the edges parallel to the y axes,

$$M_x = -D\left(\frac{\partial^2 w}{\partial x^2} + \nu \frac{\partial^2 w}{\partial y^2}\right) - M_T \tag{2.73}$$

From the displacement boundary conditions on these edges, we have $\partial^2 w/\partial y^2 = 0$, assuming, as we do, that the edge is perfectly straight. Thus, the moment boundary condition reduces to

$$M_x = -D\frac{\partial^2 w}{\partial x^2} - M_T = 0 \quad \text{on } x = 0 \text{ and } x = a \tag{2.74}$$

Similarly,

$$M_y = -D\frac{\partial^2 w}{\partial y^2} - M_T = 0 \quad \text{on } y = 0 \text{ and } y = b \tag{2.75}$$

Alternatively, we may say that the moment boundary condition on a typical edge is given by

$$D\nabla^2 w + M_T = 0 \tag{2.76}$$

where we recognize that one term in ∇^2 is always zero.

An important point that we should observe immediately is that the moment boundary conditions are non-homogeneous. The non-homogeneous boundary conditions are a signal of mathematical difficulty because separation of variables cannot be used unless the boundary conditions are homogeneous. This difficulty is often resolved by forming the solution as the sum of two or more parts such that a new non-homogeneous partial differential equation with homogeneous boundary conditions is introduced. Then the new problem is solved by separation of variables.

We use an approach that reduces the plate bending problem to that of the deflection of a membrane. The fourth-order plate equation is replaced by two second-order equations. We begin by writing Eq. (2.69) as

$$\nabla^2(D\nabla^2 w + M_T) = 0 \tag{2.77}$$

which is equivalent to two equations:

$$D\nabla^2 w + M_T = f(x, y) \quad \nabla^2 f = 0 \tag{2.78}$$

where $f(x, y)$ is a function to be determined. Each of these equations is second-order and requires only one boundary condition on each plate edge. On a plate edge, the moment boundary condition shows that the left-hand side of the first of the two preceding equations is zero. Thus, on the plate edges, the function f is zero. Now we realize from the second of the two preceding equations that we have a homogeneous partial differential equation with homogeneous boundary conditions; therefore, the solution is $f(x, y) = 0$. Thus, the problem reduces to the single second-order non-homogeneous equation,

$$D\nabla^2 w = -M_T(x, y) \tag{2.79}$$

with the homogeneous boundary condition $w = 0$ on each edge. This problem is easily solved by separation of variables using a double Fourier sine series,

$$w(x, y) = \sum_{m=1}^{\infty}\sum_{n=1}^{\infty} w_{mn} \sin\frac{m\pi x}{a} \sin\frac{n\pi y}{b} \tag{2.80}$$

where ω_{mn} are Fourier coefficients to be determined. The Fourier series automatically satisfies the displacement boundary conditions. To determine ω_{mn}, we expand the thermal moment in a similar series,

$$M_T(x, y) = \sum_{m=1}^{\infty}\sum_{n=1}^{\infty} a_{mn} \sin\frac{m\pi x}{a} \sin\frac{n\pi y}{b} \tag{2.81}$$

where

$$a_{mn} = \frac{4}{ab}\int_0^b\int_0^a M_T(x, y) \sin\frac{m\pi x}{a} \sin\frac{n\pi y}{b}\, dx\, dy \tag{2.82}$$

Substituting the infinite series for $W(x, y)$ and $M_T(x, y)$ into the second order partial differential equation, we match coefficients to find

$$w_{mn} = \frac{a_{mn}}{\pi^2 D(m^2/a^2 + n^2/b^2)} \tag{2.83}$$

and, consequently,

$$w(x, y) = \frac{1}{\pi^2 D} \sum_{m=1}^{\infty} \sum_{n=1}^{\infty} \frac{a_{mn}}{(m^2/a^2 + n^2/b^2)} \sin\frac{m\pi x}{a} \sin\frac{n\pi y}{b} \tag{2.84}$$

A solution like this one in the form of a double Fourier sine series is called a Navier solution. For a given thermal moment, we evaluate the Fourier coefficients a_{mn} and then sum the infinite series of the preceding equation to obtain $w(x, y)$. The series converge satisfactorily for the deflection w, but we encounter computational difficulty when we seek bending moments. Therefore, this solution is not practical for the determination of plate moments, and we must find an alternative solution.

Problem 2

As in the previous problem we consider a rectangular plate of dimensions a and b with all edges simply supported. The objective is to determine $w(x, y)$ when the plate is subjected to a temperature varying only through-the-thickness plate direction.

Solution

For the special case of a rectangular simply supported plate with a temperature varying only through the plate thickness, we can use the preceding solution to derive a series solution for $w(x, y)$. From the previous example, we can evaluate the coefficients a_{mn}. Since $T = T(z)$, the thermal moment is constant, and we have

$$a_{mn} = \frac{4M_T}{ab} \int_0^b \sin\frac{n\pi y}{b} dy \int_0^a \sin\frac{m\pi x}{a} dx \tag{2.85}$$

Hence,

$$\begin{aligned} a_{mn} &= \frac{16M_T}{\pi^2}\frac{1}{mn} \quad m, n \text{ odd} \\ &= 0 \quad m, n \text{ even} \end{aligned} \tag{2.86}$$

Thus, the solution is

$$w(x, y) = \frac{16}{\pi^4}\frac{M_T}{D} \sum_{m=1,3,5}^{\infty} \sum_{n=1,3,5}^{\infty} \frac{1}{mn(m^2/a^2 + n^2/b^2)} \sin\frac{m\pi x}{a} \sin\frac{n\pi y}{b} \tag{2.87}$$

As in the previous example, this series solution works well for the deflections, but bending moments derived from it are known to converge poorly.

Problem 3

Find the thermal stress for a thin plate when it is subjected to a temperature variation $\Delta T = A + Bz$.

Solution

Thermal stress is given by Eq. (2.47)

$$\sigma_{xx} = \frac{\alpha E}{1-\nu}\left[-\Delta T(z) + \frac{1}{h}\int_{-h/2}^{h/2} \Delta T(z)\,\mathrm{d}z + \frac{12z}{h^3}\int_{-h/2}^{h/2} \Delta T(z)\,z\,\mathrm{d}z\right] \tag{2.88}$$

We calculate each integral

$$\begin{aligned}\int_{-h/2}^{h/2} \Delta T(z)\,\mathrm{d}z &= \int_{-h/2}^{h/2} (A + Bz)\,\mathrm{d}z = Ah \\ \int_{-h/2}^{h/2} \Delta T(z)\,z\,\mathrm{d}z &= \int_{-h/2}^{h/2} \left(Az + Bz^2\right)\mathrm{d}z = \frac{B}{12}h^3\end{aligned} \tag{2.89}$$

Substitution of Eq. (2.89) into Eq. (2.88) leads to the final expression of the thermal stress

$$\sigma_{xx} = \frac{\alpha E}{1-\nu}\left[-(A+Bz) + \frac{1}{h}Ah + \frac{12z}{h^3}\frac{B}{12}h^3\right] = 0 \tag{2.90}$$

Thus, thermal stress does not occur.

2.1.4 Cylinders

Solid cylinders

Consider a solid circular cylinder with axis oz of radius b subjected to the radial temperature variation $\theta(r) = T(r) - T_0$ with T_0 being the reference temperature, assuming plane strain condition $\varepsilon_{zz} = 0$, $\varepsilon_{rz} = 0$, $\varepsilon_{\phi z} = 0$. The stress-strain relations are:

$$\begin{aligned}\varepsilon_{rr} &= \frac{1}{E}\left[\sigma_{rr} - \nu\left(\sigma_{\phi\phi} + \sigma_{zz}\right)\right] + \alpha\theta \\ \varepsilon_{\phi\phi} &= \frac{1}{E}\left[\sigma_{\phi\phi} - \nu\left(\sigma_{rr} + \sigma_{zz}\right)\right] + \alpha\theta \\ \sigma_{zz} &= \nu\left(\sigma_{rr} + \sigma_{\phi\phi}\right) - E\alpha\theta\end{aligned} \tag{2.91}$$

Solving for stresses in terms of strains yields:

$$\begin{aligned}\sigma_{rr} &= \frac{E}{(1+\nu)(1-2\nu)}\left[(1-\nu)\,\varepsilon_{rr} + \nu\varepsilon_{\phi\phi} - (1+\nu)\,\alpha\theta\right] \\ \sigma_{\phi\phi} &= \frac{E}{(1+\nu)(1-2\nu)}\left[(1-\nu)\,\varepsilon_{\phi\phi} + \nu\varepsilon_{rr} - (1+\nu)\,\alpha\theta\right]\end{aligned} \tag{2.92}$$

The equilibrium equation for the axial symmetry is

$$\frac{d\sigma_{rr}}{dr} + \frac{\sigma_{rr} - \sigma_{\phi\phi}}{r} = 0 \tag{2.93}$$

and the strain-displacement relations, with u_r being the radial displacement, are

$$\varepsilon_{rr} = \frac{du_r}{dr} \qquad \varepsilon_{\phi\phi} = \frac{u_r}{r} \tag{2.94}$$

Substituting Eq. (2.94) into Eq. (2.92) gives

$$\begin{aligned}\sigma_{rr} &= \frac{E}{(1+\nu)(1-2\nu)}\left[(1-\nu)\,\frac{du_r}{dr} + \nu\frac{u_r}{r} - (1+\nu)\,\alpha\theta\right] \\ \sigma_{\phi\phi} &= \frac{E}{(1+\nu)(1-2\nu)}\left[(1-\nu)\,\frac{u_r}{r} + \nu\frac{du_r}{dr} - (1+\nu)\,\alpha\theta\right]\end{aligned} \tag{2.95}$$

The use of Eqs. (2.95) into Eq. (2.93) leads to the equilibrium equation in terms of displacement u_r as

$$\frac{d}{dr}\left[\frac{1}{r}\frac{du_r r}{dr}\right] = \frac{1+\nu}{1-\nu}\alpha\frac{d\theta}{dr} \tag{2.96}$$

The integration of Eq. (2.96) yields

$$u_r = \frac{1+\nu}{1-\nu}\frac{\alpha}{r}\int_0^r \theta r dr + C_1 r + \frac{C_2}{r} \tag{2.97}$$

where C_1 and C_2 are the constants of integration. Since the displacement must be finite at $r=0$, it follows that C_2 must be zero. The strain components in Eq. (2.94) are

$$\begin{aligned}\varepsilon_{rr} &= \frac{1+\nu}{1-\nu}\frac{\alpha}{r^2}\int_0^r \theta r dr + C_1 + \frac{1+\nu}{1-\nu}\alpha\theta \\ \varepsilon_{\phi\phi} &= \frac{1+\nu}{1-\nu}\frac{\alpha}{r^2}\int_0^r \theta r dr + C_1\end{aligned} \tag{2.98}$$

and the stresses from Eq. (2.92) are

$$\begin{aligned}\sigma_{rr} &= \frac{E}{1-\nu}\frac{\alpha}{r^2}\int_0^r \theta r dr + C_1\frac{E}{(1+\nu)(1-2\nu)} \\ \sigma_{\phi\phi} &= \frac{E}{1-\nu}\frac{\alpha}{r^2}\int_0^r \theta r dr - \frac{E\alpha\theta}{1-\nu} + C_1\frac{E}{(1+\nu)(1-2\nu)}\end{aligned} \tag{2.99}$$

The constant C_1 is determined imposing the boundary condition

$$\sigma_{rr} = 0 \quad \text{at } r = b \tag{2.100}$$

which gives

$$C_1 = \frac{\alpha(1+\nu)(1-2\nu)}{(1-\nu)b^2} \int_0^b \theta r dr \tag{2.101}$$

The substitution into Eqs. (2.97) and Eq. (2.99) provides

$$\begin{aligned} u_r &= \frac{1+\nu}{1-\nu} \frac{\alpha}{r} \left[\int_0^r \theta r dr + (1-2\nu) \frac{r^2}{b^2} \int_0^r \theta r dr \right] \\ \sigma_{rr} &= \frac{E\alpha}{1-\nu} \left[\frac{1}{b^2} \int_0^r \theta r dr - \frac{1}{r^2} \int_0^r \theta r dr \right] \\ \sigma_{\phi\phi} &= \frac{E\alpha}{1-\nu} \left[\frac{1}{b^2} \int_0^r \theta r dr + \frac{1}{r^2} \int_0^r \theta r dr - \theta \right] \end{aligned} \tag{2.102}$$

The stress in the axial direction σ_{zz} is obtained from the last of Eqs. (2.91)

$$\sigma_{zz} = \frac{E\alpha}{1-\nu} \left[\frac{2\nu}{b^2} \int_0^b \theta r dr - \theta \right] \tag{2.103}$$

Hollow cylinders

Consider the hollow cylinder with inside and outside radii a and b, respectively. The same equilibrium equations hold, but the integration for the displacement u_r is carried out from a, the inside radius, to r. Thus, Eq. (2.97) assumes the following form

$$u_r = \frac{1+\nu}{1-\nu} \frac{\alpha}{r} \int_a^r \theta r dr + C_1 r + \frac{C_2}{r} \tag{2.104}$$

Substituting u_r from this equation into Eq. (2.94) and then into Eqs. (2.92) the radial stress becomes

$$\sigma_{rr} = E\left[-\frac{\alpha}{(1-\nu)r^2} \int_a^r \theta r dr + \frac{C_1}{(1+\nu)(1-2\nu)} - \frac{C_2}{(1+\nu)r^2} \right] \tag{2.105}$$

Applying the boundary conditions

$$\begin{aligned} \sigma_{rr} &= 0 \quad r = a \\ \sigma_{rr} &= 0 \quad r = b \end{aligned} \tag{2.106}$$

yields

$$\begin{aligned} C_1 &= \frac{(1+\nu)(1-2\nu)}{(1-\nu)} \frac{\alpha}{(b^2-a^2)} \int_a^b \theta r dr \\ C_2 &= \frac{(1+\nu)}{(1-\nu)} \frac{\alpha a^2}{(b^2-a^2)} \int_a^b \theta r dr \end{aligned}$$

Substituting C_1 and C_2 into Eq. (2.104), the radial displacement and the stresses become

$$
\begin{aligned}
u_r &= \frac{1+\nu}{1-\nu}\frac{\alpha}{r}\left[\frac{(1-2\nu)\,r^2+a^2}{b^2-a^2}\int_a^b \theta r dr + \int_a^r \theta r dr\right] \\
\sigma_{rr} &= \frac{E\alpha}{1-\nu}\left[\frac{1}{b^2-a^2}\left(1-\frac{a^2}{r^2}\right)\int_a^b \theta r dr - \frac{1}{r^2}\int_a^r \theta r dr\right] \\
\sigma_{\phi\phi} &= \frac{E\alpha}{1-\nu}\left[\frac{1}{b^2-a^2}\left(1+\frac{a^2}{r^2}\right)\int_a^b \theta r dr + \frac{1}{r^2}\int_a^r \theta r dr\right]
\end{aligned}
\tag{2.107}
$$

The axial stress from the last of Eqs. (2.91) is

$$
\sigma_{zz} = \frac{E\alpha}{1-\nu}\left[\frac{2\nu}{b^2-a^2}\int_a^b \theta r dr - \theta\right]
\tag{2.108}
$$

the axial force for this case, i.e., when $\varepsilon_{zz}=0$, is

$$
F_z = \int_a^b 2\pi r \sigma_{zz} dz
\tag{2.109}
$$

If the temperature of the hollow cylinder is T_a at the inner surface and T_b at the outer surface, then the temperature distribution becomes (see Subsection 2.2.1)

$$
T = \frac{T_d}{ln\left(\frac{b}{a}\right)}\left[ln\left(\frac{b}{r}\right)\right] + T_b
\tag{2.110}
$$

where $T_d = T_a - T_b$. Substituting the temperature distribution T in Eqs. (2.107), the stresses for a hollow cylinder with fixed ends become

$$
\begin{aligned}
\sigma_{rr} &= -\frac{E\alpha T_d}{2(1-\nu)\ln\left(\frac{b}{a}\right)}\left[\ln\left(\frac{b}{r}\right) + \frac{a^2}{b^2-a^2}\left(1-\frac{b^2}{r^2}\right)\ln\left(\frac{b}{a}\right)\right] \\
\sigma_{\phi\phi} &= \frac{E\alpha T_d}{2(1-\nu)\ln\left(\frac{b}{a}\right)}\left[1-\ln\left(\frac{b}{r}\right) - \frac{a^2}{b^2-a^2}\left(1+\frac{b^2}{r^2}\right)\ln\left(\frac{b}{a}\right)\right] \\
\sigma_{zz} &= \frac{E\alpha T_d}{2(1-\nu)\ln\left(\frac{b}{a}\right)}\left[1-\frac{2a^2}{b^2-a^2}\ln\left(\frac{b}{a}\right) - \frac{2}{\nu}\ln\left(\frac{b}{r}\right)\right] - E\alpha\,(T_b - T_0)
\end{aligned}
\tag{2.111}
$$

2.1.5 Spheres

The equilibrium equation for a thick sphere vessel of inside radius a and outside radius b subjected to a radial temperature change is

$$\frac{d\sigma_{rr}}{dr}+\frac{2\left(\sigma_{rr}-\sigma_{\phi\phi}\right)}{r}=0 \tag{2.112}$$

and the strain-displacement relations are

$$\varepsilon_{rr}=\frac{du_r}{dr}\qquad \varepsilon_{\phi\phi}=\frac{u_r}{r} \tag{2.113}$$

The stress-strain relations for spherical symmetry, i.e., when $\sigma_{\theta\theta}=\sigma_{\phi\phi}$, are

$$\begin{aligned}\varepsilon_{rr}&=\frac{1}{E}\left[\sigma_{rr}-2\nu\sigma_{\phi\phi}\right]+\alpha\theta\\ \varepsilon_{\phi\phi}&=\frac{1}{E}\left[(1-\nu)\,\sigma_{\phi\phi}-\nu\sigma_{rr}\right]+\alpha\theta\end{aligned} \tag{2.114}$$

Substituting Eqs. (2.113) into Eq. (2.114) and into Eq. (2.112), the equilibrium equation in terms of radial displacement u_r reduces to

$$\frac{d}{dr}\left[\frac{1}{r^2}\frac{d\left(u_r r^2\right)}{dr}\right]=\alpha\frac{1+\nu}{1-\nu}\frac{d\theta}{dr} \tag{2.115}$$

The integration of Eq. (2.115) yields

$$u_r=\frac{1+\nu}{1-\nu}\frac{\alpha}{r^2}\int_a^r\theta r^2 dr+C_1 r+\frac{C_2}{r^2} \tag{2.116}$$

Using Eq. (2.116) in Eq. (2.113) and solving as for the stresses yields

$$\begin{aligned}\sigma_{rr}&=-\frac{E}{1-\nu}\frac{2\alpha}{r^3}\int_a^r\theta r^2 dr+\frac{EC_1}{(1-2\nu)}-\frac{2EC_2}{(1+\nu)\,r^3}\\ \sigma_{\phi\phi}&=\frac{E}{1-\nu}\frac{\alpha}{r^3}\int_a^r\theta r^2 dr-\frac{E\alpha\theta}{(1-\nu)}+\frac{EC_1}{(1-2\nu)}+\frac{2EC_2}{(1+\nu)\,r^3}\end{aligned} \tag{2.117}$$

The integration constants C_1 and C_2 can be obtained by imposing the following boundary conditions

$$\begin{aligned}\sigma_{rr}&=0\quad r=a\\ \sigma_{rr}&=0\quad r=b\end{aligned} \tag{2.118}$$

Upon substitution in Eqs. (2.116) and (2.117), yields

$$
\begin{aligned}
u_r &= \frac{1+\nu}{1-\nu}\frac{\alpha}{b^3-a^3}\left[\frac{a^3}{r^2}\int_r^b \theta r^2 \mathrm{d}r + \frac{b^3}{r^2}\int_a^r \theta r^2 \mathrm{d}r + \frac{2(1-2\nu)}{1+\nu} r \int_a^r \theta r^2 \mathrm{d}\right] \\
\sigma_{rr} &= -\frac{2E}{1-\nu}\frac{\alpha}{b^3-a^3}\left[\frac{a^3}{r^3}\int_r^b \theta r^2 \mathrm{d}r + \frac{b^3}{r^3}\int_a^r \theta r^2 \mathrm{d}r - \int_a^b \theta r^2 \mathrm{d}r\right] \\
\sigma_{\phi\phi} &= \frac{E}{1-\nu}\frac{\alpha}{b^3-a^3}\left[\frac{a^3}{r^3}\int_r^b \theta r^2 \mathrm{d}r + \frac{b^3}{r^3}\int_a^r \theta r^2 \mathrm{d}r + 2\int_a^b \theta r^2 \mathrm{d}r - (b^3-a^3)\theta\right]
\end{aligned}
\tag{2.119}
$$

For constant temperature change θ all the components of stresses vanish, and the radial displacement becomes proportional to the radius of the sphere

$$u_r = \alpha\theta r \tag{2.120}$$

For a solid sphere Eqs. (2.119) are used by letting the inside radius approach zero, thus

$$
\begin{aligned}
u_r &= \frac{1+\nu}{1-\nu}\alpha\left[\frac{1}{r^2}\int_0^r \theta r^2 \mathrm{d}r + \frac{2(1-2\nu)}{1+\nu}\frac{r}{b^3}\int_0^b \theta r^2 \mathrm{d}\right] \\
\sigma_{rr} &= \frac{2E\alpha}{1-\nu}\left[\frac{1}{b^3}\int_0^b \theta r^2 \mathrm{d}r - \frac{1}{r^3}\int_0^r \theta r^2 \mathrm{d}r\right] \\
\sigma_{\phi\phi} &= \frac{E\alpha}{1-\nu}\left[\frac{1}{r^3}\int_0^r \theta r^2 \mathrm{d}r + \frac{2}{b^3}\int_0^b \theta r^2 \mathrm{d}r - \theta\right]
\end{aligned}
\tag{2.121}
$$

At $r = 0$ the value of the displacement and stresses are indeterminate, but considering that the following limits hold

$$
\begin{aligned}
&\lim_{r\to 0}\frac{1}{r^2}\int_0^r \theta r^2 \mathrm{d}r = 0 \\
&\lim_{r\to 0}\frac{1}{r^3}\int_0^r \theta r^2 \mathrm{d}r = \frac{\theta(0)}{3}
\end{aligned}
\tag{2.122}
$$

then at $r = 0$ the radial displacement is zero and the stresses are

$$\sigma_{rr} = \sigma_{\phi\phi} = \frac{2E\alpha}{1-\nu}\left[\frac{1}{b^3}\int_0^b \theta r^2 \mathrm{d}r - \frac{\theta(0)}{3}\right] \tag{2.123}$$

For an infinite body subjected to a radial temperature change, the results of the sphere can be applied by letting $b \to \infty$ and $a \to 0$. We then obtain

$$
\begin{aligned}
u_r &= \frac{1+\nu}{1-\nu}\frac{\alpha}{r^2}\int_0^r \theta r^2 \mathrm{d}r \\
\sigma_{rr} &= -\frac{2E\alpha}{1-\nu}\frac{1}{r^3}\int_0^r \theta r^2 \mathrm{d}r \\
\sigma_{\phi\phi} &= \frac{E\alpha}{1-\nu}\frac{1}{r^3}\left[\int_0^r \theta r^2 \mathrm{d}r - r^3\theta\right]
\end{aligned}
\tag{2.124}
$$

2.1.6 Disks

Solid circular and angular disks

Consider a thin circular disk subjected to a radial temperature variation $\theta(r) = T(r) - T_0$ with T_0 reference temperature. The stress and displacement components are functions of the radius, and assuming the plane stress condition, the equilibrium equation is

$$
\frac{\mathrm{d}\sigma_{rr}}{\mathrm{d}r} + \frac{2(\sigma_{rr} - \sigma_{\phi\phi})}{r} = 0 \tag{2.125}
$$

the strain-displacement relations are

$$
\varepsilon_{rr} = \frac{\mathrm{d}u_r}{\mathrm{d}r} \qquad \varepsilon_{\phi\phi} = \frac{u_r}{r} \tag{2.126}
$$

and the stress-strain relations in polar coordinates for the plane stress condition become

$$
\begin{aligned}
\varepsilon_{rr} &= \frac{1}{E}\left[\sigma_{rr} - \nu\sigma_{\phi\phi}\right] + \alpha\theta \\
\varepsilon_{\phi\phi} &= \frac{1}{E}\left[\sigma_{\phi\phi} - \nu\sigma_{rr}\right] + \alpha\theta
\end{aligned}
\tag{2.127}
$$

Solving Eqs. (2.127) for stresses gives

$$
\begin{aligned}
\sigma_{rr} &= \frac{E}{1-\nu^2}\left[\varepsilon_{rr} + \nu\varepsilon_{\phi\phi} - (1+\nu)\alpha\theta\right] \\
\sigma_{\phi\phi} &= \frac{E}{1-\nu^2}\left[\varepsilon_{\phi\phi} + \nu\varepsilon_{rr} - (1+\nu)\alpha\theta\right]
\end{aligned}
\tag{2.128}
$$

Upon introduction of Eqs. (2.127) into Eq. (2.128) and then into Eq. (2.125) the equilibrium equation in terms of displacement becomes

$$
\frac{\mathrm{d}}{\mathrm{d}r}\left[\frac{1}{r}\frac{\mathrm{d}u_r r}{\mathrm{d}r}\right] = (1+\nu)\alpha\frac{\mathrm{d}\theta}{\mathrm{d}r} \tag{2.129}
$$

Integrating this equation the radial displacement for a hollow disk of inside radius a and outside radius b becomes

$$u_r = (1-\nu)\frac{\alpha}{r}\int_a^r \theta r dr + C_1 r + \frac{C_2}{r} \tag{2.130}$$

and the stresses from Eqs. (2.126) and (2.128) are

$$\begin{aligned}\sigma_{rr} &= -\frac{E\alpha}{r^2}\int_a^r \theta r dr + \frac{EC_1}{1-\nu} - \frac{EC_2}{(1+\nu)r^2} \\ \sigma_{\phi\phi} &= \frac{E\alpha}{r^2}\int_a^r \theta r dr - E\alpha\theta + \frac{EC_1}{1-\nu} + \frac{EC_2}{(1+\nu)r^2}\end{aligned} \tag{2.131}$$

Applying the boundary conditions

$$\begin{aligned}\sigma_{rr} &= 0 \quad r = a \\ \sigma_{rr} &= 0 \quad r = b\end{aligned} \tag{2.132}$$

the constants C_1 and C_2 are found to be

$$\begin{aligned}C_1 &= \frac{(1-\nu)\alpha}{b^2-a^2}\int_a^b \theta r dr \\ C_2 &= \frac{(1+\nu)\alpha a^2}{b^2-a^2}\int_a^b \theta r dr\end{aligned} \tag{2.133}$$

Therefore, the radial displacement and the stresses for a hollow disk which is free of traction on the boundary are

$$\begin{aligned}u_r &= \frac{(1+\nu)\alpha}{r}\int_a^r \theta r dr + \frac{(1+\nu)a^2 + (1-\nu)r^2}{\left(b^2-a^2\right)r}\alpha\int_a^b \theta r dr \\ \sigma_{rr} &= -\frac{E\alpha}{r^2}\int_a^r \theta r dr + \frac{E\alpha}{\left(b^2-a^2\right)}\left(1-\frac{a^2}{r^2}\right)\int_a^b \theta r dr \\ \sigma_{\phi\phi} &= \frac{E\alpha}{r^2}\int_a^r \theta r dr - E\alpha\theta + \frac{E\alpha}{\left(b^2-a^2\right)}\left(1+\frac{a^2}{r^2}\right)\int_a^b \theta r dr\end{aligned} \tag{2.134}$$

For a solid disk free of traction on the outer boundary we take $a \to 0$ and obtain

$$\begin{aligned}u_r &= \frac{(1+\nu)\alpha}{r}\int_0^r \theta r dr + \frac{(1-\nu)\alpha r}{b^2}\int_0^b \theta r dr \\ \sigma_{rr} &= \frac{E\alpha}{b^2}\left[\int_0^b \theta r dr - \frac{b^2}{r^2}\int_0^r \theta r dr\right] \\ \sigma_{\phi\phi} &= E\alpha\left[\frac{1}{b^2}\int_0^b \theta r dr + \frac{1}{r^2}\int_0^r \theta r dr - \theta\right]\end{aligned} \tag{2.135}$$

At the centre of the disk, for continuous temperature distribution,

$$\begin{aligned} \lim_{r\to 0}\frac{1}{r}\int_0^r \theta r^2 \mathrm{d}r &= 0 \\ \lim_{r\to 0}\frac{1}{r^2}\int_0^r \theta r^2 \mathrm{d}r &= \frac{\theta(0)}{2} \end{aligned} \tag{2.136}$$

Which results in zero radial displacement and equal value for σ_{rr} and $\sigma_{\phi\phi}$

$$\sigma_{rr}\big|_{r=0} = \sigma_{\phi\phi}\big|_{r=0} = E\alpha\left[\frac{1}{b^2}\int_0^b \theta r \mathrm{d}r - \frac{\theta(0)}{2}\right] \tag{2.137}$$

Rotating disks

Consider a thin circular disk of constant thickness rotating at an angular velocity ω. The temperature distribution is assumed to be axisymmetric and a function of radius r. The body force per unit volume due to centrifugal force is $F_c = \rho r\omega^2$, where ρ is the mass density of the disk material. Since the forces are function of the radius and are axisymmetric, the shear stress is zero and the radial and tangential stresses are function of the radius. Since the disk is thin, the plane stress condition is assumed. The equilibrium equation for axisymmetric stresses in a rotating disk is

$$\frac{\mathrm{d}\sigma_{rr}}{\mathrm{d}r} + \frac{2(\sigma_{rr} - \sigma_{\phi\phi})}{r} + \rho r\omega^2 = 0 \tag{2.138}$$

the strain-displacement relations and the stress-strain relations for the plane stress condition are

$$\begin{aligned} \varepsilon_{rr} &= \frac{du_r}{dr} \qquad \varepsilon_{\phi\phi} = \frac{u_r}{r} \\ \varepsilon_{rr} &= \frac{1}{E}\left[\sigma_{rr} - \nu\sigma_{\phi\phi}\right] + \alpha\theta \\ \varepsilon_{\phi\phi} &= \frac{1}{E}\left[\sigma_{\phi\phi} - \nu\sigma_{rr}\right] + \alpha\theta \end{aligned} \tag{2.139}$$

We introduce the stress function Φ by the relations

$$\sigma_{rr} = \frac{\Phi}{r} \qquad \sigma_{\phi\phi} = \frac{d\Phi}{dr} + \rho r\omega^2 \tag{2.140}$$

The equilibrium equation (2.138) is satisfied. Eliminating u_r from Eq. (2.139) leads to

$$r\frac{\mathrm{d}\varepsilon_{\phi\phi}}{\mathrm{d}r} + \varepsilon_{\phi\phi} - \varepsilon_{rr} = 0 \tag{2.141}$$

The compatibility equation (2.141) in terms of the stress function, using Eqs. (2.140) and (2.139), becomes

$$\frac{\mathrm{d}^2\Phi}{\mathrm{d}r^2}+\frac{1}{r}\frac{\mathrm{d}\Phi}{\mathrm{d}r}-\frac{\Phi}{r^2}+(3+\nu)\,\rho r\omega^2=-E\alpha\frac{\mathrm{d}\theta}{\mathrm{d}r} \tag{2.142}$$

This differential equation is rewritten as

$$\frac{\mathrm{d}}{\mathrm{d}r}\left[\frac{1}{r}\frac{\mathrm{d}}{\mathrm{d}r}(r\Phi)\right]=-(3+\nu)\,\rho r\omega^2-E\alpha\frac{\mathrm{d}\theta}{\mathrm{d}r} \tag{2.143}$$

Integrating with respect to r gives

$$\Phi=-\frac{3+\nu}{8}\rho r^3\omega^2-\frac{E\alpha}{r}\int_a^r\theta r\mathrm{d}r+C_1\frac{r}{2}+\frac{C_2}{r} \tag{2.144}$$

where a is the inside radius of the disk. For a solid disk $a=0$. Here C_1 and C_2 are the constants of integration. The corresponding stresses from Eq. (2.140) are

$$\begin{aligned}\sigma_{rr}&=\frac{\Phi}{r}=-\frac{3+\nu}{8}\rho r^2\omega^2-\frac{E\alpha}{r^2}\int_a^r\theta r\mathrm{d}r+\frac{C_1}{2}+\frac{C_2}{r^2}\\ \sigma_{\phi\phi}&=\frac{\mathrm{d}\Phi}{\mathrm{d}r}+\rho r^2\omega^2\\ &=-\frac{1+3\nu}{8}\rho r^2\omega^2+E\alpha\left[-\theta+\frac{1}{r^2}\int_a^r\theta r\mathrm{d}r\right]+\frac{C_1}{2}+\frac{C_2}{r^2}\end{aligned} \tag{2.145}$$

For a solid disk ($a=0$) with no external force at the outside radius b the boundary condition is $\sigma_{rr}=0$ at $r=b$. Since the disk is solid, then $C_2=0$, otherwise at $r=0$ the stress becomes infinite. The constant C_1 is

$$C_1=\frac{3+\nu}{4}\rho b^2\omega^2+\frac{2E\alpha}{b^2}\int_0^b\theta r\mathrm{d}r \tag{2.146}$$

Thus the stresses become

$$\begin{aligned}\sigma_{rr}&=\frac{3+\nu}{8}\rho\omega^2\left(b^2-r^2\right)+E\alpha\left(\frac{1}{b^2}\int_0^b\theta r\mathrm{d}r-\frac{1}{r^2}\int_0^b\theta r\mathrm{d}r\right)\\ \sigma_{\phi\phi}&=\frac{1}{8}\rho\omega^2\left[(3+\nu)b^2-(1+3\nu)r^2\right]+E\alpha\left[-\theta+\frac{1}{b^2}\int_0^b\theta r\mathrm{d}r+\frac{1}{r^2}\int_0^r\theta r\mathrm{d}r\right]\end{aligned} \tag{2.147}$$

For a hollow disk of the inside radius a and outside radius b, assuming traction-free boundary conditions $\sigma_{rr}=0$ at $r=a$ and $r=b$, we obtain

$$\begin{aligned}\frac{C_1}{2}&=\frac{3+\nu}{8}\rho\omega^2\left(b^2+a^2\right)+\frac{E\alpha}{b^2-a^2}\int_a^b\theta r\mathrm{d}r\\ C_2&=-\frac{3+\nu}{8}\rho\omega^2b^2a^2-\frac{E\alpha a^2}{b^2-a^2}\int_a^b\theta r\mathrm{d}r\end{aligned} \tag{2.148}$$

Substituting the constants C_1 and C_2 into Eq. (2.145), the stresses become

$$\sigma_{rr} = \frac{3+\nu}{8}\rho r^2\omega^2\left(b^2 + a^2 - \frac{a^2b^2}{r^2} - r^2\right)$$
$$+ E\alpha\left[-\frac{1}{r^2}\int_a^r \theta r dr + \frac{1}{b^2-a^2}\int_a^b \theta r dr - \frac{a^2}{r^2\left(b^2-a^2\right)}\int_a^b \theta r dr\right]$$
$$\sigma_{\phi\phi} = \frac{3+\nu}{8}\rho r^2\omega^2\left(b^2 + a^2 + \frac{a^2b^2}{r^2} - \frac{1+3\nu}{3+\nu}r^2\right)$$
$$+ E\alpha\left[-\theta + \frac{1}{r^2}\int_a^r \theta r dr + \frac{1}{b^2-a^2}\int_a^b \theta r dr - \frac{a^2}{r^2\left(b^2-a^2\right)}\int_a^b \theta r dr\right] \quad (2.149)$$

When thermal stresses are zero, the maximum stress due to centrifugal forces occurs at the inner radius and it is

$$\sigma_{\phi\phi} = \frac{3+\nu}{4}\rho b^2\omega^2\left(1 + \frac{1-\nu}{3+\nu}\frac{a^2}{b^2}\right) \quad (2.150)$$

For a solid disk in the absence of thermal stresses the maximum stress occurs at the centre of the disk and it is

$$\sigma_{rr} = \sigma_{\phi\phi} = \frac{3+\nu}{8}\rho r^2\omega^2 \quad (2.151)$$

2.2 HEAT CONDUCTIONS PROBLEMS

The present section proposes the basic concepts and the equations of heat transfer in continuum media. The temperature distributions in one-, two- and three-dimensional problems for steady state as well as transient conditions are obtained. Classical rectangular, cylindrical and spherical coordinate systems are employed. Several examples are proposed and complete analytical solutions are provided. A comprehensive discussion on the heat transfer phenomena along with other examples can be found in textbooks focused on heat transfer such as Refs. [7–12], amongst others.

The general form of the governing differential equations of the heat transfer in structures, according to the above mentioned coordinate systems, are given as follows:

1. Rectangular Cartesian coordinate system

$$\frac{\partial}{\partial x}\left(k_x\frac{\partial T}{\partial x}\right) + \frac{\partial}{\partial y}\left(k_y\frac{\partial T}{\partial y}\right) + \frac{\partial}{\partial z}\left(k_z\frac{\partial T}{\partial z}\right) = -\mathcal{T} + \rho c\frac{\partial T}{\partial t} \quad (2.152)$$

2. Cylindrical coordinate system

$$k\left(\frac{\partial^2 T}{\partial r^2}+\frac{1}{r}\frac{\partial T}{\partial r}+\frac{1}{r^2}\frac{\partial^2 T}{\partial \phi^2}+\frac{\partial^2 T}{\partial z^2}\right)+\mathcal{T}=\rho c\frac{\partial T}{\partial t} \tag{2.153}$$

3. Spherical coordinate system

$$\frac{1}{r}\frac{\partial^2 (rT)}{\partial r^2}+\frac{1}{r^2 \sin(\theta)}\frac{\partial}{\partial \theta}\left(\sin(\theta)\frac{\partial T}{\partial \theta}\right)+\frac{1}{r^2 \sin^2(\theta)}\frac{\partial^2 T}{\partial \phi^2}+\frac{\mathcal{T}}{k}=\frac{\rho c}{k}\frac{\partial T}{\partial t} \tag{2.154}$$

where $\mathcal{T}$ is the rate of heat generated within the solid, T is the absolute temperature, k_x, k_y and k_z are the coefficient of thermal conduction along the coordinate axis x, y and z, respectively, in the rectangular Cartesian coordinate system and k is the coefficient of thermal conduction in the cylindrical and spherical coordinate systems.

2.2.1 Steady state one-dimensional

Rectangular Cartesian coordinates

When the heat flows in the x direction and $k_x = k$ is constant, the heat conduction equation (2.152) reduces to

$$\frac{d^2 T}{dx^2}=-\frac{\mathcal{T}}{k} \tag{2.155}$$

Integrating twice with respect x yields the temperature distribution in the form

$$T=\int\left(\frac{\mathcal{T}}{k}\right)dx+C_1 x+C_2 \tag{2.156}$$

where C_1 and C_2 are the constants of integration. In the following some examples are proposed.

Problem 1

Consider a flat plate of thickness h which separates two fluids at different temperatures T_1 and T_2. Determine the temperature distribution across the thickness of the plate.

Solution

The rate of heat generated within the solid is zero, then Eq. (2.152) reduces to

$$\frac{d^2 T}{dx^2}=0 \tag{2.157}$$

The boundary conditions are

$$T(0) = T_1 \quad T(h) = T_2 \tag{2.158}$$

Integrating Eq. (2.157) twice gives

$$T = C_1 x + C_2 \tag{2.159}$$

Using the boundary conditions leads to the determination of constants C_1 and C_2

$$C_2 = T \quad C_1 = \frac{T_2 - T_1}{h} \tag{2.160}$$

Thus

$$T(x) = \frac{T_2 - T_1}{h} x + T_1 \tag{2.161}$$

Equation (2.161) is linear in x and, as expected, is independent of the thermal conductivity of the material.

Problem 2

Consider a plate of thickness $2h_1$ insulated from both sides by cladding material. The coefficients of heat conduction for the plate and for the cladding are k_1 and k_2, respectively. Assume a rate of internal heat generation per internal unit volume $\mathcal{T}$ produced in the plate. Compute the temperature distribution in the plate and cladding material for the convention heat transfer coefficient l and the ambient temperature T_∞.

Solution

There are two regions in which the equations of the heat conduction have to be solved, the plate and the cladding material. The equation for the plate

$$\frac{d^2 T_1}{dx^2} + \frac{\mathcal{T}}{k_1} = 0 \tag{2.162}$$

with the boundary condition

$$\begin{gathered} \frac{dT_1(0)}{dx} = 0 \\ T_1(h_1) = T_2(h_1) \\ k_1 \frac{dT_1(h_1)}{dx} = k_2 \frac{dT_2(h_1)}{dx} \end{gathered} \tag{2.163}$$

The equation for the cladding material

$$\frac{d^2 T_2}{dx^2} = 0 \tag{2.164}$$

with the boundary conditions consisting of the second and the third of Eqs. (2.163), and

$$-k_2\frac{\mathrm{d}T_2(h_2)}{\mathrm{d}x} = l\left[T_2(h_2) - T_\infty\right] \tag{2.165}$$

The solutions of Eqs. (2.162) and (2.165) are

$$\begin{aligned} T_1 &= -\frac{\mathcal{T}x^2}{2k_1} + Ax + B \\ T_2 &= Cx + D \end{aligned} \tag{2.166}$$

Four constants A, B, C and D are determined from the four boundary conditions in Eqs. (2.163) and (2.165)

$$\begin{aligned} A &= 0 \\ -\mathcal{T}\frac{h_1^2}{2k_1} + B &= Ch_1 + D \\ -\mathcal{T}h_1 &= k_2 C \\ -k_2 C &= l(Ch_2 + D - T_\infty) \end{aligned} \tag{2.167}$$

Solving for the constants and substituting into Eqs. (2.166) leads to temperature distribution in the plate and cladding material, respectively

$$\frac{T_1 - T_\infty}{\frac{\mathcal{T}h_1^2}{2k_1}} = 1 - \left(\frac{x}{h_1}\right)^2 - 2\left(\frac{k_1}{k_2}\right) - 2\left(\frac{k_1 h_2}{k_2 h_1}\right)\left(1 + \frac{k_2}{lh_2}\right) \tag{2.168}$$

and

$$\frac{T_2 - T_\infty}{\frac{\mathcal{T}h_1^2}{2k_2}} = -\left(\frac{x}{h_1}\right) + \left(\frac{h_2}{h_1}\right)\left(1 + \frac{k_2}{lh_2}\right) \tag{2.169}$$

Cylindrical coordinates

In cylindrical coordinates the temperature distribution is a function of radius only and Eq. (2.153) reduces to

$$\frac{\mathrm{d}^2 T}{\mathrm{d}r^2} + \frac{1}{r}\frac{\mathrm{d}T}{\mathrm{d}r} + \frac{\mathcal{T}}{k} = 0 \tag{2.170}$$

This case is often encountered in practical problems and the solution is simply obtained by a simple integration.

Problem 1

Consider a hollow cylinder of inner and outer radius a and b, respectively. It is required to obtain a temperature distribution in the cylinder when the rate of heat generation is zero, and the temperature at the inner radius is different from that at the outer radius.

Solution

The differential equation of the heat conduction (2.170) becomes for this case

$$\frac{\mathrm{d}}{\mathrm{d}r}\left(\frac{r\mathrm{d}T}{\mathrm{d}r}\right)=0 \tag{2.171}$$

The boundary conditions are

$$T(a)=T_a \quad T(b)=T_b \tag{2.172}$$

Integrating Eq. (2.171) yields

$$T=A+B\ln(r) \tag{2.173}$$

where A and B are constants. Upon finding A and B from the boundary conditions, we obtain

$$T=\frac{1}{\ln\left(\frac{a}{b}\right)}\left[T_a\ln\left(\frac{b}{r}+T_b\frac{r}{a}\right)\right] \tag{2.174}$$

Now, if we assume that the inside surface is kept at $T_a=$ constant and the outside surface is exposed to convection at ambient temperature T_∞, the boundary conditions become

$$T(a)=T_a \quad k\frac{\mathrm{d}T(b)}{\mathrm{d}r}+l\left[T(b)-T_\infty\right]=0 \tag{2.175}$$

The use of these boundary conditions leads to

$$T=\frac{1}{1+mb\ln\left(\frac{b}{a}\right)}\left\{T_a\left[1+mb\ln\left(\frac{b}{r}\right)\right]+mbT_\infty\ln\left(\frac{r}{a}\right)\right\} \tag{2.176}$$

where $m=\frac{l}{k}$.

Problem 2

Consider the hollow cylinder of Problem 1, but with a constant rate of heat generation per unit volume $\mathcal{T}$. Find the temperature distribution in the cylinder.

Solution

The governing becomes

$$\frac{1}{r}\frac{\mathrm{d}}{\mathrm{d}r}\left(\frac{r\mathrm{d}T}{\mathrm{d}r}\right)+\frac{\mathcal{T}}{k}=0 \tag{2.177}$$

Assuming the boundary conditions

$$T(a) = T_a \quad T(b) = T_b \tag{2.178}$$

a solution is

$$T = A + B\ln(r) - \frac{\mathcal{T}r^2}{4k} \tag{2.179}$$

Then, substituting Eq. (2.179) into the boundary conditions Eq. (2.178) one gets

$$\begin{aligned} T_a &= A + B\ln(a) - \frac{\mathcal{T}a^2}{4k} \\ T_b &= A + B\ln(b) - \frac{\mathcal{T}b^2}{4k} \end{aligned} \tag{2.180}$$

Solving Eq. (2.180) for A and B, the constants of integration become

$$\begin{aligned} A &= T_a - \frac{\ln(a)}{\ln\left(\frac{b}{a}\right)}\left[T_b - T_a + \frac{\mathcal{T}}{4k}\left(b^2 - a^2\right)\right] + \frac{\mathcal{T}a^2}{4k} \\ B &= \frac{1}{\ln\left(\frac{b}{a}\right)}\left[T_b - T_a + \frac{\mathcal{T}}{4k}\left(b^2 - a^2\right)\right] \end{aligned} \tag{2.181}$$

Upon substitution of A and B from Eq. (2.181) in Eq (2.179), the temperature distribution is completely determined.

Spherical coordinates

The differential equation of heat conduction for radial temperature distributions in spherical coordinates in one dimensional problems reduces to

$$\frac{1}{r}\frac{\mathrm{d}^2}{\mathrm{d}r^2}(rT) + \frac{\mathcal{T}}{k} = 0 \tag{2.182}$$

If $\mathcal{T} =$ constant, a solution to this equation, including the particular solution is

$$T(r) = -\frac{\mathcal{T}}{6k}r^2 + \frac{C_1}{2} + C_2 \tag{2.183}$$

where C_1 and C_2 are the constants of integration. Some results of problems which have interest from a practical standpoint are following presented.

Problem 1

Consider a hollow thick sphere of inside radius a and outside radius b. The temperature distribution is a function of radius alone.

Solution

Consider the following cases:

(i) The rate of heat generation in the material of the sphere is zero but the temperature at the inside surface is T_a and at the outside surface is T_b. The boundary conditions are

$$T(a) = T_a, \quad T(b) = T_b \tag{2.184}$$

Substituting these conditions in Eq. (2.183) and imposing $\mathcal{T} = 0$ yields

$$T = T_a - (T_a - T_b)\frac{\left(1 - \frac{a}{r}\right)}{\left(1 - \frac{a}{b}\right)} \tag{2.185}$$

(ii) The inside temperature is T_a and the outside surface is exposed to free convection to the ambient at T_∞. We assume $\mathcal{T} = 0$. The boundary conditions are

$$T(a) = T_a, \quad \frac{\partial T(b)}{\partial r} + l_1\left[T(b) - T(\infty)\right] = 0 \tag{2.186}$$

where $l = \frac{l_1}{k}$. Substituting these conditions in Eq. (2.183) and evaluating the constants yields

$$T(r) = \frac{l_1 b^2 T_\infty (r - a) + aT_a\left[r(1 - l_1 b) + l_1 b^2\right]}{r\left[a(1 - l_1 b) + l_1 b^2\right]} \tag{2.187}$$

(iii) The sphere is solid and its surface temperature at $r = b$ is T_b, and the rate of heat generated per unit volume is a constant $\mathcal{T}$. The boundary conditions are

$$T(0) = \text{finite}, \quad T(b) = T_b \tag{2.188}$$

The temperature distribution satisfying the boundary conditions is

$$T = \frac{\mathcal{T}}{6k}\left(b^2 - a^2\right) + T_b \tag{2.189}$$

2.2.2 Steady state two-dimensional

Rectangular Cartesian coordinates

The general form of the governing partial differential equation for a steady two-dimensional problem in x and y directions is

$$\frac{\partial}{\partial x}\left(k_x\frac{\partial T}{\partial x}\right) + \frac{\partial}{\partial y}\left(k_y\frac{\partial T}{\partial y}\right) = -\mathcal{T} \tag{2.190}$$

where the thermal conductivities in x and y directions are assumed to be variable. The solution of the partial differential equation (2.190), when $k_x = k_x(x)$ and $k_y = k_y(y)$, is obtained by the method of separation of variables. This technique allows the constants of integration in each separated function to be found directly from the homogeneous boundary conditions, and the non-homogeneous boundary conditions to be treated by using the concept of expansion into a series.

The separation of variable technique is proposed as follows in several examples. Consider a general form of Eq. (2.190)

$$a_1(x)\frac{\partial^2 T}{\partial x^2} + a_2(x)\frac{\partial T}{\partial x} + a_3 T(x) + b_1(y)\frac{\partial^2 T}{\partial y^2} + b_2(y)\frac{\partial T}{\partial y} + b_3 T(y) = 0 \tag{2.191}$$

The solution of this equation may be taken in the product form as

$$T(x,y) = T_x(x)\,T_y(y) \tag{2.192}$$

where T_x is a function of x alone, and T_y is a function of y alone. Upon substitution of Eq. (2.192) into Eq. (2.191) and after dividing the whole equation by $T_x T_y$, one gets

$$\begin{aligned} &\left[a_1(x)\frac{\partial^2 T_x}{\partial x^2} + a_2(x)\frac{\partial T_x}{\partial x} + a_3 T_x(x)\right]\frac{1}{T_x} = \\ &-\left[b_1(y)\frac{\partial^2 T_y}{\partial y^2} + b_2(y)\frac{\partial T_y}{\partial y} + b_3 T_y(y)\right]\frac{1}{T_y} \end{aligned} \tag{2.193}$$

The left-hand side of Eq. (2.193) is a function of the variable x only, and the right-hand side a function of y only. Therefore, we conclude that the only way that the above equation can hold is when both sides are equal to a constant, say, $\pm\lambda^2$. This constant is called the separation constant. For what mentioned hitherto, the equations reduce to the following ordinary differential equations

$$\begin{aligned} &a_1(x)\frac{\mathrm{d}^2 T_x}{\mathrm{d}x^2} + a_2(x)\frac{\mathrm{d}T_x}{\mathrm{d}x} + \left[a_3(x) \pm \lambda^2\right]T_x = 0 \\ &b_1(y)\frac{\mathrm{d}^2 T_y}{\mathrm{d}y^2} + b_2(y)\frac{\mathrm{d}T_y}{\mathrm{d}y} + \left[b_3(y) \pm \lambda^2\right]T_y = 0 \end{aligned} \tag{2.194}$$

These equations can be solved by using the classical technique of the ordinary differential equations as two independent equations, and the constants of integration then may be found using the boundary conditions. The nature of the solution must be compatible with the given boundary conditions. A pair of the homogeneous boundary conditions in a given direction requires harmonic solution in that direction. Therefore, the sign of the separation constant selected in such a way that the solution in the direction of a pair of homogeneous boundary conditions leads to a harmonic solution. Referring to Eqs. (2.194) two boundary conditions are required in each x and y direction. If the boundary condition in x, as an example, is homogeneous, the sign of the separation constants must be selected so that the solution in x direction leads to a harmonic function. Some examples of steady state two-dimensional problems in rectangular coordinate are given as follows.

Problem 1

Consider a rectangular plate made of isotropic homogeneous material subjected to boundary conditions. Three sides of the plates are maintained at constant zero temperature, and the fourth side is exposed to a variable temperature. Find a temperature distribution within the plate when the heat generation in the plate is absent.

Solution

The two dimensional heat conduction equation is

$$\frac{\partial^2 T}{\partial x^2} + \frac{\partial^2 T}{\partial y^2} = 0 \tag{2.195}$$

and the boundary conditions are

$$\begin{aligned} T(0, y) = 0, \quad T(a, y) = 0 \\ T(x, 0) = 0, \quad T(x, b) = 0 \end{aligned} \tag{2.196}$$

By the method of separation of variables, the solution is assumed in the form

$$T(x, y) = T_x(x)\, T_y(y) \tag{2.197}$$

Substituting Eq. (2.197) in Eq. (2.195) and dividing by $T_x T_y$ yields

$$-\frac{1}{T_x}\frac{\mathrm{d}^2 T_x}{\mathrm{d}x^2} = \frac{1}{T_y}\frac{\mathrm{d}^2 T_y}{\mathrm{d}y^2} \tag{2.198}$$

Since each side of Eq. (2.198) is a function of a different independent variable, the only way that both sides may be equal is that each of them must be equal to a separation constant. Taking the separation constant as λ^2, the governing equation is

$$-\frac{1}{T_x}\frac{\mathrm{d}^2 T_x}{\mathrm{d}x^2} = \frac{1}{T_y}\frac{\mathrm{d}^2 T_y}{\mathrm{d}y^2} = \lambda^2 \tag{2.199}$$

or

$$\begin{aligned}\frac{d^2 T_x}{dx^2} + \lambda^2 T_x = 0\\ \frac{d^2 T_y}{dy^2} - \lambda^2 T_y = 0\end{aligned} \tag{2.200}$$

The value of the separation constant is determined by means of the boundary conditions, and its sign depends upon the nature of the boundary conditions involved. Three cases can be distinguished

(i) Case 1, $\lambda^2 = 0$

For this case the solution of the differential equations (2.200) leads to a simple integration

$$T_x = A_1 + A_2 x, \quad T_y = A_3 + A_4 y \tag{2.201}$$

where A_1 through A_4 are the constants of integration. From Eq. (2.197)

$$T(x, y) = (A_1 + A_2 x)(A_3 + A_4 y) \tag{2.202}$$

It is easy verified that the above solution cannot satisfy the boundary conditions. Thus, this case should be excluded.

(ii) Case 2, $\lambda^2 < 0$

Solution of Eqs. (2.200) is

$$\begin{aligned}T_x = B_1 e^{-\lambda x} + B_2 e^{\lambda x}\\ T_y = B_3 \cos \lambda y + B_4 \sin \lambda y\end{aligned} \tag{2.203}$$

where B_1 through B_4 are the constants and

$$T(x, y) = \left(B_1 e^{-\lambda x} + B_2 e^{\lambda x}\right)\left(\cos \lambda y + B_4 \sin \lambda y\right) \tag{2.204}$$

This solution cannot satisfy the nonhomogeneous boundary condition along the side $y = b$, and thus is not acceptable.

(iii) Case 3, $\lambda^2 > 0$

Integration of Eqs. (2.200) yields to the following solution

$$\begin{aligned}T_x = C_1 \cos \lambda x + C_2 \sin \lambda x\\ T_y = C_3 e^{-\lambda y} + C_4 e^{\lambda y}\end{aligned} \tag{2.205}$$

where C_1 through C_4 are the integration constants. To find these constants, the boundary conditions in Eqs. (2.196) must be used. Since three of the boundary conditions are homogeneous, they are directly applied to Eqs. (2.205), to yield

$$T_x(0) = 0, \quad T_x(a) = 0, \quad T_y(0) = 0 \tag{2.206}$$

and the nonhomogeneous boundary condition is

$$T(x, b) = f(x) \tag{2.207}$$

The first two of Eqs. (2.206) lead to

$$\begin{aligned} C_1 &= 0 \\ \sin\lambda a &= 0 \end{aligned} \tag{2.208}$$

or

$$\lambda_n = \frac{n\pi}{a} \quad n = 1, 2, 3, \cdots \tag{2.209}$$

The third of Eqs. (2.206) yields

$$C_3 = -C_4 \tag{2.210}$$

Substituting Eq. (2.210) into Eqs. (2.205) and then into Eq. (2.197) the solution of the plate problem takes the form

$$T = \sum_{n=1}^{\infty} C_n \sin\frac{n\pi}{a}x \sinh\frac{n\pi b}{a} \tag{2.211}$$

Imposing $f(x) = T_1$ at $y = b$ into a Fourier sine series gives

$$T_1 = \frac{2T_1}{\pi}\sum_{n=1}^{\infty}\frac{(-1)^{n+1}+1}{n}\sin\frac{n\pi}{a}x \tag{2.212}$$

Comparing Eqs. (2.211) and (2.212) results in

$$C_n = \frac{2T_1}{n\pi}\sum_{n=1}^{\infty}\frac{(-1)^{n+1}+1}{\sinh\frac{n\pi b}{a}} \tag{2.213}$$

and finally the solution for the temperature problem is

$$T = \sum_{n=1}^{\infty}\frac{2T_1}{n\pi}\sum_{n=1}^{\infty}\frac{(-1)^{n+1}+1}{\sinh\frac{n\pi b}{a}}\sin\frac{n\pi}{a}x \sinh\frac{n\pi b}{a} \tag{2.214}$$

Problem 2

Consider a rectangular element. It is assumed that in the element a constant amount of heat T per unit volume and per unit time is generated and that the heat convection coefficient is large. The element may represent a current-conducting electric wire of rectangular cross section. Find the temperature distribution in the cross section of the wire.

Solution

The governing equation is

$$\frac{\partial^2 T}{\partial x^2} + \frac{\partial^2 T}{\partial y^2} = -\frac{\mathcal{T}}{k} \tag{2.215}$$

and the boundary conditions are

$$\begin{gathered} T(L, y) = T_\infty, \quad T(x, l) = T_\infty \\ \frac{\partial T(0, y)}{\partial x} = 0, \quad \frac{\partial T(x, 0)}{\partial y} = 0 \end{gathered} \tag{2.216}$$

The last two boundary conditions are due to symmetry of the temperature distribution about x and y axes. By a simple transformation $\theta = T - T_\infty$ we arrive at the following equation

$$\frac{\partial^2 \theta}{\partial x^2} + \frac{\partial^2 \theta}{\partial y^2} = -\frac{\mathcal{T}}{k} \tag{2.217}$$

and the boundary conditions

$$\begin{gathered} \theta(L, y) = 0, \quad \theta(x, l) = 0 \\ \frac{\partial \theta(0, y)}{\partial x} = 0, \quad \frac{\partial \theta(x, 0)}{\partial y} = 0 \end{gathered} \tag{2.218}$$

Since the differential equation is nonhomogeneous, it is possible to assume a solution in either of the following form

$$\begin{gathered} \theta(x, y) = \psi(x, y) + \phi(x) \\ \theta(x, y) = \psi(x, y) + \phi(y) \end{gathered} \tag{2.219}$$

where ψ is a solution to the homogeneous equation and ϕ is a solution to the nonhomogeneous. Since the case of the nonhomogeneity in the differential equation, $\mathcal{T}/k$, is assumed to be constant in this case, either one of the above solutions may be selected. On the other hand, if the nonhomogeneous term were a function of x, the first of Eqs. (2.219) should be selected, and if it were a function of y, the second of Eqs. (2.219) should be selected. Upon substitution of Eq. (2.219) into Eqs. (2.217) and (2.218) we find

$$\frac{\partial^2 \psi}{\partial x^2} + \frac{\partial^2 \psi}{\partial y^2} = 0 \tag{2.220}$$

with the boundary conditions

$$\psi(L,y)=0, \quad \frac{\partial\psi(0,y)}{\partial x}=0$$
$$\psi(x,l)+\phi(x)=0, \quad \frac{\partial\psi(x,0)}{\partial y}=0 \tag{2.221}$$

and

$$\frac{\partial^2\phi}{\partial x^2}+\frac{T}{k}=0 \tag{2.222}$$

with the boundary conditions

$$\frac{\partial\psi(0)}{\partial x}=0, \quad \phi(0)=0 \tag{2.223}$$

It is noted that the non-separable form Eq. (2.217) is reduced to separable form for ψ, Eq. (2.220), and to an ordinary differential equation for ϕ, Eq. (2.222), and these equations are easily solved. The solution for θ is

$$\frac{\theta(x,y)}{RL^2/k}=\frac{1}{2}\left[1-\left(\frac{x}{L}\right)^2\right]-2\sum_{n=0}^{\infty}\frac{(-1)^n\cosh\lambda_n y}{(\lambda_n L)^3\cosh\lambda_n l}\cos\{\lambda_n x\} \tag{2.224}$$

where

$$\lambda_n L=\frac{(2n+1)\pi}{2} \quad n=1,2,3,\cdots \tag{2.225}$$

This method can be applied to many partial differential equations including problems in cylindrical coordinates and unsteady problems.

Cylindrical coordinates

In this section a solution of problems in cylindrical coordinates is presented. The problems include the flow of heat in regions bounded by cylindrical surfaces such as finite solids and hollow cylinders, and semi-cylinders. The problems depend on r and ϕ, or r and z and the method of separation of variables leads to a pair of separated equations, one being Bessel equation in r-direction. It should be bored in mind that, more generally, problem of this nature can also be solved by using the method of conformal mapping and Laplace transforms.

Problem 1

Consider a solid cylinder of outside radius b, or a hollow cylinder of inside radius a and outside radius b. The temperature is assumed to be constant along the axis of the cylinder, the z-axis but varies in r and ϕ directions. It is also assumed that the temperature reaches is steady state condition and that there is no heat generation in the cylinder. Find the temperature distribution in the cylinder.

Solution

The heat conduction equation to be solved in this case is

$$\frac{\partial^2 T}{\partial r^2}+\frac{1}{r}\frac{\partial T}{\partial r}+\frac{1}{r^2}\frac{\partial^2 T}{\partial \phi^2}=0 \tag{2.226}$$

The boundary conditions are

$$\begin{aligned}
&T(b,\phi)=F(\phi)\\
&T(0,\phi)=\text{finite for a solid cylinder}\\
&T(a,\phi)=T_i=\text{known constant for a hollow cylinder}\\
&T(r,\phi)=T(r,\phi+2\pi)\\
&\frac{1}{r}\frac{\partial T(r,\phi)}{\partial \phi}=\frac{1}{r}\frac{T(r,\phi+2\pi)}{\partial \phi}
\end{aligned} \tag{2.227}$$

The solution can be written in the following product form

$$T(r,\phi)=\mathcal{R}(r)\,\Phi(\phi) \tag{2.228}$$

upon substitution in Eq. (2.226), and knowing that the solution in ϕ direction must be periodic, the selection of a proper separation constant results in the equation

$$\frac{\mathrm{d}^2\Phi}{\mathrm{d}\phi^2}+\lambda^2\Phi=0 \tag{2.229}$$

subject to the boundary conditions

$$\Phi(\phi)=\Phi(\phi+2\pi),\quad \frac{d\Phi(\phi)}{\mathrm{d}\phi}=\frac{\mathrm{d}\Phi(\phi+2\pi)}{\mathrm{d}\phi} \tag{2.230}$$

and the separated equation in r-direction is

$$\frac{\mathrm{d}^2\mathcal{R}}{\mathrm{d}r^2}+\frac{1}{r}\frac{\mathrm{d}\mathcal{R}}{\mathrm{d}r}-\lambda^2\frac{\mathcal{R}}{r^2}=0 \tag{2.231}$$

The boundary conditions on $\mathcal{R}$ will depend on whether the cylinder is solid or hollow. Let first consider a solid cylinder which in this case leads to

$$\mathcal{R}(0)=\text{finite} \tag{2.232}$$

The solution of Eq. (2.229) is obtained as

$$\Phi=C\cos\lambda\phi+D\sin\lambda\phi \tag{2.233}$$

Applying the first condition of Eq. (2.230) defines the value of the separation constant

$$\lambda = n, \quad n = 0, 1, 2, 3, \cdots \tag{2.234}$$

The second condition of Eq. (2.230) is automatically satisfied and thus a solution for Φ reduces to

$$\Phi(\phi) = C_0 + C_n \cos n\phi + D_n \sin n\phi \tag{2.235}$$

Equation (2.231) is the Euler differential equation which easily integrated gives

$$\mathcal{R}(r) = E_n r^n + G_n r^{-n} \tag{2.236}$$

Applying the boundary condition of Eq. (2.232) for a solid cylinder gives $G_n = 0$ and then a temperature distribution after substituting in Eq. (2.228), and with the introduction of new constants, becomes

$$T(r, \phi) = A_0 + \sum_{n=1}^{\infty} r^n (A_n \cos n\phi + B_n \sin n\phi) \tag{2.237}$$

To obtain the constants A_0, A_n and B_n, we use the nonhomogeneous boundary condition on the outer boundary of the cylinder. From Eq. (2.237) and the first of Eqs. (2.227) the temperature distribution at $r = b$ is

$$F(\phi) = A_0 + \sum_{n=1}^{\infty} b^n (A_n \cos n\phi + B_n \sin n\phi) \tag{2.238}$$

Expanding $F(\phi)$ into a Fourier series with the period from 0 to 2π and equating the proper terms yields

$$\begin{aligned} A_0 &= \frac{1}{2\pi}\int_0^{2\pi} F(\phi)\,d\phi \\ A_n b^n &= \frac{1}{\pi}\int_0^{2\pi} F(\phi)\cos n\phi\,d\phi \\ B_n b^n &= \frac{1}{\pi}\int_0^{2\pi} F(\phi)\sin n\phi\,d\phi \end{aligned} \tag{2.239}$$

and upon substitution in Eq. (2.238), the temperature distribution for a solid cylinder exposed to a nonsymmetric thermal boundary condition becomes fully determined. For a hollow cylinder of inner radius a and outer radius b subject to a constant inside temperature T_i and to a variable temperature outside, the same procedure is followed. The general solution of Eq. (2.226) is obtained by substituting Eqs. (2.233) and (2.236) into Eq. (2.228), and with the introduction of new constants and with taking

$$\theta(r,\phi) = T(r,\phi) - T_i \tag{2.240}$$

it becomes

$$\begin{aligned}\theta(r,\phi) = A_0 + B_0 \ln r + \sum_{\lambda=1}^{\infty}\Bigg(&\left[A_\lambda\left(\frac{r}{a}\right)^{\lambda} + B_\lambda\left(\frac{r}{a}\right)^{-\lambda}\right]\cos\lambda\phi\\ &+\left[C_\lambda\left(\frac{r}{a}\right)^{\lambda} + D_\lambda\left(\frac{r}{a}\right)^{-\lambda}\right]\sin\lambda\phi\Bigg)\end{aligned} \tag{2.241}$$

The constants of integration A_0, B_0, A_λ, B_λ, C_λ, D_λ and the separation constants λ must be obtained using the given boundary conditions. For the sake of simplicity, it is assumed that the given boundary condition $F(\phi)$ at $r = b$, has a line of symmetry passing through the centre of the cylinder, thus the temperature distribution can be obtained for half of the cylindrical region. Measuring the angle ϕ from the line of symmetry the following boundary conditions apply

$$\frac{\partial\theta(r,0)}{\partial\phi} = 0, \quad \frac{\partial\theta(r,\pi)}{\partial\phi} = 0, \quad \theta(a,\phi) = 0, \quad \theta(b,\phi) = F(\phi) - T_i \tag{2.242}$$

From Eq. (2.241) it follows that

$$\begin{aligned}\frac{\partial\theta}{\partial\phi} = \sum_{n=1}^{\infty}\Bigg(&-\lambda\left[A_\lambda\left(\frac{r}{a}\right)^{\lambda} + B_\lambda\left(\frac{r}{a}\right)^{-\lambda}\right]\sin\lambda\phi\\ &+\lambda\left[C_\lambda\left(\frac{r}{a}\right)^{\lambda} + D_\lambda\left(\frac{r}{a}\right)^{-\lambda}\right]\cos\lambda\phi\Bigg)\end{aligned} \tag{2.243}$$

applying the first of Eq. (2.242) gives

$$\left(\frac{\partial\theta}{\partial\phi}\right)_{\phi=0} = \sum_{n=1}^{\infty}\lambda\left[C_\lambda\left(\frac{r}{a}\right)^{\lambda} + D_\lambda\left(\frac{r}{a}\right)^{-\lambda}\right] = 0 \tag{2.244}$$

which yields $C_\lambda = D_\lambda = 0$, the second condition of Eqs. (2.242) gives

$$\lambda = n \quad \text{with } n = 1, 2, 3, \cdots \tag{2.245}$$

Thus Eq. (2.241) becomes

$$\theta(r,\phi) = A_0 + B_0 \ln r + \sum_{\lambda=1}^{\infty}\left[A_n\left(\frac{r}{a}\right)^{n} + B_n\left(\frac{r}{a}\right)^{-n}\right]\cos\lambda\phi \tag{2.246}$$

This may be written as

$$A_0 + B_0 \ln(a) = 0$$
$$\sum_{n=1}^{\infty} (A_n + B_n \cos n\phi) = 0 \tag{2.247}$$

which gives

$$A_0 = -B_0 \ln(a) \quad B_n = -A_n \tag{2.248}$$

Substituting A_0 and B_n into Eq. (2.241) results in

$$\theta(r,\phi) = B_0 \ln r \frac{r}{a} + \sum_{\lambda=1}^{\infty} A_n \left[\left(\frac{r}{a}\right)^n - \left(\frac{r}{a}\right)^{-n} \right] \cos \lambda\phi \tag{2.249}$$

The constant B_0 and A_n have to be found from the last of Eqs. (2.242). The general method to compute these constants is again the expansion of $F(\phi)$ into a Fourier series and then equating it with Eq. (2.249) evaluated at $r = b$. A comparison of corresponding coefficients will result in obtaining of the constants B_0 and A_n. As an example, it is possible to assume that the temperature distribution at the outer boundary of the cylinder, $F(\phi)$, has the following form

$$\theta(b,\phi) = \begin{cases} T_1 - T_i & \text{for } 0 \le \phi \le \beta \\ T_2 - T_i & \text{for } \beta \le \phi \le \pi \end{cases} \tag{2.250}$$

where β is an unknown arbitrary angle between zero and π. Evaluating Eq. (2.249) at $r = b$ and rearranging it in a dimensionless form gives

$$\begin{aligned} \bar{\theta}(b,\phi) &= \frac{\theta(b,\phi) - (T_1 - T_i)}{(T_2 - T_i)} \\ &= \frac{B_0 \ln\left(\frac{b}{a}\right) - (T_1 - T_i)}{(T_2 - T_i)} + \sum_{n=1}^{\infty} \frac{A_n}{T_2 - T_1} \left[\left(\frac{b}{a}\right)^n - \left(\frac{b}{a}\right)^{-n} \right] \cos n\phi \end{aligned} \tag{2.251}$$

Thus

$$\bar{\theta}(b,\phi) = \begin{cases} 0 & \text{for } 0 \le \phi \le \beta \\ 1 & \text{for } \beta \le \phi \le \pi \end{cases} \tag{2.252}$$

Fourier expansion of the function $\bar{\theta}(b,\phi)$ is

$$\bar{\theta}(b,\phi) = \frac{\pi - \beta}{\pi} + \frac{2}{\pi} \sum_{n=1}^{\infty} \frac{(-1)^n}{n} \sin n(\pi - \beta) \cos n\phi \tag{2.253}$$

A comparison of Eqs. (2.253) and (2.251) yields

$$B_0 = \left[\frac{\pi (T_2 - T_i) - \beta (T_2 - T_1)}{\pi}\right] \frac{1}{\ln\left(\frac{b}{a}\right)}$$
$$A_n = \frac{2(-1)^n \sin n(\pi - \beta)}{n\pi} \frac{(T_2 - T_1)}{\left(\frac{b}{a}\right)^n - \left(\frac{b}{a}\right)^{-n}} \tag{2.254}$$

Substituting the values of the constants B_0 and A_n into Eq. (2.251), the temperature distribution in a thick-walled cylinder subjected to the boundary conditions in Eq. (2.242) becomes

$$T(r,\phi) = T_i + \frac{\pi (T_2 - T_i) - \beta (T_2 - T_1)}{\pi \ln\left(\frac{b}{a}\right)} \ln\left(\frac{r}{a}\right) + \frac{2(T_2 - T_1)}{\pi} \sum_{n=1}^{\infty} \frac{(-1)^n \sin n(\pi - \beta)}{n} \left[\frac{\left(\frac{r}{a}\right)^n - \left(\frac{r}{a}\right)^{-n}}{\left(\frac{b}{a}\right)^n - \left(\frac{b}{a}\right)^{-n}}\right] \cos n\phi \tag{2.255}$$

2.2.3 Steady state three-dimensional

Rectangular Cartesian coordinates

When a solid body is exposed to different boundary conditions in three dimension, the temperature gradient is produced in all three directions and the temperature distribution is a function of all three space coordinates. The heat conduction equation in three dimensions for a steady-state temperature distribution is

$$\frac{\partial^2 T}{\partial x^2} + \frac{\partial^2 T}{\partial y^2} + \frac{\partial^2 T}{\partial z^2} = 0 \tag{2.256}$$

A solution of Eq. (2.256) is easily obtained by using the separation of variables technique.

Problem 1

Consider a solid body in a form of rectangular parallelepiped. The surface temperature is assumed to be $T_1 =$ constant on the surface (S_1) at $x = 0$, $T_2 =$ constant on the surface (S_2) at $x = a$ and zero elsewhere. Find the temperature distribution within the body.

Solution

The differential equation for the temperature distribution is

$$\frac{\partial^2 T}{\partial x^2}+\frac{\partial^2 T}{\partial y^2}+\frac{\partial^2 T}{\partial z^2}=0 \tag{2.257}$$

The boundary conditions are

$$\begin{aligned} &T=T_1, \quad x=0\\ &T=T_2, \quad x=a\\ &T=0, \quad \text{at } y=0,\ y=b,\ z=0,\ z=c \end{aligned} \tag{2.258}$$

The solution for temperature distribution using the separation of variables method is

$$T=\sum_{m=1}^{\infty}\sum_{n=1}^{\infty}A_{mn}\frac{T_1\sinh l(a-x)+T_2\sinh lx}{\sinh la}\sin\frac{m\pi y}{b}\sin\frac{m\pi z}{c} \tag{2.259}$$

where

$$l=\left(\frac{m^2}{b^2}+\frac{n^2}{c^2}\right)\pi^2 \tag{2.260}$$

The constant coefficients A_{mn} are found from the first of Eq. (2.258) as

$$\sum_{m=1}^{\infty}\sum_{n=1}^{\infty}A_{mn}\sin\frac{m\pi y}{b}\sin\frac{n\pi z}{c}=1 \tag{2.261}$$

From the above equation it is verified that A_{mn} is zero unless m and n are both odd numbers. Thus, A_{mn} is found to be $4/\pi mn$. Finally, after substituting and taking odd number for $m=2p+1$ and $n=2q+1$, one obtains

$$T=\frac{4}{\pi^2}\sum_{p=1}^{\infty}\sum_{q=1}^{\infty}\frac{T_1\sinh la-x+T_2\sinh lx}{(2p+1)(2q+1)\sinh la}\sin\frac{(2p+1)\pi y}{b}\sin\frac{(2q+1)\pi z}{c} \tag{2.262}$$

where

$$l=\left[\frac{(2p+1)\pi^2}{b^2}+\frac{(2q+1)\pi^2}{c^2}\right]^{\frac{1}{2}} \tag{2.263}$$

Cylindrical coordinates

The problems of this nature are in general not easy to solve unless the geometry and boundary conditions are simple. Once again an effective mathematical tool to solve these problems is the separation of variable method.

Problem

Consider a half of a solid cylinder $0 \le \phi \le \pi$ of radius a and height h in which the surface temperature of all of its surfaces is zero except for the semi-circular plane at the end of the cylinder ($z = h$), which is exposed to a known temperature distribution $f(r, \phi)$. It is required to obtain the temperature distribution of the steady-state temperature in the semi-cylinder.

Solution

The steady-state temperature distribution should satisfy the Laplace equation in cylindrical coordinate

$$\frac{\partial T}{\partial r^2} + \frac{1}{r}\frac{\partial T}{\partial r} + \frac{1}{r^2}\frac{\partial^2 T}{\partial \phi^2} + \frac{\partial^2 T}{\partial z^2} = 0 \tag{2.264}$$

The boundary conditions are

$$\begin{aligned} &T(0, \phi, z) = \text{finite} \\ &T(a, \phi, z) = 0 \\ &T(r, 0, z) = 0 \\ &T(r, \pi, z) = 0 \\ &T(r, \phi, 0) = 0 \\ &T\left(r, \phi, h\right) = f(r, \phi) = \text{known} \end{aligned} \tag{2.265}$$

The solution is assumed in the product form

$$T(r, \phi, z) = \mathcal{R}(r)\, \Phi(\phi)\, \mathcal{Z}(z) \tag{2.266}$$

Substituting Eq. (2.266) into Eq. (2.264) gives

$$r^2\frac{\mathcal{R}''}{\mathcal{R}} + r\frac{\mathcal{R}'}{\mathcal{R}} + \frac{\Phi''}{\Phi} + r^2\frac{Z''}{Z} = 0 \tag{2.267}$$

or

$$\frac{\Phi''}{\Phi} = -r^2\frac{\mathcal{R}''}{\mathcal{R}} - r\frac{\mathcal{R}'}{\mathcal{R}} - r^2\frac{Z''}{Z} = -\alpha^2 \tag{2.268}$$

where α is the separation constant. The negative sign for α is chosen because the solution in ϕ-direction must be periodic, $T(r, \phi, z) = T(r, \phi + 2\pi, z)$. This choice yields

$$\Phi = A\cos\alpha\phi + B\sin\alpha\phi \tag{2.269}$$

From the third and fourth of the boundary conditions in Eq. (2.265), it follows that

$$A = 0$$
$$\alpha = n = 1, 2, 3, \cdots \tag{2.270}$$

Thus, a solution in ϕ-direction becomes

$$\Phi = B_n \sin n\phi \tag{2.271}$$

From Eq. (2.268) the separation of the function $\mathcal{R}$ and Z yields

$$\frac{Z''}{Z} = -\frac{n^2}{r^2} + \frac{\mathcal{R}''}{\mathcal{R}} - \frac{1}{r}\frac{\mathcal{R}'}{\mathcal{R}} = -\lambda^2 \tag{2.272}$$

where λ^2 is a new separation constant and the choice for the negative sign has been done because the positive sign, upon using the boundary conditions, results in a trivial solution. From Eq. (2.272)

$$\mathcal{R}'' + \frac{1}{r}\mathcal{R}' + \left(\lambda^2 - \frac{n^2}{r^2}\right)\mathcal{R} = 0 \tag{2.273}$$

This is a standard form of the Bessel equation which is readily solved to give

$$\mathcal{R} = CJ_n(\lambda r) + DY_n(\lambda r) \tag{2.274}$$

Applying the boundary conditions, the first of Eqs. (2.265) requires $D = 0$, as Y_n approaches infinity when $r \to 0$. The second of Eqs. (2.265) defines the characteristic value λ as

$$J_n(\lambda r) = 0 \tag{2.275}$$

The roots of this equation are $\lambda_m a = \rho_{mn}$, where ρ_{mn} are the roots of $J_n(\rho_{mn}) = 0$. Thus,

$$\lambda_m = \frac{\rho_{mn}}{a} \tag{2.276}$$

and the solution in r-direction reduces to

$$\mathcal{R} = CJ_n(\lambda_m r) \tag{2.277}$$

Now, the differential equation in z-direction from Eq. (2.272) becomes

$$\frac{d^2 Z}{dz^2} - \lambda_m Z = 0 \tag{2.278}$$

Integration of this equation leads to

$$Z = E\cosh\lambda_m z + F\sinh\lambda_m z \tag{2.279}$$

From the fifth of Eqs. (2.265) it follows that $E=0$ and thus

$$Z = F \sinh \lambda_m z \tag{2.280}$$

Substituting Eqs. (2.271) and (2.277) into Eq. (2.266) gives

$$T(r, \phi, z) = \sum_{m=1}^{\infty} \sum_{n=1}^{\infty} A_{mn} \sin n\phi J_n(\lambda_m r) \sinh \lambda_m z \tag{2.281}$$

The constants A_{mn} may now be determined from the nonhomogeneous boundary conditions at $z=h$

$$T(r, \phi, h) = f(r, \phi) = \sum_{m=1}^{\infty} \sum_{n=1}^{\infty} A_{mn} \sin n\phi J_n(\lambda_m r) \sinh \lambda_m h \tag{2.282}$$

The expansion of the function $f(r, \phi)$ into Fourier-Bessel series leads to the value of A_{mn} as

$$A_{mn} = \frac{2}{\pi} \frac{\int_0^a r \left[\int_0^\pi f(r, \phi) \sin n\phi J_n(\lambda_m r) \sinh \lambda_m r \right]}{\frac{a^2}{2} \sinh \lambda_m h J_{n+1}^2(\lambda_m a)} \tag{2.283}$$

Spherical coordinates

In its very general form the analytical solution of the heat conduction equation in spherical coordinates may be obtained by the use of separation of variable.

Problem

Consider heat conduction in a spherical solid body which is exposed to some kind of steady thermal fields.

Solution

The steady-state temperature will satisfy Laplace equation

$$\nabla^2 T = 0 \tag{2.284}$$

which in spherical coordinates takes the following form

$$r^2 \sin(\theta) \frac{\partial^2 T}{\partial r^2} + 2r \sin(\theta) \frac{\partial T}{\partial r} + \sin(\theta) \frac{\partial^2 T}{\partial \theta^2} + \cos(\theta) \frac{\partial T}{\partial \theta} + \frac{1}{\sin(\theta)} \frac{\partial^2 T}{\partial \phi^2} = 0 \tag{2.285}$$

Any solution of this equation is called spherical harmonic. This equation will be solved by the method of separation of variable by taking

$$T(r, \theta, \phi) = \mathcal{R}(r) F(\theta, \phi) \tag{2.286}$$

Substituting this in Eq. (2.286) yields

$$\begin{aligned} & r^2 \sin(\theta) \frac{\partial^2 \mathcal{R}}{\partial r^2} F + 2r \sin(\theta) \frac{\partial \mathcal{R}}{\partial r} F + \sin(\theta) \mathcal{R} \frac{\partial^2 F}{\partial \theta^2} \\ & + \cos(\theta) \mathcal{R} \frac{\partial F}{\partial \theta} + \frac{\mathcal{R}}{\sin(\theta)} \frac{\partial^2 F}{\partial \phi^2} = 0 \end{aligned} \tag{2.287}$$

Dividing by $\mathcal{R} F \sin(\theta)$, rearranging and equating the separated function to a constant such as λ, gives

$$\frac{r^2}{\mathcal{R}} \frac{d^2 \mathcal{R}}{dr^2} + \frac{2r}{\mathcal{R}} \frac{d\mathcal{R}}{dr} = -\left(\frac{1}{F} \frac{\partial^2 F}{\partial \theta^2} + \frac{\cos(\theta)}{F \sin(\theta)} \frac{\partial F}{\partial \theta} + \frac{1}{\sin^2(\theta)} \frac{\partial^2 F}{\partial \phi^2} \right) = \lambda \tag{2.288}$$

It will be seen later that taking the separation constant $\lambda = n(n+1)$ is more convenient. Doing so, it results in

$$\begin{aligned} & r^2 \frac{d\mathcal{R}}{dr^2} + 2r \frac{d\mathcal{R}}{dr} - n(n+1)\mathcal{R} = 0 \\ & \frac{\partial^2 F}{\partial \theta^2} + \frac{\cos(\theta)}{\sin(\theta)} \frac{\partial F}{\partial \theta} + \frac{1}{\sin^2(\theta)} \frac{\partial^2 F}{\partial \phi^2} + n(n+1) F = 0 \end{aligned} \tag{2.289}$$

The first of Eqs. (2.289) is the Euler differential equation and is readily solved to give

$$\mathcal{R}(r) = A_1 r^n + \frac{A_2}{r^{n+1}} \tag{2.290}$$

where A_1 and A_2 are constants of integration. The second of Eqs. (2.289) still has to be separated, and thus by taking

$$F(\theta, \phi) = \Theta(\theta)\ \Phi(\phi) \tag{2.291}$$

and submitting in Eq. (2.289) we find

$$\frac{d^2\Theta}{d\theta^2} \Phi + \frac{\cos(\theta)}{\sin(\theta)} \frac{d\Theta}{d\theta} \Phi + \frac{1}{\sin^2(\theta)} \Theta \frac{d^2\Phi}{d\phi^2} + n(n+1)\Theta\Phi = 0 \tag{2.292}$$

Dividing this equation by $(\Theta\Phi)/\sin^2(\theta)$ and calling the separation constant m^2, gives

$$\frac{\sin^2(\theta)}{\Theta} \frac{d^2\Theta}{d\theta^2} + \frac{\sin(\theta)\cos(\theta)}{\Theta} \frac{d\Theta}{d\theta} + n(n+1)\sin^2(\theta) = -\frac{1}{\Phi} \frac{d^2\Phi}{d\phi^2} = m^2 \tag{2.293}$$

This yields

$$\begin{aligned} & \sin^2(\theta) \frac{d^2\Theta}{d\theta^2} + \sin(\theta)\cos(\theta) \frac{d\Theta}{d\theta} + \left[n(n+1)\sin^2(\theta) - m^2 \right] \Theta = 0 \\ & \frac{d^2\Phi}{d\phi^2} + m^2 \Phi = 0 \end{aligned} \tag{2.294}$$

Solution of the second of Eqs. (2.294) is

$$\Phi = A_3 \cos(m\phi) + A_4 \sin(m\phi) \tag{2.295}$$

while the solution of the first of Eqs. (2.294) is obtained in terms of Legendre functions. In fact, the latter is the associated Legendre equation which by taking $x = \cos(\theta)$ and knowing that

$$\begin{aligned}\frac{d\Theta}{d\theta} &= \frac{d\Theta}{dx}\frac{dx}{d\theta} = -\sin(\theta)\frac{d\Theta}{dx}\\ \frac{d^2\Theta}{d\theta^2} &= \frac{d}{d\theta}\left(-\sin(\theta)\frac{d\Theta}{dx}\right) = -\cos(\theta)\frac{d\Theta}{dx} + \sin^2(\theta)\frac{d^2\Theta}{dx^2}\end{aligned}$$

is modified to

$$\left(1-x^2\right)\frac{d^2\Theta}{dx^2} - 2x\frac{d\Theta}{dx} + \left[n(n+1) - \frac{m^2}{1-x^2}\right]\Theta = 0 \tag{2.296}$$

Equation (2.297) is the associated Legendre differential equation and therefore, its solution, after substitution $x = \cos(\theta)$, is

$$\Theta = A_5 P_n^m \cos(\theta) + A_6 Q_n^m \cos(\theta) \tag{2.297}$$

Substituting Eqs. (2.290), (2.295) and (2.297) into Eqs. (2.291) and (2.286), the general solution of the Laplace equation in spherical coordinates becomes

$$\begin{aligned}T(r,\theta,\phi) = &\left(A_1 r^n + \frac{A_2}{r^{n+1}}\right)\left(A_3\cos(m\phi) + A_4\sin(m\phi)\right)\\ &\left(A_5 P_n^m \cos(\theta) + A_6 Q_n^m \cos(\theta)\right)\end{aligned} \tag{2.298}$$

There are six constants of integration, A_1 through A_6, which can be evaluated using the thermal boundary conditions.

2.2.4 Transient problems

In the present section three transient problems related to the heat conduction are proposed. In particular each examples make references to rectangular Cartesian coordinates, cylindrical coordinates and spherical coordinates.

Rectangular coordinates

When the thermal boundary conditions vary with the time, the temperature distribution is also a function of time. To formulate a transient heat conduction problems, both the initial conditions and the boundary conditions are needed. The governing differential heat conduction equation is

$$k\left(\frac{\partial^2 T}{\partial x^2}+\frac{\partial^2 T}{\partial y^2}+\frac{\partial^2 T}{\partial z^2}\right)=\rho c\frac{\partial T}{\partial t}-\mathcal{R} \tag{2.299}$$

Problem

Consider a plate of thickness $2L$ initially at uniform temperature T_0. The plate is suddenly immersed into a bath of constant temperature T_∞. Assuming the heat convection coefficient to be large, it is required to find a temperature distribution when the rate of heat generation within the plate $R=$ constant.

Solution

Taking $\theta = T - T_\infty$ the governing equation is

$$\kappa\frac{\partial^2\theta}{\partial x^2}=\frac{\partial\theta}{\partial t}-\frac{R}{\rho c} \tag{2.300}$$

where $\kappa = k/(\rho c)$ is the diffusivity. The initial and boundary conditions are

$$\begin{aligned}&\theta\left(x,0\right)=\theta_0=T_0-T_\infty\\&\frac{\partial\left(0,t\right)}{\partial x}=0,\quad \theta\left(L,t\right)=0\end{aligned} \tag{2.301}$$

Eq. (2.299) is non-homogeneous, and we assume

$$\theta\left(x,t\right)=\psi\left(x,t\right)+\phi\left(x\right) \tag{2.302}$$

where $\phi\left(x\right)$ satisfies the non-homogeneous differential equation, i.e.,

$$\frac{\partial^2\phi}{\partial x^2}+\frac{\mathcal{R}}{\kappa}=0 \tag{2.303}$$

subjected to the boundary condition

$$\phi\left(L\right)=0,\quad \frac{d\phi\left(0\right)}{dx}=0 \tag{2.304}$$

$\psi\left(x\right)$ satisfies the equation

$$\kappa\frac{\partial^2\psi}{\partial x^2}=\frac{\partial\psi}{\partial t} \tag{2.305}$$

subjected to the initial condition

$$\psi\left(x,0\right)=-\phi\left(x\right)+\theta_0 \tag{2.306}$$

and the boundary conditions

$$\psi\left(L,t\right)=0,\quad \frac{d\psi\left(0,t\right)}{dx}=0 \tag{2.307}$$

A solution of Eq. (2.303) satisfying the boundary conditions (2.304) is

$$\phi(x,) = \frac{RL^2}{2k}\left[1 - \left(\frac{x}{L}\right)^2\right] \tag{2.308}$$

To find $\psi(x, t)$ the method of separation of variables is used and a solution takes the following form

$$\psi(x,t) = X(t)\,\tau(t) \tag{2.309}$$

Substituting $\psi(x, t)$ from Eq. (2.309) into (2.305) we obtain

$$\frac{d^2X}{dx^2} + \lambda^2 X = 0 \tag{2.310}$$

with the boundary conditions

$$\frac{dX(0)}{dx} = 0, \quad X(L) = 0 \tag{2.311}$$

and

$$\frac{d\tau}{dt} + \kappa\lambda^2\tau = 0 \tag{2.312}$$

The initial condition is later applied to Eq. (2.312) using Fourier series. A solution of Eq. (2.310) is

$$X_n(x) = A_n\cos(\lambda_n x) + B_n\sin(\lambda_n x) \tag{2.313}$$

where A_n and B_n are constants. Substituting Eq. (2.313) into the boundary condition given in Eq. (2.311) leads to $B_n = 0$ and

$$\lambda_n = \frac{(2n+1)\pi}{2L} \qquad n = 0, 1, 2, 3, \cdots \tag{2.314}$$

A solution of Eq. (2.311) is also obtained by simple integration as

$$\tau_n(t) = C_n e^{(-\kappa\lambda_n^2 t)} \tag{2.315}$$

and thus Eq. (2.309) implies ψ in the form

$$\psi(x,t) = \sum_{n=0}^{\infty} a_n e^{(-\kappa\lambda_n^2 t)}\cos(\lambda_n x) \tag{2.316}$$

where a_n is a set of constants. To find the constants a_n, the initial condition provided in Eq. (2.306) is used where $\phi(x)$ is substituted from Eq. (2.309). For $\theta_0 = 0$, using Fourier series we obtain

$$a_n = -(-1)^n \frac{RL^2}{k}\frac{2}{(\lambda_n L)^3} \tag{2.317}$$

and finally

$$\frac{\theta(x,t)}{(RL^2)/k} = \frac{1}{2}\left[1 - \left(\frac{x}{L}\right)^2\right] - 2\sum_{n=0}^{\infty}\frac{(-1)^n}{(\lambda_n L)^3} e^{(-\kappa\lambda_n^2 t)}\cos(\lambda_n x) \tag{2.318}$$

Cylindrical coordinates

A general form of the governing heat conduction equation in cylindrical coordinates is obtained to be

$$\frac{1}{r}\frac{\partial}{\partial r}\left(r\frac{\partial T}{\partial r}\right)+\frac{1}{r^2}\frac{\partial^2 T}{\partial \phi^2}+\frac{\partial T}{\partial z^2}=\frac{\rho c}{k}\frac{\partial T}{\partial t} \tag{2.319}$$

Generally speaking, an analytical solution to this equation when the temperature is a function of space variables as well as the time is rather difficult to obtain, and for such problems numerical methods are recommended, especially, if the boundary conditions are complicated. In the following an analytical example is provided for a simple case.

Problem

Consider an infinite solid cylinder of radius b subject to an initial temperature $T(r,0)=g(r)$, where $g(r)$ is a known function of r. If we keep the outer surface $r=b$ at a constant temperature T_0, find the temperature distribution in the cylinder.

Solution

The temperature in this case is a function of the radius and time, and thus the heat conduction equation, taking $\theta(r,t)=T(r,t)-T_0$, becomes

$$\frac{\partial \theta}{\partial t}=\kappa\left(\frac{\partial^2\theta}{\partial r^2}+\frac{1}{r}\frac{\partial \theta}{\partial r}\right) \tag{2.320}$$

where $\kappa=k/(\rho c)$ is the diffusivity. The boundary and initial conditions are

$$\begin{aligned}&\theta(b,t)=0\\&\theta(r,0)=g(r)-T_0=f(r)\\&\theta(0,t)=\text{finite}\end{aligned} \tag{2.321}$$

A solution to Eq. (2.320) may be obtained by separation of variables, but we may proceed in a different way taking the solution in the from

$$\theta(r,t)=e^{(\kappa\lambda^2 t)}\mathcal{I}(r) \tag{2.322}$$

where $\mathcal{I}$ is a function of the radius only. Upon substitution of Eq. (2.322) into Eq. (2.320) we obtain for $\mathcal{I}$ Bessel's equation of order zero

$$\frac{d^2\mathcal{I}}{dr^2}+\frac{1}{r}\frac{d\mathcal{I}}{dr}+\lambda^2\mathcal{I}=0 \tag{2.323}$$

where λ^2 is a constant which will be found from the boundary conditions. A solution to this equation, recalling the Bessel functions of second kind, having an infinite value at $r = 0$, is

$$\mathcal{I}(r) = A J_0(\lambda r) \tag{2.324}$$

Substituting Eq. (2.324) into Eq. (2.322) gives

$$\theta(r, t) = A e^{(\kappa \lambda^2 t)} J_0(\lambda r) \tag{2.325}$$

The first boundary condition leads to the characteristic values λ as

$$J_0(\lambda_n b) = 0 \tag{2.326}$$

This equation has an infinite number of real roots $\lambda_1, \lambda_2, \lambda_3, \cdots, \lambda_n$ and thus Eq. (2.325) becomes

$$\theta(r, t) = A_n e^{(\kappa \lambda_n^2 t)} J_0(\lambda_n r) \tag{2.327}$$

From the initial condition we can find the constant coefficients A_n

$$f(r) = \sum_{n=1}^{\infty} A_n J_0(\lambda_n r) \tag{2.328}$$

Equation (2.328) is Fourier-Bessel expansion of the function $f(r)$. Multiplying both sides of Eq. (2.328) by $r J_0(\lambda_m r)$ then integrating from 0 to b and setting $n = m$ results in the proper value for the constants A_n. For $g(r) = T_1$, where T_1 is constant $\theta = T_1 - T_0$ and Fourier-Bessel expansion of θ_1, using the classical solution that can be found in several text books, yields the expression for the constants A_n

$$A_n = \frac{2\theta_1}{(\lambda_n b) J_1(\lambda_n b)} \tag{2.329}$$

The final solution for the unsteady state temperature in the solid cylinder is

$$\frac{T(r, t) - T_0}{T_1 - T_0} = 2 \sum_{n=1}^{\infty} \frac{e^{(\kappa \lambda_n^2 t)} J_0(\lambda_n r)}{(\lambda_n b) J_1(\lambda_n b)} \tag{2.330}$$

Spherical coordinates

The solutions to transient heat conduction problems in spherical coordinates are similarly obtained as those to steady-state problems. In the example that follows a problem of this nature is provided.

Problem

Consider a hot solid sphere of radius b, initially at constant temperature T_0. The surface is exposed to convective cooling as the time increases. It is required to obtain a transient temperature in the sphere.

Solution

Taking $\theta = T(r,t) - T_\infty$, the heat conduction equation is

$$\frac{\partial\theta}{\partial t} = \kappa\left(\frac{\partial^2\theta}{\partial r^2} + \frac{2}{r}\frac{\partial\theta}{\partial r}\right) \tag{2.331}$$

The boundary and initial conditions are

$$\begin{aligned} -k\frac{\partial\theta(b,t)}{\partial r} &= h\theta(b,t) \\ \frac{\partial\theta(0,t)}{\partial r} &= 0 \\ \theta(r,0) = T_0 - T_\infty &= \theta_0 \end{aligned} \tag{2.332}$$

A solution to the problem, by interchanging the variable $\theta = \tilde{\theta}/r^{\frac{1}{2}}$, and employing the method of separation of variables along with the boundary conditions is

$$\theta = r^{-\frac{1}{2}}\sum_{n=1} \infty A_n J_{\frac{1}{2}}(\lambda_n r)\, e^{(-\kappa\,\lambda_n^2\,t)} \tag{2.333}$$

which can also be written in the form

$$\theta = \sum_{n=1}^{\infty}\sqrt{\frac{2}{\pi}}A_n\left(\frac{\sin(\lambda_n r)}{r}\right) e^{(-\kappa\,\lambda_n\,t)} \tag{2.334}$$

where λ_n is the n-th root of the characteristic equation

$$\tan(b\lambda) = -\frac{b\lambda}{\frac{hb}{k} - 1} \tag{2.335}$$

Using the initial condition we have

$$\theta_0 = \sum_{n=1}^{\infty}\sqrt{\frac{2}{\pi}}A_n\left(\frac{\sin(\lambda_n r)}{r}\right) \tag{2.336}$$

Multiplying Eq. (2.336) by $r\sin(\lambda_m r)$ and integrating with respect to r from $r=0$ to $r=b$, when $m=n$, yields

$$A_n = \sqrt{\frac{\pi}{2}}\left[\frac{2\theta_0(\sin(b\lambda_n) - b\lambda_n\cos(b\lambda_n))}{\lambda_n(b\lambda_n - \sin(b\lambda_n)\cos(b\lambda_n))}\right] \tag{2.337}$$

Substituting for A_n gives

$$\theta_0 = 2\theta_0\sum_{n=1}^{\infty}\left[\frac{2\theta_0(\sin(b\lambda_n) - b\lambda_n\cos(b\lambda_n))}{\lambda_n(b\lambda_n - \sin(b\lambda_n)\cos(b\lambda_n))}\right]\left(\frac{\sin(\lambda_n r)}{r}\right) e^{(-\kappa\,\lambda_n\,t)} \tag{2.338}$$

REFERENCES

1. Thorton EA. Thermal structures for aerospace applications. First edition. Reston (Virginia, USA): AIAA Education Series; 1996.
2. Boyle BA, Weiner JH. Theory of thermal stresses. First edition. New York (USA): John Wiley & Sons; 1960.
3. Hetnarski R, Eslami RM. Thermal stresses – advanced theory and applications. First edition. Springer; 2008.
4. Eslami RM, Hetnarski R, Ignaczak J, Noda N, Sumi N, Tanigawa Y. Theory of elasticity and thermal stresses – explanations, problems and solutions. First edition. Springer; 2013.
5. Gatewood BE. Thermal stresses. First edition. New York: McGraw-Hill; 1957.
6. Johns DJ. Thermal stress analysis. London: Pergamon Press; 1965.
7. Fourier J. The analytical theory of heat. New York: Dover Publication; 1955.
8. Boelter IMK, Cherry VH, Johnson HA, Martinelli RC. Heat transfer notes. New York: McGraw-Hill; 1965.
9. Jakob M. Heat transfer. New York: John Wiley & Sons; 1949.
10. McAdams WH. Heat transmission. Third edition. New York: McGraw-Hill; 1954.
11. Carlslaw HS, Jaeger JC. Conduction of heat in solids. Second edition. New York: Oxford University Press; 1959.
12. Arpaci VS. Conduction heat transfer. Second edition. Reading (MA, USA): Addison-Wesley Publishing Company; 1966.

CHAPTER 3

Coupled and uncoupled variational formulations

3.1 CLASSICAL VARIATIONAL PRINCIPLES

3.1.1 Hamilton's principle and principle of virtual displacements

In this section the Principle of Virtual Displacements (PVD) is derived starting from Hamilton's principle. In the following theoretical formulation the viscous forces are neglected. However, in more general formulations the latter are taken into account and Rayleigh's dissipation function is employed. In its classical form Hamilton's principle [1,2] can be expressed as

$$\int_{t_1}^{t_2} \delta\,\mathcal{L}^k\,\mathrm{d}t = 0 \tag{3.1}$$

where δ is the variational operator, t denotes the time, t_1 and t_2 are the initial and generic instants respectively and $\mathcal{L}^k$ is the Lagrangian for the kth layer of the composite structure and it assumes the following form

$$\mathcal{L}^k = T^k - \Pi^k \tag{3.2}$$

Π^k and T^k are the total potential energy and the kinetic energy, respectively, and can be written as follows

$$T^k = \frac{1}{2}\int_{V^k}\left(\rho^k\,\dot{\mathbf{u}}\,\dot{\mathbf{u}}\right)\mathrm{d}V^k,$$

$$\Pi^k = \Phi^k_{se} + \Phi^k_{ef} = \frac{1}{2}\int_{V^k}\left(\boldsymbol{\varepsilon}^{k\ T}_{pG}\,\boldsymbol{\sigma}^k_{pC} + \boldsymbol{\varepsilon}^{k\ T}_{nG}\,\boldsymbol{\sigma}^k_{nC}\right)\mathrm{d}V^k + \frac{1}{2}\int_{V^k}\left(\boldsymbol{\varepsilon}^{k\ T}_{p_{nl}}\,\tilde{\boldsymbol{\sigma}}^k_{p_0}\right)\mathrm{d}V^k \tag{3.3}$$

where Φ^k_{se} and Φ^k_{ef} identify the potential strain energy and the potential energy related to the external forces, respectively. In this particular case Φ^k_{ef} is given as the product of the non-linear strains $\left(\boldsymbol{\varepsilon}_{p_{nl}}\right)$ and the algebraic summation of the initial mechanical and thermal stresses $\left(\tilde{\boldsymbol{\sigma}}_{p_0}\right)$. Substituting Eq. (3.2) in Eq. (3.1), Hamilton's principle becomes

$$\delta\int_{t_1}^{t_2}\Pi^k\,\mathrm{d}t - \delta\int_{t_1}^{t_2} T^k\,\mathrm{d}t = 0 \tag{3.4}$$

Thermal Stress Analysis of Composite Beams, Plates and Shells.
DOI: http://dx.doi.org/10.1016/B978-0-12-849892-7.00004-2

In particular, the kinetic energy can be manipulated as follows

$$\delta \int_{t_1}^{t_2} T^k \,\mathrm{d}t = \delta \int_{t_1}^{t_2} \left[\frac{1}{2} \int_{V^k} \left(\rho^k \dot{\mathbf{u}}\, \dot{\mathbf{u}} \right) \mathrm{d}V^k \right] \mathrm{d}t = \int_{t_1}^{t_2} \int_{V^k} \left(\rho^k \dot{\mathbf{u}}\, \delta \dot{\mathbf{u}} \right) \mathrm{d}V^k \,\mathrm{d}t$$
$$= \int_{V^k} \left(\rho^k \dot{\mathbf{u}}\, \delta \dot{\mathbf{u}} \right) \mathrm{d}V^k \Big|_{t_1}^{t_2} - \int_{t_1}^{t_2} \int_{V^k} \left(\rho^k \ddot{\mathbf{u}}\, \delta \mathbf{u} \right) \mathrm{d}V^k \mathrm{d}t \tag{3.5}$$

V^k is the volume of the kth layer, $\mathbf{u}$ is the displacement vector and dot denotes differentiation with respect to the time. More specifically, $\delta \mathbf{u}$ is equal to zero in $t = t_1$ and $t = t_2$, so that

$$\delta \int_{t_1}^{t_2} T^k \,\mathrm{d}t = - \int_{t_1}^{t_2} \int_{V^k} \left(\rho^k \ddot{\mathbf{u}}\, \delta \mathbf{u} \right) \mathrm{d}V^k \mathrm{d}t \tag{3.6}$$

Which implies that

$$\delta \int_{t_1}^{t_2} T^k \,\mathrm{d}t = - \int_{t_1}^{t_2} \delta L^k_{F_{in}} \,\mathrm{d}t \tag{3.7}$$

where $L^k_{F_{in}}$ is the work done by the inertial forces at layer level. The first variation of the potential energy Π^k can be written as the sum of the first variation of Φ^k_{se} and Φ^k_{ef} as follows

$$\delta \int_{t_1}^{t_2} \Pi^k \,\mathrm{d}t = \delta \int_{t_1}^{t_2} \left(\Phi^k_{se} + \Phi^k_{ef} \right) \mathrm{d}t \tag{3.8}$$

For elastic systems subjected to conservative forces, it is possible to write

$$\delta L^k_{int} = \delta \Phi^k_{se}, \quad \delta L^k_{ext} = - \delta \Phi^k_{ef} \tag{3.9}$$

with L^k_{int} being the internal work and L^k_{ext} the work done from the external forces. Substituting Eq. (3.9) in Eq. (3.8), it follows

$$\delta \int_{t_1}^{t_2} \Pi^k \,\mathrm{d}t = \int_{t_1}^{t_2} \left(\delta L^k_{int} - \delta L^k_{ext} \right) \mathrm{d}t \tag{3.10}$$

Now, substituting Eq. (3.7) and Eq. (3.10) in Eq. (3.4),

$$\int_{t_1}^{t_2} \delta L^k_{int} \,\mathrm{d}t = \int_{t_1}^{t_2} \delta L^k_{ext} \,\mathrm{d}t - \int_{t_1}^{t_2} \delta L^k_{F_{in}} \,\mathrm{d}t \tag{3.11}$$

Eliminating the time integral, the PVD for the kth layer is obtained

$$\delta L^k_{int} = \delta L^k_{ext} - \delta L^k_{F_{in}} \tag{3.12}$$

or equivalently

$$\int_{V^k} \left(\delta \boldsymbol{\varepsilon}^{k\,T}_{pG} \boldsymbol{\sigma}^k_{pC} + \delta \boldsymbol{\varepsilon}^{k\,T}_{nG} \boldsymbol{\sigma}^k_{nC} \right) \mathrm{d}V^k = \delta L^k_{ext} - \delta L^k_{F_{in}} \tag{3.13}$$

3.1.2 Reissner's mixed variational theorem

In solid mechanics, it is well known that the Principle of Virtual Displacements (PVD) involves only a compatible displacement field as a variable and has for its Euler-Lagrange equations the conditions of balance of momenta and traction boundary conditions. Likewise, the dual form of PVD, i.e., the Principle of Virtual Forces (PVF), involves a stress field which is equilibrated and satisfies the traction boundary conditions, alone as a variable and has the kinematic compatibility conditions and displacement boundary conditions as its Euler-Lagrange equations. If in PVD kinematic compatibility and displacement boundary conditions are introduced as conditions of constraint through Lagrange multipliers, which turn out to be stresses and surface traction respectively, one then obtains the so-called Hu-Washizu Variational Principle [3–6]. Likewise, if the conditions of equilibrium of stresses are introduced as a constraint condition through a Lagrange multipliers field which turns out to be displacements into PVF, one is led to the so-called Hellinger-Reissner principle [7–10]. Thus, the Hu-Washizu and Hellinger-Reissner principles, which involve that one field is the continuum as variables (some of which play the role of Lagrange multipliers to enforce certain constraint conditions), are often referred to as mixed variational principles. This is the scenario in which Reissner's mixed variational theorem (RMVT) [11–16] can be simply interpreted as a particular case of the previously mentioned mixed principles in which only compatibility of transverse strain $\boldsymbol{\varepsilon}_n = \{\gamma_{xz}, \gamma_{yz}, \varepsilon_{zz}\}$ is enforced by means of Lagrange multipliers which, in this case, turn out to be transverse shear and normal stresses $\boldsymbol{\sigma}_n = \{\tau_{xz}, \tau_{yz}, \sigma_{zz}\}$. In this framework Reissner's intuition, related to composite structures, was the restriction of the mixed assumptions to $\boldsymbol{\sigma}_n$. It is for such stresses that an independent field is in fact required to a priori and completely fulfil the C_z^0-Requirements [18]. Then, RMVT can be easily developed by simply adding the Lagrange multipliers for the transverse shear and normal stresses to the PVD (3.13). These constraint equations can be formulated by evaluating the transverse shear and normal strains in two different ways:

1. Geometrical relationships ($\boldsymbol{\varepsilon}_{nG}$);
2. Hooke's Law ($\boldsymbol{\varepsilon}_{nH}$).

The Lagrange multipliers are:

$$\boldsymbol{\varepsilon}_{nG} - \boldsymbol{\varepsilon}_{nH} = 0 \tag{3.14}$$

By considering a k-layer, and the integral on the volume V^k of the layer k as an integral on the in-plane domain Ω_k plus the integral in the thickness direction domain A_k, it is possible to write the RMVT in the following form

$$\int_{\Omega^k}\int_{A^k}\left[\delta\boldsymbol{\varepsilon}_{pG}^{k^T}\boldsymbol{\sigma}_{pH}^{k} + \delta\boldsymbol{\varepsilon}_{nG}^{k^T}\boldsymbol{\sigma}_{nM}^{k} + \overbrace{\delta\boldsymbol{\sigma}_{nM}^{k^T}\left(\boldsymbol{\varepsilon}_{nG} - \boldsymbol{\varepsilon}_{nH}\right)}^{\text{Compatibility variationally enforced}}\right] d\Omega^k\, dz$$
$$= \delta L_{ext}^{k} - L_{F_{in}}^{k}\, dz \tag{3.15}$$

The third *mixed* term variationally enforces the compatibility of the transverse strain components. Subscript M underlines that transverse stresses are those of the assumed model. Depending on the application one may decide to restrict the modelling of the transverse stresses. In other words, it is possible to derive the following two alternative versions of the RMVT:

- Compatibility variationally enforced only on the *transverse normal stress* σ_{zz}. Thus the considered virtual variations as well as primary variable in the formulation are: δu_x, δu_y, δu_z and $\delta\sigma_{zz}$, and the RMVT takes the following form

$$\int_{\Omega^k}\int_{A^k}\left[\delta\boldsymbol{\varepsilon}_{pG}^{k^T}\boldsymbol{\sigma}_{pH}^{k} + \delta\boldsymbol{\varepsilon}_{nG}^{k^T}\boldsymbol{\sigma}_{nM}^{k} + \delta\sigma_{zzM}^{k^T}\left(\boldsymbol{\varepsilon}_{zzG} - \boldsymbol{\varepsilon}_{zzH}\right)\right] d\Omega^k\, dz =$$
$$\delta L_{ext}^{k} - L_{F_{in}}^{k}\, dz \tag{3.16}$$

- Compatibility variationally enforced only on the *transverse shear stresses* τ_{xz} and τ_{yz}. Thus the considered virtual variations as well as primary variable in the formulation are: δu_x, δu_y, δu_z, $\delta\tau_{xz}$ and $\delta\tau_{yz}$, and the RMVT takes the following form

$$\int_{\Omega^k}\int_{A^k}\left[\delta\boldsymbol{\varepsilon}_{pG}^{k^T}\boldsymbol{\sigma}_{pH}^{k} + \delta\boldsymbol{\varepsilon}_{nG}^{k^T}\boldsymbol{\sigma}_{nM}^{k} + \delta\tau_{xzM}^{k^T}\left(\boldsymbol{\gamma}_{xzG} - \boldsymbol{\gamma}_{xzH}\right)\right.$$
$$\left. + \delta\delta\tau_{yzM}^{k^T}\left(\boldsymbol{\gamma}_{yzG} - \boldsymbol{\gamma}_{yzH}\right)\right] d\Omega^k\, dz = \delta L_{ext}^{k} - L_{F_{in}}^{k}\, dz \tag{3.17}$$

3.2 THERMOELASTIC VARIATIONAL FORMULATIONS

In this section two different extensions of the PVD and RMVT are proposed. The first is related to the partially coupled thermoelastic formulation, the second refers to the coupled thermoelastic formulation. In the first case, the thermal stresses are algebraically added to the mechanical ones, in the second case, the virtual internal thermal work is added to the virtual internal mechanical work [17,18], and the temperature is considered to be a primary variable in the analysis along with the displacement components. The constitutive equations in the case of thermoelastic problems are provided in Chapter 6.

3.2.1 Partially coupled thermoelastic variational statements

Thermo-mechanical problems where the temperature and deformation fields are discounted are here referred to as partially coupled thermoelastic problems.

PVD-based formulation

In the case of the thermal stress analysis, a possible extension of the PVD considers the temperature as an external load without any coupling between the mechanical and thermal fields. In the variational statement obtained in Eq. (3.13) the stresses are seen as an algebraic addition of mechanical (d) and thermal (t) contributions

$$\begin{aligned}\sigma_{pC}^k &= \sigma_{pd}^k - \sigma_{pt}^k \\ \sigma_{nC}^k &= \sigma_{nd}^k - \sigma_{nt}^k\end{aligned} \tag{3.18}$$

According to this assumption, it is possible to write the thermoelastic variational statement as

$$\int_{\Omega_k}\int_{A_k}\left[\delta\boldsymbol{\varepsilon}_{pG}^{k^T}\left(\sigma_{pd}^k - \sigma_{pt}^k\right) + \delta\boldsymbol{\varepsilon}_{nG}^{k^T}\left(\sigma_{nd}^k - \sigma_{nt}^k\right)\right] \mathrm{d}\Omega_k\,\mathrm{d}z = \delta L_{F_{ext}}^k - \delta L_{F_{in}}^k \tag{3.19}$$

RMVT-based formulation

In the mixed case, equilibrium and compatibility are both formulated in terms of the $\mathbf{u}^k$ and $\boldsymbol{\sigma}_n^k$ unknowns via Reissner's variational equation (see Eq. (3.15)),

$$\begin{aligned}&\int_{\Omega_k}\int_{A_k}\left\{\delta\boldsymbol{\varepsilon}_{pG}^{k^T}\left(\sigma_{pd}^k - \sigma_{pt}^k\right) + \delta\boldsymbol{\varepsilon}_{nG}^{k^T}\sigma_{nM}^k + \delta\sigma_{nM}^k\left[\boldsymbol{\varepsilon}_{pG} - (\boldsymbol{\varepsilon}_{nH} - \boldsymbol{\varepsilon}_{nT})\right]\right\} \mathrm{d}\Omega_k\,\mathrm{d}z \\ &= \delta L_{F_{ext}}^k - \delta L_{F_{in}}^k\end{aligned} \tag{3.20}$$

The left-hand side of Eq. (3.20) includes the variations of the internal work: the first two terms come from the displacement formulation and lead to variationally consistent equilibrium conditions; the third mixed term variationally enforces the compatibility of the transverse strain components.

3.2.2 Coupled thermoelastic variational statements

In the coupled thermoelastic formulations the temperature is a primary variable along with the displacement components.

PVD-based formulation

A PVD-based coupled thermoelastic variational principle is obtained by adding to the classical version of the PVD the internal thermal virtual work. Therefore, the variational statement can be written as

$$\int_{\Omega_k}\int_{A_k}\left(\delta\boldsymbol{\varepsilon}_{pG}^{k^T}\boldsymbol{\sigma}_{pC}^{k}+\delta\boldsymbol{\varepsilon}_{nG}^{k^T}\boldsymbol{\sigma}_{nC}^{k}-\delta\theta^{k}\,\eta_{C}^{k}-\delta\boldsymbol{\vartheta}_{pG}^{k^T}\boldsymbol{h}_{pC}^{k}-\delta\boldsymbol{\vartheta}_{nG}^{k^T}\boldsymbol{h}_{nC}^{k}\right)\mathrm{d}\Omega_k\,\mathrm{d}z$$
$$=\delta L_{ext}^{k}-\delta L_{F_{in}}^{k} \tag{3.21}$$

where θ is the temperature, η_C is the entropy per unit volume, $\boldsymbol{\vartheta}_{pG}$ and $\boldsymbol{\vartheta}_{nG}$ are the spatial temperature gradients and $\boldsymbol{h}_{pC}$ and $\boldsymbol{h}_{nC}$ are the heat fluxes. In the case of free vibration analysis, in the variational statement in Eq. (3.21) the terms $\left(\delta\boldsymbol{\vartheta}_{pG}^{k^T}\boldsymbol{h}_{pC}^{k}\right)$ and $\left(\delta\boldsymbol{\vartheta}_{nG}^{k^T}\boldsymbol{h}_{nC}^{k}\right)$ are neglected because temperature gradients variation does not exist. Under these hypotheses the coupled thermoelastic variational statement becomes

$$\int_{\Omega_k}\int_{A_k}\left(\delta\boldsymbol{\varepsilon}_{pG}^{k^T}\boldsymbol{\sigma}_{pC}^{k}+\delta\boldsymbol{\varepsilon}_{nG}^{k^T}\boldsymbol{\sigma}_{nC}^{k}-\delta\theta^{k}\,\eta_{C}^{k}\right)\mathrm{d}\Omega_k\,\mathrm{d}z=\delta L_{ext}^{k}-\delta L_{F_{in}}^{k} \tag{3.22}$$

RMVT-based formulation

In the case of mixed formulation (RMVT) the coupled thermoelastic variational principle takes the following form

$$\int_{\Omega^k}\int_{A^k}\Big[\delta\boldsymbol{\varepsilon}_{pG}^{k^T}\boldsymbol{\sigma}_{pH}^{k}+\delta\boldsymbol{\varepsilon}_{nG}^{k^T}\boldsymbol{\sigma}_{nM}^{k}+\delta\boldsymbol{\sigma}_{nM}^{k^T}\left(\boldsymbol{\varepsilon}_{nG}-\boldsymbol{\varepsilon}_{nH}\right)-\delta\theta^{k}\,\eta_{C}^{k}-\delta\boldsymbol{\vartheta}_{pG}^{k^T}\boldsymbol{h}_{pC}^{k}$$
$$-\delta\boldsymbol{\vartheta}_{nG}^{k^T}\boldsymbol{h}_{nC}^{k}\Big]d\Omega^k\,dz=\delta L_{ext}^{k}-L_{F_{in}}^{k} \tag{3.23}$$

and in the case of only mechanical load and/or free vibration, the variational statement becomes

$$\int_{\Omega^k}\int_{A^k}\Big[\delta\boldsymbol{\varepsilon}_{pG}^{k^T}\boldsymbol{\sigma}_{pH}^{k}+\delta\boldsymbol{\varepsilon}_{nG}^{k^T}\boldsymbol{\sigma}_{nM}^{k}+\delta\boldsymbol{\sigma}_{nM}^{k^T}\left(\boldsymbol{\varepsilon}_{nG}-\boldsymbol{\varepsilon}_{nH}\right)-\delta\theta^{k}\,\eta_{C}^{k}\Big]\,d\Omega^k\,dz$$
$$=\delta L_{ext}^{k}-L_{F_{in}}^{k} \tag{3.24}$$

REFERENCES

1. Reddy JN. Energy principles and variational methods in applied mechanics. 2nd edition. New Jersey: John Wiley & Sons; 2002.
2. Reddy JN. Variational methods in theoretical mechanics. 2nd edition. Berlin: Springer-Verlag; 1982.
3. Washitsu K. The relations between two energy theorems applicable in structural theory. Phil Mag J Sci 7th Ser 1938;26:617–35.
4. Washitsu K. On the variational principles of elasticity and plasticity. Aeroelastic research laboratory, MIT Tech. Rep. 25-18. Cambridge (MA): MIT; 1955.
5. Washitsu K. Variational principles in continuum mechanics. Seattle (WA): Department of Aeronautical Engineering, 62-2, University of Washington; 1962.
6. Washitsu K. Variational methods in elasticity and plasticity. 3rd edition. New York: Pergamon Press; 1982.
7. Hellinger H. Die allegemeinen Ansätze der Mechanik der Kontinua. In: Klein F, Muller C, editors. Encyclopädie der Mathematischen Wissenschafter, vol. 4. Leipzig: Teubner; 1914. p. 654–5 [Art 30].
8. Hu H-C. On some variational principles in the theory of elasticity and the theory of plasticity. Sci Sin 1955;4:33–54.
9. Reissner E. On a variational theorem in elasticity. J Math Phys 1950;29:90–5.
10. Reissner E. On variational principles in elasticity. Proc Symp Appl Math 1958;8:1–6.
11. Reissner E. On a certain mixed variational theory and a proposed application. Int J Numer Methods Eng 1984;20:1366–8.
12. Reissner E. On a mixed variational theorem and on a shear deformable plate theory. Int J Numer Methods Eng 1986;23:193–8.
13. Reissner E. On a certain mixed variational theorem and on laminated elastic shell theory. In: Proc of Euromech-Colloquium, No. 219. 1986. p. 17–27.
14. Carrera E. Developments, ideas, and evaluations based upon Reissner's Mixed Variational Theorem in the modelling of multilayered plates and shells. Appl Mech Rev 2001;54(4):301–29.
15. Fazzolari FA. Reissner's Mixed Variational Theorem and variable kinematics in the modelling of laminated composite and FGM doubly-curved shells. Composites, Part B, Eng 2016;89:408–23.
16. Fazzolari FA, Banerjee JR. Axiomatic/asymptotic PVD/RMVT-based shell theories for free vibrations of anisotropic shells using an advanced Ritz formulation and accurate curvature descriptions. Compos Struct 2014;108:91–110.
17. Fazzolari FA, Carrera E. Coupled thermoelastic effect in free vibration analysis of anisotropic multilayered plates and FGM plates by using a variable-kinematics Ritz formulation. Eur J Mech A, Solids 2014;44:157–74.
18. Carrera E, Brischetto S, Nali P. Variational statements and computational models for multifield problems and multilayered structures. Mech Adv Mat Struct 2008;15(3–4):182–98.

PART 2

Classical and Advanced Modelling of Thermal Structures

CHAPTER 4

Fundamental of mechanics of beams, plates and shells

The fundamental equations of continuum mechanics accounting for thermal effects were introduced in Chapter 1. The exact closed-form solution of these equations is generally available only for a few sets of geometries and boundary conditions. Approximated solutions of the general 3D problem are required in most cases. This has led, during the last two centuries, to the development of a significative number of structural theories that provide approximated solutions of the 3D problem. This chapter provides a short description of the two main methods adopted to derive structural theories, namely

- *The asymptotic method.*
- *The axiomatic method.*

These two approaches are usually exploited to reduce the 3D problem to a 2D or 1D problem. In a 3D problem, each variable (displacements u, stresses σ and strains ε) is defined at each point $P(x, y, z)$ of volume V. In a 2D or 1D problem, each variable is defined by means of additional functions. These functions are defined above a surface Ω (2D problem) or along a line s (1D problem). In other words, in a 2D problem a generic variable (f) is defined as a function of two coordinates ($f = f(x, y)$), whereas in a 1D problem $f = f(z)$.

The choice and the development of a 2D or 1D model is closely related to the structural component that has to be analyzed. In practical applications, for instance, both 2D and 1D models are commonly adopted. In some cases, e.g. when local effects have to be accurately investigated, a direct solution of the 3D problem is often required.

4.1 TYPICAL STRUCTURES

The typical structures that are used in civil, automotive, mechanical, ship and aerospace constructions are, generally, composed of three-dimensional, two-dimensional or one-dimensional structural elements.

Thermal Stress Analysis of Composite Beams, Plates and Shells.
DOI: http://dx.doi.org/10.1016/B978-0-12-849892-7.00006-6

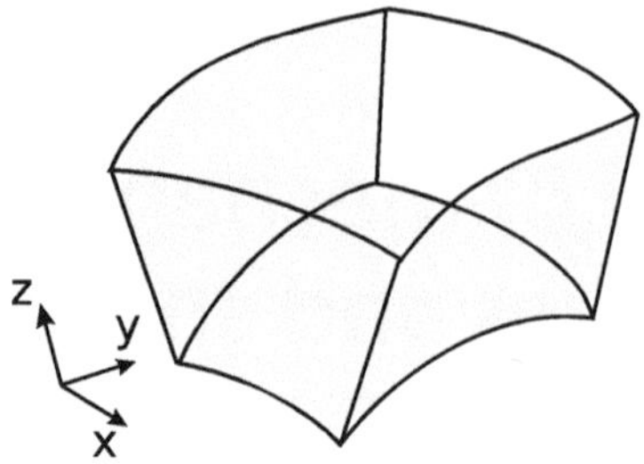

Figure 4.1 A 3D body.

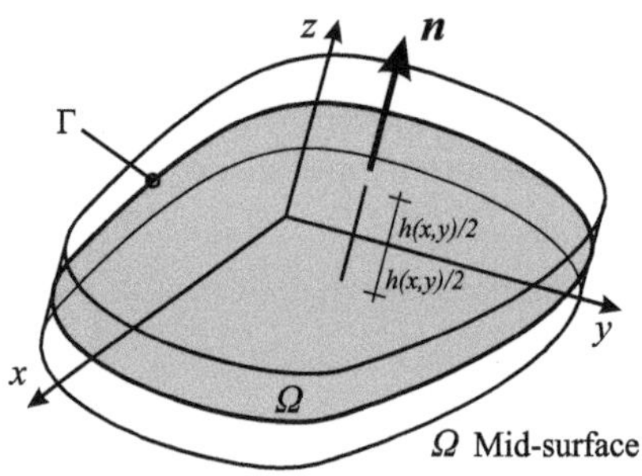

Figure 4.2 A plate element.

4.1.1 Three-dimensional structures, 3D (solids)

A structural component can be considered as a 3D body if the characteristic lengths along three different directions have the same order of magnitude, see Fig. 4.1.

4.1.2 Two-dimensional structures, 2D (plates, shells and membranes)

Plates are two-dimensional structural elements. They can be considered as special solids generated by segments of length h (see Fig. 4.2), whose midpoints belong to a surface Ω. This mid-surface is composed of all the midpoints, and each h is perpendicular to Ω. h is the *thickness* of the plate and, in general, the thickness can vary from point to point. Ω is called the *mid-surface*, or *reference surface* of the plate. The two surfaces generated by the top and bottom points of h are the top and bottom surfaces (or faces) of

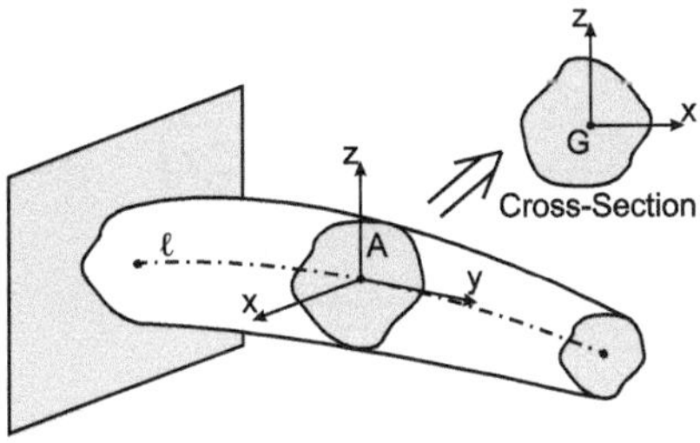

Figure 4.3 A beam element.

the plate. Λ is a closed line that represents the contour of Ω. The surface normal to Ω and passing through Λ is called the *edge* of the plate.

Let us define t and L as the characteristic lengths of the plate, measured along the thickness and above Ω, respectively. The two-dimensionality of the plate becomes more valid when L is much larger than t,

$$\frac{L}{t} >> 1$$

Like plates, shells are two-dimensional elements with a curved mid-surface. The nomenclature adopted for plates can also be exploited for shells.

Plates and shells are referred to as membranes if they are only loaded by forces that produce in-plane stresses which are constant along h.

4.1.3 One-dimensional structures, 1D (beams and bars)

Let us consider a section A whose centroid G moves along a line l where A is kept normal to l. The one-dimensional solid generated in this manner is called a *beam*, see Fig. 4.3. Let us define a and b as the characteristic lengths of A and let L be the length of s. The one dimensionality of a solid body becomes more valid as

$$\frac{L}{a} >> 1, \quad \frac{L}{b} >> 1$$

If a beam has no rigidity along the directions perpendicular to s, it is referred to as a *bar*.

4.2 AXIOMATIC METHOD

Axiomatic theories are developed on the basis of a number of hypotheses that *cannot* be mathematically proved. Despite this lack of mathematical

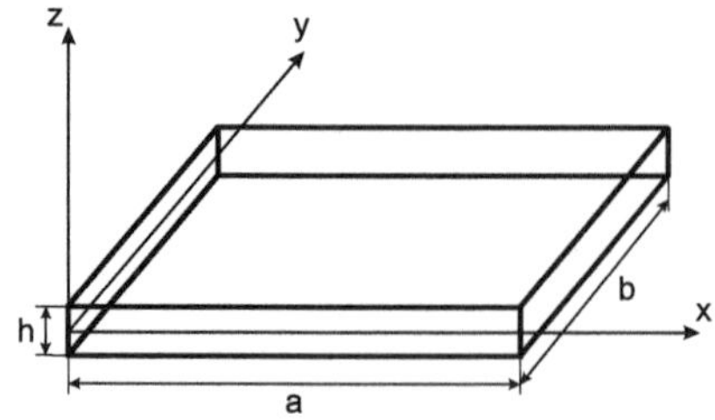

Figure 4.4 Reference frame for a plate element.

strength, axiomatic structural theories have been exploited to design most of the structures built over the last few decades.

In the development of an axiomatic theory, the starting point is the introduction of a number of *hypotheses* on the behaviour of the unknown functions. These hypotheses are introduced to reduce the mathematical complexity of the 3D elasticity differential equations. In other words, these hypotheses allow us to reduce the complexity of the problem in order to derive a new set of governing differential equations that can be solved in a more comfortable manner, and in some specific cases admit closed-form solutions. Axiomatic theories have been introduced over the last two centuries by eminent scientists whose intuition allowed them to detect and retain the most important aspects of structural problems while discarding the marginal features.

4.2.1 2D case

Let us introduce an orthogonal Cartesian reference frame, as shown in Fig. 4.4, where x and y lie on the mid-surface of the plate and z is defined along the thickness of the plate. Let $f(x, y, z)$ be an unknown variable of the 3D structural problem (e.g. a displacement u, stress σ or strain ε component). The 3D structural problem becomes a 2D problem when the unknown variables are defined above a reference surface. The 2D variables can be explicitly defined as $f_\Omega(x, y)$. In other words, it is necessary to define an explicit expression that relates variables (f) to the third coordinate (z). In the most general case, each variable f can require one or more f_Ω functions in order to reduce the 3D problem to a 2D one. The axiomatic approach can be seen as a tool that can be used to introduce an expansion of a 3D variable f along thickness z. The choice of the unknown variables that

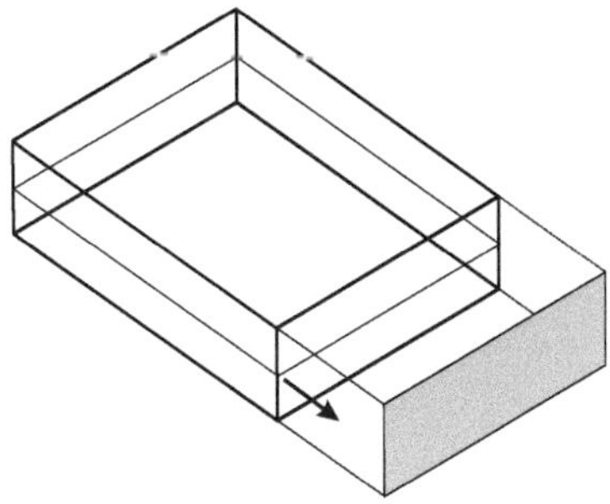

Figure 4.5 An axiomatic theory with only constant terms, the membranal case.

has to be expanded and the choice of the type of expansion are based on scientist's intuition.

Membrane behaviour

Let us consider the basic expansion case, that is, an expansion in which the generic function f is expanded along z via a constant term,

$$f(x, y, z) = f_{\Omega}(x, y) \tag{4.1}$$

This means f is assumed independent of z, or in other words, f is constant along the thickness of the plate (see Fig. 4.5). If f represents a displacement component (along x or y, for instance), the previous expression leads to an axiomatic theory in which constant displacement distributions are assumed along the thickness of the plate. These kinds of theories (i.e. those with constant displacements along x and y) are referred to as *membranal*, since the mechanical behaviour of the plate is, in this case, equivalent to that of a membrane.

Bending behaviour

The membrane case can be extended to a more sophisticated plate model by adding a term that is linearly dependent on z,

$$f(x, y, z) = f_{\Omega}(x, y) + z f_{\Omega}^{*}(x, y) \tag{4.2}$$

The behaviour of a generic function f along z is now assumed to be given by the sum of a constant and a linear term (see Fig. 4.6). Therefore, f at a generic point $P(x, y, z)$ is given by two functions (f_{Ω} and f_{Ω}^{*}) defined above the mid-surface of the plate. If f is a normal stress (e.g. σ_{xx}), the membranal (constant) term describes a constant stress distribution along the thickness

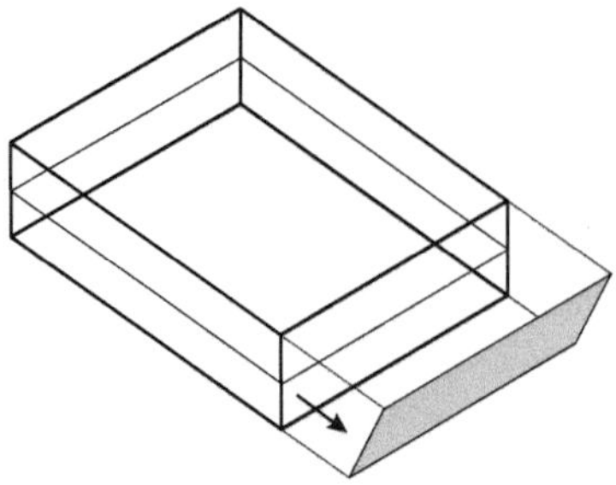

Figure 4.6 An axiomatic theory with constant and linear terms, the plate case.

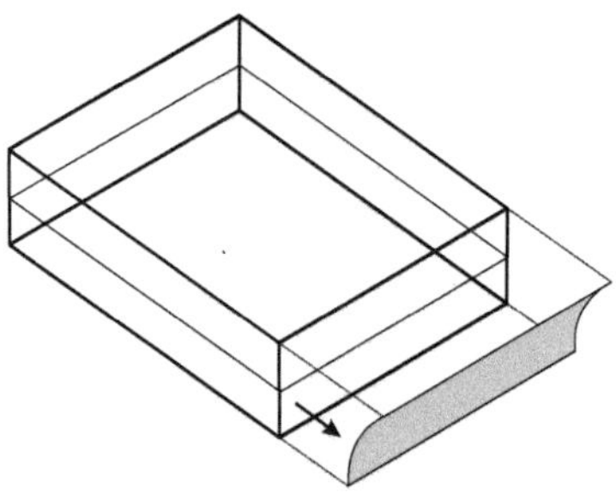

Figure 4.7 An axiomatic theory with constant, linear and higher-order terms.

of the plate, whereas the linear term introduces a variation of the stress distribution along the thickness. This implies the presence of compressed and stretched portions along the thickness, which are required to properly describe the *bending* behaviour of the plate.

Higher-order behaviour

Further refinements can similarly be carried out in order to introduce more complicated expansions of f. For instance (see Fig. 4.7),

$$f(x, y, z) = f_\Omega(x, y) + z f_\Omega^*(x, y) + z^2 f_\Omega^\star(x, y) \tag{4.3}$$

where $f^\star$ is a new unknown variable. This expansion can also be considered as a series expansion of the unknown variables. However, it is important to underline that, from a practical point of view, the physical meaning of the 2D variables does not always coincide with the derivatives of f. This means that the new 2D variables, related to a certain expansion order, might only have a mathematical meaning. Most of the plate theories exploited in real applications are based on constant or linear distributions of the unknown variables (membrane and bending behaviour).

The procedure described above can easily be extended to shells. This extension does not imply any conceptual difficulties. However, some geometrical complications can arise due to the curvature of the shell.

4.2.2 The N-order case for 2D theories, an introduction to a unified description of the displacement field

The displacement field refinement process can be extended to any expansion order. It is convenient to formulate the displacement expansion as

$$\mathbf{u} = F_\tau \, \mathbf{u}_\tau, \quad \tau = 1, ..., M \tag{4.4}$$

where F_τ is the expansion function, $\mathbf{u}_\tau$ is the vector of the unknown displacements and M is the number of the expansion terms. The following aspects are of fundamental importance:

- F_τ can be of any-type. This means that one can assume polynomial expansions, Lagrange/Legendre polynomials, harmonics, exponentials, combinations of different expansions types, etc.
- M can be arbitrary. This means that the number of terms can be increased to any extent. According to [3], as $M \to \infty$, the 2D (or 1D) model solution coincides with the exact 3D solution independently of the problem characteristics.

If, for instance, a Taylor-like polynomial expansion is adopted, Eq. (4.4) can be rewritten as

$$\mathbf{u} = z^{\tau-1} \, \mathbf{u}_\tau, \quad \tau = 1, ..., M \tag{4.5}$$

A second-order model ($N = 2$, $M = 3$) is then described by the following displacement field:

$$\begin{aligned} u_x &= u_{x_1} + z\, u_{x_2} + z^2\, u_{x_3} \\ u_y &= u_{y_1} + z\, u_{y_2} + z^2\, u_{y_3} \\ u_z &= u_{z_1} + z\, u_{z_2} + z^2\, u_{z_3} \end{aligned} \tag{4.6}$$

In this case, the 2D model has 9 unknown displacement variables.

4.2.3 1D case

The development of 1D structural models in an axiomatic framework requires a 2D expansion of the generic function f above the cross-section

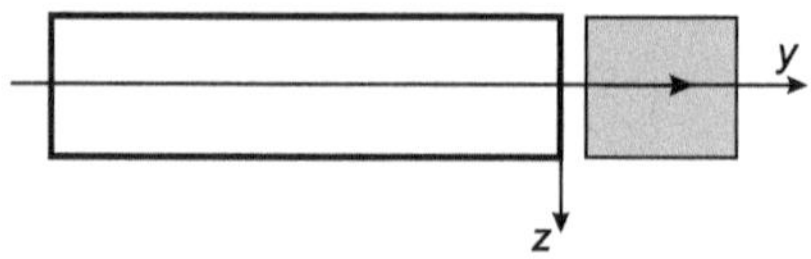

Figure 4.8 An axiomatic theory with constant terms, 1D case above the $y - z$ plane.

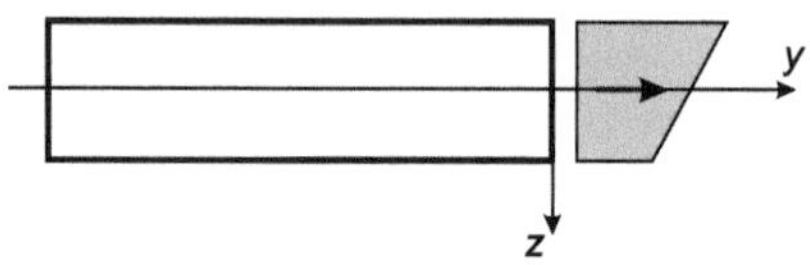

Figure 4.9 An axiomatic theory with constant and linear terms, 1D case above the $y - z$ plane.

domain. Let y be the coordinate along the axis (l) of the beam. The simplest expansion (constant term only) will therefore be

$$f(x, y, z) = f_s(y) \tag{4.7}$$

This expansion implies a constant distribution of f above the cross-section of the beam (see Fig. 4.8). Such a distribution can vary from one section to another. If f is the axial tension σ_{yy}, a membrane stress state will be obtained. In this particular case, a beam is referred to as *rod* or *bar*. The bending behaviour cannot be obtained in a bar and the axial tension is usually constant along the axis. In some cases, such as in stringers for aerospace applications, σ_{yy} is constant above the cross-section whereas it varies linearly along the axis. An improvement in the expansion can be obtained by adding a linear term (see Fig. 4.9),

$$f(x, y, z) = f_s(y) + x f_s^a(y) \tag{4.8}$$

Or,

$$f(x, y, z) = f_s(y) + z f_s^b(y) \tag{4.9}$$

And, by combining the two cases,

$$f(x, y, z) = f_s(y) + x f_s^a(y) + z f_s^b(y) \tag{4.10}$$

These theories are usually adopted to include bending behaviour in a beam element.

Further improvements of the beam theory can be obtained through the addition of higher-order terms to the axiomatic expansion, for instance (see Fig. 4.10)

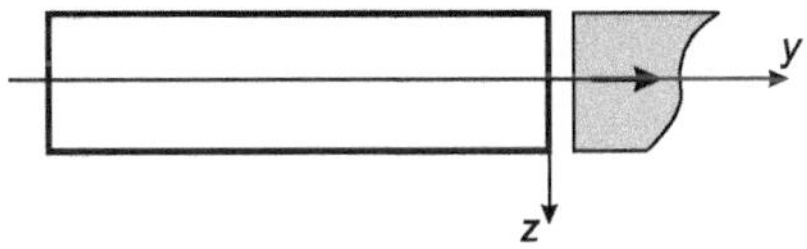

Figure 4.10 An axiomatic theory with higher-order terms, 1D case above the $y-z$ plane.

$$f(x, y, z) = f_s(y) + x f^a(y) + z f_s^b(y) + x^2 f_s^c(y) + z^2 f_s^d(y) + xz f_s^e(y) \tag{4.11}$$

4.2.4 The N-order case for 1D theories, an introduction to a unified description of the displacement field

The displacement field refinement process can be extended to any expansion order even for the 1D case and the compact notation (see Eq. (4.4)) does not change,

$$\mathbf{u} = F_\tau \, \mathbf{u}_\tau, \quad \tau = 1, \ldots, M \tag{4.12}$$

where, again, F_τ is the expansion function and it can be of any type (Taylor, Lagrange, harmonics, etc.), $\mathbf{u}_\tau$ is the vector of the unknown displacements and M is the number of the expansion terms and it can be arbitrary. A second-order model based on Taylor-like expansions ($N = 2$, $M = 6$) is described by the following displacement fields:

$$\begin{aligned} u_x &= u_{x_1} + x\, u_{x_2} + z\, u_{x_3} + x^2\, u_{x_4} + xz\, u_{x_5} + z^2\, u_{x_6} \\ u_y &= u_{y_1} + x\, u_{y_2} + z\, u_{y_3} + x^2\, u_{y_4} + xz\, u_{y_5} + z^2\, u_{y_6} \\ u_z &= u_{z_1} + x\, u_{z_2} + z\, u_{z_3} + x^2\, u_{z_4} + xz\, u_{z_5} + z^2\, u_{z_6} \end{aligned} \tag{4.13}$$

In this case, the 1D model has 18 unknown displacement variables.

4.3 ASYMPTOTIC METHOD

As mentioned in the previous sections, axiomatic theories have been introduced by distinguished scientists who have been able to understand the most important mathematical terms that need to be considered for a given structural problem. Further improvements are generally obtained by adding terms to the series expansions. An important drawback of axiomatic methods is the lack of information about the accuracy of the approximated theory with respect to the exact 3D solution. In other words, it is not usually possible to evaluate a-priori the accuracy of an axiomatic theory. In this respect, knowledge and experience play a crucial role.

A valid alternative to the axiomatic approach is the asymptotic method [1,2]. The latter can be seen as a step towards the development of approximated theories with known accuracy with respect to the 3D exact solution. Let us introduce two generic theories, A and B and let B be a theory adopted to enhance A through additional terms in the expansion. An important issue is the effectiveness of the additional terms in B. In other words, do the additional terms improve the accuracy of A? In order to deal with this issue, let us assume an exact solution and its series expansion,

$$f_{exact} = f_0(x, y) + f_1(x, y)\, z + \frac{1}{2} f_2(x, y)\, z^2 + \frac{1}{6} f_3(x, y)\, z^3 + \cdots + \frac{1}{n!} f_n(x, y)\, z^n \tag{4.14}$$

To better understand this aspect, let us assume that A contains all the terms that have the same effectiveness as $f_0(x, y)$ and $f_1(x, y)$ in the solution. It is necessary to evaluate whether B has all the terms with the same effectiveness as (for instance) $f_0(x, y)$, $f_1(x, y)$ and $f_2(x, y)$. During the last few decades, it has been shown that, in the development of axiomatic theories, some fundamental terms have erroneously been omitted from the series expansions. This means that many axiomatic theories are based on expansions that lack of effective terms. The asymptotic method overcomes this drawback by introducing some controls on the order of magnitude of each term introduced into an expansion.

In order to build an asymptotic theory, a reference exact solution must first be introduced. This exact solution can usually be obtained considering the limits of a function with respect to a characteristic feature of the problem (for instance a characteristic length). This characteristic length is the *parameter* of the asymptotic method. For instance, in a 2D theory, this parameter could be the ratio between the thickness and the length of a plate,

$$\delta = \frac{h}{L} = \frac{\text{thickness}}{\text{reference length}} \tag{4.15}$$

When $\delta \rightarrow 0$, the 3D solid plate becomes a 2D surface and the 2D plate theory is exact.

A typical procedure that is followed to build an asymptotic theory is:

1. An infinite expansion of an unknown function is introduced, for instance

$$f(x, y, z) = \sum_{i=1}^{\infty} f_i(x, y) z^i.$$

2. These expansions are introduced into the problem governing equations and the thickness parameter is isolated.
3. The 3D equations are then written as a series expansion with respect to the thickness parameter δ (this step is usually extremely difficult).
4. All the terms in the equations that multiply δ's by exponents that are lower or equal to n are retrieved for a given value of the exponent.

The development of asymptotic theories is generally more difficult than the development of axiomatic ones. The main advantage of these theories is that they contain all the terms whose effectiveness is of the same order of magnitude. Moreover, these theories are exact as $\delta \rightarrow 0$.

4.4 BEAM

The unknown variables of a one-dimensional (1D) model depend on one coordinate which is generally the axial coordinate of the structure. In this book, if not differently stated, the axial coordinate of a 1D model is y, while x and z are the so-called cross-section coordinates. The behaviour of the unknowns of a 1D structural problem can be axiomatically assumed. For instance, polynomial expansions of the unknown variables can be adopted above the cross-section, this means that for a given y the distribution of the unknowns above the cross-section will be given by a 2D polynomial in x and z. The choice of a particular expansion characterizes the capabilities of a structural model. The Richer is the expansion the higher is the results accuracy (see Ref. [3]). The number of the expansion terms to be retained to fulfil a given accuracy requirement is strictly related to the characteristics of the structural problem to be analyzed (e.g. geometry, boundary conditions, material, etc.). The main feature of the present unified formulation is due to the possibility of arbitrarily choosing the number of terms of the expansion. Therefore, the number of terms to be retained can be evaluated through a convergence analysis.

This chapter presents 1D elements based on Taylor-like expansions of the displacement variables, these elements are hereafter referred to as TE elements. These expressions are valid in a Cartesian orthogonal reference frame. If, for instance, a curvilinear system is adopted, the explicit expressions of the fundamental nucleus components can vary. In what follows, classical models (da Vinci-Euler-Bernoulli and Timoshenko) are briefly recalled along with the more general complete linear expansion case. Higher-order models are presented and the unified formulation is introduced. The

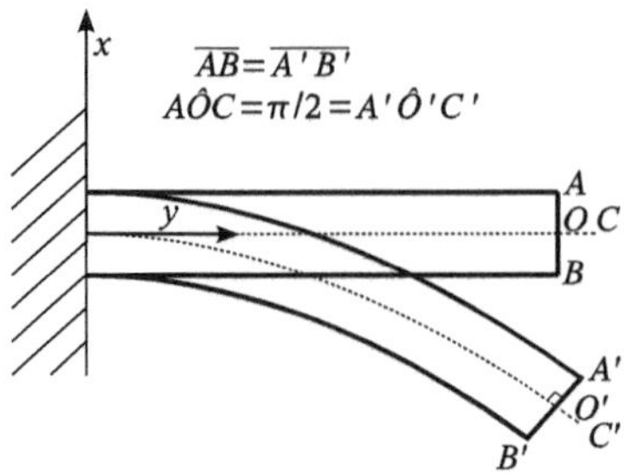

Figure 4.11 Euler-Bernoulli beam model.

principle of virtual displacements is employed to derive the governing differential equations and the finite element formulation. Locking phenomena and their corrections are discussed.

4.5 CLASSICAL MODELS AND THE COMPLETE LINEAR EXPANSION CASE

The mechanics of a beam under bending was first understood and described by Leonardo da Vinci as stated by [4] and [5]. The mathematical formulations of beams under bending were provided by da Vinci-Euler-Bernoulli [6] and [7,8], their models represent the classical beam theories. They are the reference models to analyze slender homogeneous structures under bending loads or to compute bending natural modes. These theories will be briefly described hereafter, more details can be found in Ref. [9].

4.5.1 The da Vinci-Euler-Bernoulli beam model (DEBBT)

The da Vinci-Euler-Bernoulli beam theory, hereafter referred to as DEBBT, was derived from the following a-priori assumptions (see Fig. 4.11):

1. the cross-section is rigid on its plane,
2. the cross-section rotates around a neutral surface remaining plane,
3. the cross-section remains perpendicular to the neutral surface during deformation.

According to the first hypothesis, no in-plane deformations are accounted for and, therefore, the in-plane displacements u_x and u_z depend upon the axial coordinate y only,

$$\begin{cases} \epsilon_{xx} = u_{x,x} = 0 \\ \epsilon_{zz} = u_{z,z} = 0 \\ \gamma_{xz} = u_{x,z} + u_{z,x} = 0 \end{cases} \Rightarrow \begin{cases} u_x\left(x, y, z\right) = u_{x_1}\left(y\right) \\ u_z\left(x, y, z\right) = u_{z_1}\left(y\right) \end{cases} \tag{4.16}$$

On the basis of the second hypothesis, the out-of-plane (or axial displacement) u_y is linear versus the in-plane coordinates,

$$u_y\left(x, y, z\right) = u_{y_1}\left(y\right) + \phi_z\left(y\right) x + \phi_x\left(y\right) z \tag{4.17}$$

where ϕ_z and ϕ_x are the rotation angles along the z- and the x-axis, respectively. ϕ_z is positive when, according to the "right-hand grip rule", the thumb points in the positive direction of the z-axis, whereas the thumb points in the negative direction of the x-axis for positive values of ϕ_x. On the basis of the third hypothesis and according to the definition of shear strains, shear deformations γ_{yz} and γ_{yx} are disregarded,

$$\gamma_{yz} = \gamma_{yx} = 0 \tag{4.18}$$

Equations (4.16), (4.17) and (4.18) allow obtaining the rotation angles as functions of the derivatives of the in-plane displacements,

$$\begin{cases} \gamma_{xy} = u_{y,x} + u_{x,y} = \phi_z + u_{x1,y} = 0 \\ \gamma_{yz} = u_{y,z} + u_{z,y} = \phi_x + u_{z1,y} = 0 \end{cases} \Rightarrow \begin{cases} \phi_z = -u_{x1,y} \\ \phi_x = -u_{z1,y} \end{cases} \tag{4.19}$$

The displacement field under the DEBBT assumption assumes the following form

$$\begin{aligned} u_x &= u_{x_1} \\ u_y &= u_{y_1} - u_{x1,y}\, x - u_{z1,y}\, z \\ u_z &= u_{z_1} \end{aligned} \tag{4.20}$$

From a mathematical point of view, the DEBBT displacement field can be seen as a McLaurin-like series expansion in which a zeroth-order approximation is used for the in-plane components and an expansion order (N) equal to one is adopted for the axial displacement. The relations amongst the unknowns have been derived from kinematic considerations. DEBBT presents three unknown variables.

According to the kinematic hypotheses, DEBBT accounts for the axial strain only. On the basis of its definition and of the DEBBT displacement

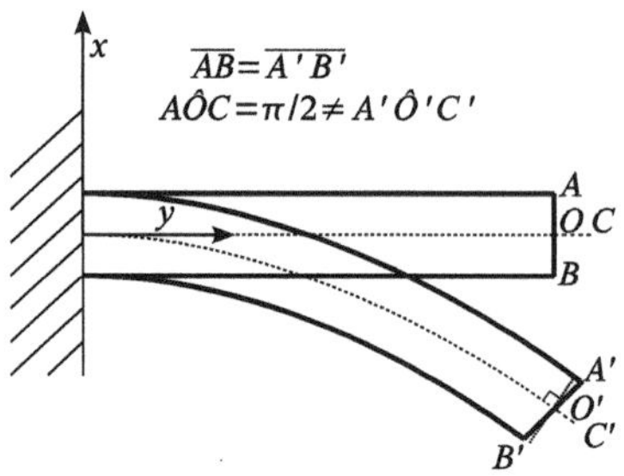

Figure 4.12 Timoshenko beam model.

field, the axial strain ε_{yy} is:

$$\varepsilon_{yy} = u_{y,y} = \underbrace{\frac{\partial u_{y_1}}{\partial y}}_{k^y_y} \underbrace{- \frac{\partial^2 u_{x_1}}{\partial y^2}}_{k^x_{yy}} x \underbrace{- \frac{\partial^2 u_{z_1}}{\partial y^2}}_{k^z_{yy}} z = k^y_y + k^x_{yy} x + k^z_{yy} z \tag{4.21}$$

k^y_y has the physical meaning of membrane deformation, whereas k^x_{yy} and k^x_{yy} are the second-order derivatives of the transverse displacements, they represent the curvatures in the case of infinitesimal deformations and small rotations. The axial stress, σ_{yy}, is obtained from the axial strain by means of the reduced constitutive equations:

$$\sigma_{yy} = E\varepsilon_{yy} = E\left(k^y_y + k^x_{yy} x + k^z_{yy} z\right) \tag{4.22}$$

4.5.2 The Timoshenko beam theory (TBT)

In a TBT model, the cross-section is still rigid on its plane, it rotates around a neutral surface remaining plane but is no longer constrained to remain perpendicular to it (see Fig. 4.12). Shear deformations γ_{xy} and γ_{yz} are now accounted for. According to the previous a-priori kinematic assumptions, the displacement field of TBT is

$$\begin{aligned} u_x(x, y, z) &= u_{x_1}(y) \\ u_y(x, y, z) &= u_{y_1}(y) + \phi_z(y)\, x + \phi_x(y)\, z \\ u_z(x, y, z) &= u_{z_1}(y) \end{aligned} \tag{4.23}$$

The strain components are obtained by substituting the displacement field in Eq. (4.23) into the geometrical relations. Only the non-null strain com-

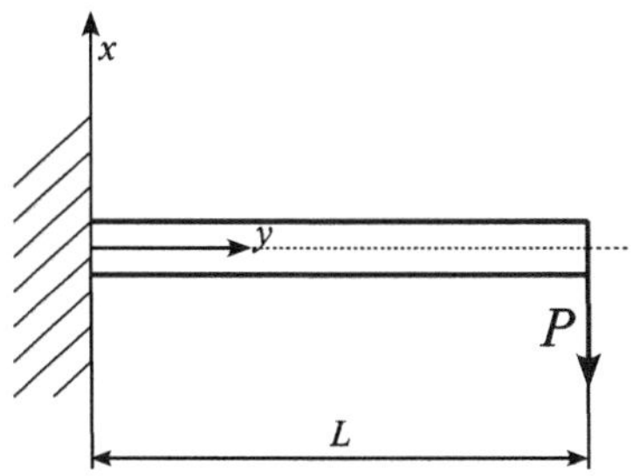

Figure 4.13 Cantilever beam bent by a vertical force.

ponents are reported,

$$\begin{aligned}\varepsilon_{yy} &= u_{y,y} = u_{y1,y} + \phi_{z,y}\, x + \phi_{x,y}\, z \\ \gamma_{xy} &= u_{y,x} + u_{x,y} = \phi_z + u_{x1,y} \\ \gamma_{yz} &= u_{y,z} + u_{z,y} = \phi_x + u_{z1,y}\end{aligned} \tag{4.24}$$

The constitutive relations are used to obtain the axial stress and the shear stress components,

$$\begin{aligned}\sigma_{yy} &= E\varepsilon_{yy} = E\left(u_{y1,y} + \phi_{z,y}\, x + \phi_{x,y}\, z\right) \\ \sigma_{xy} &= \kappa G\left(\phi_z + u_{x1,y}\right) \\ \sigma_{yz} &= \kappa G\left(\phi_x + u_{z1,y}\right)\end{aligned} \tag{4.25}$$

where κ is the shear correction factor. The shear predicted by TBT should be corrected since the model yields a constant shear distribution above the cross-section, whereas the shear distribution has to be at least parabolic in order to satisfy the stress free boundary conditions on the unloaded edges of the cross-section. The shear correction factor is mainly related to the cross-section geometry. In literature there are many methods to compute κ, see, for instance, [7,10–13]. A discussion on the shear correction factor is beyond the scope of this book. It will be shown that the adoption of higher-order models represents a general approach to avoid the introduction of shear correction factors.

4.5.3 Example

A cantilever beam is considered to highlight the differences between EBBT and TBT solutions. Figure 4.13 shows the structure that is loaded by a vertical force applied at the free-tip. Let us consider the elastica equation,

$$\frac{\partial^2 u_{x1}}{\partial y^2} = \frac{M_z(y)}{EI_z} = -\frac{PL}{EI_z}\left(\frac{L-y}{L}\right) \tag{4.26}$$

By integrating Eq. (4.26) the following relations are obtained:

$$
\begin{aligned}
u_{x_1,y} &= -\frac{P\,L}{E\,I_z}\left(y - \frac{y^2}{2\,L}\right) + C_1 \\
u_{x_1} &= -\frac{P\,L}{E\,I_z}\left(\frac{y^2}{L} - \frac{y^3}{6\,L}\right) + C_1\,y + C_2
\end{aligned} \tag{4.27}
$$

The following BCs are applied in order to evaluate C_1 and C_2:

$$
\begin{aligned}
u_{x_1,y}\big|_{y=0} &= 0 \quad \Rightarrow \quad C_1 = 0 \\
u_{x_1}\big|_{y=1} &= 0 \quad \Rightarrow \quad C_2 = 0
\end{aligned} \tag{4.28}
$$

The value of the maximum vertical displacement is then given by

$$
u_{x_1}\big|_{y=1} = -\frac{P\,L^3}{3\,E\,I_z} \tag{4.29}
$$

In the case of TBT the elastica equation becomes:

$$
\begin{cases}
\phi_{z,y} = -\dfrac{M_z(y)}{EI_z} = \dfrac{P\,L}{E\,I_z}\left(\dfrac{L-y}{L}\right) \\
u_{x_1,y} = -\dfrac{P}{\kappa\,G\,A} - \phi_z
\end{cases} \tag{4.30}
$$

By integrating the first relation of Eq. (4.30) the following relations are obtained:

$$
\begin{cases}
\phi_z = \dfrac{P\,L}{E\,I_z}\left(y - \dfrac{y^2}{2\,L}\right) + C_1 \\
u_{x_1,y} = -\dfrac{P}{\kappa\,G\,A} - \dfrac{P\,L}{E\,I_z}\left(y - \dfrac{y^2}{2\,L}\right) - C_1
\end{cases} \tag{4.31}
$$

A further integration leads to

$$
\begin{cases}
\phi_z = \dfrac{P\,L}{E\,I_z}\left(y - \dfrac{y^2}{2\,L}\right) + C_1 \\
u_{x_1} = -\dfrac{P\,y}{\kappa\,G\,A} - \dfrac{P\,L}{E\,I_z}\left(\dfrac{y^2}{2\,L} - \dfrac{y^3}{6\,L}\right) - C_1\,y + C_2
\end{cases} \tag{4.32}
$$

The following BCs are applied in order to evaluate C_1 and C_2:

$$
\begin{aligned}
\phi_z\big|_{y=0} &= 0 \quad \Rightarrow \quad C_1 = 0 \\
u_{x_1}\big|_{y=0} &= 0 \quad \Rightarrow \quad C_2 = 0
\end{aligned} \tag{4.33}
$$

The value of the maximum vertical displacement is then given by:

$$
u_{x_1}\big|_{y=0} = -\underbrace{\frac{P\,L}{\kappa\,G\,A}}_{TBT,\ shear\ contribution} - \underbrace{\frac{P\,L^3}{3\,E\,I_z}}_{DEBBT,\ bending\ contribution} \tag{4.34}
$$

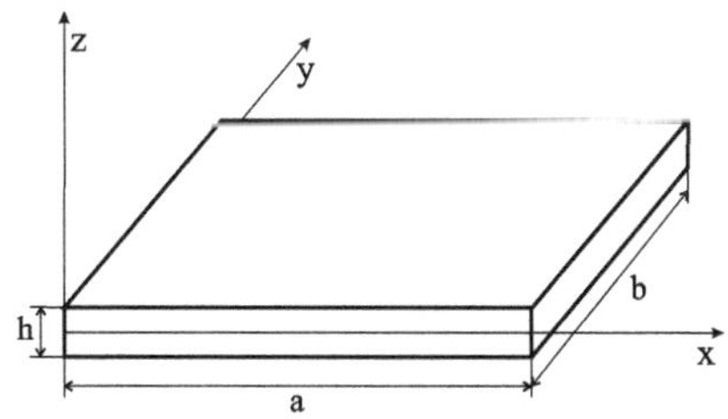

Figure 4.14 Plate geometry and reference system.

By comparing this last equation with Eq. (4.29) it is clear that TBT solution provides the DEBBT solution enhanced by the shear deformation contribution. The TBT contribution is significant whenever short beams or composite materials are considered. In an orthotropic material, for instance, the E/G ratio is much larger than in isotropic materials, this leads to higher shear deformability [14]. Similar considerations are still valid when the Classical Laminated Theory, CLT, and the First-Order Shear Deformability Theory, FSDT, for plates are considered.

4.6 PLATE

Plates are two-dimensional structures in which one dimension, in general the thickness h, is at least one order of magnitude lower than the in-plane dimensions a and b (Fig. 4.14). This fact permits the reduction of the 3D problem to a 2D one. Such a reduction can be seen as a transformation of the problem defined at each point $Q_V(x, y, z)$ of the 3D continuum plate into a problem defined at each point $Q_\Omega(x, y)$ of the plate surface Ω. 2D modelling of plates is a classical problem of the theory of structures. The elimination of the z-coordinate can be performed through several methodologies that lead to a significant number of approaches and techniques. For instance, the unknown variables can be axiomatically assumed along z. This means that, for a given point $Q_\Omega(x, y)$ in the plane, the distribution of the unknowns along the thickness will be given by a polynomial expansion in z. The main feature of the unified formulation is the possibility of arbitrarily choosing the kind of expansion and the number of terms.

This chapter presents 2D flat elements based on Taylor expansions of the displacement variables. First of all, classical models (Kirchoff and Reissner-Mindlin) will be briefly described together with the more general complete linear expansion case. Higher-order models will then be presented and the unified formulation introduced. The principle of virtual displacements will be employed to derive governing equations and the finite element formulation.

4.7 CLASSICAL MODELS AND THE COMPLETE LINEAR EXPANSION

The first mathematical formulations of plates under stretching and bending were provided by [15] and Reissner-Mindlin [16,17]. Their models represent the classical plate theories. They are reference models to analyze two-dimensional flat structures under stretching and bending loads or to compute natural modes. These theories will be briefly described in this section together with the case of complete linear expansion.

4.7.1 Classical plate theory

The Kirchhoff plate model, hereafter referred to as CPT (Classical plate theory), was derived from the following a-priori assumptions:

1. straight lines perpendicular to the mid-surface (i.e., transverse normals) before deformation remain straight after deformation,
2. the transverse normals do not experience elongation (i.e., they are inextensible),
3. the transverse normals rotate such that they remain perpendicular to the mid-surface after deformation.

According to the first hypothesis, the in-plane displacements u_x and u_y are linear versus the thickness coordinate z

$$\begin{aligned} u_x(x, y, z) &= u_{x_0}(x, y) + \phi_x(x, y)z \\ u_y(x, y, z) &= u_{y_0}(x, y) + \phi_y(x, y)z \end{aligned} \tag{4.35}$$

where ϕ_x and ϕ_y are the rotations of a transverse normal about the y- and x-axes, respectively. The notation that ϕ_x denotes the rotation of a transverse normal about the y-axis and ϕ_y denotes the rotation about the x-axis may be confusing to some, and they do not follow the right-hand rule. However, the notation has been used extensively in literature, and we will not depart from it. If (β_x, β_y) denote the rotations about the x and y axes, respectively, that follow the right-hand rule, then

$$\beta_x = \phi_y, \quad \beta_y = -\phi_x \tag{4.36}$$

On the basis of the second hypothesis, the transverse displacement u_z is independent of the transverse (or thickness) coordinate and the transverse normal strain ε_{zz} is disregarded

$$u_z(x, y, z) = u_{z_0}(x, y) \quad \Rightarrow \quad \varepsilon_{zz} = \frac{\partial u_z}{\partial z} = 0 \tag{4.37}$$

On the basis of the third hypothesis and according to the definition of shear strains, shear deformations γ_{xz} and γ_{yz} are disregarded

$$\gamma_{xz} = \gamma_{yz} = 0 \tag{4.38}$$

Equations (4.35), (4.37) and (4.38) allow obtaining the rotation angles as function of the derivatives of the transverse displacement

$$\begin{cases} \gamma_{xz} = \dfrac{\partial u_z}{\partial x} + \dfrac{\partial u_x}{\partial z} = \dfrac{\partial u_{z0}}{\partial x} + \phi_x = 0 \\ \gamma_{xz} = \dfrac{\partial u_z}{\partial y} + \dfrac{\partial u_y}{\partial z} = \dfrac{\partial u_{z0}}{\partial y} + \phi_y = 0 \end{cases} \Rightarrow \begin{cases} \phi_x = -\dfrac{\partial u_{z0}}{\partial x} \\ \phi_y = -\dfrac{\partial u_{z0}}{\partial y} \end{cases} \tag{4.39}$$

The displacement field of CPT is then

$$\begin{aligned} u_x &= u_{x0} - \frac{\partial u_{z0}}{\partial x} z \\ u_y &= u_{y0} - \frac{\partial u_{z0}}{\partial y} z \\ u_z &= u_{z0} \end{aligned} \tag{4.40}$$

where $(u_{x_0}, u_{y_0}, u_{z_0})$ are the displacements along the coordinate lines of a material point on the xy-plane. Note that the form of the displacement field in Eq. (4.40) allows the reduction of the 3-D problem to one of studying the deformation of the reference plane $z=0$ (or midplane). Once the midplane displacements $(u_{x_0}, u_{y_0}, u_{z_0})$ are known, the displacements of any arbitrary point (u_x, u_y, u_z) in the 3-D continuum can be determined using Eq. (4.40). From a mathematical point of view, the CPT displacement field can be seen as a McLaurin-like series expansion in which a zero-order approximation is used for the transversal component and an expansion order N equal to one is adopted for the in-plane displacements. CPT presents three unknown variables and the relations amongst them have been derived from kinematic considerations. Figure 4.15 shows the typical distribution of displacement components according to CPT: linear for u_x and u_y and constant for u_z. Also the physical meaning of the derivatives of transversal displacement, $u_{z,x}$ and $u_{z,y}$, is represented.

According to the kinematics hypotheses, CPT accounts for the in-plane strains only. On the basis of their definition, and of CPT displacement field,

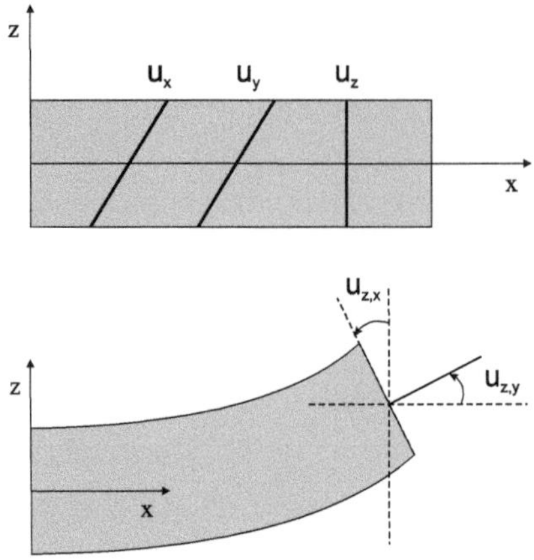

Figure 4.15 Distribution of displacements in Classical Plate Theory.

these strains are

$$
\begin{aligned}
\varepsilon_{xx} &= \frac{\partial u_x}{\partial x} = \frac{\partial u_{x0}}{\partial x} - \frac{\partial^2 u_{z0}}{\partial x^2} z = k_x^x + k_{xx}^z z \\
\varepsilon_{yy} &= \frac{\partial u_y}{\partial y} = \frac{\partial u_{y0}}{\partial y} - \frac{\partial^2 u_{z0}}{\partial y^2} z = k_y^y + k_{yy}^z z \\
\gamma_{xy} &= \frac{\partial u_x}{\partial y} + \frac{\partial u_y}{\partial x} = \frac{\partial u_{x0}}{\partial y} + \frac{\partial u_{y0}}{\partial x} - 2\frac{\partial^2 u_{z0}}{\partial xy} z = k_y^x + k_x^y + 2k_{xy}^z z
\end{aligned}
\tag{4.41}
$$

k_x^x, k_y^y, k_y^x and k_x^y have the physical meaning of membrane deformation, whereas k_{xx}^z, k_{yy}^z and k_{xy}^z, being the second order derivatives of the transverse displacement, represent the curvatures in the case of infinitesimal deformations and small rotations. The corresponding in-plane stresses are obtained by means of the reduced constitutive equations

$$
\begin{Bmatrix} \sigma_{xx} \\ \sigma_{yy} \\ \tau_{xy} \end{Bmatrix} = \frac{E}{1-\nu^2} \begin{bmatrix} 1 & \nu & 0 \\ \nu & 1 & 0 \\ 0 & 0 & \frac{1-\nu}{2} \end{bmatrix} \begin{Bmatrix} \varepsilon_{xx} \\ \varepsilon_{yy} \\ \gamma_{xy} \end{Bmatrix}
\tag{4.42}
$$

4.7.2 First-order shear deformation theory

In the Reissner-Mindlin theory, also called First-order Shear Deformation Theory (FSDT), the third part of Kirchhoff hypothesis is removed, so the transverse normals do not remain perpendicular to the midsurface after deformation. In this way, transverse shear strains γ_{xz} and γ_{yz} are included in the theory. However, the inextensibility of transverse normal remains, so

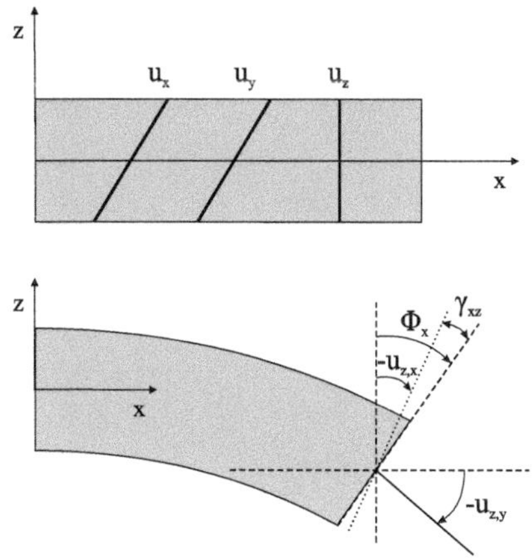

Figure 4.16 Distribution of displacements in FSDT.

displacement u_z is constant in the thickness direction z. The displacement field in the case of FSDT is

$$\begin{aligned} u_x(x,y,z) &= u_{x_0}(x,y) + \phi_x(x,y)z \\ u_y(x,y,z) &= u_{y_0}(x,y) + \phi_y(x,y)z \\ u_z(x,y,z) &= u_{z_0}(x,y) \end{aligned} \tag{4.43}$$

The quantities $(u_{x_0}, u_{y_0}, u_{z_0}, \phi_x, \phi_y)$ will be the unknowns. For thin plates, i.e., when the plate in-plane characteristic dimension to thickness ratio is on the order 50 or greater, the rotation functions ϕ_x and ϕ_y should approach the respective slopes of the transverse deflection $-\frac{\partial u_{z_0}}{\partial x}$ and $-\frac{\partial u_{z_0}}{\partial y}$. Figure 4.16 shows the typical distribution of displacement components according to FSDT: linear for u_x and u_y and constant for u_z. Also the physical meaning of the rotations, ϕ_x and ϕ_y, is represented.

The strain components are obtained by substituting the displacement field (Eqs. (4.43)) in the geometrical relations. Only strain ε_{zz} is zero, so the not-null strains are

$$\begin{aligned} \varepsilon_{xx} &= \frac{\partial u_x}{\partial x} = u_{x0,x} + \phi_{x,x}z \\ \varepsilon_{yy} &= \frac{\partial u_y}{\partial y} = u_{y0,y} + \phi_{y,y}z \\ \gamma_{xy} &= \frac{\partial u_x}{\partial y} + \frac{\partial u_y}{\partial x} = u_{x0,y} + u_{y0,x} + \phi_{x,y}z + \phi_{y,x}z \end{aligned} \tag{4.44}$$

$$\gamma_{xz} = \frac{\partial u_x}{\partial z} + \frac{\partial u_z}{\partial x} = \phi_x + u_{z0,x}$$
$$\gamma_{xy} = \frac{\partial u_x}{\partial y} + \frac{\partial u_y}{\partial x} = \phi_y + u_{z0,y}$$

The constitutive relations are used to obtain the in-plane stresses and the shear stress components

$$\left\{\begin{array}{c}\sigma_{xx}\\ \sigma_{yy}\\ \tau_{xy}\end{array}\right\} = \frac{E}{1-\nu^2}\left[\begin{array}{ccc}1 & \nu & 0\\ \nu & 1 & 0\\ 0 & 0 & \frac{1-\nu}{2}\end{array}\right]\left\{\begin{array}{c}\varepsilon_{xx}\\ \varepsilon_{yy}\\ \gamma_{xy}\end{array}\right\} \qquad (4.45)$$
$$\tau_{xz} = \kappa\, G\gamma_{xz} \quad \tau_{yz} = \kappa\, G\gamma_{yz}$$

Where κ is the shear correction factor. Similarly to the TBT, the shear predicted by FSDT should be corrected since the model yields a constant value along the thickness, whereas it is at least parabolic in order to satisfy the stress free boundary conditions on the unloaded top and bottom faces of the plate. In literature, there are many methods to compute κ for the FSDT, see, for instance, [18] and [19]. The shear correction factor is not here discussed, but it will be shown that the adoption of higher-order models represents a general approach to avoid shear correction factors.

4.8 2D SHELL MODELS WITH N-ORDER DISPLACEMENT FIELD, THE TAYLOR EXPANSION CLASS

A thin shell is defined as a three-dimensional body bounded by two closely spaced curved surfaces, where the distance between the two surfaces should be small compared to the other dimensions. The middle surface of the shell is the locus of the points that lie midway between these surfaces. The distance between the surfaces, measured along the normal to the middle surface, is the thickness of the shell at that point. Shells may be considered as generalizations of a flat plate; conversely, a flat plate can be considered as a special case of a shell with no curvature.

The governing equations are derived from the PVD and RMVT in chapter 7. Membrane and shear locking phenomena and their correction are discussed and numerical results are given.

4.9 GEOMETRY DESCRIPTION

The shell can be considered as a solid medium geometrically defined by a midsurface, given by the coordinates α, β, immersed in the physical space and a parameter representing the thickness z of the medium around this surface.

The Cartesian coordinates (x, y, z) can be expressed in function of the curvilinear coordinates α, β, z by means of the so-called 3D chart $\boldsymbol{\Phi}$, that is

$$\boldsymbol{\Phi}(\alpha, \beta, z) = \boldsymbol{\phi}(\alpha, \beta) + z\boldsymbol{a}_z(\alpha, \beta). \tag{4.46}$$

So, the 3D chart is defined by means of the 2D chart $\boldsymbol{\phi}$ and a unit vector $\boldsymbol{a}_z$, that is introduced below.

Starting from $\boldsymbol{\Phi}$, one can calculate the 3D *covariant basis* $(\boldsymbol{g}_\alpha, \boldsymbol{g}_\beta, \boldsymbol{g}_z)$ as follows:

$$\boldsymbol{g}_m = \frac{\partial \boldsymbol{\Phi}(\alpha, \beta, z)}{\partial m} \qquad m = \alpha, \beta, z, \tag{4.47}$$

$\boldsymbol{g}_m$ is a local basis and it is defined in each point of the shell volume. The 3D *contravariant basis* $(\boldsymbol{g}^\alpha, \boldsymbol{g}^\beta, \boldsymbol{g}^z)$ can be inferred from the 3D covariant basis by the relations:

$$\boldsymbol{g}_m \cdot \boldsymbol{g}^n = \delta_m^n \qquad m, n = \alpha, \beta, z, \tag{4.48}$$

where δ denotes the Kronecker symbol ($\delta_m^n = 1$ if $m = n$ and 0 otherwise).

On the other hand, the vectors $\boldsymbol{a}_l$ $(l = \alpha, \beta)$ and $\boldsymbol{a}_z$, that form the covariant basis of the plane tangent to the midsurface at each point, are calculated from the 2D chart as follows:

$$\boldsymbol{a}_l = \frac{\partial \boldsymbol{\phi}(\alpha, \beta)}{\partial l} \qquad l = \alpha, \beta, \qquad \boldsymbol{a}_z = \frac{\boldsymbol{a}_\alpha \wedge \boldsymbol{a}_\beta}{\|\boldsymbol{a}_\alpha \wedge \boldsymbol{a}_\beta\|}. \tag{4.49}$$

Similarly to the 3D case, a contravariant basis of the tangent plane $(\boldsymbol{a}^\alpha, \boldsymbol{a}^\beta)$ can be defined by the relations:

$$\boldsymbol{a}_l \cdot \boldsymbol{a}^r = \delta_l^r \qquad l, r = \alpha, \beta. \tag{4.50}$$

The 3D base vectors can be defined by means of the basis of the tangent plane, using the following relations:

$$\begin{aligned} \boldsymbol{g}_l &= (\delta_l^r - z b_l^r)\boldsymbol{a}_r = \mu_l^r \boldsymbol{a}_r \qquad l, r = \alpha, \beta, \\ \boldsymbol{g}_z &= \boldsymbol{a}_z, \end{aligned} \tag{4.51}$$

where the tensors b_l^r and μ_l^r have been introduced. b_l^r takes into account the curvature of the shell and it is:

$$b_l^r = \boldsymbol{a}^r_{,l} \cdot \boldsymbol{a}_z.$$

In the case of the contravariant basis, one has:

$$\boldsymbol{g}^l = m_r^l \boldsymbol{a}^r \qquad l, r = \alpha, \beta, \tag{4.52}$$

where m_r^l is the inverse of the tensor μ_l^r introduced in the previous relations (4.51):

$$m_r^l = (\mu^{-1})_r^l \, .$$

One can conclude that, knowing the 2D chart $\boldsymbol{\phi}$, it is possible to define each point in the volume of the shell and all the quantities (3D base vectors, 2D base vectors and so on) here presented. These last are used to write the strain-displacement relations of the shell.

4.10 CLASSICAL MODELS AND UNIFIED FORMULATION

The most common mathematical models used to describe shell structures can be classified in two classes, on the basis of different physical assumptions. The Koiter model is based on the Kirchhoff-Love hypothesis and it is hereafter indicated as the Classical Shell Theory (CST). The Naghdi model is based on the Reissner-Mindlin assumptions that take into account the transverse shear deformation, and it is hereafter indicated as the First-order Shear Deformation Theory. In the first model, normals to the reference surface remain normal in the deformed states and do not change in length. This means that transverse shear and transverse normal strains are negligible compared to the other strains. In the second model, one or more CST postulates are removed and, for example, the effects of transverse shear stresses can be taken into account. For more details on the assumptions of the CST and FSDT models, the reader can refer to the works by [20] and [21], respectively. The formulation of these models is very similar to that of the CPT and FSDT presented for the plates in Section 4.7. The difference is that, in the case of shells, the displacement components are expressed in curvilinear coordinates.

Being (u_α, u_β, u_z) the displacement components along the coordinates (α, β, z), respectively, the displacement field for CST model is:

$$\begin{aligned}
u_\alpha(\alpha,\beta,z) &= u_{\alpha_0}(\alpha,\beta) - \frac{\partial u_{z_0}}{\partial \alpha}(\alpha,\beta) z \\
u_\beta(\alpha,\beta,z) &= u_{\beta_0}(\alpha,\beta) - \frac{\partial u_{z_0}}{\partial \beta}(\alpha,\beta) z \\
u_z(\alpha,\beta,z) &= u_{z_0}(\alpha,\beta)
\end{aligned} \tag{4.53}$$

On the other hand, the displacement field of FSDT model is:

$$\begin{aligned} u_\alpha(\alpha,\beta,z) &= u_{\alpha_0}(\alpha,\beta)+\phi_\alpha(\alpha,\beta)z \\ u_\beta(\alpha,\beta,z) &= u_{\beta_0}(\alpha,\beta)+\phi_\beta(\alpha,\beta)z \\ u_z(\alpha,\beta,z) &= u_{z_0}(\alpha,\beta) \end{aligned} \tag{4.54}$$

where ϕ_α and ϕ_β are the rotations of a transverse normal about the β- and α-axes, respectively.

As in the case of plates, the CST and FSDT models can be seen as particular cases of the $N=1$ model. By acting on the full linear expansion, they can be obtained applying penalty techniques as shown in Section 5.16.2. Generalizing the displacement field according to the unified formulation, the displacement field is obtained through a formal expression regardless of the order of the theory (N), which is considered as an input of the analysis.

Like in the plates, the unified formulation of the through-the-thickness displacement field $\mathbf{u}=(u_\alpha, u_\beta, u_z)$ is described by an expansion of generic functions (F_τ)

$$\mathbf{u}(\alpha,\beta,z)=F_\tau(z)\mathbf{u}_\tau(\alpha,\beta), \qquad \tau=0,1,\ldots,N \tag{4.55}$$

but, for shells, the thickness functions F_τ depend on the thickness coordinate z and the displacement components $\mathbf{u}_\tau$ depend on the coordinates α,β. N is the order of expansion of the model and the Taylor polynomials, consisting of the base z^i, with $i=0,1,\ldots,N$, are chosen as thickness functions. Therefore, the complete expansion of an N-order model is:

$$\begin{aligned} u_\alpha &= u_{\alpha_0}+z\,u_{\alpha_1}+ \quad \ldots \quad \Big\| \quad +z^N\,u_{\alpha N} \\ u_\beta &= u_{\beta_0}+z\,u_{\beta_1}+ \quad \ldots \quad \Big\| \quad +z^N\,u_{\beta N} \\ u_z &= u_{z_0}+z\,u_{z_1}+ \quad \ldots \quad \Big\| \quad +z^N\,u_{zN} \end{aligned} \tag{4.56}$$

Usually, orders of expansion up to $N=4$ are sufficient to reproduce the 3D solution of shell problems.

REFERENCES

1. Cicala P. Systematic approximation approach to linear shell theory. Torino: Levrotto e Bella; 1965.
2. Gol'denweizer AL. Derivation of an approximate theory of bending of a plate by the method of asymptotic integration of the equations of the theory of elasticity. Prikl Mat Meh 1962;26:1000–25.
3. Washizu K. Variational methods in elasticity and plasticity. Oxford: Pergamon Press; 1968.
4. Reti L. The unknown Leonardo. New York: McGraw-Hill Book Co.; 1974.

5. Ballarini R. The Da Vinci-Euler-Bernoulli beam theory? Mech Engrg Mag 2003. http://www.memagazine.org/.
6. Euler L. De curvis elasticis. In: Methodus Inveniendi Lineas Curvas Maximi Minimive Proprietate Gaudentes, Sive Solutio Problematis Isoperimetrici Lattissimo Sensu Accepti. Lausanne and Geneva: Bousquet; 1744.
7. Timoshenko SP. On the corrections for shear of the differential equation for transverse vibrations of prismatic bars. Philos Mag 1921;41:744–6.
8. Timoshenko SP. On the transverse vibrations of bars of uniform cross-section. Philos Mag 1922;43:125–31.
9. Carrera E, Giunta G, Petrolo M. Beam structures: classical and advanced theories. Wiley; 2011.
10. Cowper GR. The shear coefficient in Timoshenko's beam theory. J Appl Mech 1966;33(2):335–40.
11. Pai PF, Schulz MJ. Shear correction factors and an energy-consistent beam theory. Int J Solids Struct 1999;36:1523–40.
12. Gruttmann F, Sauer R, Wagner W. Shear stresses in prismatic beams with arbitrary cross-sections. Int J Numer Methods Eng 1999;45:865–89.
13. Gruttmann F, Wagner W. Shear correction factors in Timoshenko's beam theory for arbitrary shaped cross-sections. Comput Mech 2001;27:199–207.
14. Kapania K, Raciti S. Recent advances in analysis of laminated beams and plates, part I: shear effects and buckling. AIAA J 1989;27(7):923–35.
15. Kirchhoff G. Über das Gleichgewicht und die Bewegung einer elastishen Scheibe. J Reine Angew Math 1850;40:51–88.
16. Reissner E. The effect of transverse shear deformation on the bending of elastic plates. J Appl Mech 1945;12(2):69–77.
17. Mindlin RD. Influence of rotatory inertia and shear in flexural motions of isotropic elastic plates. J Appl Mech 1951;18:1031–6.
18. Babuška I, d'Harcourt JM, Schwab C. Optimal shear correction factors for hierarchical plate models. Math Modell Sci Comput 1994;I:1–30.
19. Rössle A. On the derivation of an asymptotically correct shear correction factor for the Reissner-Mindlin plate model. C R Acad Sci, Ser 1 Math 1999;328(3):269–74.
20. Koiter WT. On the foundations of the linear theory of thin elastic shell. Proc K Ned Akad 1970;73:169–95.
21. Naghdi WT. The theory of shells and plates. Handb Phis 1972;6:425–640.

CHAPTER 5

Advanced theories for composite beams, plates and shells

5.1 INTRODUCTION TO THE UNIFIED FORMULATION

The present chapter proposes a new approach for the derivation of FE matrices. This approach is based on the Unified Formulation (UF), which allows FE matrices to be derived in terms of fundamental nuclei. The UF is introduced in this chapter by extending the indicial notation (indices i and j), which is also found in finite element procedures, to the theory of structures (indices τ ans s). As a result, a fundamental nucleus (FN), expressed in terms of four indices (τ, s, i and j), is obtained. This FN is a 3×3 array and its form does not change for 1D, 2D or 3D problems.

5.2 STIFFNESS MATRIX OF A BAR AND THE RELATED FUNDAMENTAL NUCLEUS

In this section, the stiffness matrix of a bar is derived using a compact indicial approach.

In order to distinguish the displacements from their virtual variations, two different indices are introduced,

$$u_y(y) = N_i(y)\, u_{y_i} \tag{5.1}$$

$$\delta u_y(y) = N_j(y)\, \delta u_{y_j} \tag{5.2}$$

where i denotes the displacements and j the virtual variations. Repeated indices indicate a sum, therefore

$$u_y(y) = N_i(y)\, u_{y_i} = \sum_{i=1}^{N_{NE}} N_i(y) u_{y_i} = N_1 u_{y_1} + N_2 u_{y_2} + \ldots + N_{N_{NE}} u_{y_{N_{NE}}} \tag{5.3}$$

$$\delta u_y(y) = N_j(y)\, \delta u_j = \sum_{j=1}^{N_{NE}} N_i(y) \delta u_{y_j} = N_1 \delta u_{y_1} + N_2 \delta u_{y_2} + \ldots + N_{N_{NE}} \delta u_{y_{N_{NE}}} \tag{5.4}$$

Thermal Stress Analysis of Composite Beams, Plates and Shells.
DOI: http://dx.doi.org/10.1016/B978-0-12-849892-7.00007-8

where N_{NE} is the number of nodes in the finite element. The strains and the relative virtual variations are

$$\varepsilon = N_{i,y}\, u_{y_i} \tag{5.5}$$

$$\delta\varepsilon = N_{j,y}\, \delta u_{y_j} \tag{5.6}$$

where $N_{i,y} = \dfrac{dN}{dy}$. The virtual variation of the internal work, according to the indicial formulation, becomes

$$\begin{aligned} \delta L_{int} &= \int_V \delta\varepsilon^T\, \sigma\, dV \\ &= \int_V \delta\varepsilon^T\, E\, \varepsilon\, dV \\ &= \delta u_{y_j} \left(\int_V N_{j,y}\, E\, N_{i,y}\, dV \right) u_{y_i} \\ &= \delta u_{y_j}\, k^{ij}\, u_{y_i} \end{aligned} \tag{5.7}$$

This is what we call the 'fundamental nucleus' of a bar. Its form does not depend on the N formulation (see next paragraph). It is invariant with respect to

- the number of element nodes, N_{NE};
- the choice of the shape functions, N_i and N_j.

The explicit form of the nucleus is

$$k^{11} = \int_V N_{1,y}\, E\, N_{1,y}\, dV = \frac{1}{L}\, E\, \frac{1}{L}\, AL = EA/L \tag{5.8}$$

$$k^{12} = \int_V N_{2,y}\, E\, N_{1,y}\, dV = -\frac{1}{L}\, E\, \frac{1}{L}\, AL = -EA/L \tag{5.9}$$

$$k^{21} = \int_V N_{1,y}\, E\, N_{2,y}\, dV = -\frac{1}{L}\, E\, \frac{1}{L}\, AL = -EA/L \tag{5.10}$$

$$k^{22} = \int_V N_{2,y}\, E\, N_{2,y}\, dV = \frac{1}{L}\, E\, \frac{1}{L}\, AL = EA/L \tag{5.11}$$

It appears evident that the fundamental nucleus can easily be used in a computer program:

- two loops are made on i and j;
- the stiffness term for each i,j is calculated and assembled in the global matrix through an appropriate identification of their position in the stiffness matrix.

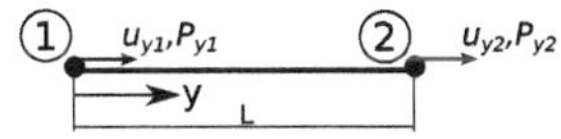

Figure 5.1 Bar element with 2 nodes in its local reference system.

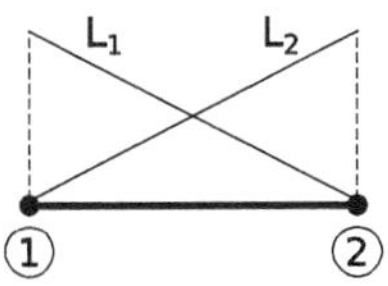

Figure 5.2 Example of two node bar, $N_{NE} = 2$, linear shape functions.

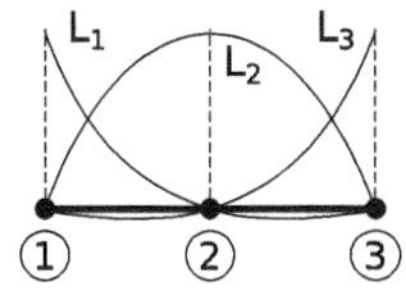

Figure 5.3 Example of three node bar, $N_{NE} = 3$, parabolic shape functions.

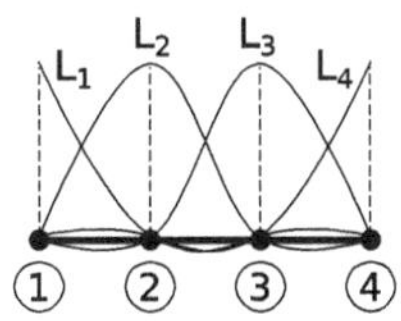

Figure 5.4 Example of four node bar, $N_{NE} = 4$, cubic shape functions.

The stiffness matrix derived using the fundamental nucleus can be written as

$$[K] = \begin{bmatrix} k^{11} & k^{12} \\ k^{21} & k^{22} \end{bmatrix} = \frac{EA}{L} \begin{bmatrix} 1 & -1 \\ -1 & 1 \end{bmatrix} \tag{5.12}$$

If the considered bar element has 2 nodes ($N_{NE} = 2$), as shown in Fig. 5.1, the virtual variation of the internal virtual work can be written in explicit form, according to Eq. (5.7),

$$\delta L_{int} = \delta u_{y_1}\, k^{11}\, u_{y_1} + \delta u_{y_2}\, k^{12}\, u_{y_1} + \delta u_{y_1}\, k^{21}\, u_{y_2} + \delta u_{y_2}\, k^{22}\, u_{y_2} \tag{5.13}$$

5.3 FUNDAMENTAL NUCLEUS FOR THE CASE OF A BAR ELEMENT WITH INTERNAL NODES

The indicial notation allows additional nodes to be introduced into the bar without any difficulty. Figures 5.2, 5.3 and 5.4 show different bar element configurations. The first (Fig. 5.2) has 2 nodes and the shape functions are linear. The second and the third cases (Fig. 5.3 and 5.4) have three and four nodes and the shape functions are quadratic and cubic, respectively. The stiffness matrix, for the linear case, was derived in Section 5.2. The stiffness matrix can be written for refined elements using the same approach. The displacement and its virtual variation can be written as

$$u_y(y) = N_i\, u_{y_i} \tag{5.14}$$

$$\delta u_y(y) = N_j\, \delta u_{y_j} \tag{5.15}$$

with $i,j = 1 \ldots N_{NE}$, where N_{NE} is the number of nodes of the bar element. Let us consider the case in which $N_{NE} = 3$, the shape functions are quadratic and, in the physical coordinates, their explicit expressions are

$$N_1(y) = 1 - 3\left(\frac{y}{L}\right) + 2\left(\frac{y}{L}\right)^2 \tag{5.16}$$

$$N_2(y) = 0 + 4\left(\frac{y}{L}\right) - 4\left(\frac{y}{L}\right)^2 \tag{5.17}$$

$$N_3(y) = 0 - 1\left(\frac{y}{L}\right) + 2\left(\frac{y}{L}\right)^2 \tag{5.18}$$

The indicial notation allows the stiffness matrix to be written independently of the number of nodes N_{NE},

$$k^{ij} = \int_V N_{j,y}\, E\, N_{i,y}\, dV \tag{5.19}$$

This is once again what we call the fundamental nucleus of a bar as it is in the notation of the UF. Its form does not depend on N_i or N_j and it is exactly the same as the one computed in the $N_{NE} = 2$ case. Once again, this equation is invariant with respect to:

- the number of element nodes;
- the choice of the shape functions.

The derivatives of the shape functions are:

$$N_1(y)_{,y} = -\left(\frac{3}{L}\right) + 4\left(\frac{y}{L^2}\right) \tag{5.20}$$

$$N_2(y)_{,y} = +\left(\frac{4}{L}\right) - 8\left(\frac{y}{L^2}\right) \tag{5.21}$$

$$N_3(y)_{,y} = -\left(\frac{1}{L}\right) + 4\left(\frac{y}{L^2}\right) \tag{5.22}$$

The components of the stiffness matrix can be calculated easily using an iterative procedure,

$$k^{11} = EA \int_z N_{1,y}\, N_{1,y}\, dy = +\frac{7}{3}\frac{EA}{L}$$

$$k^{12} = EA \int_z N_{2,y}\, N_{1,y}\, dy = -\frac{8}{3}\frac{EA}{L}$$

$$k^{13} = EA \int_z N_{3,y}\, N_{1,y}\, dy = +\frac{1}{3}\frac{EA}{L}$$

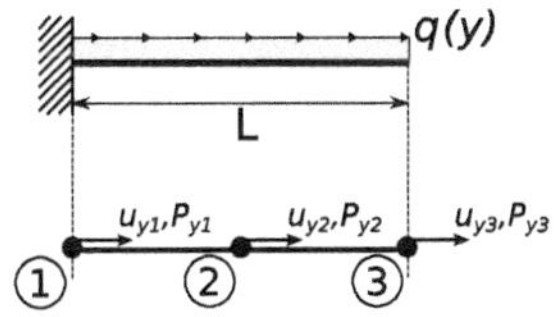

Figure 5.5 Example of bar axially loaded, physical and FEM model.

$$
\begin{aligned}
k^{21} &= EA\int_z N_{1,y}\, N_{2,y}\, dy = -\frac{8}{3}\frac{EA}{L} \\
k^{22} &= EA\int_z N_{2,y}\, N_{2,y}\, dy = +\frac{16}{3}\frac{EA}{L} \\
k^{23} &= EA\int_z N_{3,y}\, N_{2,y}\, dy = -\frac{8}{3}\frac{EA}{L} \\
k^{31} &= EA\int_z N_{1,y}\, N_{3,y}\, dy = +\frac{1}{3}\frac{EA}{L} \\
k^{32} &= EA\int_z N_{2,y}\, N_{3,y}\, dy = -\frac{8}{3}\frac{EA}{L} \\
k^{33} &= EA\int_z N_{3,y}\, N_{3,y}\, dy = +\frac{7}{3}\frac{EA}{L}
\end{aligned}
\tag{5.23}
$$

Finally, the stiffness matrix is

$$
[K] = \frac{EA}{3L}\begin{bmatrix} 7 & -8 & 1 \\ -8 & 16 & -8 \\ 1 & -8 & 7 \end{bmatrix}
\tag{5.24}
$$

5.3.1 Example

The $N_{NE} = 2$ case (two node bar) does not allow a constant distribution of axial loading ($q = cost$) to be applied. At least the $N_{NE} = 3$ case has to be adopted. In this example, a three-node bar loaded by a distributed load $q(y)$ is considered. Figure 5.5 shows the structure. Node 1 is clamped while load $q(y)$ is constant and distributed along the whole axis of the bar. The external work made by load $q(y)$ can be written as

$$
\delta L_{ext} = \int_0^L \delta u_y\, q(y) dy = \bar{q}\int_0^L \delta u\, dy
\tag{5.25}
$$

Load $q(y)$ is considered constant and it can therefore be excluded from the integral. The indicial notation introduced by the UF allows the external work to be written through the introduction of the shape functions,

$$\delta L_e = q\int_0^L N_j \delta u_j dy = \delta u_j\, q\int_0^L N_j dy \tag{5.26}$$

The integral of the shape functions is equal to,

$$\int_0^L N_1 dy = \frac{L}{6}; \quad \int_0^L N_2 dy = \frac{2L}{3}; \quad \int_0^L N_3 dy = \frac{L}{6} \tag{5.27}$$

therefore,

$$\delta L_e = q\frac{L}{6}\delta u_{y1} + q\frac{2L}{3}u_{y2} + q\frac{L}{6}u_{y3} \tag{5.28}$$

The system that has to be solved becomes

$$\frac{EA}{3L}\begin{bmatrix} 7 & -8 & 1 \\ -8 & 16 & -8 \\ 1 & -8 & 7 \end{bmatrix}\begin{Bmatrix} u_{y1} \\ u_{y2} \\ u_{y3} \end{Bmatrix} = \begin{Bmatrix} P_{y1} \\ P_{y2} \\ P_{y3} \end{Bmatrix} \tag{5.29}$$

where:

$$P_{y1} = q\frac{L}{6}, \quad P_{y2} = q\frac{2L}{3}, \quad P_{y3} = q\frac{L}{6} \tag{5.30}$$

If node 1 is clamped and no other loads are applied, the boundary condition imposes a displacement equal to zero at node 1,

$$u_{y1} = 0 \tag{5.31}$$

The system can be partitioned, as shown in the previous sections and the new system that has to be solved becomes

$$\frac{EA}{3L}\left[\begin{array}{c|cc} 7 & -8 & 1 \\ \hline -8 & 16 & -8 \\ 1 & -8 & 7 \end{array}\right]\left\{\begin{array}{c} 0 \\ \hline u_{y2} \\ u_{y3} \end{array}\right\} = \left\{\begin{array}{c} R_{y1} \\ \hline q\frac{2L}{3} \\ q\frac{L}{6} \end{array}\right\} \tag{5.32}$$

It is possible to compute the displacements u_{y2} and u_{y3} from the second and third equations,

$$\frac{EA}{3L}\begin{bmatrix} 16 & -8 \\ -8 & 7 \end{bmatrix}\begin{Bmatrix} u_{y2} \\ u_{y3} \end{Bmatrix} = Lq\begin{Bmatrix} 2/3 \\ 1/6 \end{Bmatrix} \tag{5.33}$$

The solution of the problem gives the displacements

$$u_{y1} = \frac{3}{8}\frac{qL^2}{EA}, \quad u_{y2} = \frac{1}{2}\frac{qL^2}{EA} \tag{5.34}$$

The solution of the first equation gives the reaction forces at node 1, R_{y1},

$$R_{y1} = \frac{EA}{3L}\begin{bmatrix} -8 & 1 \end{bmatrix}\frac{qL^2}{EA}\begin{Bmatrix} 3/8 \\ 1/2 \end{Bmatrix} = -0.83333\, q\, L \tag{5.35}$$

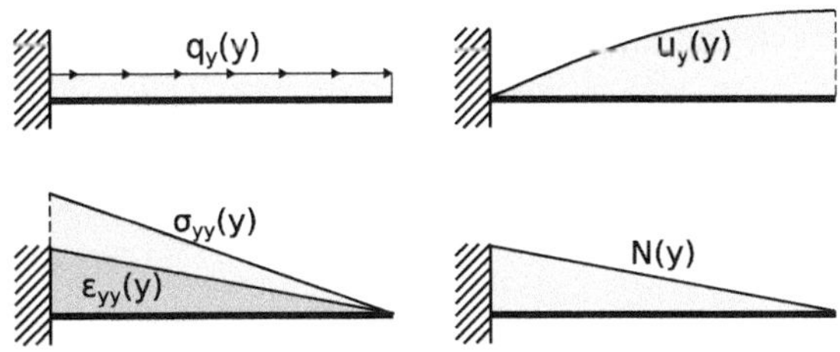

Figure 5.6 Qualitative solutions of a bar with a uniform axial load.

From these results, it is possible to recover the displacements along the whole axis,

$$u_y(y) = N_i u_{y_i} = \left(y\,L - \frac{y^2}{2}\right)\frac{q}{AE} \tag{5.36}$$

The axial strain is

$$\varepsilon_{yy}(y) = N_{i,y} u_{y_i} = (L - y)\,\frac{q}{AE} \tag{5.37}$$

Using Hooke's law, it is possible to derive the axial stress,

$$\sigma_{yy}(y) = E\varepsilon_y(y) = N_{i,y} u_{y_i} E = (L - y)\,\frac{q}{A} \tag{5.38}$$

Finally, the axial force becomes

$$N(y) = \int_A \sigma_{yy}(y)dA = \sigma_{yy}(y)\int_A 1dA = (L - y)\,q \tag{5.39}$$

The results obtained in Eqs. (5.36)–(5.39) are reported in Fig. 5.6. The displacement has quadratic behaviour, and, in the case of a distributed load, at least a three-node element is needed. The stresses, strains and resultant axial force are linear. They are equal to zero at $y = L$ and have the maximum value at $y = 0$, where the constraint was placed.

5.3.2 The case of an arbitrary defined number of nodes

The results presented in sections 5.2 and 5.3 show how to build stiffness matrices for a bar element with two or three nodes. The indicial formulation allows this formulation to be extended to a bar with any number of nodes.

Because of the one-dimensionality nature of the bar problem, the fundamental nucleus has only one term,

$$k^{ij} = \int_V N_{j,y}\,E\,N_{i,y}\,dV \tag{5.40}$$

Indices i and j represent the displacement and its virtual variation, respectively.

There are no limitations to the choice of i and j, and the UF therefore offers a tool that can be used to create the stiffness matrix of a bar element with any number of nodes[1]

$$\mathbf{K} = \begin{bmatrix} k^{11} & k^{12} & \dots & k^{1i} & \dots & k^{1\,N_{NE}} \\ k^{21} & k^{22} & & & & \\ \vdots & & \ddots & & & \vdots \\ k^{j1} & & & k^{ij} & & \vdots \\ \vdots & & & & \ddots & \\ k^{N_{NE}1} & & \dots & \dots & & k^{N_{NE}N_{NE}} \end{bmatrix} \tag{5.41}$$

Whatever the choice of N_{NE}, the stiffness matrix can be derived using only the fundamental nucleus, as shown in Eq. (5.41). The same approach can be used to define the load vector $\mathbf{P}$. The nucleus of the load vector is

$$P_j = \int_0^L q(y) N_j dy \tag{5.42}$$

The vector P_j can be used to build the load vector for a bar element with any number of nodes,

$$\mathbf{P} = \begin{bmatrix} P_{y_1} \\ P_{y_2} \\ \vdots \\ P_{y_j} \\ \vdots \\ P_{y_{N_{NE}}} \end{bmatrix} \tag{5.43}$$

5.4 FEM AND THE THEORY OF STRUCTURE: A FOUR INDICES FUNDAMENTAL NUCLEUS

In the previous section, the axial displacement, u_y, was considered to be constant over the cross-section and the transversal displacements were neglected. In the UF framework, it is very easy to overcome this limitation

[1] Indicial notations are already known in FE application. The UF extends them to the theory of structures as will be shown in the next chapters.

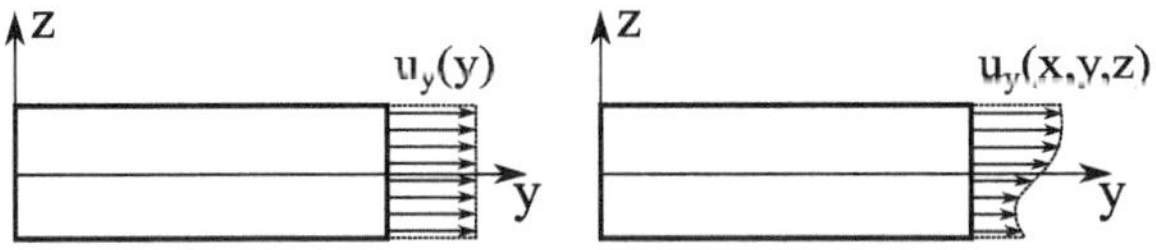

Figure 5.7 An example of a bar with a rigid cross-section, $u_y(y)$, on the left-hand and a bar with a variable axial displacement over the cross-section, $u_y(x, y, z)$, on the right-hand.

and include the effects of a non-constant axial displacement over the cross-section of a bar,

$$u_y(y) \longrightarrow u_y(x, y, z) \tag{5.44}$$

it is also easy to include all three components of the displacement field (which cannot be avoided if Eq. (5.44) is considered)

$$u_y(x, y, z) \longrightarrow \mathbf{u}^T = (u_x, u_y, u_z) \tag{5.45}$$

The fundamental nuclei for these cases are derived in the following sections.

5.4.1 Fundamental nucleus for a bar with a variable axial displacement over the cross-section

If the cross-section of a bar is considered rigid, the problem can be considered one-dimensional. In other words, the value of the displacement on the axis is enough to describe the deformation of the whole cross-section. In order to overcome this assumption, the axial displacement, u_y, should not be considered constant over the cross-section,

$$u_y(y) \longrightarrow u_y(x, y, z) \tag{5.46}$$

The displacement field that was originally defined in a one-dimensional domain now becomes a three-dimensional, 3D, domain, as shown in Fig. 5.7. A bar element with a variable displacement field over the cross-section does not have many practical applications in structural analysis but is introduced here as an intermediate step, before the next section in which a refined beam model is introduced.

The bar element introduced in Section 5.2 can be used to approximate this displacement field, thus:

$$u_y(x, y, z) = N_i(y)\, u_{y_i}(x, z) \tag{5.47}$$

The shape functions, $N_i(y)$, are used to approximate the displacements along the y axis, while the u_{y_i} coefficients are now functions of the cross-

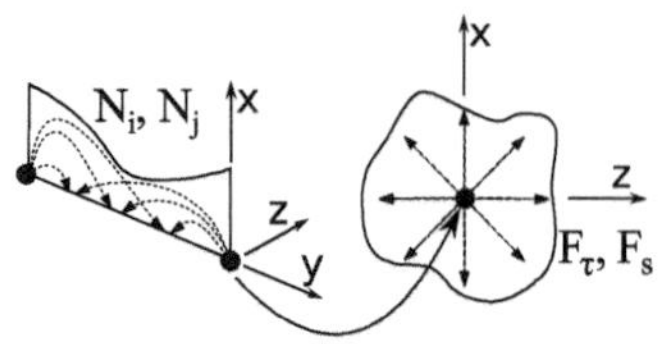

Figure 5.8 Representation of the axial approximation (using N_i, N_j), and the cross-section expansion (using F_τ, F_s).

section coordinates. The term $u_{y_i}(x, z)$ can be approximated by introducing a generic expansion to the cross-section. In general, u_{y_i}, can be written as the sum of the generic functions $F_\tau(x, z)$,

$$\begin{aligned} u_{y_i}(x, z) &= F_\tau(x, z) u_{y_{\tau i}} = \\ &= F_1(x, z) u_{y_{i1}} + F_2(x, z) u_{y_{i2}} + \ldots + F_\tau(x, z) u_{y_{\tau i}} + \ldots + F_m(x, z) u_{y_{iM}} \end{aligned} \tag{5.48}$$

where M is the number of terms in the expansion. Figure 5.8 shows the two approximations. The shape functions N_i and N_j expand the solution from the nodes to the axis. Expansions F_τ ad F_s expand the solution from the nodes to the cross-section of the bar. The complete displacement field can be obtained by introducing Eq. (5.48) into Eq. (5.47),

$$u_y(x, y, z) = N_i(y)\, u_{y_i}(x, z) = N_i(y)\, F_\tau(x, z)\, u_{y_{\tau i}} \tag{5.49}$$

The introduction of the new expansion, indicated with the indices τ and s, does not change the approach used to derive the FEM matrices introduced above. It is possible to introduce the virtual variation using the s index instead of τ,

$$u_y(x, y, z) = N_i(y)\, F_\tau(x, z)\, u_{y_{\tau i}} \tag{5.50}$$

$$\delta u_y(x, y, z) = N_j(y)\, F_s(x, z)\, \delta u_{y_{sj}} \tag{5.51}$$

The stress and strain become

$$\boldsymbol{\varepsilon}(x, y, z) = \mathbf{b} N_i(y)\, F_\tau(x, z)\, u_{y_{\tau i}} \tag{5.52}$$

$$\boldsymbol{\sigma}(x, y, z) = \mathbf{C}\mathbf{b} N_i(y)\, F_\tau(x, z)\, u_{y_{\tau i}} \tag{5.53}$$

In this case, the stress and strain become vectors because, if the axial displacement is considered not constant over the cross-section, the shear stress and strain appear in the formulation.[2]

[2] If u_y has a three-dimensional formulation one has:

$$\varepsilon_{xy} = \frac{\partial u_y}{\partial x} + \frac{\partial u_x}{\partial y} \neq 0, \quad \varepsilon_{zy} = \frac{\partial u_y}{\partial z} + \frac{\partial u_z}{\partial y} \neq 0 \tag{5.54}$$

5.4.2 Fundamental nucleus for a 1D structure with a complete displacement field: the case of a refined beam model

The formulation introduced in Section 5.4.1 considers a variable axial displacement on the cross-section of the bar. Engineering problems usually deal with three-dimensional problems and the structural solution has the aim of studying the displacements in all three directions. If all the components of the displacement have to be considered, the displacement field becomes

$$u_y(x, y, z) \longrightarrow \mathbf{u}^T = (u_x, u_y, u_z) \tag{5.58}$$

If the approximations introduced into Eqs. (5.47) and (5.48) are applied to $[u]$, the displacement field can be written in compact form as,

$$\mathbf{u} = N_i(y)\, F_\tau(x, z)\, \mathbf{u}_{\tau i} \tag{5.59}$$

or,

$$u_x(x, y, z) = N_i(y)\, F_\tau(x, z)\, u_{x_{\tau i}} \tag{5.60}$$

$$u_y(x, y, z) = N_i(y)\, F_\tau(x, z)\, u_{y_{\tau i}} \tag{5.61}$$

$$u_z(x, y, z) = N_i(y)\, F_\tau(x, z)\, u_{z_{\tau i}} \tag{5.62}$$

and the stress and strain are expressed as vectors. These vectors are derived using matrix **b**. The virtual variation of the displacements is expressed using the j and s indices,

$$\delta\boldsymbol{\varepsilon}(x, y, z) = \mathbf{b}N_j(y)\, F_s(x, z)\, \delta u_{y_{sj}} \tag{5.55}$$

The virtual variation of the internal work can be written as

$$\delta L_{int} = \int_V \delta\boldsymbol{\varepsilon}^T \boldsymbol{\sigma}\, dV = \delta u_{y_{sj}} \int_V F_s(x, z) N_j(y) \mathbf{b}^T \mathbf{C} \mathbf{b} N_i(y)\, F_\tau(x, z)\, dV u_{y_{\tau i}} \tag{5.56}$$

Thus, the fundamental nucleus becomes

$$\mathbf{k}^{\tau sij} = \int_V F_s(x, z) N_j(y) \mathbf{b}^T \mathbf{C} \mathbf{b} N_i(y)\, F_\tau(x, z)\, dV \tag{5.57}$$

Since only the axial displacement, u_y, is still considered, the fundamental nucleus is still a 1×1 matrix, but it now has two new indices. The four indices are

- i, the displacement in the FEM approximation, see Eq. (5.47);
- j, the virtual variation of the displacement in the FEM approximation, see Eq. (5.47);
- τ, the displacement in the cross-section approximation, see Eq. (5.48);
- s, the virtual variation of the displacement in the cross-section approximation, see Eq. (5.48).

Indices i and j vary from 1 to the number of nodes, N_{NE}. Indices τ and s vary from 1 to the number of terms of the expansion, M.

The introduction of the complete displacement field[3] does not lead to any changes in the approach to build the fundamental nucleus or the stiffness matrix. The virtual variation can be introduced

$$\delta \mathbf{u} = N_j(y)\, F_s(x, z)\, \delta \mathbf{u}_{sj} \tag{5.63}$$

The stress and strain can be derived using the differential operator $\mathbf{b}$ and the material coefficient matrix $[C]$,

$$\boldsymbol{\varepsilon}(x, y, z) = \mathbf{b} N_i(y)\, F_\tau(x, z)\, \mathbf{u}_{\tau i} \tag{5.64}$$

$$\boldsymbol{\sigma}(x, y, z) = \mathbf{C}\mathbf{b} N_i(y)\, F_\tau(x, z)\, \mathbf{u}_{\tau i} \tag{5.65}$$

The virtual variation becomes

$$\delta\boldsymbol{\varepsilon}(x, y, z) = \mathbf{b} N_j(y)\, F_s(x, z)\, \delta \mathbf{u}_{sj} \tag{5.66}$$

In this case, $\boldsymbol{\varepsilon}$ and $\boldsymbol{\sigma}$ are vectors that contain all the stress and stain components, $\mathbf{C}$ is a 6×6 matrix with all the material coefficients, $\mathbf{b}$ is a differential operator matrix of 6×3 size which contains the geometrical relation between the displacements and strains. All these quantities were introduced in the previous chapter of the book.

The virtual variation of the internal work can be derived in the same way as for the previous case,

$$\delta L_{int} = \int_V \delta\boldsymbol{\varepsilon}^T \boldsymbol{\sigma}\, dV = \delta \mathbf{u}_{sj} \mathbf{k}^{\tau sij} \mathbf{u}_{\tau s} = \tag{5.67}$$

$$= \delta \mathbf{u}_{sj}^T \underbrace{\int_V \Big[F_s(x, z) N_j(y) \underbrace{\overbrace{[3 \times 6]}^{\mathbf{b}^T} \overbrace{\Big[6 \times 6 \Big]}^{\mathbf{C}} \overbrace{\Big[6 \times 3 \Big]}^{\mathbf{b}}}_{[3\times 3]} N_i(y) F_\tau(x, z) \Big] dV}_{\text{Fundamental Nucleus } \mathbf{k}^{\tau sij}} \mathbf{u}_{\tau i} \tag{5.68}$$

If all the components of the displacement are considered, the nucleus becomes a 3×3 matrix.

5.5 THE ASSEMBLY PROCEDURE

Whatever the dimension of the nucleus, 3×3 or 1×1, the nucleus is denoted with the same notation: $\mathbf{k}^{\tau sij}$. The use of the UF makes the assembly

[3] Functions N_i and F_τ can be different for the various displacement components, but for the sake of simplicity, they have here been assumed to be the same.

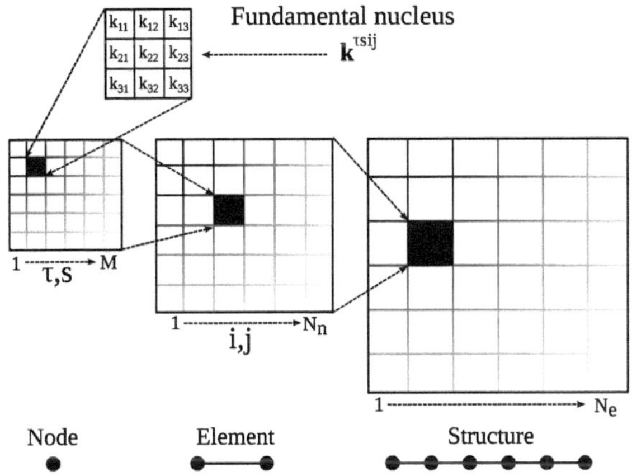

Figure 5.9 Representation of the assembling procedure: the fundamental nucleus is the core, the loops on τ and s build the matrix for a given pair of i and j, the loops on i and j give the matrix of the elements, the loop on the element gives the global stiffness matrix.

of the matrices a trivial operation that can easily be implemented in a computer code. The assembly of the matrix consists of four loops on indices i, j, τ and s, and a fundamental nucleus is calculated for each combination of these indices. A representation of this procedure is shown in Fig. 5.9. The picture shows how it is possible to build a matrix of the node, of the element and, finally, of the global stiffness matrix by exploiting the nucleus.

$$
\begin{array}{ll|ccc c ccc}
 & & \multicolumn{3}{c}{i=1} & & \multicolumn{3}{c}{i=N_{NE}} \\
 & & \tau=1 & \cdots & \tau=M & & \tau=1 & \cdots & \tau=M \\
\hline
 & s=1 & [k]^{1111} & \cdots & [k]^{1M11} & & [k]^{111N_n} & \cdots & [k]^{1M1N_n} \\
j=1 & \vdots & \vdots & & \vdots & \cdots & \vdots & & \vdots \\
 & s=M & [k]^{M111} & \cdots & [k]^{MM11} & & [k]^{M11N_n} & \cdots & [k]^{MM1N_n} \\
 & & & \vdots & & \ddots & & \vdots & \\
 & s=1 & [k]^{11N_n1} & \cdots & [k]^{1MN_n1} & & [k]^{11N_nN_n} & \cdots & [k]^{1MN_nN_n} \\
j=N_n & \vdots & \vdots & & \vdots & \cdots & \vdots & & \vdots \\
 & s=M & [k]^{M1N_n1} & \cdots & [k]^{MMN_n1} & & [k]^{M1N_nN_n} & \cdots & [k]^{MMN_nN_n}
\end{array}
\tag{5.69}
$$

Equation (5.69) shows the general formulation of the stiffness matrix. Each fundamental nucleus is reported as $\mathbf{k}^{\tau sij}$ and it works as the core of the matrix construction. The indices indicate the position of the nucleus that has to be identified in the global matrix.

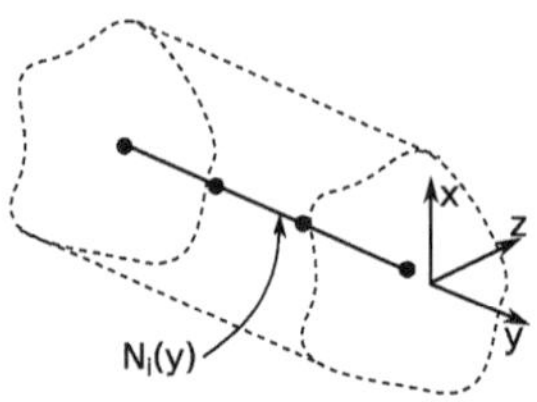

Figure 5.10 Example of 1D FEM model.

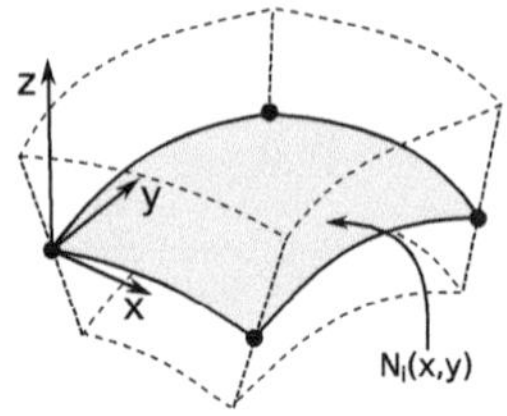

Figure 5.11 Example of 2D FEM model.

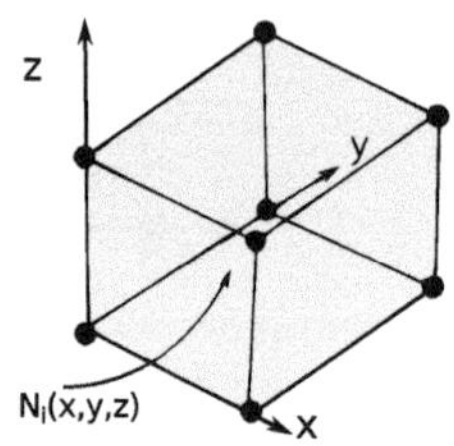

Figure 5.12 Example of 3D FEM model.

5.6 A UNIFIED APPROACH FOR ONE-, TWO- AND THREE-DIMENSIONAL STRUCTURES

The displacement field of 1D models is described in Section 5.4.2 using the classical FE approach ($N_i(y)$) along the axis and functions $F_\tau(x, z)$ over the cross-section,

$$\mathbf{u} = N_i(y)\, F_\tau(x, z)\, \mathbf{u}_{\tau i} \tag{5.70}$$

The indicial formulation used in the UF offers the possibility of extending the formulation introduced in the previous sections to any other structural theory, such as plate and shell models. The shape functions, N_i, introduced by the FEM model, can be used to approximate a domains with 2D and 3D dimensions,

$$1D\colon \longrightarrow N_i(y) \tag{5.71}$$

$$2D\colon \longrightarrow N_i(x, y) \tag{5.72}$$

$$3D\colon \longrightarrow N_i(x, y, z) \tag{5.73}$$

Figures 5.10, 5.11 and 5.12 show examples of one- two- and three-dimensional FE models, respectively. The introduction of expansions F_τ,

$$1D\colon \longrightarrow F_\tau(y, z) \tag{5.74}$$

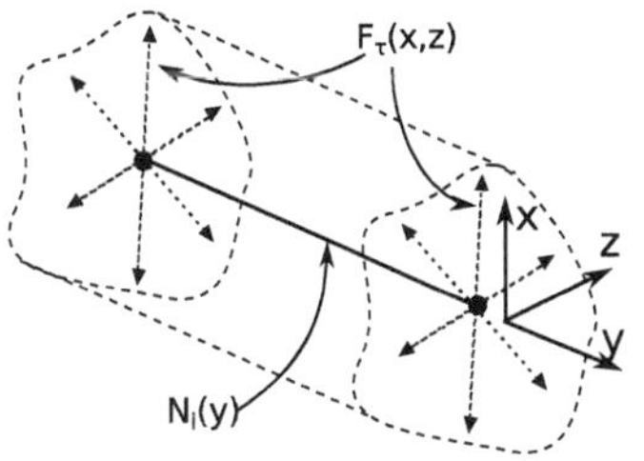

Figure 5.13 Example of beam model.

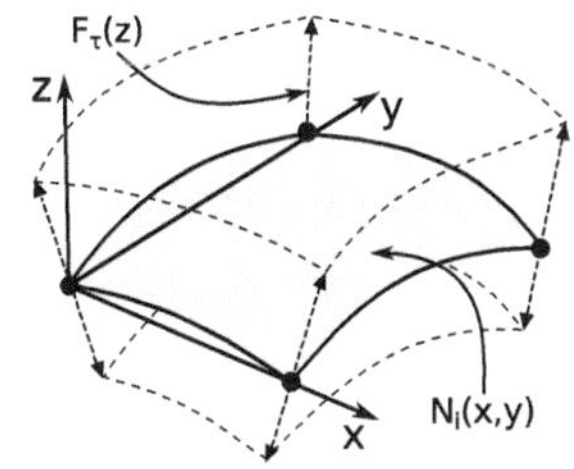

Figure 5.14 Example of plate/shell model.

$$2D: \longrightarrow F_\tau(z) \tag{5.75}$$

$$3D: \longrightarrow 1 \tag{5.76}$$

completes the description of the theoretical problem for the 1D, 2D and 3D case. The combination of the FEM and the theory of structure approximations leads to

$$1D: \quad \mathbf{u}(x, y, z) = N_i(y)\, F_\tau(x, z) \tag{5.77}$$

$$2D: \quad \mathbf{u}(x, y, z) = N_i(x, y)\, F_\tau(z) \tag{5.78}$$

$$3D: \quad \mathbf{u}(x, y, z) = N_i(x, y, z) \tag{5.79}$$

Figures 5.13, 5.14 and 5.15 show how functions F_τ approximate the solution on the cross-section of a beam, or along the thickness of a plate/shell. The 1D approach is described in Eq. (5.77) and shown in Fig. 5.13. The FE model is used to approximate the problem along the y axis, while the expansion, $F_\tau(x, z)$, is used to approximate the displacement on the cross-section. Any beam model can be derived using this formulation and classical models, such as da Vinci-Euler-Bernoulli or Timoshenko models, can be obtained as particular cases. The 2D model, see Eq. (5.78), uses the FEM to solve the problem on a reference surface, while expansion, $F_\tau(z)$, is used to describe the displacements along the thickness of the element. An example is shown in Fig. 5.14. This approach can be used to derive any plate or shell element. Finally, the 3D approach is shown in Eq. (5.79). In this case, the whole domain is approximated using the FEM approach and the F_τ expansion is not used, see Fig. 5.15. The 3D FEM models can be derived using this approach. Whatever model is used, the displacement field has a three dimensional formulation. Details of the 3D, 2D and 1D models will be introduced in the following Chapters.

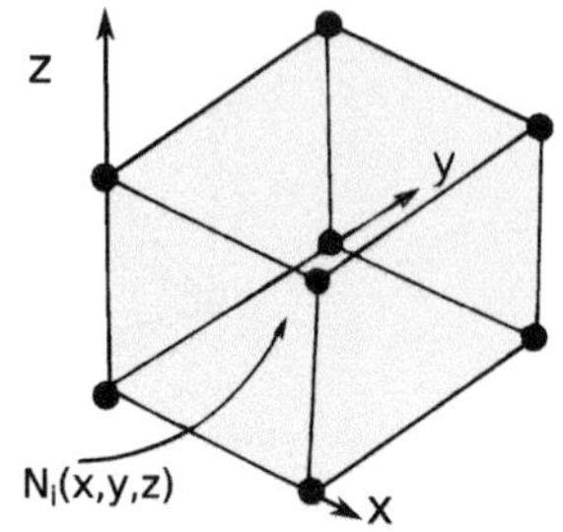

Figure 5.15 Example of solid model.

5.7 BEAM

5.7.1 The complete linear expansion case

The complete linear expansion model involves a first-order ($N = 1$) Taylor-like polynomial to describe the cross-section displacement field,

$$\begin{aligned} u_x &= \underbrace{u_{x_1}}_{N=0} \;\Big\|\; \underbrace{+x\, u_{x_2} + z\, u_{x3}}_{} \\ u_y &= u_{y_1} \;\Big\|\; +x\, u_{y_2} + z\, u_{y3} \\ u_z &= u_{z_1} \;\Big\|\; \underbrace{+x\, u_{z_2} + z\, u_{z_3}}_{N=1} \end{aligned} \tag{5.80}$$

The beam model given in Eq. (5.80) has 9 displacement variables (or unknowns): three constant ($N = 0$) and six linear ($N = 1$). Strain components are given by

$$\begin{aligned} \varepsilon_{xx} &= u_{x,x} = u_{x_2} \\ \varepsilon_{yy} &= u_{y,y} = u_{y1,y} + x\, u_{y2,y} + z\, u_{y3,y} \\ \varepsilon_{zz} &= u_{z,z} = u_{z3} \\ \gamma_{xy} &= u_{x,y} + u_{y,x} = u_{x1,y} + x\, u_{x2,y} + z\, u_{x3,y} + u_{y_2} \\ \gamma_{xz} &= u_{x,z} + u_{z,x} = u_{x_3} + u_{z_2} \\ \gamma_{yz} &= u_{y,z} + u_{z,y} = u_{y_3} + u_{z1,y} + x\, u_{z2,y} + z\, u_{z3,y} \end{aligned} \tag{5.81}$$

The linear model leads to a constant distribution of ε_{xx}, ε_{zz}, and γ_{xz} above the cross-section, and a linear distribution of ε_{yy}, γ_{xy}, and γ_{yz}.

The adoption of the $N = 1$ model is necessary to introduce the in-plane stretching of the cross-section. A simple example is presented in the following to underline the importance of the in-plane stretching terms (u_{x_2}, u_{x_3}, u_{z_2} and u_{z_3}). Figure 5.16 shows a rectangular cross-section loaded by two point forces which provide a pure torsion load. The adoption of DEBBT or TBT to analyze this problem would provide null displacements since

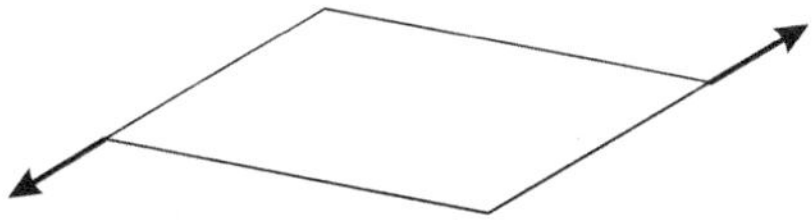

Figure 5.16 Torsion of a square beam.

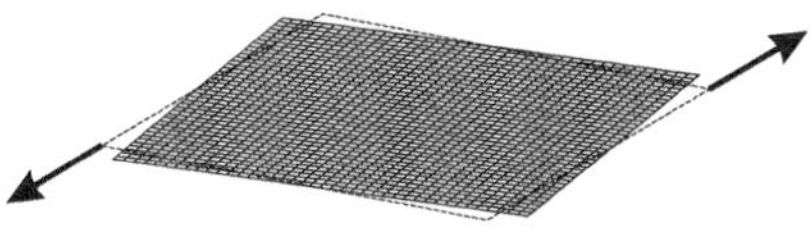

Figure 5.17 Deformed cross-section due to torsion, $N=1$ model.

only constant in-plane translations can be detected. The $N=1$ model can provide a linear distribution of the in-plane stretching as shown in Fig. 5.17.

5.7.2 A finite element based on $N=1$

The $N=1$ model has nine displacement variables, this implies that, in a finite element formulation, each node has nine generalized displacement variables. The aim of this section is to provide the FE formulation for $N=1$. The derivation of the governing finite element equations begins with the definition of the nodal displacement vector

$$\mathbf{u}_i = \left\{ \begin{array}{ccccccccc} u_{x1} & u_{y1} & u_{z1} & u_{x2} & u_{y2} & u_{z2} & u_{x3} & u_{y3} & u_{z3} \end{array} \right\}^T \tag{5.82}$$

The displacement variables are interpolated along the beam axis by means of the shape functions, N_i:

$$\mathbf{u} = N_i \mathbf{u}_i \tag{5.83}$$

Beam elements with two, B2, nodes are considered here, with the following shape functions

$$N_1 = \frac{1}{2}(1-r), \quad N_2 = \frac{1}{2}(1+r), \quad \begin{cases} r_1 = -1 \\ r_2 = +1 \end{cases} \tag{5.84}$$

where the natural coordinate, r, varies from -1 and $+1$ and r_i indicates the position of the node within the natural boundaries of the beam. The total number of degrees of freedom of the structural model will be given by

$$DOFs = \underbrace{3 \times 3}_{\text{number of DOFs per node}} \times [(\underbrace{2}_{\text{number of nodes per element}} - 1) \times \underbrace{N_{BE}}_{\text{total number of beam elements}} + 1] \tag{5.85}$$

The principle of virtual displacements, PVD, is employed to compute the FE matrices

$$\delta L_{int} = \delta L_{ext} \tag{5.86}$$

where

$$\delta L_{int} = \int_V \delta \boldsymbol{\varepsilon}^T \boldsymbol{\sigma} \, dV \tag{5.87}$$

L_{int} stands for the strain energy, L_{ext} is the work of the external loadings, and δ stands for the virtual variation. Using Eq. (5.84), a compact form of the virtual variation of the internal work can be obtained, as known in the FE procedure,

$$\delta L_{int} = \delta \mathbf{u}_j^T \mathbf{k}^{ij} \mathbf{u}_i \tag{5.88}$$

where $\mathbf{k}^{ij}$ is the stiffness matrix. For a given i, j pair, the stiffness matrix has the form

$$\left[\begin{array}{ccc|ccc|ccc}
k_{xx}^{1,1} & k_{xy}^{1,1} & k_{xz}^{1,1} & k_{xx}^{1,x} & k_{xy}^{1,x} & k_{xz}^{1,x} & k_{xx}^{1,z} & k_{xy}^{1,z} & k_{xz}^{1,z} \\
k_{yx}^{1,1} & k_{yy}^{1,1} & k_{yz}^{1,1} & k_{yx}^{1,x} & k_{yy}^{1,x} & k_{yz}^{1,x} & k_{yx}^{1,z} & k_{yy}^{1,z} & k_{yz}^{1,z} \\
k_{zx}^{1,1} & k_{zy}^{1,1} & k_{zz}^{1,1} & k_{zx}^{1,x} & k_{zy}^{1,x} & k_{zz}^{1,x} & k_{zx}^{1,z} & k_{zy}^{1,z} & k_{zz}^{1,z} \\
\hline
k_{xx}^{x,1} & k_{xy}^{x,1} & k_{xz}^{x,1} & k_{xx}^{x,x} & k_{xy}^{x,x} & k_{xz}^{x,x} & k_{xx}^{x,z} & k_{xy}^{x,z} & k_{xz}^{x,z} \\
k_{yx}^{x,1} & k_{yy}^{x,1} & k_{yz}^{x,1} & k_{yx}^{x,x} & k_{yy}^{x,x} & k_{yz}^{x,x} & k_{yx}^{x,z} & k_{yy}^{x,z} & k_{yz}^{x,z} \\
k_{zx}^{x,1} & k_{zy}^{x,1} & k_{zz}^{x,1} & k_{zx}^{x,x} & k_{zy}^{x,x} & k_{zz}^{x,x} & k_{zx}^{x,z} & k_{zy}^{x,z} & k_{zz}^{x,z} \\
\hline
k_{xx}^{z,1} & k_{xy}^{z,1} & k_{xz}^{z,1} & k_{xx}^{z,x} & k_{xy}^{z,x} & k_{xz}^{z,x} & k_{xx}^{z,z} & k_{xy}^{z,z} & k_{xz}^{z,z} \\
k_{yx}^{z,1} & k_{yy}^{z,1} & k_{yz}^{z,1} & k_{yx}^{z,x} & k_{yy}^{z,x} & k_{yz}^{z,x} & k_{yx}^{z,z} & k_{yy}^{z,z} & k_{yz}^{z,z} \\
k_{zx}^{z,1} & k_{zy}^{z,1} & k_{zz}^{z,1} & k_{zx}^{z,x} & k_{zy}^{z,x} & k_{zz}^{z,x} & k_{zx}^{z,z} & k_{zy}^{z,z} & k_{zz}^{z,z}
\end{array}\right]_{ij} \tag{5.89}$$

where the superscripts indicate the expansion functions that are involved in each component of the stiffness matrix, i.e. 1, x, or z. For the sake of clarity, the explicit expression of two components is reported hereafter

$$
\begin{aligned}
k_{xx}^{1,1} &= C_{44} \int_A 1 \cdot 1 \, dxdz \int_l N_{i,y} N_{j,y} \, dy \\
k_{yx}^{x,z} &= C_{23} \int_A \frac{\partial x}{\partial x} \cdot z \, dxdz \int_l N_i N_{j,y} \, dy + C_{44} \int_A x \cdot \frac{\partial z}{\partial x} \, dxdz \int_l N_{i,y} N_j \, dy \\
&= C_{23} \int_A 1 \cdot z \, dxdz \int_l N_i N_{j,y} \, dy
\end{aligned}
\tag{5.90}
$$

where A indicates the cross-section area and l the element length.

5.8 DEBBT, TBT AND $N = 1$ IN UNIFIED FORM

The DEBBT, TBT and $N = 1$ models can be obtained in a unified manner via a condensed notation that represents the basic step towards the UF. In this section, the $N = 1$ will be reformulated by means of a new notation, then TBT and DEBBT will be obtained as particular cases.

5.8.1 Unified formulation of $N = 1$

The $N = 1$ stiffness matrix was given in Eq. (5.89), that matrix can be considered as being composed of nine 3×3 sub-matrices,

$$
\begin{array}{cccc}
 & 1 & x & z \\
1 & \begin{bmatrix} k_{xx}^{1,1} & k_{xy}^{1,1} & k_{xz}^{1,1} \\ k_{yx}^{1,1} & k_{yy}^{1,1} & k_{yz}^{1,1} \\ k_{zx}^{1,1} & k_{zy}^{1,1} & k_{zz}^{1,1} \end{bmatrix} & \cdots & \cdots \\
x & \cdots & \cdots & \begin{bmatrix} k_{xx}^{x,z} & k_{xy}^{x,z} & k_{xz}^{x,z} \\ k_{yx}^{x,z} & k_{yy}^{x,z} & k_{yz}^{x,z} \\ k_{zx}^{x,z} & k_{zy}^{x,z} & k_{zz}^{x,z} \end{bmatrix} \\
z & \cdots & \begin{bmatrix} k_{xx}^{z,x} & k_{xy}^{z,x} & k_{xz}^{z,x} \\ k_{yx}^{z,x} & k_{yy}^{z,x} & k_{yz}^{z,x} \\ k_{zx}^{z,x} & k_{zy}^{z,x} & k_{zz}^{z,x} \end{bmatrix} & \cdots
\end{array}
\tag{5.91}
$$

Each sub-matrix has a fixed pair of expansion functions that are used in the explicit computation of the integrals, as shown in Eq. (5.90). It is extremely important to point out that the formal expression of each component of the sub-matrices does not depend on the expansion functions, that is, corresponding components of different sub-matrices have the same formal expression, as shown hereafter:

$$
\begin{aligned}
k_{xx}^{1,1} &= C_{44} \int_A 1 \cdot 1 \; dxdz \int_l N_{i,y} N_{j,y} dy \\
k_{xx}^{x,z} &= C_{44} \int_A x \cdot z \; dxdz \int_l N_{i,y} N_{j,y} dy \\
k_{xx}^{1,z} &= C_{44} \int_A 1 \cdot z \; dxdz \int_l N_{i,y} N_{j,y} dy
\end{aligned}
\tag{5.92}
$$

This implies that the sub-matrix can be considered as a fundamental invariant nucleus which can be used to build the global stiffness matrix. Let us introduce the following notation for the expansion functions, F_τ

$$
\begin{aligned}
F_{\tau=1} &= 1 \\
F_{\tau=2} &= x \\
F_{\tau=3} &= z
\end{aligned}
\tag{5.93}
$$

The displacement field in Eq. (5.80) becomes

$$
\begin{aligned}
u_x &= F_1 \, u_{x_1} + F_2 \, u_{x_2} + F_3 \, u_{x_3} = F_\tau \, u_{x_\tau} \\
u_y &= F_1 \, u_{y_1} + F_2 \, u_{y_2} + F_3 \, u_{y_3} = F_\tau \, u_{y_\tau} \\
u_z &= F_1 \, u_{z_1} + F_2 \, u_{z_2} + F_3 \, u_{z_3} = F_\tau \, u_{z_\tau}
\end{aligned}
\tag{5.94}
$$

where the repeated indexes indicate summation according to the Einstein notation. The displacement vector can be written as

$$
\mathbf{u} = F_\tau \mathbf{u}_\tau, \qquad \tau = 1, 2, 3 \tag{5.95}
$$

If a finite element formulation is introduced and two-node elements are adopted, the nodal unknown vector is given by

$$
\mathbf{u}_{\tau i} = \left\{ \begin{array}{ccc} u_{x_{\tau i}} & u_{y_{\tau i}} & u_{z_{\tau i}} \end{array} \right\}^T \qquad \tau = 1, 2, 3 \qquad i = 1, 2 \tag{5.96}
$$

The compact form of the internal work seen in Eq. (5.88) becomes

$$
\delta L_{int} = \delta \mathbf{u}_{sj}^T \mathbf{k}^{\tau sij} \mathbf{u}_{\tau i} \tag{5.97}
$$

where

- τ and s are the expansion function indexes;
- i and j are the shape function indexes.

Coherently with the notation introduced, the matrix in Eq. (5.91) can be expressed as

$$
\begin{array}{cccc}
 & \tau=1 & \tau=2 & \tau=3 \\
s=1 & \begin{bmatrix} k_{xx}^{11} & k_{xy}^{11} & k_{xz}^{11} \\ k_{yx}^{11} & k_{yy}^{11} & k_{yz}^{11} \\ k_{zx}^{11} & k_{zy}^{11} & k_{zz}^{11} \end{bmatrix} & \cdots & \cdots \\
s=2 & \cdots & \cdots & \begin{bmatrix} k_{xx}^{32} & k_{xy}^{32} & k_{xz}^{32} \\ k_{yx}^{32} & k_{yy}^{32} & k_{yz}^{32} \\ k_{zx}^{32} & k_{zy}^{32} & k_{zz}^{32} \end{bmatrix} \\
s=3 & \cdots & \begin{bmatrix} k_{xx}^{23} & k_{xy}^{23} & k_{xz}^{23} \\ k_{yx}^{23} & k_{yy}^{23} & k_{yz}^{23} \\ k_{zx}^{23} & k_{zy}^{23} & k_{zz}^{23} \end{bmatrix} & \cdots
\end{array}
\tag{5.98}
$$

Each 3×3 block is the fundamental nucleus of the stiffness matrix. A component of the nucleus is given for different combinations of the expansion functions,

$$
\begin{aligned}
k_{xx}^{ij11} &= C_{44} \int_A F_1 \cdot F_1 \, dxdz \int_l N_{i,y} N_{j,y} dy \\
k_{xx}^{ij23} &= C_{44} \int_A F_2 \cdot F_3 \, dxdz \int_l N_{i,y} N_{j,y} dy \\
k_{xx}^{ij13} &= C_{44} \int_A F_1 \cdot F_3 \, dxdz \int_l N_{i,y} N_{j,y} dy
\end{aligned}
\tag{5.99}
$$

The unified formulation introduced is of particular interest when a computational implementation is considered. The exploitation of the four indexes in the formal expression of the fundamental nucleus makes it possible to compute the stiffness matrix by means of four nested *FOR*-cycles.

5.8.2 DEBBT and TBT as particular cases of $N = 1$

The DEBBT and TBT are particular cases of the $N = 1$, and they can be obtained by acting on the full linear expansion. As far as TBT is concerned, the displacement field is given by

$$
\begin{aligned}
u_x &= u_{x_1} \\
u_y &= u_{y_1} + x\, u_{y_2} + z\, u_{y_3} \\
u_z &= u_{z_1}
\end{aligned}
\tag{5.100}
$$

A linear out-of-plane warping distribution is considered and constant in-plane displacement distributions are accounted for. Starting from the $N = 1$

case, two possible techniques can be used to obtain TBT: 1. the rearranging of rows and columns of the stiffness matrix; 2. penalization of the stiffness terms related to u_{x_2}, u_{x_3}, u_{z_2}, and u_{z_3} (the latter is here preferred in the numerical applications). The main diagonal terms have to be considered, that is, $i=j$ and $\tau=s$; moreover, only the component with $\tau, s=2,3$ has to be penalized, therefore

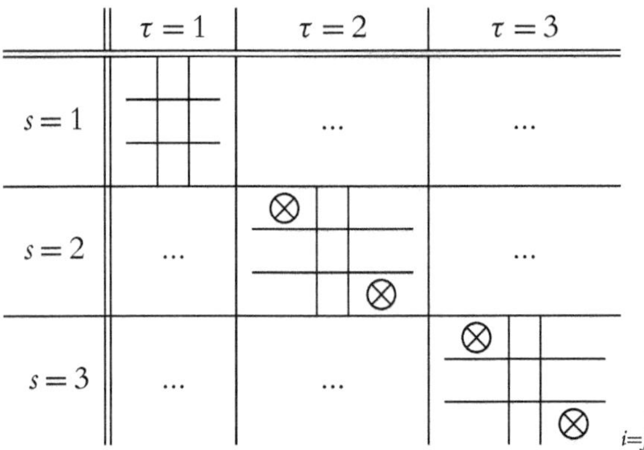

(5.101)

EBBT can be obtained through the penalization of γ_{xy} and γ_{zy}. This condition can be imposed using a penalty value χ in the following constitutive equations

$$
\begin{aligned}
\tau_{xy} &= \chi C_{55}\gamma_{xy} + \chi C_{45}\gamma_{zy} \\
\tau_{zy} &= \chi C_{45}\gamma_{xy} + \chi C_{44}\gamma_{zy}
\end{aligned}
\tag{5.102}
$$

5.9 HIGHER-ORDER MODELS

Classical 1D models can properly deal with the bending of a compact cross-section beam, on the other hand, refined models are mandatory to describe the mechanical response of more complex boundary (e.g. torsion) or geometrical (e.g. thin-walls) conditions. Refinements of classical theories are possible as shown by many authors, however, most of the times, refinements are problem dependent. This means that, as the structural problem changes, a new model has to be developed and adopted.

This section presents a novel unified approach to deal with any-order structural models. In the present unified formulation, in fact, displacement fields are obtained through a single formal expression in a unified manner regardless of the order of the theory (N) which is considered as an input of the analysis.

Table 5.1 Taylor-like polynomials. This table presents the compact form of the Taylor-like polynomials

N	M	F_τ
0	1	$F_1 = 1$
1	3	$F_2 = x \; F_3 = z$
2	6	$F_4 = x^2 \; F_5 = xz \; F_6 = z^2$
3	10	$F_7 = x^3 \; F_8 = x^2 z \; F_9 = xz^2 \; F_{10} = z^3$
...	...	...
N	$\frac{(N+1)(N+2)}{2}$	$F_{\frac{(N^2+N+2)}{2}} = x^N \; \ldots\ldots \; F_{\frac{(N+1)(N+2)}{2}} = z^N$

The unified formulation of the cross-section displacement field is described by an expansion of generic functions (F_τ),

$$\mathbf{u} = F_\tau \mathbf{u}_\tau, \qquad \tau = 1, 2, ..., M \tag{5.103}$$

where F_τ are functions of the cross-section coordinates x and z, $\mathbf{u}_\tau$ is the displacement vector and M stands for the number of terms of the expansion. According to the Einstein notation, the repeated subscript τ indicates summation. The choice of F_τ and M is arbitrary, different base functions of any-order can be taken into account to model the displacement field of a structure above its cross-section. One possible choice deals with the adoption of Taylor-like polynomials consisting of the 2D base $x^i z^j$, where i and j are positive integers. Table 5.1 presents M and F_τ as functions of the order of the beam model, N. Each row shows the expansion terms of an N-order theory. The second-order model ($N = 2$, $M = 6$) exploits a parabolic expansion of the Taylor-like polynomials,

$$\begin{array}{ll|l|l} u_x = & \underbrace{u_{x_1}}_{N=0} & +x\,u_{x_2} + z\,u_{x_3} & +x^2\,u_{x_4} + xz\,u_{x_5} + z^2\,u_{x_6} \\ u_y = & u_{y_1} & +x\,u_{y_2} + z\,u_{y_3} & +x^2\,u_{y_4} + xz\,u_{y_5} + z^2\,u_{y_6} \\ u_z = & \underbrace{u_{z_1}}_{N=0} & \underbrace{+x\,u_{z_2} + z\,u_{z_3}}_{N=1} & \underbrace{+x^2\,u_{z_4} + xz\,u_{z_5} + z^2\,u_{z_6}}_{N=2} \end{array} \tag{5.104}$$

The 1D model given by Eq. (5.104) has 18 displacement variables: three constant ($N = 0$), six linear ($N = 1$), and nine parabolic ($N = 2$). The strain components can be obtained through partial derivations, for instance, the normal components are

$$\begin{aligned} \varepsilon_{xx} &= u_{x,x} = \quad u_{x_2} + 2\,x\,u_{x_4} + z\,u_{x_5} \\ \varepsilon_{yy} &= u_{y,y} = u_{y_1,y} + x\,u_{y_2,y} + z\,u_{y_3,y} + x^2\,u_{y_4,y} + xz\,u_{y_5,y} + z^2\,u_{y_6,y} \\ \varepsilon_{zz} &= u_{z,z} = \quad u_{z_3} + z\,u_{z_5} + 2\,z\,u_{z_6} \end{aligned} \tag{5.105}$$

5.9.1 Example

Let us consider a two-node 1D elements (B2) modelled via the $N = 2$ theory. The normal strain components have to be computed at node 1 (i.e. $y_p = 0$) in a generic point of coordinates (x_p, z_p). For an $N = 2$ model the expansion functions (and their partial derivatives) are

$$
\begin{aligned}
&\tau, s = 1, 2, 3, 4, 5, 6 \quad \Rightarrow F_1 = 1, F_2 = x, F_3 = z, F_4 = x^2, F_5 = x\,z, F_6 = z \\
&\tau, s = 1, 2, 3, 4, 5, 6 \Rightarrow \\
&\quad F_{1,x} = 0, F_{2,x} = 1, F_{3,x} = 0, F_{4,x} = 2\,x, F_{5,x} = z, F_{6,x} = 0 \\
&\tau, s = 1, 2, 3, 4, 5, 6 \Rightarrow \\
&\quad F_{1,z} = 0, F_{2,z} = 0, F_{3,z} = 1, F_{4,z} = 0, F_{5,z} = x, F_{6,z} = 2\,z
\end{aligned}
$$

The use of a B2 element implies that

$$
\begin{aligned}
i, j = 1, 2 &\Rightarrow N_1 = 1 - \frac{y}{L}, N_2 = \frac{y}{L} \\
i, j = 1, 2 &\Rightarrow N_{1,_y} = -\frac{1}{L}, N_{2,_y} = \frac{1}{L}
\end{aligned}
$$

The strain components are given by

$$
\begin{aligned}
\varepsilon_{xx} &= u_{x,x} = (F_\tau\, N_i)_x\, u_{x_{\tau i}} = F_{\tau,_x}(x_p, z_p)\, N_i(y_p)\, u_{x_{\tau i}} \\
\varepsilon_{yy} &= u_{y,y} = (F_\tau\, N_i)_y\, u_{y_{\tau i}} = F_\tau(x_p, z_p)\, N_{i,_y(y_p)}\, u_{y_{\tau i}} \\
\varepsilon_{zz} &= u_{z,z} = (F_\tau\, N_i)_z\, u_{z_{\tau i}} = F_{\tau,_z}(x_p, z_p)\, N_i(y_p)\, u_{z_{\tau i}}
\end{aligned}
$$

Thus

$$
\begin{aligned}
\varepsilon_{xx}(x_p, y_p, z_p) &= u_{x_{21}} + 2\,x_p\, u_{x_{41}} + z_p\, u_{x_{51}} \\
\varepsilon_{yy}(x_p, y_p, z_p) &= -\frac{1}{L}(1\, u_{y_{11}} + x_p\, u_{y_{21}} + z_p\, u_{y_{31}} + x_p^2\, u_{y_{41}} + x_p\, z_p\, u_{y_{51}} + z_p^2\, u_{y_{61}}) \\
&\quad + \frac{1}{L}(1\, u_{y_{12}} + x_p\, u_{y_{22}} + z_p\, u_{y_{32}} + x_p^2\, u_{y_{42}} + x_p\, z_p\, u_{y_{52}} + z_p^2\, u_{y_{62}}) \\
\varepsilon_{zz}(x_p, y_p, z_p) &= u_{z_{31}} + x_p\, u_{z_{51}} + 2\, z_p\, u_{z_{61}}
\end{aligned}
$$

The UF can also deal with *reduced* models by exploiting a lower number of variables. An example of reduced models is given by

$$
\begin{aligned}
u_x &= u_{x_1} + x\, u_{x_2} + \qquad\qquad + x^2\, u_{x_4} + xz\, u_{x_5} \\
u_y &= u_{y_1} + x\, u_{y_2} + z\, u_{y_3} + x^2\, u_{y_4} + \qquad\qquad + z^2\, u_{y_6} \\
u_z &= u_{z_1} + \qquad\qquad + z\, u_{z_3} \; + \qquad\qquad + xz\, u_{z_5} + z^2\, u_{z_6}
\end{aligned}
\tag{5.106}
$$

The 1D model given in Eq. (5.106) has 13 displacement variables: three constant, four linear, and six parabolic. More details about this kind of reduced models will be provided in the next chapters.

5.9.2 $N=3$ and $N=4$

The third-order model ($N=3$, $M=10$) exploits a cubic expansion of the Taylor-like polynomials,

$$\begin{array}{ll} u_x = \dots & + x^3\, u_{x_7} + x^2\, z\, u_{x_8} + x\, z^2\, u_{x_9} + z^3\, u_{x10} \\ u_y = \dots & + x^3\, u_{y_7} + x^2\, z\, u_{y_8} + x\, z^2\, u_{y_9} + z^3\, u_{y10} \\ u_z = \underbrace{\dots}_{N\leq 2} & \underbrace{+ x^3\, u_{z_7} + x^2\, z\, u_{z_8} + x\, z^2\, u_{z_9} + z^3\, u_{z_{10}}}_{N=3} \end{array} \tag{5.107}$$

The 1D model given by Eq. (5.107) has 30 displacement variables: 12 cubic ($N=3$) and 18 lower-order terms ($N \leq 2$). Shear strain components are

$$\begin{array}{ll} \gamma_{xy} = u_{x,y} + u_{y,x} = \overbrace{\dots}^{N\leq 2} & + x^3\, u_{x7,y} + x^2\, z\, u_{x8,y} + x\, z^2\, u_{x9,y} + z^3\, u_{x10,y} + \\ & + 3\, x^2\, u_{y_7} + 2\, x\, z\, u_{y_8} + z^2\, u_{y_9} \\ \gamma_{yz} = u_{y,z} + u_{z,y} = \dots & + x^2\, u_{y_8} + 2\, x\, z\, u_{y_9} + 3\, z^2\, u_{y10} + \\ & + x^3\, u_{z7,y} + x^2\, z\, u_{z8,y} + x\, z^2\, u_{z9,y} + z^3\, u_{z10,y} \\ \gamma_{xz} = u_{x,z} + u_{z,x} = \dots & + x^2\, u_{x_8} + 2\, x\, z\, u_{x_9} + 3\, z^2\, u_{x10} + \\ & + 3\, x^2\, u_{z_7} + 2\, x\, z\, u_{z_8} + z^2\, u_{z_9} \end{array} \tag{5.108}$$

The fourth-order model ($N=4$) exploits a quartic expansion of the Taylor-like polynomials

$$\begin{array}{ll} u_x = \dots & + x^4\, u_{x11} + x^3\, z\, u_{x12} + x^2\, z^2\, u_{x13} + x\, z^3\, u_{x14} + z^4\, u_{x15} \\ u_y = \dots & + x^4\, u_{y11} + x^3\, z\, u_{y12} + x^2\, z^2\, u_{y13} + x\, z^3\, u_{y14} + z^4\, u_{y15} \\ u_z = \underbrace{\dots}_{N\leq 3} & \underbrace{+ x^4\, u_{z11} + x^3\, z\, u_{z12} + x^2\, z^2\, u_{z13} + x\, z^3\, u_{z14} + z^4\, u_{z15}}_{N=4} \end{array} \tag{5.109}$$

The beam model given by Eq. (5.109) has 45 displacement variables: 15 quartic ($N=4$) and 30 lower-order terms ($N \leq 3$).

5.9.3 N-order

The present 1D formulation is able to implement any-order theory by setting the order (N) as an input. An arbitrary refined model can be obtained by the following compact expression:

$$
\begin{array}{rcl|l}
u_x = & \cdots & & + \sum\limits_{M=0}^{N} x^{N-M} z^{M} \, u_{x \, \frac{N(N+1)}{2}+M+1} \\
u_y = & \cdots & & + \sum\limits_{M=0}^{N} x^{N-M} z^{M} \, u_{y \, \frac{N(N+1)}{2}+M+1} \\
u_z = & \underbrace{\cdots}_{1,\ldots,N-1} & & \underbrace{+ \sum\limits_{M=0}^{N} x^{N-M} z^{M} \, u_{z \, \frac{N(N+1)}{2}+M+1}}_{N\text{-order}}
\end{array}
\tag{5.110}
$$

The total number of displacement variables of the model (N_{DV}) is related to N,

$$
N_{DV} = 3 \times \frac{(N+1)(N+2)}{2} \tag{5.111}
$$

In the case of the finite element formulation, N_{DV} indicates the number of degrees of freedom per node. The strain components can be expressed in a compact manner,

$$
\begin{array}{rcl|l}
\varepsilon_{xx} = & \cdots & & + \sum\limits_{M=0}^{N-1} (N-M) \, x^{N-M-1} z^{M} \, u_{x \, \frac{N(N+1)}{2}+M+1} \\
\varepsilon_{yy} = & \cdots & & + \sum\limits_{M=0}^{N} x^{N-M} z^{M} \left(u_{y \, \frac{N(N+1)}{2}+M+1} \right)_{,y} \\
\varepsilon_{zz} = & \cdots & & + \sum\limits_{M=1}^{N} M \, x^{N-M} z^{M-1} \, u_{z \, \frac{N(N+1)}{2}+M+1} \\
\gamma_{xy} = & \cdots & & + \sum\limits_{M=0}^{N} x^{N-M} z^{M} \left(u_{x \, \frac{N(N+1)}{2}+M+1} \right)_{,y} + \\
 & & & + \sum\limits_{M=0}^{N-1} (N-M) \, x^{N-M-1} z^{M} \, u_{y \, \frac{N(N+1)}{2}+M+1} \\
\gamma_{yz} = & \cdots & & + \sum\limits_{M=0}^{N} x^{N-M} z^{M} \left(u_{z \, \frac{N(N+1)}{2}+M+1} \right)_{,y} + \\
 & & & + \sum\limits_{M=1}^{N} M \, x^{N-M} z^{M-1} \, u_{y \, \frac{N(N+1)}{2}+M+1} \\
\gamma_{xz} = & \cdots & & + \sum\limits_{M=0}^{N-1} (N-M) \, x^{N-M-1} z^{M} \, u_{z \, \frac{N(N+1)}{2}+M+1} + \\
 & \underbrace{\cdots}_{1,\ldots,N-1} & & \underbrace{+ \sum\limits_{M=1}^{N} M \, x^{N-M} z^{M-1} \, u_{x \, \frac{N(N+1)}{2}+M+1}}_{N\text{-order}}
\end{array}
\tag{5.112}
$$

5.10 1D MODELS WITH A PHYSICAL VOLUME/SURFACE-BASED GEOMETRY AND PURE DISPLACEMENT VARIABLES, THE LAGRANGE EXPANSION CLASS (LE)

Different expansion functions can be implemented through the unified formulation. In a displacement-based approach, the displacement field of the structure cross-section can be defined via different classes of functions, such as polynomials, harmonics and exponentials. In the previous chapter, TE models based on Taylor-like polynomial expansions were described. In this chapter a second class of 1D UF models based on Lagrange polynomial expansions is described. The Lagrange-expansion 1D models – hereafter referred to as LEs – have the following main characteristics:

1. LE model variables and boundary conditions can be located above the physical surfaces of the structure. This feature is particular relevant in a CAD-FEM coupling scenario.
2. The unknown variables of the problem are pure displacement components. No rotations or higher-order variables are exploited to describe the displacement field of an LE model.
3. Locally refined models can be easily built since Lagrange polynomial sets can be arbitrarily spread above the cross-section.

Each of these capabilities will be outlined in the following sections of this chapter and numerical examples will be provided in order to highlight the enhanced capabilities of 1D UF LEs in terms of 3D-like accuracies and very low computational costs.

5.11 PHYSICAL VOLUME/SURFACE APPROACH

Finite element analyses (FEA) are typically conducted on structural models whose geometries are derived from computer aided design tools (CAD). CAD and FEA representations of the geometry may differ significantly and the proper finite element modelling of a CAD-based structure is a critical and lengthy task which can also affect the accuracy of the results.

There are two main aspects that should be carefully considered in a CAD-FEA scenario:

1. The finite element discretization is, by definition, a process that leads to a modified geometry of a structure. Mesh refinements or higher-order shape functions are typical remedies to this problem.
2. Many structural elements (e.g. beams, plates and shells) require the definition of reference surfaces or axes where these elements and the

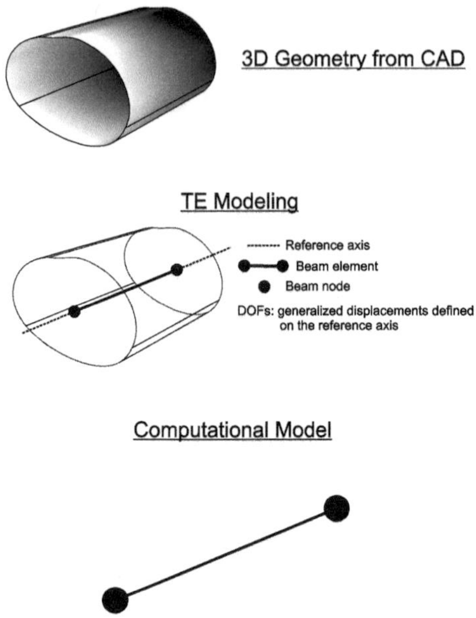

Figure 5.18 Geometrical considerations about the TE modelling approach, a line is used to model the entire 3D volume.

problem unknowns lie. The definition of reference surfaces/axes is particularly critical when 3D CAD geometries are provided. This difficulty increases if FEA is, for instance, exploited to conduct the topological optimization of the geometry of a structure.

The present 1D finite element formulation offers significant advantages related to the second point listed above, as LE elements can directly deal with the 3D geometry given by a CAD model.

Figure 5.18 shows the typical steps required to implement a 1D beam element, both classical (DEBBT, TBT) and refined (TE) models are considered. Starting from the 3D geometry of the structure, an axis is defined and used to create the finite element discretization. The problem unknowns are defined along this axis. The geometrical characteristics of the cross-section are retained by means of the surface integration of the expansion functions within the finite element matrices. This process can be particularly critical when multiple CAD-FEA iterative processes are required (e.g. in an optimization problem) since it can be difficult to redefine a 3D geometry starting from a 1D FE model. Figure 5.19 shows the LE modelling approach. In this case, the cross-section nodes can be directly located along the surface contour of the 3D structure. This implies that the finite element

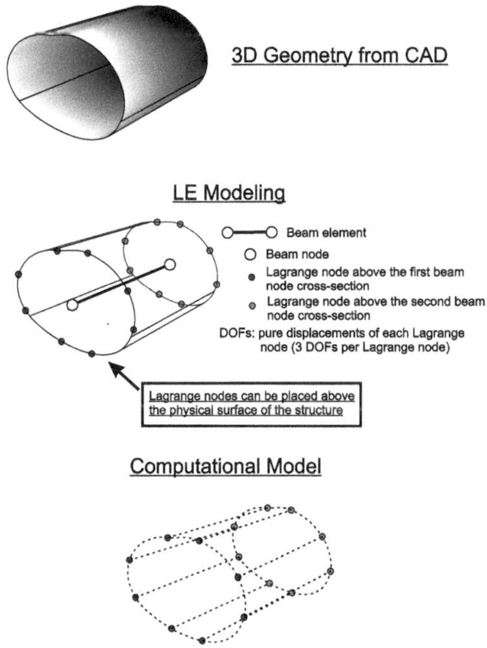

Figure 5.19 Geometrical considerations about the LE modelling approach, lines and nodes lie on the external surface of the physical body.

unknowns lie above the physical surface of the structure. The definition of reference surfaces or axes is not required and a 3D CAD geometry can be directly exploited for the finite element analysis. In other words, a 1D finite element can be used for a 3D geometrical description. This important feature makes LE models extremely attractive to:

1. Easily create finite element models derived from 3D CAD.
2. Improve the FEA-CAD coupling capabilities in an iterative design scenario.

Figure 5.20 offers a summary of the main differences between the different modelling approaches described above.

5.12 LAGRANGE POLYNOMIALS AND ISOPARAMETRIC FORMULATION

The Lagrange expansion 1D models (LEs) represent the second class of 1D models developed in the framework of UF. In an LE model, the F_τ expansion functions coincide with the Lagrange polynomials. The use of LEs as F_τ does not imply a reformulation of the problem equations and

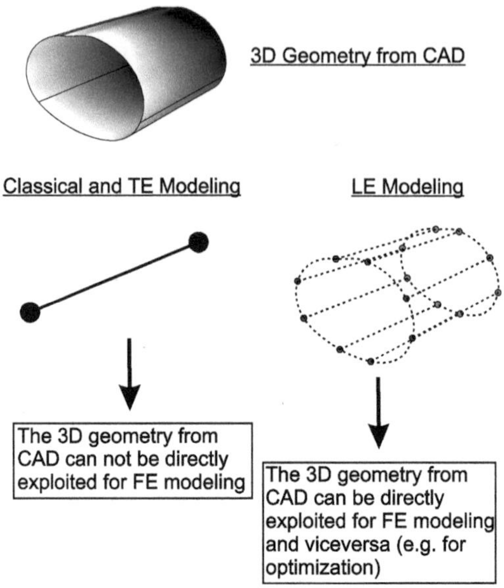

Figure 5.20 Different geometrical modelling approaches, TE vs. LE; it is clear that TE formulations introduce fictitious entities.

matrices. LEs, on the other hand, use a different approach with respect to TE since the isoparametric formulation is exploited. This section will first describe LE polynomials and then LE models will be given together with a number of numerical examples. Descriptions provided in the following will focus on the specific issues related to the LE models. Details on Lagrange polynomials and the isoparametric formulation not directly related to LEs are covered in [1,2]. In these books this formulation is usually provided for 2D finite elements.

5.12.1 Lagrange polynomials

Lagrange polynomials are usually given in terms of normalized – or natural – coordinates. This choice is not compulsory since LEs can also be implemented in terms of actual coordinates. However, the normalized formulation was preferred since it offers many advantages. Quadrilateral and triangular sets are presented in this book and each set is named according to its number of nodes. Only 2D polynomials are hereafter presented since only these are employed on the cross-section of the model.

The simplest quadrilateral Lagrange polynomial is the four-point (L4) set as shown in Fig. 5.21 and the polynomials are given by Eq. (5.113)

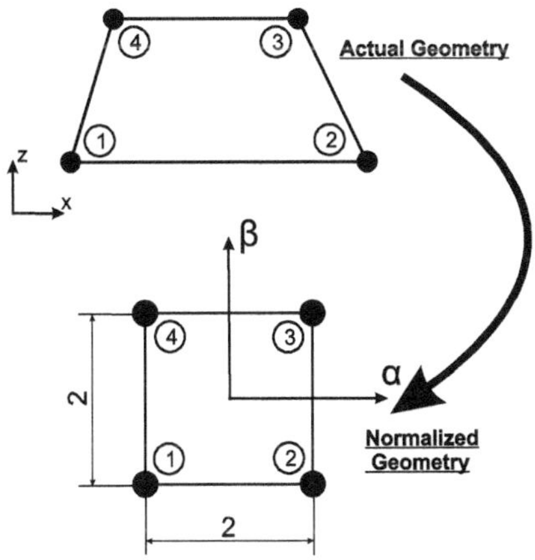

Figure 5.21 Four-node Lagrange element (L4) in actual and normalized geometry.

Table 5.2 L4 point normalized coordinates, see Fig. 5.21. This table presents the normalized coordinates of the four points of an L4 element

Point	α_τ	β_τ
1	−1	−1
2	1	−1
3	1	1
4	−1	1

where α and β are the normalized coordinates and α_τ and β_τ are the coordinates of the four nodes given in Table 5.2.

$$F_\tau = \frac{1}{4}(1 + \alpha\,\alpha_\tau)(1 + \beta\,\beta_\tau) \qquad \tau = 1, 2, 3, 4 \tag{5.113}$$

An L4 can be seen as a linear expansion (terms 1, α and β) plus a bilinear term ($\alpha\beta$). The second L-set is given by the L9 in Fig. 5.22. L9 polynomials and point coordinates are given by Eq. (5.114) and Table 5.3.

$$\begin{aligned}
F_\tau &= \frac{1}{4}(\alpha^2 + \alpha\,\alpha_\tau)(\beta^2 + \beta\,\beta_\tau) \qquad \tau = 1, 3, 5, 7 \\
F_\tau &= \frac{1}{2}\beta_\tau^2(\beta^2 + \beta\,\beta_\tau)(1 - \alpha^2) + \frac{1}{2}\alpha_\tau^2(\alpha^2 + \alpha\,\alpha_\tau)(1 - \beta^2) \\
&\quad \tau = 2, 4, 6, 8 \\
F_\tau &= (1 - \alpha^2)(1 - \beta^2) \qquad \tau = 9
\end{aligned} \tag{5.114}$$

An L9 can be seen as a parabolic expansion plus two cubic terms ($\alpha\beta^2$ and $\alpha^2\beta$) and a quartic term ($\alpha^2\beta^2$).

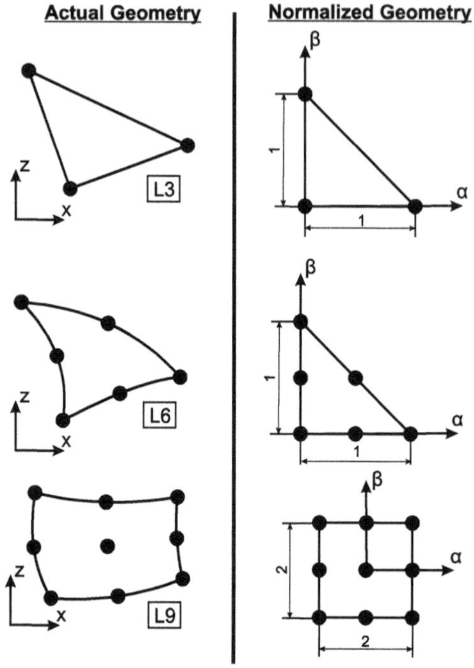

Figure 5.22 Three-, six- and nine-node Lagrange elements (L3, L6 and L9) in actual and normalized geometry.

Table 5.3 L9 point normalized coordinates, see Fig. 5.22. This table presents the normalized coordinates of the nine points of an L9 element

Point	α_τ	β_τ
1	−1	−1
2	0	−1
3	1	−1
4	1	0
5	1	1
6	0	1
7	−1	1
8	−1	0
9	0	0

The simplest triangular set is the L3 shown in Fig. 5.22. L3 polynomials are given by Eq. (5.115) and the node coordinates are shown in Table 5.4.

$$F_1 = 1 - \alpha - \beta \qquad F_2 = \alpha \qquad F_3 = \beta \tag{5.115}$$

Table 5.4 L3 point normalized coordinates, see Fig. 5.22. This table presents the normalized coordinates of the three points of an L3 element

Point	α_τ	β_τ
1	0	0
2	1	0
3	0	1

Table 5.5 L6 point normalized coordinates, see Fig. 5.22. This table presents the normalized coordinates of the six points of an L6 element

Point	α_τ	β_τ
1	0	0
2	0.5	0
3	1	0
4	0.5	0.5
5	0	1
6	0	0.5

An L3 has only linear terms. The L6 set is shown in Fig. 5.22, polynomials and points are shown in Eq. (5.116) and Table 5.5 respectively.

$$\begin{aligned} &F_1 = 1 - 3\,(\alpha + \beta) + 2\,(\alpha^2 + 2\,\alpha\,\beta + \beta^2) && F_2 = 4\,\alpha\,(1 - \alpha - \beta) \\ &F_3 = \alpha\,(2\,\alpha - 1) && F_4 = 4\,\alpha\,\beta \\ &F_5 = s\,(2\,\beta - 1) && F_6 = 4\,\beta\,(1 - \alpha - \beta) \end{aligned} \tag{5.116}$$

An L6 has only linear and parabolic terms.

5.12.2 Isoparametric formulation

Isoparametric formulations can be 1D, 2D or 3D. The isoparametric formulation is exploited in the LE finite elements to deal with

1. 1D shape functions along the longitudinal axis of the structure.
2. 2D expansion functions to describe the displacement field of the structure on its cross-section.

The 2D isoparametric formulation is described hereafter, as it is adopted to implement LEs. The same 2D formulation is usually employed for 2D finite elements (plate models).

As shown in the chapter on TE, the computation of the fundamental nucleus in a 1D formulation requires an evaluation of the surface integrals on the Cartesian coordinates (x and z), so that:

$$\int_A F_{\tau,x} F_{s,z}\, dxdz = \int_A F_{\tau,x} F_{s,z}\, dxdz \tag{5.117}$$

where the 2D integration domain (A) can be arbitrary-shaped. These integrals are formally independent of the adopted class of F_τ functions, i.e. if Lagrange polynomials are adopted – instead of Taylor expansions – the surface integrals will not formally change. If normalized coordinates (α and β) are accounted for, the integrals can be computed above a fixed 2D domain, regardless of the actual geometry. For instance, if quadrilateral domains are considered, the following will hold

$$\int_A F_{\tau,x}(x,z) F_{s,z}(x,z)\, dxdz = \int_{-1}^{+1}\int_{-1}^{+1} F_{\tau,x}(\alpha,\beta) F_{s,z}(\alpha,\beta) |J(\alpha,\beta)|\, d\alpha\, d\beta \tag{5.118}$$

where $|J|$ is the Jacobian determinant of the transformation. In some cases, the new integral in the normalized coordinates can be computed analytically, but numerical techniques often have to be employed.

Partial derivatives have to be computed, with respect to the normalized coordinates, according to the chain rule,

$$\begin{aligned} F_{\tau,x} &= F_{\tau,\alpha}\, \alpha_{,x} + F_{\tau,\beta}\, \beta_{,x} \\ F_{\tau,z} &= F_{\tau,\alpha}\, \alpha_{,z} + F_{\tau,\beta}\, \beta_{,z} \end{aligned} \tag{5.119}$$

The evaluation of Eq. (5.119) requires the following explicit relationships:

$$\alpha = \alpha(x,z), \quad \beta = \beta(x,z) \tag{5.120}$$

These explicit relationships are often difficult to establish and, for this reason, it is preferable to use the chain rule as follows:

$$\begin{aligned} F_{\tau,\alpha} &= F_{\tau,x}\, x_{,\alpha} + F_{\tau,z}\, z_{,\alpha} \\ F_{\tau,\beta} &= F_{\tau,x}\, x_{,\beta} + F_{\tau,z}\, z_{,\beta} \end{aligned} \tag{5.121}$$

Equation (5.121) also holds for F_τ and can be rewritten in matrix form,

$$\begin{Bmatrix} F_{\tau,\alpha} \\ F_{\tau,\beta} \end{Bmatrix} = \underbrace{\begin{bmatrix} x_{,\alpha} & z_{,\alpha} \\ x_{,\beta} & z_{,\beta} \end{bmatrix}}_{\mathbf{J}} \begin{Bmatrix} F_{\tau,x} \\ F_{\tau,z} \end{Bmatrix} \tag{5.122}$$

The inverse relationship is given by

$$\begin{Bmatrix} F_{\tau,x} \\ F_{\tau,z} \end{Bmatrix} = \frac{1}{|J|} \begin{bmatrix} z_{,\beta} & -z_{,\alpha} \\ -x_{,\alpha} & x_{,\beta} \end{bmatrix} \begin{Bmatrix} F_{\tau,\alpha} \\ F_{\tau,\beta} \end{Bmatrix} \tag{5.123}$$

The four terms of the Jacobian matrix can be computed if a known relation between the Cartesian and normalized coordinates exists,

$$x = x(\alpha, \beta), \quad z = z(\alpha, \beta) \tag{5.124}$$

The isoparametric formulation is generally used to obtain this relation and, therefore, to compute the Jacobian and to associate the actual geometries to the normalized geometry of the LEs. It is important to note that Eq. (5.123) requires that the inverse of $\mathbf{J}$ exists. In order to fulfil this requirement, attention should be paid to the geometry of the Lagrange element. Major distortions, or folded back elements, can cause a singularity problem.

The term *isoparametric* means that the same functions are adopted to interpolate the displacement field and the geometry of a structural element. In LEs, Lagrange polynomials are employed,

$$\begin{gathered} \mathbf{u} = F_\tau \mathbf{u}_\tau \\ x = F_\tau\, x_\tau \qquad\qquad z = F_\tau\, z_\tau \end{gathered} \tag{5.125}$$

where x_τ and z_τ are the actual coordinates of the Lagrange nodes. If an L4 element is considered, the geometry will be interpolated as

$$\begin{aligned} x &= F_1\, x_1 + F_2\, x_2 + F_3\, x_3 + F_4\, x_4 \\ z &= F_1\, z_1 + F_2\, z_2 + F_3\, z_3 + F_4\, z_4 \end{aligned} \tag{5.126}$$

Typical derivatives for the Jacobian matrix are given by

$$\begin{aligned} x_{,\alpha} &= F_{1,\alpha}\, x_1 + F_{2,\alpha}\, x_2 + F_{3,\alpha}\, x_3 + F_{4,\alpha}\, x_4 \\ z_{,\beta} &= F_{1,\beta}\, z_1 + F_{2,\beta}\, z_2 + F_{3,\beta}\, z_3 + F_{4,\beta}\, z_4 \end{aligned} \tag{5.127}$$

The integrals in Eq. (5.118) can have analytical solutions but, in practice, numerical integrations are required. Gauss quadrature formulas are extensively adopted in FEA for these purposes. In 1D UF, these formulas are employed to compute 1D shape function integrals. In 1D LE UF, a Gauss quadrature is also implemented to compute the F_τ-integrals on the cross-section domain,

$$\int_{-1}^{+1}\int_{-1}^{+1} F_\tau F_s |J|\, d\alpha\, d\beta = \sum_{h,k} w_h\, w_k\, F_\tau(\alpha_h, \beta_k)\, F_\tau(\alpha_h, \beta_k)\, |J(\alpha_h, \beta_k)| \tag{5.128}$$

where w_h and w_k are the integration weights and α_h and β_k are the integration points. The weights and points depend on the adopted set of Lagrange polynomials – three-, four-, six- or nine-node in the present.

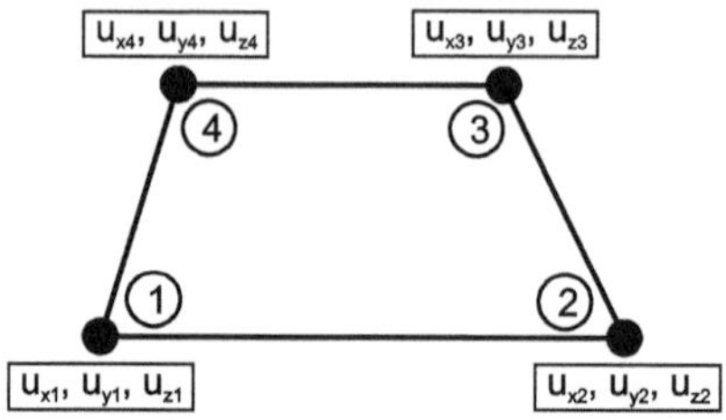

Figure 5.23 L4 element DOFs, only pure displacement unknowns are employed.

5.13 LE DISPLACEMENT FIELDS AND CROSS-SECTION ELEMENTS

Cross-section elements are introduced in this section. A cross-section element is described by a set of Lagrange polynomials defined on a given number of points. Cross-section elements are used to define the cross-section displacement field. As stated above, three-, four-, six- and nine-node elements are described hereafter and they are referred to as L3, L4, L6 and L9 according to the Lagrange polynomials they are based on. This means, for instance, that the L4 cross-section element is based on the four-point Lagrange polynomials given in Eq. (5.113). L-elements based on other polynomial sets (e.g. L16) could also be implemented easily.

L4 leads to the following displacement field:

$$\begin{aligned} u_x &= F_1\, u_{x_1} + F_2\, u_{x_2} + F_3\, u_{x_3} + F_4\, u_{x_4} \\ u_y &= F_1\, u_{y_1} + F_2\, u_{y_2} + F_3\, u_{y_3} + F_4\, u_{y_4} \\ u_z &= F_1\, u_{z_1} + F_2\, u_{z_2} + F_3\, u_{z_3} + F_4\, u_{z_4} \end{aligned} \tag{5.129}$$

Figure 5.23 shows an L4 element and its nodes. The unknown variables – $u_{x_1}, \ldots, u_{z_4}$ – are the three displacements components of each node. This means that

1. The problem unknowns are only physical translational displacements.
2. The problem unknowns can be placed on the physical surfaces of the body.

These two fundamental characteristics are valid for each L-element, regardless of the number of nodes. A typical L-element modelling approach is shown in Fig. 5.24, where the following modelling steps are pointed out:

1. The 3D body is discretized at the cross-section level by means of L-elements. The number of L-elements depends on the geometry of the structure and on the boundary conditions (geometrical or mechanical). For the sake of simplicity, only one L4 was adopted in Fig. 5.24

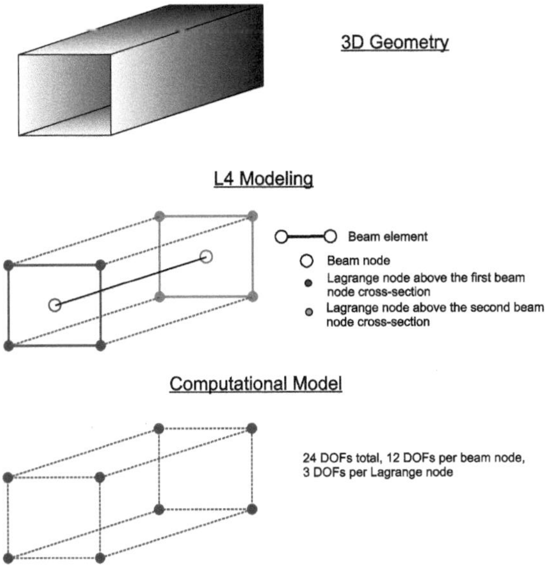

Figure 5.24 An example of LE modelling via an L4 element.

and a two-node (B2) beam element was considered. More complex discretizations are described in the following sections.

2. In the finite element LE formulation, the cross-section discretization determines the number of degrees of freedom (DOFs) of each beam node. If an L4 element is used, 12 DOFs per beam node will be exploited.
3. A classical finite element approach, based on beam elements, is employed to build finite element matrices. It is important to underline that – unlike classical and TE models – the unknown variables of the computational model do not lie on the beam element axis.

The L3 displacement field is given by

$$\begin{aligned} u_x &= F_1\, u_{x_1} + F_2\, u_{x_2} + F_3\, u_{x3} \\ u_y &= F_1\, u_{y_1} + F_2\, u_{y_2} + F_3\, u_{y3} \\ u_z &= F_1\, u_{z_1} + F_2\, u_{z_2} + F_3\, u_{z_3} \end{aligned} \tag{5.130}$$

Nine unknowns are employed, as shown in Fig. 5.25. L6 is based on the following 18 unknown fields:

$$\begin{aligned} u_x &= F_1\, u_{x_1} + F_2\, u_{x_2} + F_3\, u_{x3} + F_4\, u_{x_4} + F_5\, u_{x_5} + F_6\, u_{x_6} \\ u_y &= F_1\, u_{y_1} + F_2\, u_{y2} + F_3\, u_{y3} + F_4\, u_{y_4} + F_5\, u_{y_5} + F_6\, u_{y_6} \end{aligned} \tag{5.131}$$

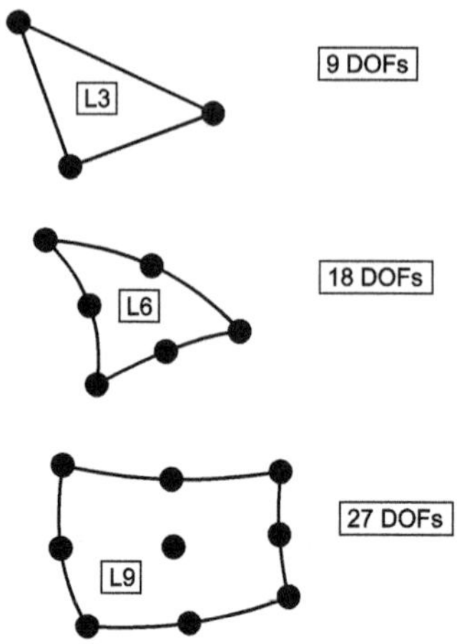

Figure 5.25 L3, L6 and L9 element DOFs.

$$u_z = F_1\, u_{z_1} + F_2\, u_{z_2} + F_3\, u_{z_3} + F_4\, u_{z_4} + F_5\, u_{z_5} + F_6\, u_{z_6}$$

Eventually, the L9 element is

$$\begin{aligned} u_x &= F_1\, u_{x_1} + F_2\, u_{x_2} + F_3\, u_{x_3} + \cdots + F_7\, u_{x_7} + F_8\, u_{x_8} + F_9\, u_{x_9} \\ u_y &= F_1\, u_{y_1} + F_2\, u_{y_2} + F_3\, u_{y_3} + \cdots + F_7\, u_{y_7} + F_8\, u_{y_8} + F_9\, u_{y_9} \\ u_z &= F_1\, u_{z_1} + F_2\, u_{z_2} + F_3\, u_{z_3} + \cdots + F_7\, u_{z_7} + F_8\, u_{z_8} + F_9\, u_{z_9} \end{aligned} \quad (5.132)$$

5.13.1 Example

Let us consider the L4 element in Fig. 5.23, a strain component (ε_{xx}) has to be computed at a generic point (α_k, β_k). The inputs are the coordinates of the four nodes (x_i, z_i, $i = 1, 2, 3, 4$) and their displacements.

The strain component that has to be computed is given by

$$\varepsilon_{xx} = u_{x,x}$$

where u_x is given by

$$u_x = F_1\, u_{x_1} + F_2\, u_{x_2} + F_3\, u_{x_3} + F_4\, u_{x_4}$$

According to Eq. (5.123), the derivative of u_x is

$$u_{x,x} = \frac{1}{|J(\alpha_k, \beta_k)|}\left(z_{,\beta}\, u_{x,\alpha} - z_{,\alpha}\, u_{x,\beta}\right)_{\alpha=\alpha_k, \beta=\beta_k}$$

where

$$
\begin{aligned}
z_{,\beta} &= -\frac{1}{4}(1-\alpha_k)\, z_1 - \frac{1}{4}(1+\alpha_k)\, z_2 + \frac{1}{4}(1+\alpha_k)\, z_3 + \frac{1}{4}(1-\alpha_k)\, z_4 \\
u_{x,\alpha} &= -\frac{1}{4}(1-\beta_k)\, u_{x1} - \frac{1}{4}(1+\beta_k)\, u_{x2} + \frac{1}{4}(1+\beta_k)\, u_{x3} + \frac{1}{4}(1-\beta_k)\, u_{x4} \\
z_{,\alpha} &= -\frac{1}{4}(1-\beta_k)\, z_1 - \frac{1}{4}(1+\beta_k)\, z_2 + \frac{1}{4}(1+\beta_k)\, z_3 + \frac{1}{4}(1-\beta_k)\, z_4 \\
u_{x,\beta} &= -\frac{1}{4}(1-\alpha_k)\, u_{x1} - \frac{1}{4}(1+\alpha_k)\, u_{x2} + \frac{1}{4}(1+\alpha_k)\, u_{x3} + \frac{1}{4}(1-\alpha_k)\, u_{x4}
\end{aligned}
$$

and $|J(\alpha_k, \beta_k)|$ can be computed according to Eq. (5.122).

This procedure is usually carried out, for example, to compute strain components at given points for the result postprocessing, starting from the displacement vector.

5.14 CROSS-SECTION MULTI-ELEMENTS AND LOCALLY REFINED MODELS

Cross-sections can be discretized by means of multiple LEs. This is a fundamental function of LEs. Multi-elements are generally adopted for three main purposes:

1. To refine the cross-section displacement field without increasing the polynomial expansion order.
2. To impose the geometrical discontinuities above the cross-section.
3. To locally refine the structural model.

Figure 5.26 shows a typical example in which two L4s are used to model the cross-section displacement field. A refined model is obtained by combining two piecewise linear elements. The assembly of the FE matrices requires the definition of a local and global connectivity of the cross-section nodes, as shown in Fig. 5.27. The definition of the global connectivity should take into account the band matrix assembly, an aspect that has not been analyzed in this book. The stiffness matrices of each L4 element are computed and assembled on the basis of the global connectivity, see Figs. 5.28, 5.29 and 5.30. This assembly technique is analogous to the one that is commonly used for 2D and 3D finite elements. Continuity of the displacements is imposed at the interface nodes. Different L-elements can be assembled simultaneously, e.g. a combination of L4s and L9s can be used in a cross-section.

Geometrical discontinuities can easily be modelled. Figure 5.31 shows an example in which nodes 6 and 7 are physically disconnected – although

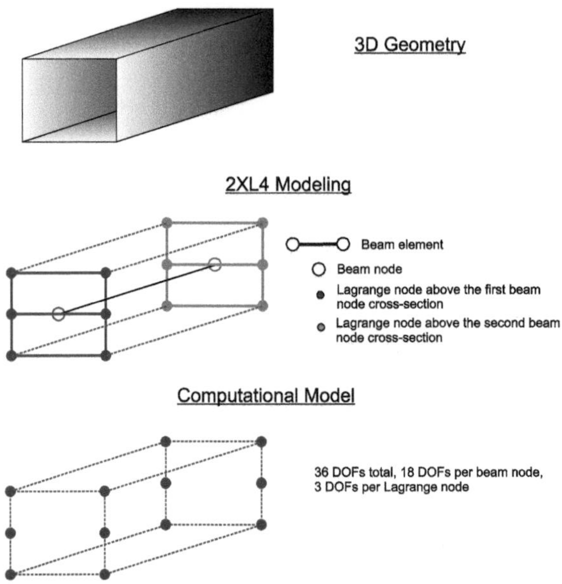

Figure 5.26 An example of LE modelling via two L4 elements.

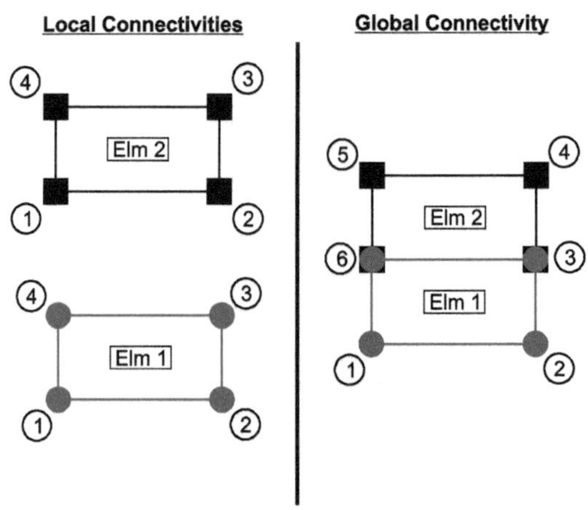

Figure 5.27 An example of two L4 element assembly within a beam node.

located in the same position – and Fig. 5.32 shows the new stiffness matrix. The overall number of DOFs per cross-section (N_{cs}) (i.e. per beam node) is given by

$$N_{cs} = 3 \times N_{cn} \tag{5.133}$$

where N_{cn} is the number of cross-section nodes.

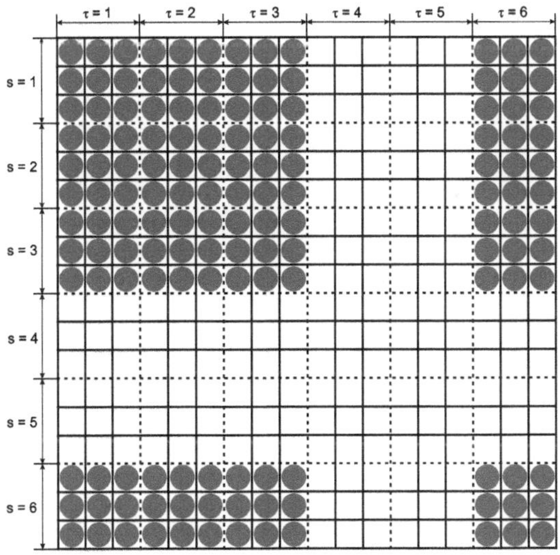

Figure 5.28 Element 1 assembly within a beam node.

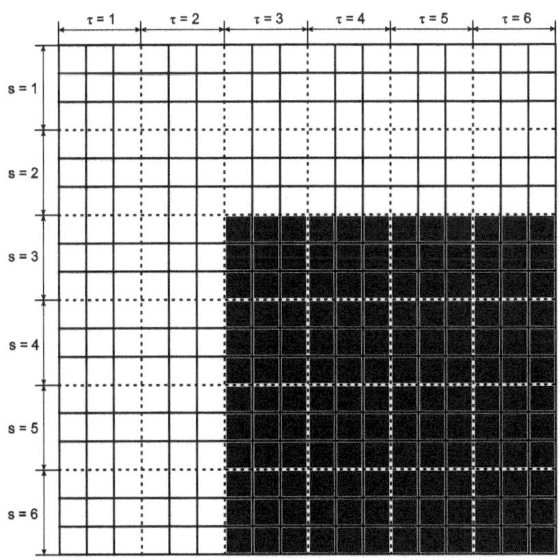

Figure 5.29 Element 2 assembly within a beam node.

Locally refined models are of particular interest when local effects play an important role in a structural problem. A typical example is that of a thin-walled structure under point loads. Figure 5.33 shows a thin-walled cross-section and two point loads acting on the top and bottom edges.

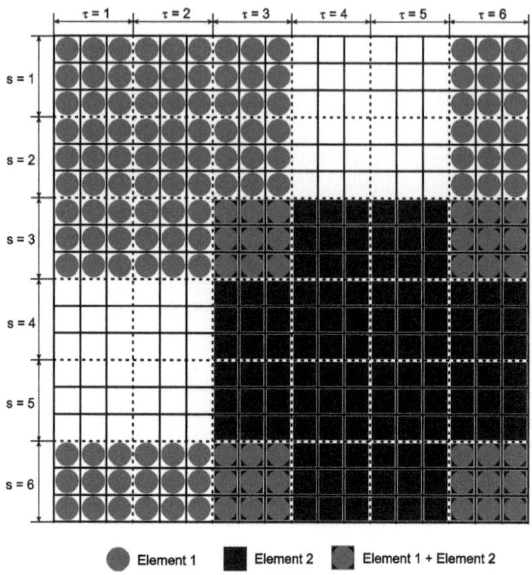

Figure 5.30 2 L4 element assembly within a beam node.

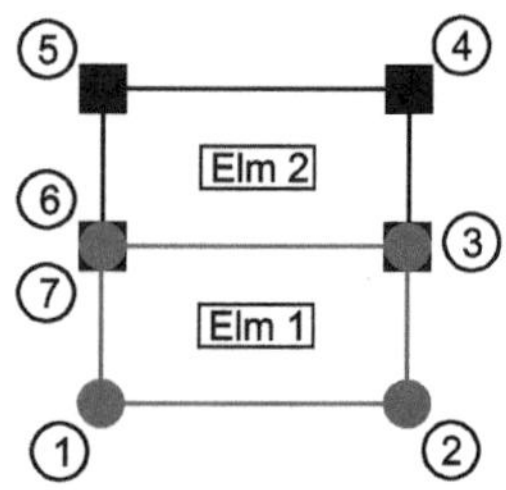

Figure 5.31 An example of two L4 element assembly within a beam node, two cross-section nodes are disconnected.

In this case, the deformed displacement field is affected to a great extent by local effects close to the loading points, while the loads have much less influence on the vertical edges. A proper detection of the in-plane deformed configuration requires higher-order polynomials, since a fairly complex deformed shape has to be modelled. If TE models are adopted, higher-order expansions are compulsory. The drawback of TEs is that a single expansion set can be adopted, which means that both the highly deformed and the barely deformed zones of the cross-section are modelled by means of higher-order models. It would be preferable to locally tune the refinement in order to optimize computational costs. Local re-

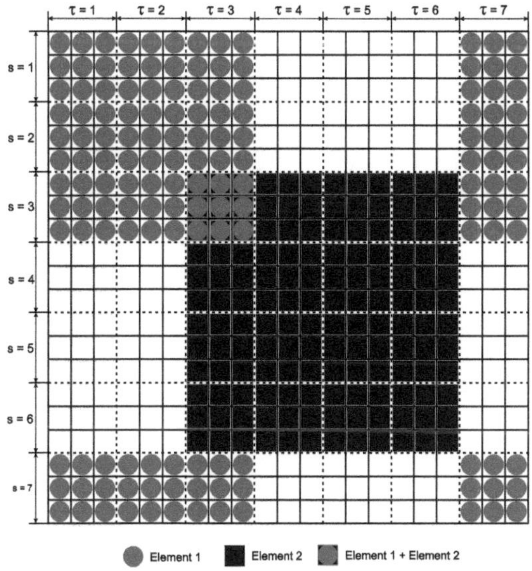

Figure 5.32 2 L4 element assembly, two cross-section nodes are disconnected.

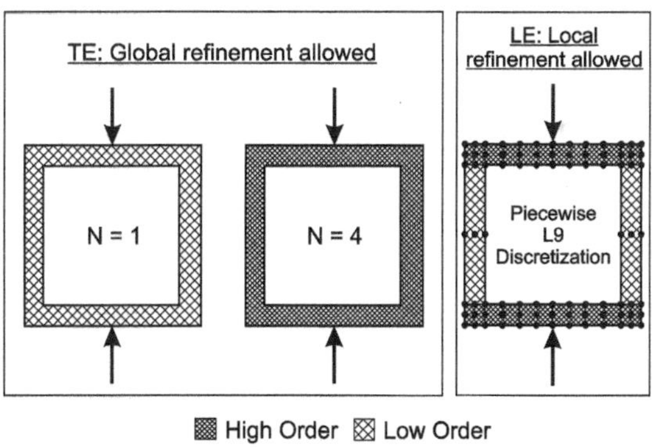

Figure 5.33 Global vs. local refinements on the cross-section.

finements can be implemented straightforwardly via LEs, since a finer cross-section mesh (or higher-order sets) can be used when needed. For the sake of completeness, it should be noted that locally refined models could also be obtained with TEs, but this process would require the imposition of compatibility conditions through, for instance, Lagrange multipliers.

5.15 PLATE

5.15.1 The complete linear expansion case

The complete linear expansion model involves a first-order ($N = 1$) Taylor polynomial to describe the through-the-thickness displacement field

$$
\begin{aligned}
u_x &= \underbrace{u_{x0}}_{} && + z\, u_{x1} \\
u_y &= u_{y0} && + z\, u_{y1} \\
u_z &= \underbrace{u_{z0}}_{N=0} && \underbrace{+ z\, u_{z1}}_{N=1}
\end{aligned}
\tag{5.134}
$$

The plate model given in Eq. (5.134) has 6 displacement variables: three constant ($N = 0$) and three linear ($N = 1$). Strain components are given by

$$
\begin{aligned}
\varepsilon_{xx} &= \frac{\partial u_x}{\partial x} &&= u_{x0,x} + z\, u_{x1,x} \\
\varepsilon_{yy} &= \frac{\partial u_y}{\partial y} &&= u_{y0,y} + z\, u_{y1,y} \\
\varepsilon_{zz} &= \frac{\partial u_z}{\partial z} &&= u_{z1} \\
\gamma_{xy} &= \frac{\partial u_x}{\partial y} + \frac{\partial u_y}{\partial x} &&= u_{x0,y} + u_{y0,x} + z\, u_{x1,y} + z\, u_{y1,x} \\
\gamma_{xz} &= \frac{\partial u_x}{\partial z} + \frac{\partial u_z}{\partial x} &&= u_{x1} + u_{z0,x} + z\, u_{z1,x} \\
\gamma_{yz} &= \frac{\partial u_y}{\partial z} + \frac{\partial u_z}{\partial y} &&= u_{y1} + u_{z0,y} + z\, u_{z1,y}
\end{aligned}
\tag{5.135}
$$

The linear model leads to a constant distribution of ε_{zz} along the thickness, and a linear distribution of other strain components.

The adoption of the $N = 1$ model is necessary to introduce the through-the-thickness stretching of the plate, given by ε_{zz}. As demonstrated by the results provided below, the thickness stretching cannot be negligible when the plate is very thick.

5.15.2 A finite element based on $N = 1$

The $N = 1$ model has six displacement variables, which implies that, in a finite element formulation, each node has six generalized displacement variables. The aim of this section is to provide the FE formulation for the complete linear expansion case. The derivation of the governing finite element equations begins with the definition of the nodal displacement

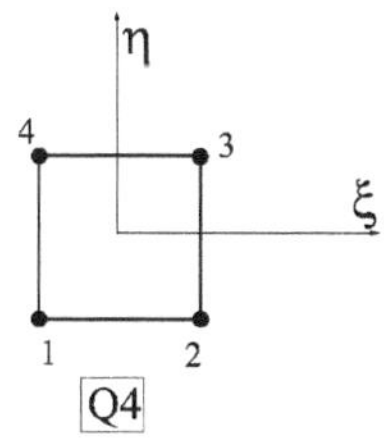

Figure 5.34 Q4 element.

vector

$$\mathbf{u}_i = \left\{ \begin{array}{cccccc} u_{x_0} & u_{y_0} & u_{z_0} & u_{x_1} & u_{y_1} & u_{z_1} \end{array} \right\}^T \tag{5.136}$$

The displacement variables are interpolated in the plane of the plate by means of the shape functions N_i

$$\mathbf{u} = N_i \mathbf{u}_i \tag{5.137}$$

Plate elements with 4 nodes, Q4, are here considered (see Fig. 5.34), with the following shape functions

$$\begin{aligned} N_1 &= \frac{1}{4}(1-\xi)(1-\eta) \\ N_2 &= \frac{1}{4}(1+\xi)(1-\eta) \\ N_3 &= \frac{1}{4}(1+\xi)(1+\eta) \\ N_4 &= \frac{1}{4}(1-\xi)(1+\eta) \end{aligned} \tag{5.138}$$

where the natural coordinates ξ and η vary from -1 to $+1$. Considering a rectangular plate, these are defined as follows

$$\begin{aligned} \xi &= \frac{2(x-x_1)-a}{a} \\ \eta &= \frac{2(y-y_1)-b}{b} \end{aligned} \tag{5.139}$$

x_1, y_1 are the global coordinates of the node 1 of the element and a, b are the dimensions of the plate along the x and y axis, respectively.

The total number of degrees of freedom of the structural model will be given by

$$DOFs = \underbrace{3\times 2}_{\text{DOFs per node}} \times [(\underbrace{2}_{\text{nodes per edge}} -1)\times \underbrace{N_{Ex}}_{\text{elements along x}} +1]\times [(\underbrace{2}_{\text{nodes per edge}} -1)\times \underbrace{N_{Ey}}_{\text{elements along y}} +1] \tag{5.140}$$

The principle of virtual displacements, PVD, is employed to compute the FEM matrices

$$\delta L_{int} = \delta L_{ext} \tag{5.141}$$

where

$$\delta L_{int} = \int_V (\delta \boldsymbol{\varepsilon}_p^T [\sigma]_p + \delta [\varepsilon]_n^T [\sigma]_n) dV \tag{5.142}$$

L_{int} stands for the strain energy, L_{ext} is the work of the external loadings, and δ stands for the virtual variation. Using Eq. (5.138), a compact form of the virtual variation of the internal work can be obtained following the well-known FE procedure

$$\delta L_{int} = \delta \mathbf{u}_j^T \mathbf{K}^{ij} \mathbf{u}_i \tag{5.143}$$

where $\mathbf{K}^{ij}$ is the stiffness matrix. For a given i, j pair, the stiffness matrix has the form

$$\left[\begin{array}{ccc|ccc} k_{xx}^{1,1} & k_{xy}^{1,1} & k_{xz}^{1,1} & k_{xx}^{1,z} & k_{xy}^{1,z} & k_{xz}^{1,z} \\ k_{yx}^{1,1} & k_{yy}^{1,1} & k_{yz}^{1,1} & k_{yx}^{1,z} & k_{yy}^{1,z} & k_{yz}^{1,z} \\ k_{zx}^{1,1} & k_{zy}^{1,1} & k_{zz}^{1,1} & k_{zx}^{1,z} & k_{zy}^{1,z} & k_{zz}^{1,z} \\ \hline k_{xx}^{z,1} & k_{xy}^{z,1} & k_{xz}^{z,1} & k_{xx}^{z,z} & k_{xy}^{z,z} & k_{xz}^{z,z} \\ k_{yx}^{z,1} & k_{yy}^{z,1} & k_{yz}^{z,1} & k_{yx}^{z,z} & k_{yy}^{z,z} & k_{yz}^{z,z} \\ k_{zx}^{z,1} & k_{zy}^{z,1} & k_{zz}^{z,1} & k_{zx}^{z,z} & k_{zy}^{z,z} & k_{zz}^{z,z} \end{array}\right]_{ij} \tag{5.144}$$

where the superscripts indicate the expansion functions that are involved in each component of the stiffness matrix, i.e. 1 and z. For the sake of clarity, the explicit expression of two components is reported hereafter

$$\begin{aligned} k_{xx}^{1,1} &= \tilde{C}_{11} \int_A 1 \cdot 1\, dz \int_\Omega N_{i,x} N_{j,x} d\Omega + C_{66} \int_A 1 \cdot 1\, dz \int_\Omega N_{i,y} N_{j,y} d\Omega \\ k_{yz}^{1,z} &= C_{23} \int_A 1 \cdot \frac{\partial z}{\partial z} dz \int_\Omega N_{i,y} N_j d\Omega = C_{23} \int_A 1 \cdot 1\, dz \int_\Omega N_{i,y} N_j d\Omega \end{aligned} \tag{5.145}$$

where Ω indicates the in-plane domain and A the through-the-thickness domain.

5.16 CPT, FSDT AND $N = 1$ MODEL IN UNIFIED FORM

The CPT, FSDT and $N = 1$ models can be obtained in a unified manner via a condensed notation that represents the basic step towards the UF. In this section, the $N = 1$ will be reformulated by means of a new notation, the FSDT and CPT will be obtained as particular cases.

5.16.1 Unified formulation of $N = 1$ model

The $N = 1$ stiffness matrix was given in Eq. (5.144). That matrix can be considered as being composed of four 3×3 sub-matrices, as in the following

$$
\begin{array}{ccc}
 & 1 & z \\
1 & \begin{bmatrix} k_{xx}^{1,1} & k_{xy}^{1,1} & k_{xz}^{1,1} \\ k_{yx}^{1,1} & k_{yy}^{1,1} & k_{yz}^{1,1} \\ k_{zx}^{1,1} & k_{zy}^{1,1} & k_{zz}^{1,1} \end{bmatrix} & \begin{bmatrix} k_{xx}^{1,z} & k_{xy}^{1,z} & k_{xz}^{1,z} \\ k_{yx}^{1,z} & k_{yy}^{1,z} & k_{yz}^{1,z} \\ k_{zx}^{1,z} & k_{zy}^{1,z} & k_{zz}^{1,z} \end{bmatrix} \\
z & \begin{bmatrix} k_{xx}^{z,1} & k_{xy}^{z,1} & k_{xz}^{z,1} \\ k_{yx}^{z,1} & k_{yy}^{z,1} & k_{yz}^{z,1} \\ k_{zx}^{z,1} & k_{zy}^{z,1} & k_{zz}^{z,1} \end{bmatrix} & \begin{bmatrix} k_{xx}^{z,z} & k_{xy}^{z,z} & k_{xz}^{z,z} \\ k_{yx}^{z,z} & k_{yy}^{z,z} & k_{yz}^{z,z} \\ k_{zx}^{z,z} & k_{zy}^{z,z} & k_{zz}^{z,z} \end{bmatrix}
\end{array}
\tag{5.146}
$$

Each sub-matrix has a fixed pair of expansion functions that are used in the explicit computation of the integrals along the thickness, as shown in Eq. (5.145). As highlighted for 1D models, the formal expression of each component of the sub-matrices does not depend on the thickness functions. For example, if one considers the following components, they have the same formal expression

$$
\begin{aligned}
k_{xx}^{1,1} &= C_{11} \int_A 1 \cdot 1 \, dz \int_\Omega N_{i,x} N_{j,x} \, d\Omega + C_{66} \int_A 1 \cdot 1 \, dz \int_\Omega N_{i,y} N_{j,y} \, d\Omega \\
k_{xx}^{1,z} &= C_{11} \int_A 1 \cdot z \, dz \int_\Omega N_{i,x} N_{j,x} \, d\Omega + C_{66} \int_A 1 \cdot z \, dz \int_\Omega N_{i,y} N_{j,y} \, d\Omega
\end{aligned}
\tag{5.147}
$$

This implies that the sub-matrix can be considered as a fundamental invariant nucleus which can be used to build the global stiffness matrix. Let us introduce the following notation for the expansion functions, F_τ,

$$
\begin{aligned}
F_{\tau=0} &= 1 \\
F_{\tau=1} &= z
\end{aligned}
\tag{5.148}
$$

The displacement field in Eq. (5.134) becomes

$$
\begin{aligned}
u_x &= F_0\, u_{x_0} + F_1\, u_{x_1} = F_\tau\, u_{x_\tau} \\
u_y &= F_0\, u_{y_0} + F_1\, u_{y_1} = F_\tau\, u_{y_\tau} \\
u_z &= F_0\, u_{z_0} + F_1\, u_{z_1} = F_\tau\, u_{z_\tau}
\end{aligned}
\tag{5.149}
$$

where the repeated index indicates summation according to the Einstein notation. The displacement vector can be written as

$$
\mathbf{u} = F_\tau \mathbf{u}_\tau, \qquad \tau = 0, 1 \tag{5.150}
$$

If a finite element formulation is introduced and four-node elements are adopted, the nodal unknown vector is given by

$$
\mathbf{u}_{\tau i} = \left\{ \begin{array}{ccc} u_{x_{\tau i}} & u_{y_{\tau i}} & u_{z_{\tau i}} \end{array} \right\}^T \qquad \tau = 0, 1 \qquad i = 1, \ldots, 4 \tag{5.151}
$$

The compact form of the internal work seen in Eq. (5.143) becomes

$$
\delta L_{int} = \delta \mathbf{u}_{sj}^T \mathbf{K}^{\tau sij} \mathbf{u}_{\tau i} \tag{5.152}
$$

where

- τ and s are the thickness function indexes;
- i and j are the shape function indexes.

Coherently with the notation introduced, the matrix in Eq. (5.146) can be expressed as

$$
\begin{array}{ccc}
 & 0 & 1 \\
0 & \begin{bmatrix} k_{xx}^{0,0} & k_{xy}^{0,0} & k_{xz}^{0,0} \\ k_{yx}^{0,0} & k_{yy}^{0,0} & k_{yz}^{0,0} \\ k_{zx}^{0,0} & k_{zy}^{0,0} & k_{zz}^{0,0} \end{bmatrix} & \begin{bmatrix} k_{xx}^{0,1} & k_{xy}^{0,1} & k_{xz}^{0,1} \\ k_{yx}^{0,1} & k_{yy}^{0,1} & k_{yz}^{0,1} \\ k_{zx}^{0,1} & k_{zy}^{0,1} & k_{zz}^{0,1} \end{bmatrix} \\
1 & \begin{bmatrix} k_{xx}^{1,0} & k_{xy}^{1,0} & k_{xz}^{1,0} \\ k_{yx}^{1,0} & k_{yy}^{1,0} & k_{yz}^{1,0} \\ k_{zx}^{1,0} & k_{zy}^{1,0} & k_{zz}^{1,0} \end{bmatrix} & \begin{bmatrix} k_{xx}^{1,1} & k_{xy}^{1,1} & k_{xz}^{1,1} \\ k_{yx}^{1,1} & k_{yy}^{1,1} & k_{yz}^{1,1} \\ k_{zx}^{1,1} & k_{zy}^{1,1} & k_{zz}^{1,1} \end{bmatrix}
\end{array}
\tag{5.153}
$$

Each 3×3 block is the fundamental nucleus of the stiffness matrix. A component of the nucleus is given for different combinations of thickness functions. For example, the components of Eq. (5.145) become

$$
\begin{aligned}
k_{xx}^{1,1} &= C_{11} \int_A F_0 \cdot F_0\, dz \int_\Omega N_{i,x} N_{j,x}\, d\Omega + C_{66} \int_A F_0 \cdot F_0\, dz \int_\Omega N_{i,y} N_{j,y}\, d\Omega \\
k_{yz}^{1,z} &= C_{23} \int_A F_0 \cdot \frac{\partial F_1}{\partial z}\, dz \int_\Omega N_{i,y} N_j\, d\Omega
\end{aligned}
\tag{5.154}
$$

5.16.2 CPT and FSDT as particular cases of $N = 1$

The CPT and FSDT are particular cases of $N = 1$, and they can be obtained by acting on the full linear expansion. As far as FSDT is concerned, the displacement field is given by

$$
\begin{aligned}
u_x &= u_{x_0} + z\, u_{x_1} \\
u_y &= u_{y_0} + z\, u_{y_1} \\
u_z &= u_{z_0}
\end{aligned}
\tag{5.155}
$$

that is, a linear distribution of in-plane displacements and constant transversal displacement distribution are accounted for. Starting from the $N = 1$ case, two possible techniques can be used to obtain FSDT: 1. the rearranging of rows and columns of the stiffness matrix; 2. penalization of the stiffness terms related to u_{z_1} (the latter is here preferred in the numerical applications). The main diagonal terms have to be considered, that is, $i = j$ and $\tau = s$; moreover, only the component with $\tau, s = 1$ has to be penalized, therefore

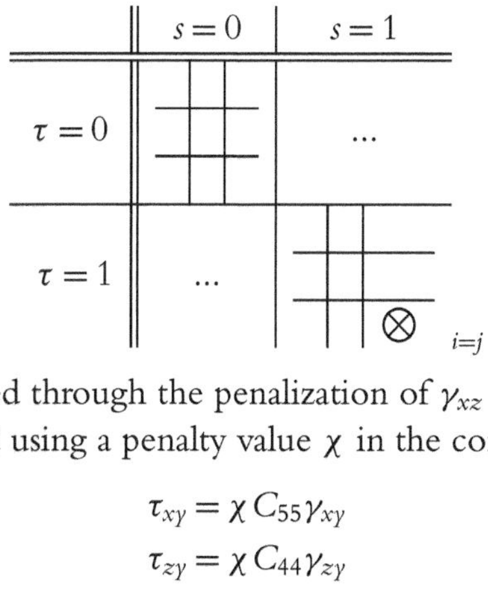

(5.156)

CPT can be obtained through the penalization of γ_{xz} and γ_{yz}. The condition can be imposed using a penalty value χ in the constitutive equations

$$
\begin{aligned}
\tau_{xy} &= \chi\, C_{55} \gamma_{xy} \\
\tau_{zy} &= \chi\, C_{44} \gamma_{zy}
\end{aligned}
\tag{5.157}
$$

5.17 UNIFIED FORMULATION OF N-ORDER

The classical plate theory and the first-order shear deformation theory are the simplest theories that permit the kinematic behaviour of most thin plates to be adequately described. Refined theories can represent the kinematics better, may not require shear correction factors, and can yield more accurate results in the case of thick plates. However, they involve considerably more computational effort. Therefore, such theories should be chosen

Table 5.6 Taylor polynomials. This table presents the compact form of the Taylor polynomials

N	M	F_τ
0	1	$F_0 = 1$
1	2	$F_1 = z$
2	3	$F_2 = z^2$
3	4	$F_3 = z^3$
...	...	...
N	$N+1$	$F_N = z^N$

case by case, depending on the structural problem analyzed, in order to consider only higher-order terms that are necessary.

This section presents a unified approach to dealing with plate models of any order. In the present unified formulation, the displacement field is obtained through a formal expression regardless of the order of the theory (N), which is considered as an input of the analysis.

The unified formulation of the through-the-thickness displacement field is described by an expansion of generic functions (F_τ)

$$\mathbf{u}(x, y, z) = F_\tau(z)\mathbf{u}_\tau(x, y), \qquad \tau = 0, 1, ..., N \tag{5.158}$$

where F_τ are functions of the thickness coordinate z, $\mathbf{u}_\tau$ is the displacement vector depending on the in-plane coordinates x, y and N is the order of expansion of the model. According to Einstein notation, the repeated subscript τ indicates summation. Note that τ assumes values from 0 to N (differently from 1D UF) because, in two-dimensional modelling, it coincides with the polynomial order of the corresponding thickness function. The choice of F_τ and N is arbitrary, different base-functions of any order can be taken into account to model the displacement field of a structure along the thickness. One possible choice is the adoption of Taylor polynomials consisting of the base z^i, where i is a positive integer. Table 5.6 presents F_τ, as functions of the order of expansion N, and M, which is the number of terms in the expansion. Each row shows the expansion terms of an N-order theory. The second-order model ($N = 2$) exploits a parabolic expansion of the Taylor polynomials

$$\begin{array}{ll|l|l} u_x = & \underbrace{u_{x_0}} & + z\, u_{x_1} & + z^2\, u_{x_2} \\ u_y = & u_{y_0} & + z\, u_{y_1} & + z^2\, u_{y_2} \\ u_z = & \underbrace{u_{z_0}}_{N=0} & \underbrace{+ z\, u_{z_1}}_{N=1} & \underbrace{+ z^2\, u_{z_2}}_{N=2} \end{array} \tag{5.159}$$

The 2D model given by Eq. (5.159) has 9 displacement variables. The strain components can be derived through partial derivations, for instance, the shear and normal components are:

$$\begin{aligned} \gamma_{xz} &= u_{x,z} + u_{z,x} = u_{x_1} + 2\,z\,u_{x_2} + u_{z_0,x} + z\,u_{z_1,x} + z^2\,u_{z_2,x} \\ \varepsilon_{zz} &= \quad u_{z,z} \quad = \qquad u_{z_1} + 2\,z\,u_{z_2} \end{aligned} \tag{5.160}$$

The UF can also deal with *reduced* models by exploiting a lower number of variables. An example of a reduced model is given by:

$$\begin{aligned} u_x &= u_{x_0} + z\,u_{x_1} \\ u_y &= u_{y_0} + z\,u_{y_1} \\ u_z &= u_{z_0} + z\,u_{z_1} + z^2\,u_{z_2} \end{aligned} \tag{5.161}$$

The 2D model given in Eq. (5.161) has 7 displacement variables: three constants, three linear and one parabolic. More details about reduced models will be provided in the next chapters.

5.17.1 $N = 3$ and $N = 4$

The third-order model ($N = 3$) exploits a cubic expansion of the Taylor polynomials

$$\begin{aligned} u_x &= \ \ldots \ && \Big\| + z^3\,u_{x_3} \\ u_y &= \ \ldots \ && \Big\| + z^3\,u_{y_3} \\ u_z &= \underbrace{\ldots}_{N\le 2} && \Big\| \underbrace{+ z^3\,u_{z_3}}_{N=3} \end{aligned} \tag{5.162}$$

The 2D model given by Eq. (5.162) has 12 displacement variables. The shear and normal strain components are:

$$\begin{aligned} \gamma_{xz} &= u_{x,z} + u_{z,x} \\ &= u_{x_1} + 2\,z\,u_{x_2} + 3\,z^2\,u_{x_3} + u_{z_0,x} + z\,u_{z_1,x} + z^2\,u_{z_2,x} + z^3\,u_{z_3,x} \\ \varepsilon_{zz} &= u_{z,z} \\ &= u_{z_1} + 2\,z\,u_{z_2} + 3\,z^2\,u_{z_3} \end{aligned} \tag{5.163}$$

Similarly, the fourth-order model ($N = 4$) exploits a quartic expansion of the Taylor polynomials

$$\begin{aligned} u_x &= \ \ldots \ && \Big\| + z^4\,u_{x_4} \\ u_y &= \ \ldots \ && \Big\| + z^4\,u_{y_4} \\ u_z &= \underbrace{\ldots}_{N\le 3} && \Big\| \underbrace{+ z^4\,u_{z_4}}_{N=4} \end{aligned} \tag{5.164}$$

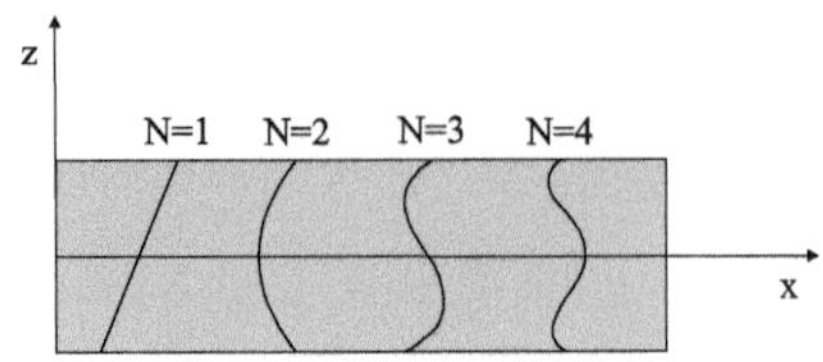

Figure 5.35 Distribution of displacements according to linear and higher-order theories in unified formulation.

The plate model given by Eq. (5.164) has 15 displacement variables. Usually in 2D modelling, a fourth-order model is sufficient for reaching the convergence solution. Figure 5.35 shows the typical distribution of the displacements for different orders of expansions.

5.18 2D MODELS WITH PHYSICAL VOLUME/SURFACE-BASED GEOMETRY AND PURE DISPLACEMENT VARIABLES, THE LAGRANGE EXPANSION CLASS (LE)

The unified formulation permits one to implement different models for the analysis of bi-dimensional structures (plates and shells) by varying the class of the thickness functions and the order of expansion along the thickness. Models based on a Taylor-like polynomial expansion were presented in the previous chapter. The second class of UF models for plates and shells is based on Lagrange polynomials. Lagrange-expansion 2D models – hereafter referred to as LEs – are described in this chapter. As mentioned in section 5.10, LEs have the following main characteristics:

1. The variables and boundary conditions can be located above the physical surfaces of the structure. This feature is particularly relevant for a CAD-FEM coupling scenario.
2. The unknown variables of the problem are the pure displacement components. No rotations or higher-order variables are exploited to describe the displacement field.

These features are described in the following sections of this chapter and numerical examples are provided in order to highlight the enhanced capabilities of 2D UF LEs, in terms of 3D-like accuracies and very-low computational costs.

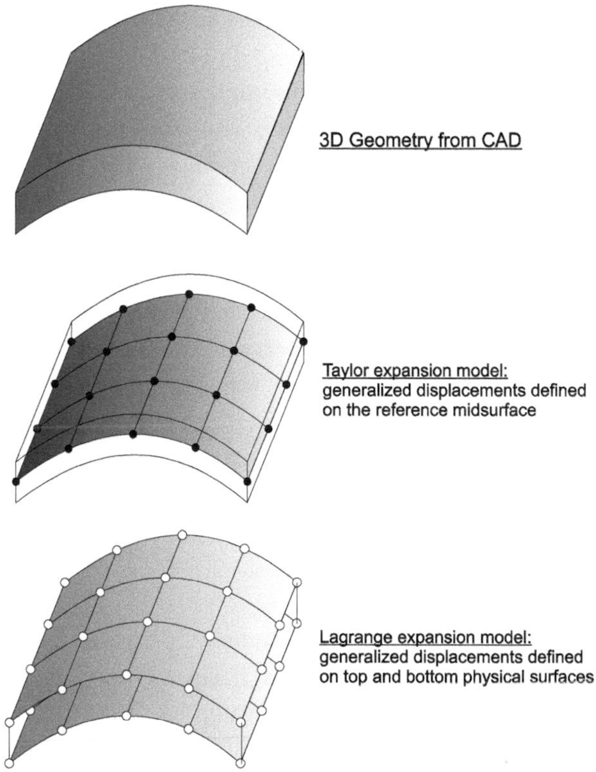

Figure 5.36 Different geometrical modelling approaches, Taylor expansion model vs. Lagrange expansion model.

5.19 PHYSICAL VOLUME/SURFACE APPROACH

In the frame of FEA (Finite Element Analysis) and CAD (Computer Aided Design), it should be considered that many structural elements require the definition of the reference surfaces on which these elements – and the problem unknowns – lie. The definition of the reference surfaces is particularly critical when 3D CAD geometries are considered. The present 2D finite element formulation, based on a physical volume/surface approach, offers significant advantages as far as this point is concerned because LE elements can deal directly with the 3D geometry given by a CAD model.

Figure 5.36 shows the finite element models of a shell structure based on both Taylor and Lagrange polynomial expansions. For the case of the Taylor expansion model and starting from the 3D geometry of the structure, the finite element discretization requires that a reference midsurface of the shell is defined. The problem unknowns are defined above this surface. Instead, in the LE modelling approach, the nodes can be located directly on the top and bottom surfaces of the 3D structure. This implies that the finite element unknowns lie above the physical surface of the structure. It is not necessary to define reference surfaces and a 3D CAD geometry can be exploited directly for the finite element analysis. In other words, a 2D finite element can be used for a 3D geometrical description. Moreover, the LE model can be particularly advantageous because it permits boundary conditions to be applied directly to pure displacement components. For example, it is easy to manage problems in which mechanical loads act on the top/bottom surface of the structure. The LE model also becomes very attractive in view of the analysis of multilayered structures. A layer-wise description of a multilayer presupposes the use of LE models that allow compatibility conditions to be easily imposed between the different layers.

5.20 LAGRANGE EXPANSION MODEL

The Lagrange expansion 2D models (LEs) represent the second class of 2D models developed in the framework of UF. Descriptions provided in this section are tuned on the specific issues related to the UF. Details on Lagrange polynomials not directly related to UF can be found in many excellent books [1–3].

In an LE model, the F_τ thickness functions coincide with the Lagrange polynomials. Therefore, the through-the-thickness displacement field is written as:

$$\begin{aligned} u_\alpha &= F_t\, u_{\alpha_t} + F_b\, u_{\alpha_b} \\ u_\beta &= F_t\, u_{\beta_t} + F_b\, u_{\beta_b} \\ u_z &= F_t\, u_{z_t} + F_b\, u_{z_b} \end{aligned} \tag{5.165}$$

where $(u_{\alpha_t}, u_{\beta_t}, u_{z_t})$ and $(u_{\alpha_b}, u_{\beta_b}, u_{z_b})$ are the values of displacement components at top and bottom surfaces of the shell, respectively. F_t and F_b are linear Lagrange polynomials and they read:

$$\begin{aligned} F_t &= \frac{1+\zeta}{2} \\ F_b &= \frac{1-\zeta}{2} \end{aligned} \tag{5.166}$$

Lagrange polynomials are usually given in terms of normalized – also known as natural – thickness coordinate $-1 < \zeta < 1$. This choice is not

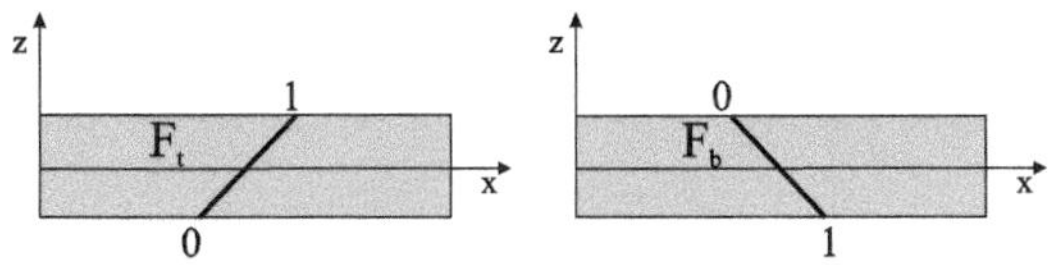

Figure 5.37 Lagrange thickness functions.

compulsory since LEs could also be implemented in terms of actual coordinate z, however the normalized formulation was preferred since it offers many advantages. Only linear expansions are considered in this book, even though higher-order theories can be formulated on the basis of Lagrange polynomials.

These thickness functions have the following particular properties:

$$\begin{aligned} &\zeta = 1 : F_t = 1 \quad \text{and} \quad F_b = 0 \quad \text{at the top surface} \\ &\zeta = 1 : F_b = 1 \quad \text{and} \quad F_t = 0 \quad \text{at the bottom surface} \end{aligned} \tag{5.167}$$

as shown in Fig. 5.37. Therefore, the boundary surface values of the displacements are considered as variable unknowns. This fact permits to apply the boundary conditions directly to pure displacement components and, in the case of multilayered structures, one can easily impose the following compatibility conditions:

$$\mathbf{u}_t^k = \mathbf{u}_b^{k+1} \tag{5.168}$$

where k is the layer index.

In the framework of unified formulation, the displacement field preserves the formal expression of Eq. (4.55):

$$\mathbf{u} = F_\tau \mathbf{u}_\tau, \qquad \tau = 0, 1, ..., N \tag{5.169}$$

The use of LEs as F_τ does not imply a reformulation of the problem equations and matrices, as typical in the UF environment. The governing equations, the finite element formulation and the fundamental nuclei are exactly the same for plates and shells, respectively. Note that the LEs are here presented referring to shell displacement components, but they remain valid for plates that are shells with $R = \infty$.

5.21 EXTENSION TO MULTILAYERED STRUCTURES

The aim of this chapter is to offer some ideas about the capabilities of the UF in the analysis of multilayered structures. The UF was, in fact, conceived by the first author in order to provide a powerful tool for the study

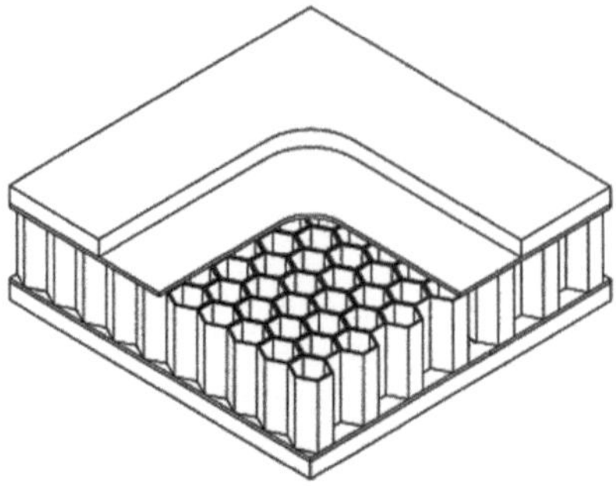

Figure 5.38 Honeycomb core sandwich.

of new advanced materials. For the sake of brevity, only two-dimensional multilayered theories will be discussed here. However, the UF can also be used for beam structures, as shown in the previous chapters.

5.22 MULTILAYERED STRUCTURES

The previous chapters dealt with traditional structures made of metallic materials with isotropic behaviour. Most of structure theories have, in fact, been developed to obtain a better understanding of the mechanical behaviour of one-layered beams, plates and shells.

However, there are a number of applications in which the use of a single material is not convenient. These are briefly listed below.

- Sandwich structures:
  sandwich[4] beams, plates and shells (see Fig. 5.38) are composite structures that consist of at least three different layers. Two or more high-strength stiff layers (faces) are bonded to one or more low-density flexible layer (core). The core, which is usually the cheapest material, mainly has the task of keeping the two faces away from the neutral axis, thus improving bending resistance. Sandwich structures can be defined as composite structures, since they are composed of at least two different materials at a microscopic level. There are many types of sandwich

[4] This word is derived from Lord Sandwich who lived a few centuries ago. A very conversant gambler, Lord Sandwich did not take the time to have a meal during his long hours playing at the card table. Consequently, he would ask his servants to bring him slices of meat between two slices of bread; a habit well known among his gambling friends. Since John Montagu was the Earl of Sandwich, others began to order "the same as Sandwich!" – and the word "sandwich" was thus created.

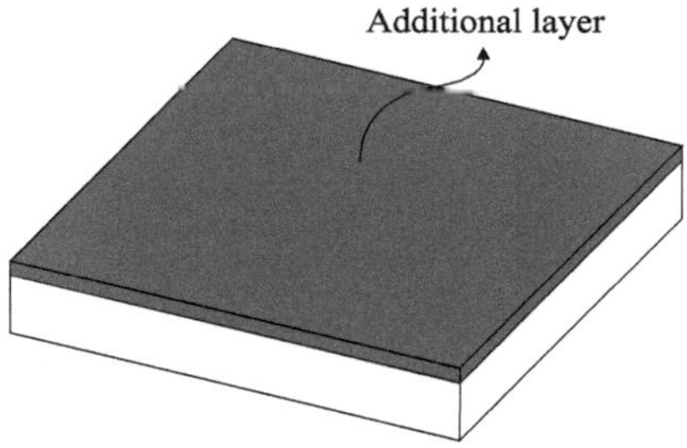

Figure 5.39 Layered plate made of isotropic layers for thermal protection.

structures in aerospace construction that have not been dealt with in detail here.

- Layered structures for thermal protection purposes:
  engine components, reentry vehicles and supersonic aircraft often require adequate thermal protection. In many cases, it is not possible to obtain a material that can stand both mechanical and thermal loadings at the same time. The solution, in this case, is to build a composite multilayered structure: the mechanical structure is protected by an additional layer that leads to high resistance with respect to thermal loadings, as shown in Fig. 5.39.
- Piezo-layered materials for smart structures:
  the phenomenon of piezoelectricity is a particular feature of certain classes of crystalline materials. The piezoelectric effect consists of a linear energy conversion between the mechanical and electrical fields in both directions that defines a direct or converse piezoelectric effect. The direct piezoelectric effect generates an electric polarization by applying mechanical stresses. The converse piezoelectric effect instead induces mechanical stresses or strains by applying an electric field. Multilayered structures are also obtained when piezo-layers are bonded as sensors or actuators to a given structure, as can be seen in Fig. 5.40.
- Laminates:
  the laminate structure is the most common case of composite materials (see Fig. 5.41). Composites are multilayered structures (made mostly of flat and curved panels) constituted by several layers or laminae that are bonded perfectly together. Each lamina is composed of fibers embedded in a matrix. These fibers are produced according to a specific technological process that confers high mechanical properties in the longitudinal direction (L) of the fibers (see Fig. 5.41). The matrix has

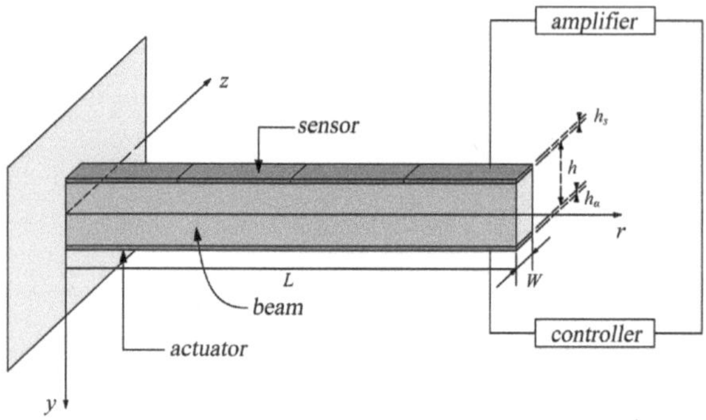

Figure 5.40 Example of a smart structure: sensor-actuator network for a plate.

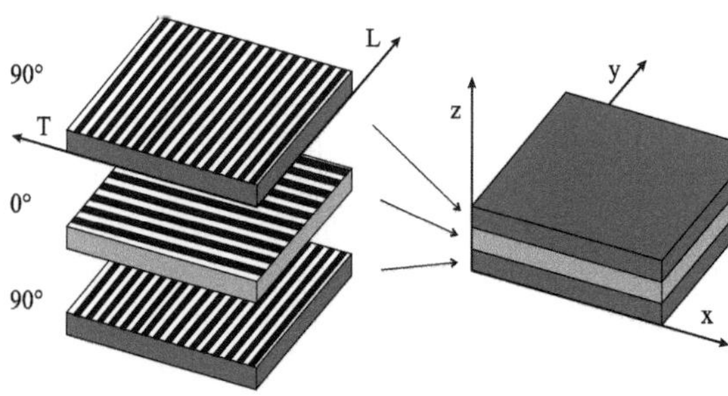

Figure 5.41 Layered plate made of unidirectional laminae.

the role of holding the fibers together. Carbon, boron and glass fibers are used above, all along with organic products. The matrices are mostly of an epoxy type. There are several possible ways of putting the fibers and matrix together. Uni-directional laminae or laminates made of differently oriented laminae are used in most applications related to the construction of aerospace, automotive or sea vehicles. The laminae are placed one over the other, according to a given lay-out, as shown in Fig. 5.41. Such a possibility, which is known as "tailoring", permits one to optimize the use of the material for a given set of design requirements.

- Functionally graded materials:

  FGMs are advanced composite materials, within which the composition of each constituent material varies gradually with respect to spatial coordinates. Therefore, the macroscopic material properties in FGMs vary continuously, thus distinguishing them from laminated composite materials in which the abrupt change of the material properties across

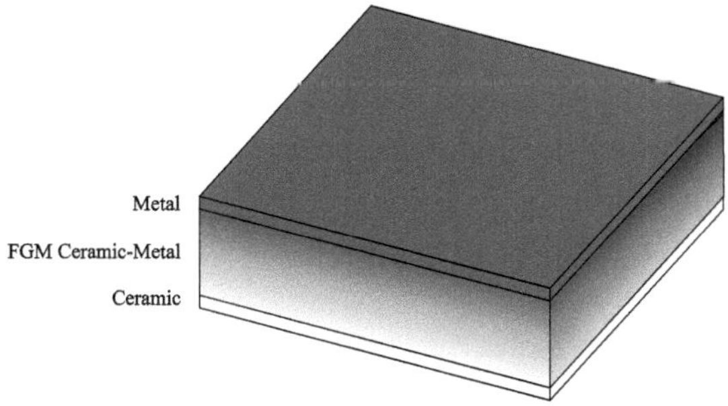

Figure 5.42 Multilayered plate embedding a FGM layer.

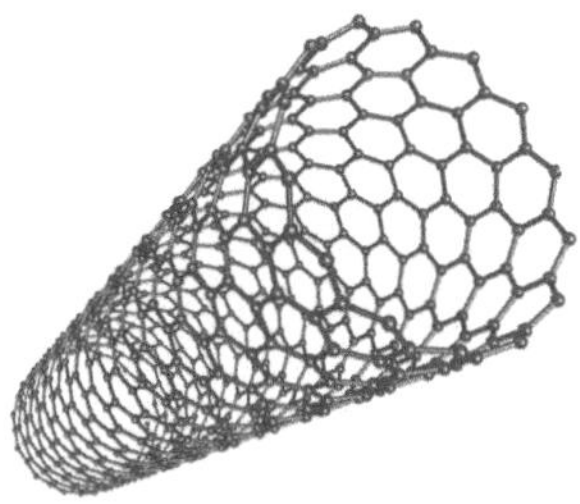

Figure 5.43 Example of single-walled carbon nanotube.

the layer interfaces leads to large interlaminar stresses which can lead to damage development. As in the case of laminated composite materials, FGMs combine the desirable properties of the constituent phases to obtain a superior performance. To this aim, functionally graded layers can be embedded in multilayered structures (see Fig. 5.42).

- Multiwalled carbon-nanotubes:
  carbon nanotubes (CNT) have exceptional mechanical properties (Young's modulus, tensile strength, toughness, etc.), which are due to their molecular structure, which consists of single or multiple sheets of graphite wrapped into seamless hollow cylinders (see Fig. 5.43). Owing to the great stiffness, strength and high aspect ratio of CNTs, it is expected that, by dispersing them evenly throughout a polymer matrix, it is possible to produce composites with considerably improved overall effective mechanical properties. Furthermore, CNTs have a relatively low density of about 1.75 g/cm^3 and nanotube reinforced polymers (NRPs), therefore, excel due to their extremely high specific stiffness, strength and toughness.

5.23 THEORIES ON MULTILAYERED STRUCTURES

Theories on multilayered structures could be developed by making an appropriate choice concerning the following points.

1. The choice of the unknown variables:
   - displacement formulation;
   - mixed formulation.
2. The choice of the variable description:
   - equivalent single layer (ESL) models;
   - layer-wise (LW) models.

Many other choices which have already been adopted for traditional metallic structures can be made. Nevertheless, it will be clear, from reading this chapter, that the "sic and simpliciter" extension of the modellings that are already known for monocoque structures would not lead to satisfactory results. Complicating effects, such as the so-called C_z^0 Requirements, should be taken into account for this purpose.

5.23.1 C_z^0-Requirements

Layered structures are said to be "transversely anisotropic" because they exhibit different mechanical-physical properties in the thickness direction. Transverse discontinuous mechanical properties cause a displacement field, $\boldsymbol{u}$, in the thickness direction, which can exhibit a rapid change in its slope in correspondence to each layer interface. This is known as the Zig-Zag, or ZZ, form of displacement field in the thickness direction (see Fig. 5.44(b)). In-plane stresses $\boldsymbol{\sigma}_p = (\sigma_{xx}, \sigma_{yy}, \sigma_{xy})$ can in general be discontinuous at each layer interface. Nevertheless, transverse stresses $\boldsymbol{\sigma}_n = (\sigma_{xz}, \sigma_{yz}, \sigma_{zz})$, for equilibrium reasons, i.e. the Cauchy theorem, must be continuous at each layer interface. This is often referred to in literature as Interlaminar Continuity, IC, of the transverse stresses. Figure 5.44 shows, from a qualitative point of view, the possible displacements and transverse stresses distribution scenarios of a multilayered structure, which depend on the variables description. It appears evident that displacements and transverse stresses are C^0-continuous functions in the thickness z direction, due to compatibility and equilibrium reasons respectively. It is also evident that the displacement $\boldsymbol{u}$ has discontinuous first derivative corresponding to each interface, where the mechanical properties change. In Refs. [8,9], ZZ and IC were referred to as C_z^0-Requirements. The fulfilment of the C_z^0-Requirements is a crucial point in the development of any theory suitable for multilayered structures.

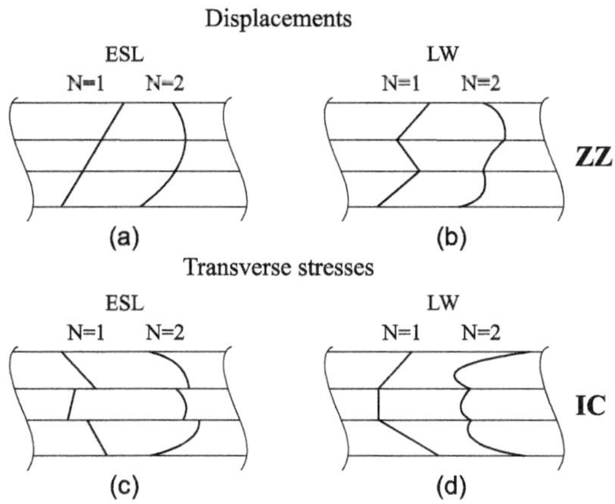

Figure 5.44 C_z^0 requirements.

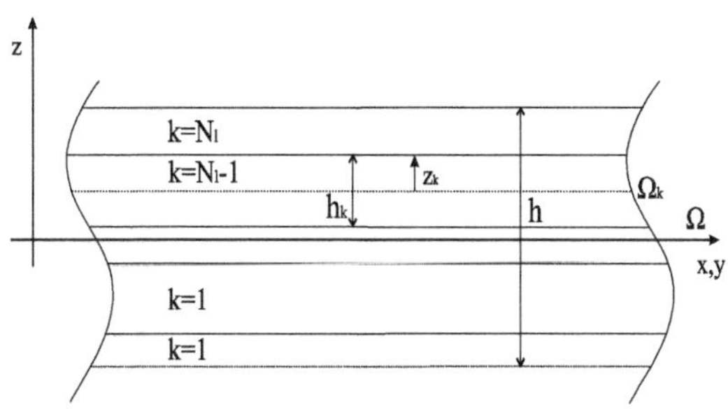

Figure 5.45 Notation for multilayered structures.

5.23.2 Refined theories

The theories that were originally developed for single-layer "monocoque" structures made of traditional isotropic materials, can conveniently be grouped into two cases: Love First Approximation Theories, LFAT, or Love Second Approximation Theories, LSAT. Applications of LFAT to multilayered structures are often referred to as Classical Lamination Theories, CLTs, see Ref. [31]. CLTs assume that the normals to the reference surface Ω (see Fig. 5.45) remain normal in the deformed states and do not change in length, that is, transverse shear as well as normal strains are postulated to be negligible with respect to the other strains. Extensions of the so-called Reissner–Mindlin (see Refs. [19,27]) LSAT type model, which includes transverse shear strains, to layered structures are known as the Shear Deformation Theory SDT (or First-order SDT, FSDT) see Ref. [39].

A simple refinement of the Reissner-Mindlin theory was conducted by Vlasov [38] for monocoque structures. Vlasov's FSDT type-theory permits the homogeneous conditions to be fulfilled for the transverse shear stresses in correspondence of the top and bottom shell/plate surfaces $\sigma_{iz}(\pm h/2) = 0$. In Refs. [24,26] has been shown that such a simple inclusion can lead to significant improvements, compared to FSDT analysis, in tracing the static and dynamic response of thick laminated structures.

Further refinements of FSDT are known as Higher-Order Theories, HOT. In general, higher-order theories are based on displacement models of the following type:

$$u_i(x, y, z) = u_0 + z\, u_1 + z^2\, u_2 + \dots + z^N\, u_N \tag{5.170}$$

where N is the order of expansion used for the displacement variables. Displacement models related to HOT have been traced in Fig. 5.44. Examples of the application of these types of models to flat and curved laminated structures can be found in Refs. [14,15,18,23,33,35–37].

5.23.3 Zig-zag theories

The extension of CLT, FSDT and HOT to multilayered plates does not permit the C_z^0-Requirements to be fulfilled. Refined theories have therefore been introduced to resolve this problem. Owing to the form of the displacement field in the thickness direction, see Fig. 5.44, these types of theories are referred to as Zig-Zag theories.

The fundamental idea behind zig-zag theories is that a certain displacement and/or stress model is assumed in each layer and then compatibility and equilibrium conditions are used at the interface to reduce the number of unknown variables.

The first, most significant contributions to Zig-Zag theories came from the Russian school. The first Zig-Zag theory was proposed by Lekhnitskii [17] for beam geometries. Other outstanding contributions to plates and shells have been given by Ambartsumian [4–6]. An independent manner of formulating zig-zag plate/shell theories has been provided in the Western by Reissner [28–30]. Many other zig-zag theories based on the above mentioned models have been collected and discussed in Ref. [11].

5.23.4 Layer-wise theories

The theories mentioned before consider a number of unknown variables that is independent of the number of constitutive layers N_l. These all are known as Equivalent Single Layer Models or ESLMs. Although these kinematic theories can describe transverse shear and normal strains, including transverse warping of the cross-section, their approach is "kinematically homogeneous" in the sense that the kinematics is insensitive to individual layers, unless zig-zag models are used. If a detailed response of the individual layers is required, and if significant variations in the displacements gradients between layers exist, as is the case of a local phenomena description, this approach needs the use of special higher-order theories in each of the constitutive layers along with a concomitant increase in the number of unknowns in the solution process, which leads to an increase in the complexity of the analysis. In other words, a possible "natural" way of including the ZZ effect in the framework of a classical model with only displacement variables, could be to apply CLT, FSDT or HOT at a layer level, that is each layer is considered as an independent plate, and compatibility of the displacement components, corresponding to each interface, is then imposed as a constraint (see Fig. 5.44(b)). In these cases, Layer Wise Models or LWMs are obtained. Examples of these types of theories can be found in Refs. [12,13,16,32,34]. These displacement models require constraint conditions to be included in order to enforce the compatibility conditions at each interface. Generalizations of these types of theories were given in Refs. [21,25], in which the displacement variables in the thickness direction are expressed in terms of Lagrange polynomials:

$$u_i^k(x, y, z) = L_1(z_k)u_i^k|_{h/2} + L_2(z_k)u_i^k|_{-h/2} + L_3(z_k)u_{i3}^k + \ldots + L_N(z_k)u_{iN}^k \tag{5.171}$$

with $k = 1, ..., N_l$. Interface values are used as the unknown variables (the first two terms of the expansion $u_i^k|_{h/2}$ and $u_i^k|_{-h/2}$) and this permits easy linkage of the compatibility conditions at each interface. $L1$, $L2$, in fact, coincide with linear Lagrangian polynomials, while $L3$, ...L_N should be an independent base of polynomials that start from the parabolic one ($L3$).

5.23.5 Mixed theories

A third approach for laminated structures was presented in the two papers by Reissner [28,29], in which a mixed variational equation, namely

Reissner's Mixed Variational Theorem, RMVT, was proposed. The displacement $\boldsymbol{u}$ and transverse stress $\boldsymbol{\sigma}_n$ variables are assumed independently in the framework of RMVT, which states:

$$\sum_{k=1}^{N_l} \int_{\Omega_k} \int_{A_k} \left\{ \delta\boldsymbol{\epsilon}_{pG}^{k\ T} \boldsymbol{\sigma}_{pH}^k + \delta\boldsymbol{\epsilon}_{nG}^{k\ T} \boldsymbol{\sigma}_{nM}^k + \delta\boldsymbol{\sigma}_{nM}^{k\ T} (\boldsymbol{\varepsilon}_{nG}^k - \boldsymbol{\varepsilon}_{nH}^k) \right\} d\Omega_k dz = \sum_{k=1}^{N_l} \delta L_{ext}^k \tag{5.172}$$

The subscript H underlines that stresses are computed via Hooke's law. The variation in the internal work has been split into in-plane and out-of-plane parts and involves stresses from Hooke's law and strains from geometrical relations (subscript G). δL_{ext} is the virtual variation of the work made by the external layer forces. The third "mixed" term variationally enforces compatibility of the transverse strain components. Subscript M underlines that the transverse stresses are those of the assumed model, see the discussion reported in Ref. [10].

Full mixed methods have also been developed, in which all six stress components and the three displacements are assumed as the unknown variables. A detailed discussion has been provided in Ref. [22]. An interesting discussion on the possible improvement of FSDT type models, using mixed and partially mixed formulation, has been given in Ref. [7]. The Hellingher-Reissner mixed principle has been employed in this work to determine the FSDT governing equations. Two FSDT type models, both of which describe transverse shear stresses as independent variables, have been discussed and implemented.

5.24 UNIFIED FORMULATION FOR MULTILAYERED STRUCTURES

As discussed in the previous section, the introduced assumptions for displacements and stresses can be made at layer or at multilayer level. Layer-Wise (LW) description is obtained in the first case, while Equivalent Single Layer (ESL) description is acquired in the latter one. If LW description is employed, $\boldsymbol{u}_\tau$ and $\boldsymbol{\sigma}_{nM_\tau}$ are layer variables. These are different in each layer. If one refers to ESL description, $\boldsymbol{u}_\tau$ and $\boldsymbol{\sigma}_{nM_\tau}$ are plate/shell variables. These are the same for the whole multilayer.

Examples of ESL and LW assumptions are shown in Fig. 5.44.

5.24.1 ESL models

In the most general case, higher-order ESL models appear in the following form:

$$
\begin{aligned}
u_x(x, y, z) &= u_{x0}(x, y) + z\, u_{x1}(x, y) + \cdots + z^N\, u_{xN}(x, y) \\
u_y(x, y, z) &= u_{y0}(x, y) + z\, u_{y1}(x, y) + \cdots + z^N\, u_{yN}(x, y) \\
u_z(x, y, z) &= u_{z0}(x, y) + z\, u_{z1}(x, y) + \cdots + z^N\, u_{zN}(x, y)
\end{aligned}
\tag{5.173}
$$

According to unified formulation, they can be written in the following compact form:

$$
\mathbf{u} = F_0\, \mathbf{u}_0 + F_1\, \mathbf{u}_1 + ..F_N\, \mathbf{u}_N = F_\tau\, \mathbf{u}_\tau, \qquad \tau = 0, 1, 2, \cdots N \tag{5.174}
$$

where N is the order of the expansion and

$$
F_0 = 1, \quad F_1 = z, \quad F_0 = z^2, \; \cdots \; F_N = z^N \tag{5.175}
$$

These higher-order theories are herein denoted by acronyms ED1, ED2, ED3, $\cdots$, EDN. Further letter d means that ED1d model is obtained by ED1 one by neglecting the linear term in u_z (ED1d, in fact, coincides with FSDT).

5.24.2 Inclusion of Murakami's zig-zag function

Murakami [20] introduced in a first order ESL displacement field a zig-zag function able to describe zig-zag form for the displacements. He modified the FSDT theories according to the following model,

$$
\begin{aligned}
u_x(x, y, z) &= u_{0x} + z\, u_{x1}(x, y) + (-1)^k \zeta_k\, u_{xZ} \\
u_y(x, y, z) &= u_{0y} + z\, u_{x1}(x, y) + (-1)^k \zeta_k\, u_{yZ} \\
u_z(x, y, z) &= u_{0z}
\end{aligned}
\tag{5.176}
$$

Subscript Z refers to the introduced Murakami's zig-zag function. $\zeta_k = 2z_k/h_k$ is a non-dimensioned layer coordinate (z_k is the physical coordinate of the k-layer whose thickness is h_k, as indicated in Fig. 5.45). The exponent k changes the sign of the zig-zag term in each layer. Such an artifice permits the discontinuity of the first derivative of the displacement variables to be reproduced in the z-directions, which physically comes from the intrinsic transverse anisotropy of multilayer structures.

Transverse normal strain/stress effects can be included as done for the ED1 models, leading to

$$\begin{aligned} u_x(x,y,z) &= u_{0x} + z\,u_{x1}(x,y) + (-1)^k \zeta_k\, u_{xZ} \\ u_y(x,y,z) &= u_{0y} + z\,u_{y1}(x,y) + (-1)^k \zeta_k\, u_{yZ} \\ u_z(x,y,z) &= u_{0z} + z\,u_{z1}(x,y) + (-1)^k \zeta_k\, u_{zZ} \end{aligned} \tag{5.177}$$

By introducing the following notations,

$$\tau = 0, 1, 2 \quad F_0 = 1, \quad F_1 = z, \quad F_2 = F_Z = (-1)^k \zeta_k \tag{5.178}$$

Equation (5.177) can be written in the following array form

$$\mathbf{u} = \mathbf{u}_0 + (-1)^k \zeta_k\, \mathbf{u}_Z + z\,\mathbf{u}_1 = F_\tau\, \mathbf{u}_\tau, \quad \tau = 0, 1, Z \tag{5.179}$$

Such a model is denoted by the acronym EDZ1 in which Z underlines that a zig-zag function has also been used for the expansion in z.

Higher-order models appear in the following form:

$$\begin{aligned} u_x(x,y,z) &= u_{x0}(x,y) + z\,u_{x1}(x,y) + \cdots + z^{N-1}\, u_{xN}(x,y) + (-1)^k \zeta_k\, u_{xZ} \\ u_y(x,y,z) &= u_{y0}(x,y) + z\,u_{y1}(x,y) + \cdots + z^{N-1}\, u_{yN}(x,y) + (-1)^k \zeta_k\, u_{yZ} \\ u_z(x,y,z) &= u_{z0}(x,y) + z\,u_{z1}(x,y) + \cdots + z^{N-1}\, u_{zN}(x,y) + (-1)^k \zeta_k\, u_{zZ} \end{aligned} \tag{5.180}$$

Or in compact form

$$\mathbf{u} = \quad \mathbf{u}_0 + (-1)^k \zeta_k\, \mathbf{u}_Z + z^r\, \mathbf{u}_r = \quad F_\tau\, \mathbf{u}_\tau, \quad \tau = 0, 1, 2, \cdots N \tag{5.181}$$

N is the order of the expansion and

$$F_0 = 1, \quad F_1 = z, \quad F_0 = z^2, \ \cdots \ F_{N-1} = z^{N-1}, \quad F_N = F_Z = (-1)^k \zeta_k \tag{5.182}$$

These higher-order theories are herein denoted by acronyms EDZ1, EDZ2, EDZ3, $\cdots$, EDZN.

5.24.3 Layer-wise theory and Legendre expansion

In the case of layer-wise models, LWMs, each layer is seen as independent and compatibility of displacement components at each interface is then imposed. To this aim, Lagrange polynomials could be used, like in Ref. [25]. However, the use of thickness functions in terms of Legendre polynomials is here preferred. The following expansion is employed,

$$\begin{aligned} u_x^k &= F_t\, u_{xt}^k + F_b\, u_{xb}^k \\ u_y^k &= F_t\, u_{yt}^k + F_b\, u_{yb}^k \\ u_z^k &= F_t\, u_{zt}^k + F_b\, u_{zb}^k \end{aligned} \tag{5.183}$$

It is now intended that the subscripts t and b denote values related to the top and bottom layer-surface, respectively. These two terms consist of the linear part of the expansion. The thickness functions $F_\tau(\zeta_k)$ have now been defined at the k-layer level,

$$F_t = \frac{P_0 + P_1}{2}, \quad F_b = \frac{P_0 - P_1}{2}, \tag{5.184}$$

in which $P_j = P_j(\zeta_k)$ is the Legendre polynomial of the j-order defined in the ζ_k-domain: $\zeta_k = \frac{2z_k}{h_k}$ and $-1 \leq \zeta_k \leq 1$. For instance, the first five Legendre Polynomials are $P_0 = 1$, $P_1 = \zeta_k$, $P_2 = (3\zeta_k^2 - 1)/2$, $P_3 = \frac{5\zeta_k^3}{2} - \frac{3\zeta_k}{2}$, $P_4 = \frac{35\zeta_k^4}{8} - \frac{15\zeta_k^2}{4} + \frac{3}{8}$. The chosen functions have the following interesting properties:

$$\zeta_k = \begin{cases} 1 & : \quad F_t = 1; \quad F_b = 0; \quad F_r = 0 \\ -1 & : \quad F_t = 0; \quad F_b = 1; \quad F_r = 0, \end{cases} \tag{5.185}$$

That permits to have interface values as unknown variables, avoiding therefore the inclusion of constraint equations to impose C_z^0-requirements.

In a unified form

$$\mathbf{u}^k = F_t\,\mathbf{u}_t^k + F_b\,\mathbf{u}_b^k + = F_\tau\,\mathbf{u}_\tau^k \quad \tau = t, b, \tag{5.186}$$

This layer-wise theory will be denoted as LD1: Layer-Wise theory, with displacement unknowns and first order expansion.

Higher-order layer-wise theories are written by adding higher-order terms,

$$\begin{aligned} u_x^k &= F_t\,u_{xt}^k + F_b\,u_{xb}^k + F_2\,u_{x2}^k + .. + F_N\,u_{xN}^k \\ u_y^k &= F_t\,u_{yt}^k + F_b\,u_{yb}^k + F_2\,u_{y2}^k + .. + F_N\,u_{yN}^k \\ u_z^k &= F_t\,u_{zt}^k + F_b\,u_{zb}^k + F_2\,u_{z2}^k + .. + F_N\,u_{zN}^k \end{aligned} \tag{5.187}$$

where

$$F_r = P_r - P_{r-2}, \quad r = 2, 3, .., N \tag{5.188}$$

In a unified form

$$\mathbf{u}^k = F_t\,\mathbf{u}_t^k + F_b\,\mathbf{u}_b^k + F_r\,\mathbf{u}_r^k = F_\tau\,\mathbf{u}_\tau^k \quad \tau = t, b, \; r = 2, 3, ..N, \tag{5.189}$$

These higher-order expansions have been denoted by the acronyms LD2, LD3, $\cdots$, LDN.

5.24.4 Mixed models with displacement and transverse stress variables

The ESL models, described in the previous section, can be still used for displacement variables in the framework of RMVT applications. Anyway, an appropriate transverse stress field is required. A first choice could be using for $\boldsymbol{\sigma}_n$ the same expansions used for $\boldsymbol{u}$. Of course interlaminar continuity should be fulfilled. To meet such a requirement, a layer-wise description is required for the transverse stresses. The Legendre expansion already used for displacements field seems to be very appropriate to this aim. For the generic higher-order case, one has

$$
\begin{aligned}
\sigma_{xz}^k &= F_t\, \sigma_{txz}^k + F_b\, \sigma_{xzb}^k + F_2\, \sigma_{xz2}^k + \cdots + F_N\, \sigma_{xzN}^k \\
\sigma_{yz}^k &= F_t\, \sigma_{tyz}^k + F_b\, \sigma_{yzb}^k + F_2\, \sigma_{yz2}^k + \cdots + F_N\, \sigma_{yzN}^k \\
\sigma_{zz}^k &= F_t\, \sigma_{tzz}^k + F_b\, \sigma_{zzb}^k + F_2\, \sigma_{zz2}^k + \cdots + F_N\, \sigma_{zzN}^k
\end{aligned}
\tag{5.190}
$$

or in compact form

$$
\boldsymbol{\sigma}_n^k = F_t\, \boldsymbol{\sigma}_{nt}^k + F_b\, \boldsymbol{\sigma}_b^k + = F_\tau\, \boldsymbol{\sigma}_{n\tau}^k \qquad \tau = t, b, \tag{5.191}
$$

Two possible choices can be made for displacements: ESL or LW description.

In the first case, even though it is not extremely appropriate, the mixed models are denoted as EMN. In case interlaminar continuity is imposed in the theories, that is

$$
\boldsymbol{\sigma}_{nt}^k = \boldsymbol{\sigma}_{nb}^{(k+1)}, \qquad k = 1, N_l - 1 \tag{5.192}
$$

and/or top/bottom surface stress values are prescribed (zero or imposed values), the following additional equilibrium conditions should be accounted for:

$$
\boldsymbol{\sigma}_{nb}^1 = \bar{\boldsymbol{\sigma}}_{nb}, \qquad \boldsymbol{\sigma}_{nt}^{N_l} = \bar{\boldsymbol{\sigma}}_{nt} \tag{5.193}
$$

where the over-bar indicates the imposed values at the plate boundary surfaces.

The resulting models are denoted by the acronym EMCN. Letter C is included to underline the fulfilment of interlaminar continuity.

Zig-zag form of the displacement field can be included by using the EDZ type model along with related Murakami's zig-zag function. Resulting models are denoted by acronyms EMZN. Letter Z has been included to underline the description of zig-zag effects.

Full layer-wise description is acquired by combining both layer-wise description for displacement and transverse stresses. Resulting models are denoted by LM1, LM2, $\cdots$, LMN acronyms.

5.25 UF IN TERMS OF 1 x 1 SECONDARY NUCLEI

5.25.1 Beam models

The UF can alternatively be formulated in a way in which each displacement component can be expanded independently from the others and with respect to the results accuracy and the computational cost. In this case, the most general displacement field can be written as follows

$$
\begin{aligned}
u_x\left(x, y, z, t\right) &= F_{\tau_{ux}}\left(x, y\right) u_{x\tau_{ux}}\left(z, t\right), \qquad && \tau_{u_x} = 0, 1, \cdots, N_{u_x} \\
u_y\left(x, y, z, t\right) &= F_{\tau_{uy}}\left(x, y\right) u_{y\tau_{uy}}\left(z, t\right), && \tau_{u_y} = 0, 1, \cdots, N_{u_y} \\
u_z\left(x, y, z, t\right) &= F_{\tau_{uz}}\left(x, y\right) u_{z\tau_{uz}}\left(z, t\right), && \tau_{u_z} = 0, 1, \cdots, N_{u_z}
\end{aligned}
\tag{5.194}
$$

or in compact form

$$
\mathbf{u}\left(x, y, z, t\right) = \mathbf{F}_\tau\left(x, y\right) \mathbf{u}_\tau\left(z, t\right) \tag{5.195}
$$

with

$$
\begin{aligned}
\mathbf{u}_\tau\left(z, t\right) &= \begin{Bmatrix} u_{x\tau_{ux}}\left(z, t\right) \\ u_{y\tau_{uy}}\left(z, t\right) \\ u_{z\tau_{uz}}\left(z, t\right) \end{Bmatrix}, \\
\mathbf{F}_\tau\left(x, y\right) &= \begin{bmatrix} F_{\tau_{ux}}\left(x, y\right) & 0 & 0 \\ 0 & F_{\tau_{uy}}\left(x, y\right) & 0 \\ 0 & 0 & F_{\tau_{uz}}\left(x, y\right) \end{bmatrix}
\end{aligned}
\tag{5.196}
$$

where $F_{\tau_{ux}}$, $F_{\tau_{uy}}$, $F_{\tau_{uz}}$ are the cross-section functions; $u_{x\tau_{ux}}$, $u_{y\tau_{uy}}$, $u_{z\tau_{uz}}$ are the displacement vector components and N_{u_x}, N_{u_y} and N_{u_z} are the orders of expansion. According to Einstein's notation, the repeated subscripts τ_{u_x}, τ_{u_y}, τ_{u_z} indicate summation.

When the cross-section functions are chosen to be Taylor's series expansion then Eq. (5.194) can be rewritten as

$$
\begin{aligned}
u_x(x,y,z,t) &= \sum_{n_{ux}=0}^{N_{ux}} \left[\sum_{n^*_{ux}=0}^{n_{ux}} x^{\left(n_{ux}-n^*_{ux}\right)} y^{n^*_{ux}} u_{x\tilde{N}_{ux}}(z,t) \right] \\
u_y(x,y,z,t) &= \sum_{n_{uy}=0}^{N_{uy}} \left[\sum_{n^*_{uy}=0}^{n_{uy}} x^{\left(n_{uy}-n^*_{uy}\right)} y^{n^*_{uy}} u_{y\tilde{N}_{uy}}(z,t) \right] \\
u_z(x,y,z,t) &= \sum_{n_{uz}=0}^{N_{uz}} \left[\sum_{n^*_{uz}=0}^{n_{uz}} x^{\left(n_{uz}-n^*_{uz}\right)} y^{n^*_{uz}} u_{z\tilde{N}_{uz}}(z,t) \right]
\end{aligned}
\tag{5.197}
$$

where $\tilde{N}_u = \frac{\left[n_u(n_u+1)+2\left(n^*_u+1\right)\right]}{2}$. Equation (5.197) can alternatively be formulated as follows

$$
\begin{aligned}
u_x &= u_{x_1} + xu_{x_2} + yu_{x_3} + \cdots + x^{N_{ux}} u_{x\frac{\left(N^2_{ux}+N_{ux}+2\right)}{2}} + \cdots + y^{N_{ux}} u_{xN^*_{ux}} \\
u_y &= u_{y_1} + xu_{y_2} + yu_{y_3} + \cdots + x^{N_{uy}} u_{y\frac{\left(N^2_{uy}+N_{uy}+2\right)}{2}} + \cdots + y^{N_{uy}} u_{yN^*_{uy}} \\
u_z &= u_{z_1} + xu_{z_2} + yu_{z_3} + \cdots + x^{N_{uz}} u_{z\frac{\left(N^2_{uz}+N_{uz}+2\right)}{2}} + \cdots + y^{N_{uz}} u_{zN^*_{uz}}
\end{aligned}
\tag{5.198}
$$

the total number of unknowns for each displacement component is indicated as N^*_u. The latter along with the cross-section functions F_τ are related to N_u by means of the Pascal's triangle. The orders N_{u_x}, N_{u_y} and N_{u_z} of the expansion are arbitrary and are set as an input of the analysis. An example of a possible displacement field according to the unified formulation in Eq. (5.197) and with the expansion orders $N_{u_x}=2$, $N_{u_y}=2$ and $N_{u_z}=1$ is given as follows

$$
\begin{aligned}
u_x &= u_{x_1} + xu_{x_2} + yu_{x_3} + x^2 u_{x_4} + xyu_{x_5} + y^2 u_{x_6} \\
u_y &= u_{y_1} + xu_{y_2} + yu_{y_3} + x^2 u_{y_4} + xyu_{y_5} + y^2 u_{y_6} \\
u_z &= u_{z_1} + xu_{z_2} + yu_{z_3}
\end{aligned}
\tag{5.199}
$$

The acronyms used to define the theories are indicated as $TE_{N_{ux}N_{uy}N_{uz}}$ where TE means that a Taylor expansion has been employed and N_{u_x}, N_{u_y}, N_{u_z} are the three expansion orders used in the displacement field.

5.25.2 Plate and shell models

Therefore, following the proposed approach, the displacement models, for the displacement components and their virtual variation, can be written as:

$$\begin{aligned} u_\alpha(\alpha,\beta,z) &= F_{\tau_{u_\alpha}}(z)\, u_{\alpha\tau_{u_\alpha}}(\alpha,\beta), \quad \tau_{u_\alpha}, s_{u_\alpha} = 0,1,\ldots,N_{u_\alpha} \\ u_\beta(\alpha,\beta,z) &= F_{\tau_{u_\beta}}(z)\, u_{\beta\tau_{u_\beta}}(\alpha,\beta), \quad \tau_{u_\beta}, s_{u_\beta} = 0,1,\ldots,N_{u_\beta} \\ u_z(\alpha,\beta,z) &= F_{\tau_{u_z}}(z)\, u_{z\tau_{u_z}}(\alpha,\beta), \quad \tau_{u_z}, s_{u_z} = 0,1,\ldots,N_{u_z} \end{aligned} \tag{5.200}$$

where (α, β, z) represent a curvilinear orthogonal coordinates system in the case of shell, when plate structures are analyzed the simple substitution of the in-plan coordinates $(\alpha, \beta) = (x, y)$ can be easily applied. Equation (5.200) can be alternatively written in compact form as

$$\mathbf{u} = \mathbf{F}_\tau\, \mathbf{u}_\tau, \quad \tau = \tau_{u_\alpha}, \tau_{u_\beta}, \tau_{u_z}, \quad s = s_{u_\alpha}, s_{u_\beta}, s_{u_z}; \tag{5.201}$$

where

$$\mathbf{F}_\tau = \begin{bmatrix} F_{\tau_{u_\alpha}} & 0 & 0 \\ 0 & F_{\tau_{u_\beta}} & 0 \\ 0 & 0 & F_{\tau_{u_z}} \end{bmatrix}, \quad \mathbf{u}_\tau = \begin{bmatrix} u_{\alpha\tau_{u_\alpha}} \\ u_{\beta\tau_{u_\beta}} \\ u_{z\tau_{u_z}} \end{bmatrix} \tag{5.202}$$

$F_{\tau_{u_\alpha}}$, $F_{\tau_{u_\beta}}$, $F_{\tau_{u_z}}$ are functions of z and identify the kinematics description used for each displacement component. The functions $u_{\alpha\tau_{u_\alpha}}$, $u_{\beta\tau_{u_\beta}}$, $u_{z\tau_{u_z}}$ are displacement variables. According to Einstein's notation, the repeated subscripts τ_{u_α}, τ_{u_β}, τ_{u_z} indicate summation. Three examples of possible displacement fields are given, hereafter, according to the hierarchical ESL, ZZ and LW plate models. In the first two approaches (ESL, ZZ) the thickness functions are expressed in Taylor series, in the last one (LW) Legendre polynomials P_i have been effectively employed to satisfy the interlaminar continuity of the displacement components, then by using the expansion indexes $N_{u_\alpha} = 3$, $N_{u_\beta} = 1$, $N_{u_z} = 2$:

Equivalent Single Layer Displacement Field

$$\begin{aligned} u_\alpha &= u_{\alpha_0} + z\, u_{\alpha_1} + z^2\, u_{\alpha_2} + z^3\, u_{\alpha_3} \\ u_\beta &= u_{\beta_0} + z\, u_{\beta_1} \\ u_z &= u_{z_0} + z\, u_{z_1} + z^2\, u_{z_2} \end{aligned}$$

Equivalent Single Layer Displacement Field including Murakami's Zig-Zag Function

$$\begin{aligned} u_\alpha &= u_{\alpha_0} + z\, u_{\alpha_1} + z^2\, u_{\alpha_2} + z^3\, u_{\alpha_3} + (-1)^k\, \zeta_k\, u_{Z_{u_\alpha}} \\ u_\beta &= u_{\beta_0} + z\, u_{\beta_1} + (-1)^k\, \zeta_k\, u_{Z_{u_\beta}} \\ u_z &= u_{z_0} + z\, u_{z_1} + z^2\, u_{z_2} + (-1)^k\, \zeta_k\, u_{Z_{u_z}} \end{aligned} \tag{5.203}$$

Layer-wise Displacement Field

$$
\begin{aligned}
u_\alpha^k &= \frac{P_0 - P_1}{2}\, u_{b_{u\alpha}}^k + (P_2 - P_0)\, u_{\alpha_1}^k + (P_3 - P_1)\, u_{\alpha_2}^k + (P_4 - P_2)\, u_{\alpha_3}^k \\
&\quad + \frac{P_0 + P_1}{2}\, u_{t_{u\alpha}}^k \\
u_\beta^k &= \frac{P_0 - P_1}{2}\, u_{b_{u\beta}}^k + (P_2 - P_0)\, u_{\beta_1}^k + \frac{P_0 + P_1}{2}\, u_{t_{u\beta}}^k \\
u_z^k &= \frac{P_0 - P_1}{2}\, u_{b_{uz}}^k + (P_2 - P_0)\, u_{z_1}^k + (P_3 - P_1)\, u_{z_2}^k + \frac{P_0 + P_1}{2}\, u_{t_{uz}}^k
\end{aligned}
$$

In the classical expansion and assembly procedures carried out in the UF each 3×3 is expanded according to the thickness function F_τ which identifies the used kinematic description, the expansion order N and then if ESL or ZZ models are employed the expansion continues in terms of M and N which are the half-wave numbers used in the analysis, and the fundamental nucleus has dimension $3\,(N+1)\,P \times 3\,(N+1)\,P$. The presence of several layers does not affect, in this case, the expansion procedure, because they can be accounted for by simply adding the matrices related at each layer. On the contrary if LW modes are considered the assembly procedure on the k-layers influences the expansion and as has to be performed before the expansion on the half-wave numbers, and the final fundamental nucleus has dimension, $3\,(N\,N_l+1)\,P \times 3\,(N\,N_l+1)\,P$. In the presented hierarchical plate models the assembly procedure undergoes to a significant change (see Ref. [41]). Both for the ESL, ZZ and for the LW models the logical-consequential steps change with respect to the classical models, according to the thickness functions $F_{\tau_{u\alpha}}$, $F_{\tau_{u\beta}}$, $F_{\tau_{uz}}$ and the expansion indexes N_{u_α}, N_{u_β} and N_{u_z} used. In the case of ESL and ZZ models the primary fundamental nucleus has dimension

$$
\begin{aligned}
\mathcal{D}_{\mathrm{ESL}} = &\left[(N_{u_\alpha}+1)+(N_{u_\beta}+1)+(N_{u_z}+1)\right] P \times \\
&\left[(N_{u_\alpha}+1)+(N_{u_\beta}+1)+(N_{u_z}+1)\right] P
\end{aligned}
\tag{5.204}
$$

and like in the case of classical models the expansion is independent of the layers number, in the case of LW models the fundamental primary nucleus dimension is

$$
\begin{aligned}
\mathcal{D}_{\mathrm{LW}} = &\left[(N_{u_\alpha}\,N_l+1)+(N_{u_\beta}\,N_l+1)+(N_{u_z}\,N_l+1)\right] P \times \\
&\left[(N_{u_\alpha}\,N_l+1)+(N_{u_\beta}\,N_l+1)+(N_{u_z}\,N_l+1)\right] P
\end{aligned}
\tag{5.205}
$$

5.26 DISCUSSION ON POSSIBLE BEST BEAM, PLATE AND SHELL DIAGRAMS

The advanced structural models introduced through the unified formulation are generally based on full expansions of a given order; this means that all the displacement variables of a given order theory are employed. The contribution of each term of a theory varies depending on the structural problem and some variables can be more important than others in predicting the mechanical behaviour of a structure. Moreover, some terms might not have any influence, since their absence does not corrupt the accuracy of the solution.

The unified formulation can be exploited to investigate the influence of each variable of a refined theory on the solution. This capability is hereafter referred to as the mixed axiomatic-asymptotic approach (MAAA). MAAA can be considered a powerful tool to build:

- Reduced refined theories which have the same accuracies as full expansion models but fewer unknown variables.
- The best theory diagram (BTD), where the accuracy of a model can be evaluated against the number of variables.

This chapter describes the MAAA, gives general guidelines to determine the most adequate model for a given problem and introduces the BTD. Numerical examples are presented to evaluate reduced models and to highlight the influence of characteristic parameters, such as the slenderness ratio, the thickness and the loading conditions.

5.27 THE MIXED AXIOMATIC/ASYMPTOTIC METHOD

Theories of structures can be built on the basis of different approaches, in which the common target of each approach is to evaluate the minimum number of unknown variables in order to solve a given problem against a given accuracy. Axiomatic and asymptotic methods are the main tools to build structural models. More details on the development of the MAAA can be found in Ref. [42]. The choice of the name is due to the fact that MAAA is capable of obtaining asymptotic-like results, starting from axiomatic-like hypothesis. The influence of a variable can in fact be investigated, against the variation of various parameters (e.g. thickness, boundary conditions, etc.) using the MAAA and this analysis is conducted starting from axiomatic-like hypotheses. The smallest number of variables required to fulfil a given accuracy requirement can thus be determined. The basic

Table 5.7 Positions of the displacement variables within the table layout. This table shows the layout notation that defines the position of each displacement variable in the case of a 1D second-order model, $N=2$

$N=0$	$N=1$		$N=2$		
u_{x1}	u_{x2}	u_{x3}	u_{x4}	u_{x5}	u_{x6}
u_{y1}	u_{y2}	u_{y3}	u_{y4}	u_{y5}	u_{y6}
u_{z1}	u_{z2}	u_{z3}	u_{z4}	u_{z5}	u_{z6}

Table 5.8 Symbolic representation of the reduced kinematic model with u_{y_3} deactivated. This table shows the graphic representation of a second-order 1D model that has u_{y_3} deactivated

▲	▲	▲	▲	▲	▲
▲	▲	△	▲	▲	▲
▲	▲	▲	▲	▲	▲

procedure is described hereafter. This procedure is exactly the same for 1D and 2D models. A graphic notation is introduced to improve the readability of the results. Table 5.7 shows the locations held by each second-order 1D model term within the tabular layout. The first column presents the constant terms ($N=0$), the second and third columns the linear terms ($N=1$), and the last three columns show the parabolic terms ($N=2$). Each term can be activated (black) or deactivated (white), as shown in Table 5.8. On the basis of the adopted notation, the 1D model given in Table 5.8 refers to the following cross-section displacement field:

$$\begin{aligned}
u_x &= u_{x_1} + x\, u_{x_2} + z\, u_{x_3} + x^2\, u_{x_4} + xz\, u_{x_5} + z^2\, u_{x_6} \\
u_y &= u_{y_1} + x\, u_{y_2} + \qquad\qquad + x^2\, u_{y_4} + xz\, u_{y_5} + z^2\, u_{y_6} \\
u_z &= u_{z_1} + x\, u_{z_2} + z\, u_{z_3} + x^2\, u_{z_4} + xz\, u_{z_5} + z^2\, u_{z_6}
\end{aligned} \tag{5.206}$$

The 2D model notation is given in Tables 5.9 and 5.10. The 2D model given in Table 5.10 refers to the following displacement field:

$$\begin{aligned}
u_x &= u_{x_1} + z\, u_{x_2} + z^2\, u_{x_3} + z^3\, u_{x_4} + z^4\, u_{x_5} \\
u_y &= u_{y_1} + z\, u_{y_2} + \qquad\qquad + z^3\, u_{y_4} + z^4\, u_{y_5} \\
u_z &= u_{z_1} + z\, u_{z_2} + z^2\, u_{z_3} + z^3\, u_{z_4} + z^4\, u_{z_5}
\end{aligned} \tag{5.207}$$

The reduced models can be obtained by opportunely rearranging the rows and columns of the FE matrices or through penalty techniques. The latter is the approach that has been adopted in this book.

Table 5.9 Locations of the displacement variables within the table layout for 2D models. This table shows the layout notation that defines the position of each displacement variable in the case of a fourth-order model for plates, $N = 4$

$N=0$	$N=1$	$N=2$	$N=3$	$N=4$
u_{x1}	u_{x2}	u_{x3}	u_{x4}	u_{x5}
u_{y1}	u_{y2}	u_{y3}	u_{y4}	u_{y5}
u_{z1}	u_{z2}	u_{z3}	u_{z4}	u_{z5}

Table 5.10 Symbolic representation of the reduced kinematic model with u_{y_3} deactivated, 2D models. This table shows the graphic representation of a fourth-order 2D model that has u_{y_3} deactivated

▲	▲	▲	▲	▲
▲	▲	△	▲	▲
▲	▲	▲	▲	▲

Table 5.11 Reduced fourth-order model for the torsional analysis of a thin-walled cylinder. This table presents a reduced fourth-order 1D model where the u_{z_3} variable is deactivated

●	●	●	●	●	●	●	●	●	●	●	●	●	●	●
●	●	●	●	●	●	●	●	●	●	●	●	●	●	●
●	●	○	●	●	●	●	●	●	●	●	●	●	●	●

5.27.1 Example

Let us evaluate the influence of the u_{z_3} term of a fourth-order 1D model in the torsional analysis of a thin-walled cylinder with L/d equal to 10. The torsion is investigated through the application of two concentrated forces (P_z) at $[\pm\frac{d}{2}, L, 0]$. Table 5.11 shows the reduced 1D model considered. The influence of the term is evaluated by means of percentage variations

$$\delta_u = \frac{u}{u_{N=4}} \times 100, \quad \delta_\sigma = \frac{\sigma}{\sigma_{N=4}} \times 100 \tag{5.208}$$

computed with respect to the full fourth-order ($N = 4$). Table 5.12 shows the results referring to the three displacement variables computed at $[\frac{d}{2}, L, 0]$. It can be seen that u_{z_3} influences u_x and u_z, but it does not affect u_y. Example 5.27.1 illustrates the procedure used to investigate the effect of a single term on different output variables. This procedure is used to determine all the inactive terms of a refined model for a given structural problem.

Table 5.12 Effect of the absence of u_{z_3} on different output variables. This table presents the effects caused by the absence of u_{z_3} on different output variables

δ_{u_x} %	δ_{u_y} %	δ_{u_z} %
96.9	100.0	118.8

Table 5.13 Symbols that indicate the loading cases and the presence of a displacement variable. This table shows the symbols that are used to indicate the presence of a displacement variable and the considered loading case in the static analysis of beams

Loading case	Active term	Inactive term
Bending	▲	△
Torsion	●	○
Axial	▼	▽

5.28 STATIC ANALYSIS OF BEAMS

In this section, MAAA is exploited to determine reduced 1D models for the static analysis of cantilever beams. For the sake of clarity, different symbols, which are shown in Table 5.13, are adopted to distinguish the loading cases.

5.28.1 Influence of the loading conditions

Bending, torsion, and traction load cases are considered for a compact square cantilever beam. An $N = 4$ 1D model is considered as a reference solution in order to evaluate the effectiveness of each displacement variable in detecting the displacement components.

Bending is considered as the first loading case. Table 5.14 shows the set of terms that are needed to detect fourth-order accuracy, it can be seen that 11 out of 45 terms are needed and the explicit expression of the 1D model is

$$\begin{aligned} u_x &= xz\, u_{x_5} + x^3 z\, u_{x_{12}} + xz^3\, u_{x_{14}} \\ u_y &= z\, u_{y_3} + x^2 z\, u_{y_8} + z^3\, u_{y_{10}} \\ u_z &= u_{z_1} + z^2\, u_{z_6} + x^4\, u_{z_{11}} + x^2 z^2\, u_{z_{13}} + z^4\, u_{z_{15}} \end{aligned} \tag{5.209}$$

The second loading deals with torsion. The related reduced model is presented in Table 5.15. In this case, 9 out of 45 terms are needed and the 1D model is given by

$$\begin{aligned} u_x &= z\, u_{x_3} + x^2 z\, u_{x_8} + z^3\, u_{x_{10}} \\ u_y &= xz\, u_{y_5} + x^3 z\, u_{y_{11}} + xz^3\, u_{y_{14}} \\ u_z &= x\, u_{z_2} + x^3\, u_{z_7} + xz^2\, u_{z_9} \end{aligned} \tag{5.210}$$

Table 5.14 Set of active displacement variables for the bending analysis of a square cross-section beam. This table presents the set of active terms to detect the bending behaviour of a square cross-section beam, 11 out of 45 terms are needed

$M_{eff}/M = 11/45$

△	△	△	△	▲	△	△	△	△	△	△	▲	△	▲	△
△	△	▲	△	△	△	△	▲	△	▲	△	△	△	△	△
▲	△	△	△	△	▲	△	△	△	△	▲	△	▲	△	▲

Table 5.15 Set of active displacement variables for the torsional analysis of a square cross-section beam. This table presents the set of active terms to detect the torsional behaviour of a square cross-section beam, 9 out of 45 terms are needed

$M_{eff}/M = 9/45$

○	○	●	○	○	○	○	●	○	●	○	○	○	○	○
○	○	○	○	●	○	○	○	○	○	○	●	○	●	○
○	●	○	○	○	○	●	○	●	○	○	○	○	○	○

Table 5.16 Set of active displacement variables for the traction analysis of a square cross-section beam. This table presents the set of active terms to detect the traction behaviour of a square cross-section beam, 12 out of 45 terms are needed

$M_{eff}/M = 12/45$

▽	▼	▽	▽	▽	▽	▼	▽	▼	▽	▽	▽	▽	▽	▽
▼	▽	▽	▼	▽	▼	▽	▽	▽	▽	▼	▽	▼	▽	▼
▽	▽	▼	▽	▽	▽	▽	▼	▽	▼	▽	▽	▽	▽	▽

Traction is considered as a third loading case. The results are given in Table 5.16 and the reduced model is

$$\begin{aligned} u_x &= x\, u_{x_2} + x^3\, u_{x_7} + x z^2\, u_{x_9} \\ u_y &= u_{y_1} + x^2\, u_{y_4} + z^2\, u_{y_6} + x^4\, u_{y_{11}} + x^2 z^2\, u_{y_{13}} + z^4\, u_{y_{15}} \\ u_z &= z\, u_{z_3} + x^2 z\, u_{z_8} + z^3\, u_{z_{10}} \end{aligned} \tag{5.211}$$

It is important to underline that the 1D models in Eqs. (5.209), (5.210) and (5.211) are substantially different from each other. This means that each loading case needs its own reduced beam model. The combined reduced 1D model necessary to detect the fourth-order solution for the bending, torsion and traction loads is presented in Table 5.17.

5.28.2 Influence of the cross-section

The cross-section geometry is another important parameter that is necessary to determine a refined 1D model. General guidelines state that

Table 5.17 Combined set of active displacement variables for the bending, torsional and traction analysis of a square cross-section beam. This table presents the combined set of active terms to detect the bending, torsional and traction behaviour of a square cross-section beam, 32 out of 45 terms are needed

$M_{eff}/M = 32/45$

	▼	●		▲		▼	●	▼	●		▲		▲	
▼		▲	▼	●	▼		▲		▲	▼	●	▼	●	▼
▲	●	▼			▲	●	▼	●	▼	▲		▲		▲

Table 5.18 Set of active displacement variables for the torsional analysis of a thin-walled annular beam. This table presents the set of active terms to detect the torsional behaviour of a thin-walled annular beam, 21 out of 45 terms are needed

$M_{eff}/M = 21/45$

▽	▼	▼	▽	▽	▽	▼	▼	▼	▼	▽	▽	▽	▽	▽
▼	▽	▽	▼	▼	▼	▽	▽	▽	▽	▼	▼	▼	▼	▼
▽	▼	▼	▽	▽	▽	▼	▼	▼	▼	▽	▽	▽	▽	▽

compact beams need fewer variables than thin-walled beams. Other key aspects concern the symmetry/asymmetry and the presence of closed/open sections. Two cross-section geometries are investigated hereafter: annular and airfoil-shaped. In both cases, a torsional load is applied to the free-tip. Table 5.18 shows the reduced 1D model for the annular cross-section that is equivalent to a full fourth-order model, this model is

$$
\begin{aligned}
u_x &= x\, u_{x_2} + z\, u_{x_3} + x^3\, u_{x_7} + x^2 z\, u_{x_8} + x z^2\, u_{x_9} + z^3\, u_{x_{10}} \\
u_y &= u_{y_1} + x^2\, u_{y_4} + xz\, u_{y_5} + z^2\, u_{y_6} + x^4\, u_{y_{11}} + x^3 z\, u_{y_{12}} + \\
&\quad + x^2 z^2\, u_{y_{13}} + x z^3\, u_{y_{14}} + z^4\, u_{y_{15}} \\
u_z &= x\, u_{z_2} + z\, u_{z_3} + x^3\, u_{z_7} + x^2 z\, u_{z_8} + x z^2\, u_{z_9} + z^3\, u_{z_{10}}
\end{aligned}
\tag{5.212}
$$

Table 5.19 shows the equivalent result for an airfoil-shaped cantilever beam. In this case, all the 45 displacement variables are needed, that is, each term of the fourth-order expansion plays a role in detecting the mechanical behaviour of the structure.

5.28.3 Reduced models vs accuracy

Reduced models that were equivalent to full fourth-order expansions were considered in the previous sections. Another important option offered by the present formulation is that it is possible to choose the accuracy range of the refined model. In other words, a reduced model that offers a given

Table 5.19 Set of active displacement variables for the torsional analysis of an airfoil-shaped beam. This table presents the set of active terms to detect the torsional behaviour of an airfoil-shaped beam, 45 out of 45 terms are needed

$M_{eff}/M = 45/45$

•	•	•	•	•	•	•	•	•	•	•	•	•	•	•
•	•	•	•	•	•	•	•	•	•	•	•	•	•	•
•	•	•	•	•	•	•	•	•	•	•	•	•	•	•

accuracy can be determined. The analysis is conducted on an airfoil-shaped cantilever beam under torsional and bending loads. The input parameter is the error with respect to the fourth-order solution:

$$\overline{\delta_u} = \left\| \frac{u - u_{N=4}}{u_{N=4}} \right\| \times 100, \tag{5.213}$$

For instance, $\overline{\delta_u} = 0$ implies that a full $N = 4$ solution is necessary. Table 5.20 shows the torsion-related results, while the bending case is addressed in Table 5.21. In both cases, significant reductions in the total number of variables can be observed for the totally different reduced models requested for the torsion and the bending loading cases. These results confirm that the development of reduced higher-order models is decidedly problem-dependent. The possibility of dealing with full arbitrary order models that is offered by the present unified formulation, is fundamental to analyze structures of engineering interest in which different loads, geometries and BCs are usually present simultaneously. It should be underlined that the present analysis has only isotropic materials. The case of composite materials, other important parameters such as the orthotropic ratio, and the stacking sequence, can be expected to play the same role in determining the reduced models as those seen above.

5.29 MODAL ANALYSIS OF BEAMS

In this section, MAAA is applied to the modal analysis of beams in order to build reduced 1D models that are able to detect the natural modes and frequencies of beams.

The accuracy of a reduced model is evaluated on the natural frequency through E_f defined as

$$E_f = \left\| \frac{f - f_{\text{ref}}}{f_{\text{ref}}} \right\| \times 100 \tag{5.214}$$

Table 5.20 Set of active displacement variables that offers a given accuracy for an airfoil-shaped beam under torsion. This table presents the set of active terms to detect the torsional behaviour of an airfoil-shaped beam with a given accuracy

$\overline{\delta_u} = 0\,\%, M_{eff}/M = 45/45$

●	●	●	●	●	●	●	●	●	●	●	●	●	●	●
●	●	●	●	●	●	●	●	●	●	●	●	●	●	●
●	●	●	●	●	●	●	●	●	●	●	●	●	●	●

$\overline{\delta_u} \leq 15\,\%, M_{eff}/M = 42/45$

●	●	●	●	●	●	●	●	●	●	●	●	○	○	●
●	●	●	●	○	●	●	●	●	●	●	●	●	●	●
●	●	●	●	●	●	●	●	●	●	●	●	●	●	●

$\overline{\delta_u} \leq 35\,\%, M_{eff}/M = 25/45$

●	●	●	○	●	○	○	●	○	●	○	●	○	○	○
●	●	○	●	○	●	●	○	○	○	●	○	○	○	○
●	●	●	●	●	○	●	●	●	●	●	●	○	●	○

Table 5.21 Set of active displacement variables that offers a given accuracy requirement for an airfoil-shaped beam under bending. This table presents the set of active terms to detect the bending behaviour of an airfoil-shaped beam with a given accuracy

$\overline{\delta_u} = 0\,\%, M_{eff}/M = 45/45$

▲	▲	▲	▲	▲	▲	▲	▲	▲	▲	▲	▲	▲	▲	▲
▲	▲	▲	▲	▲	▲	▲	▲	▲	▲	▲	▲	▲	▲	▲
▲	▲	▲	▲	▲	▲	▲	▲	▲	▲	▲	▲	▲	▲	▲

$\overline{\delta_u} \leq 15\,\%, M_{eff}/M = 23/45$

▲	▲	▲	▲	▲	△	△	▲	△	△	▲	▲	△	△	△
▲	▲	△	▲	▲	△	△	△	△	△	▲	△	△	△	△
▲	▲	▲	▲	▲	△	▲	△	▲	▲	△	△	▲	▲	△

$\overline{\delta_u} \leq 35\,\%, M_{eff}/M = 9/45$

▲	△	▲	△	▲	△	△	△	△	△	△	△	△	△	△
▲	▲	△	△	△	△	△	△	△	△	△	△	△	△	△
△	▲	▲	△	▲	△	△	△	△	▲	△	△	△	△	△

where f_{ref} denotes the frequency computed by means of the reference model ($N = 4$).

5.29.1 Influence of the cross-section

Reduced 1D models for the modal analysis of beams with various cross-sections are dealt with in this section using the MAAA. Four geometries

Table 5.22 Reduced models for the modal analysis of a cantilever rectangular cross-section beam. This table presents the reduced models required to detect the first bending and torsional modes of a cantilever rectangular cross-section beam

1st bending z, $M_{eff}/M=7/45$														
▽	▽	▽	▽	▼	▽	▽	▽	▽	▽	▽	▼	▽	▽	▽
▽	▽	▼	▽	▽	▽	▽	▽	▽	▽	▽	▽	▽	▽	▽
▼	▽	▽	▼	▽	▼	▽	▽	▽	▽	▼	▽	▽	▽	▽
1st bending x, $M_{eff}/M=5/45$														
▼	▽	▽	▼	▽	▽	▽	▽	▽	▽	▽	▽	▽	▽	▽
▽	▼	▽	▽	▽	▽	▼	▽	▽	▽	▽	▽	▽	▽	▽
▽	▽	▽	▽	▼	▽	▽	▽	▽	▽	▽	▽	▽	▽	▽
1st torsional, $M_{eff}/M=9/45$														
▽	▽	▼	▽	▽	▽	▽	▼	▽	▼	▽	▽	▽	▽	▽
▽	▽	▽	▽	▼	▽	▽	▽	▽	▽	▽	▼	▽	▼	▽
▽	▼	▽	▽	▽	▽	▼	▽	▼	▽	▽	▽	▽	▽	▽

are considered: rectangular, rectangular thin-walled, C-shaped and annular. Three natural modes are considered to evaluate the influence of each generalized variables on the solution:

1. The frequency related to the first bending mode along the z-direction, hereafter referred to as *Bending z*.
2. The frequency related to the first bending mode along the x-direction, hereafter referred to as *Bending x*.
3. The frequency related to the first torsional mode, hereafter referred to as *Torsional*.

Table 5.22 shows the reduced models for the rectangular beam. These models were obtained by retaining the active terms for a given mode. The explicit expression of the beam model needed to detect the first bending mode along z is

$$
\begin{aligned}
u_x &= xz\, u_{x_5} + x^3 z\, u_{x_{12}} \\
u_y &= z\, u_{y_3} + x\, u_{y_2} + xz\, u_{y_5} \\
u_z &= u_{z_1} + x^2\, u_{z_4} + z^2\, u_{z_6} + x^4\, u_{z_{11}}
\end{aligned}
\qquad (5.215)
$$

This model requires only seven terms out of 45, that is, seven degrees of freedom per node. It should, however, be underlined that the reduced model related to the bending along x requires fewer terms than those in Eq. (5.215) because of the higher flexibility of the rectangular cross-section along z. The torsional mode is the most cumbersome for this cross-section

Table 5.23 Reduced models for the modal analysis of a cantilever rectangular thin-walled cross-section beam. This table presents the reduced models required to detect the first bending and torsional modes of a cantilever rectangular thin-walled cross-section beam

1st bending z, $M_{eff}/M = 10/45$														
○	○	○	○	●	○	○	○	○	○	○	●	○	●	○
○	○	●	○	○	○	○	●	○	●	○	○	○	○	○
●	○	○	●	○	●	○	○	○	○	●	○	○	○	○
1st bending x, $M_{eff}/M = 6/45$														
●	○	○	●	○	○	○	○	○	○	○	○	○	○	●
○	●	○	○	○	○	●	○	○	○	○	○	○	○	○
○	○	○	○	●	○	○	○	○	○	○	○	○	○	○
1st torsional, $M_{eff}/M = 8/45$														
○	○	●	○	○	○	○	●	○	●	○	○	○	○	○
○	○	○	○	●	○	○	○	○	○	○	●	○	●	○
○	●	○	○	○	○	○	○	●	○	○	○	○	○	○

configuration,

$$\begin{aligned} u_x &= z\, u_{x_3} + x^2 z\, u_{x_8} + z^3\, u_{x10} \\ u_y &= xz\, u_{y_5} + x^3 z\, u_{y12} + xz^3\, u_{y14} \\ u_z &= x\, u_{z_2} + x^3\, u_{z_7} + xz^2\, u_{z_9} \end{aligned} \tag{5.216}$$

The beam model that is necessary to detect all these three modes is

$$\begin{aligned} u_x &= u_{x_1} + z\, u_{x_3} + x^2\, u_{x_4} + xz\, u_{x_5} + x^2 z\, u_{x_8} + z^3\, u_{x10} + x^3 z\, u_{x12} \\ u_y &= x\, u_{y_2} + z\, u_{y_3} + xz\, u_{y_5} + x^3\, u_{y_7} + x^3 z\, u_{y12} + xz^3\, u_{y14} \\ u_z &= u_{z_1} + x\, u_{z_2} + x^2\, u_{z_4} + xz\, u_{z_5} + z^2\, u_{z_6} + x^3\, u_{z_7} + xz^2\, u_{z_9} + x^4\, u_{z11} \end{aligned} \tag{5.217}$$

Tables 5.23, 5.24 and 5.25 present the reduced models for the remaining three geometries.The symmetry of the annular cross-section makes the reduced beam models for bending symmetric. Table 5.26 shows the reduced models needed to detect all the considered modes. For instance, the beam model needed to detect the first bending and torsional modes of an annular cross-section beam is

$$\begin{aligned} u_x &= u_{x_1} + z\, u_{x_3} + x^2\, u_{x_4} + xz\, u_{x_5} + z^2\, u_{x_6} + x^3 z\, u_{x12} + xz^3\, u_{x14} \\ u_y &= x\, u_{y_2} + z\, u_{y_3} + x^3\, u_{y_7} + x^2 z\, u_{y_8} + xz^2\, u_{y_9} + z^3\, u_{y10} \\ u_z &= u_{z_1} + x\, u_{z_2} + x^2\, u_{z_4} + xz\, u_{z_5} + z^2\, u_{z_6} + x^3 z\, u_{z12} + xz^3\, u_{z14} \end{aligned} \tag{5.218}$$

These results suggest:

Table 5.24 Reduced models for the modal analysis of a cantilever C-shaped cross-section beam. This table presents the reduced models required to detect the first bending and torsional modes of a cantilever C-shaped cross-section beam

1st bending z, $M_{eff}/M = 20/45$														
△	△	▲	△	▲	△	△	▲	△	▲	△	▲	△	▲	△
△	△	▲	△	▲	△	△	▲	△	▲	△	▲	△	▲	△
▲	▲	△	▲	△	▲	▲	△	▲	△	▲	△	▲	△	△
1st bending x, $M_{eff}/M = 17/45$														
▲	▲	△	▲	△	▲	△	△	△	△	△	△	△	△	▲
▲	▲	△	▲	△	▲	▲	△	△	△	▲	△	▲	△	▲
△	△	▲	△	▲	△	△	△	△	▲	△	△	△	▲	△
1st torsional, $M_{eff}/M = 19/45$														
△	△	▲	△	▲	△	△	▲	△	▲	△	▲	△	▲	△
△	△	▲	△	▲	△	△	▲	△	▲	△	▲	△	▲	△
▲	▲	△	▲	△	▲	▲	△	▲	△	▲	△	△	△	△

Table 5.25 Reduced models for the modal analysis of a cantilever annular cross-section beam. This table presents the reduced models required to detect the first bending and torsional modes of a cantilever annular cross-section beam

1st bending z, $M_{eff}/M = 9/45$														
□	□	□	□	■	□	□	□	□	□	□	■	□	■	□
□	□	■	□	□	□	□	■	□	■	□	□	□	□	□
■	□	□	■	□	■	□	□	□	□	□	□	□	□	□
1st bending x, $M_{eff}/M = 9/45$														
■	□	□	■	□	■	□	□	□	□	□	□	□	□	□
□	■	□	□	□	□	■	□	■	□	□	□	□	□	□
□	□	□	□	■	□	□	□	□	□	□	■	□	■	□
Torsional, $M_{eff}/M = 3/45$														
□	□	■	□	□	□	□	□	□	□	□	□	□	□	□
□	□	□	□	□	□	□	□	□	□	□	□	□	□	□
■	■	□	□	□	□	□	□	□	□	□	□	□	□	□

1. Different sets of displacement variables are needed to detect different modes, as previously stated.
2. Thin walls and the asymmetry of the cross-section play a fundamental role in determining the number of terms needed to detect a given mode. The asymmetry, in particular, seems to be of primary importance.

Table 5.26 Combined reduced models for the modal analysis of cantilever beams with various cross-sections. This table presents the reduced models required to detect the first bending and torsional modes of cantilever beams with various cross-sections

Rectangular, $M_{eff}/M = 21/45$

▼	▽	▼	▼	▼	▽	▽	▼	▽	▼	▽	▼	▽	▽	▽
▽	▼	▼	▽	▼	▽	▼	▽	▽	▽	▽	▼	▽	▼	▽
▼	▼	▽	▼	▼	▼	▼	▽	▼	▽	▼	▽	▽	▽	▽

Rectangular thin-walled, $M_{eff}/M = 24/45$

●	○	●	●	●	○	○	●	○	●	○	●	○	●	●
○	●	●	○	●	○	●	●	○	●	○	●	○	●	○
●	●	○	●	●	●	○	○	●	○	●	○	○	○	○

C-shaped, $M_{eff}/M = 37/45$

▲	▲	▲	▲	▲	▲	△	▲	△	▲	△	▲	△	▲	▲
▲	▲	▲	▲	▲	▲	▲	▲	△	▲	▲	▲	▲	▲	▲
▲	▲	▲	▲	▲	▲	▲	△	▲	▲	▲	△	▲	▲	△

Annular, $M_{eff}/M = 20/45$

■	□	■	■	■	■	□	□	□	□	□	■	□	■	□
□	■	■	□	□	□	■	■	■	■	□	□	□	□	□
■	■	□	■	■	■	□	□	□	□	□	■	□	■	□

3. As a general guideline, it can be stated that if asymmetric cross-sections are considered, full models should be adopted, whereas symmetric geometries should be analyzed by means of reduced models, since great reductions in the computational costs are possible.

5.29.2 Influence of the boundary conditions

The influence of the boundary conditions is investigated in this section through the MAAA. A rectangular beam is considered and three different boundary conditions are considered: clamped at both ends, hinged at both ends and simply-supported.

Tables 5.27, 5.28 and 5.29 show the reduced models for the three modes considered for all the boundary conditions under investigation. Table 5.30 presents the combined models for all the modes. The following comments can be made:

1. The boundary conditions play a significant role in the construction of reduced models. Their effect is equivalent to that observed for the cross-section geometry.

Table 5.27 Reduced models for the modal analysis of a clamped-clamped rectangular cross-section beam. This table presents the reduced models required to detect the first bending and torsional modes of a clamped-clamped rectangular cross-section beam

1st bending z, $M_{eff}/M = 6/45$

○	○	○	○	○	●	○	○	○	○	○	○	○	○	○
○	○	●	○	○	○	○	○	○	●	○	○	○	○	○
●	○	○	●	○	●	○	○	○	○	○	○	○	○	○

1st bending x, $M_{eff}/M = 5/45$

●	○	○	●	○	●	○	○	○	○	○	○	○	○	○
○	●	○	○	○	○	○	○	○	○	○	○	○	○	○
○	○	○	○	●	○	○	○	○	○	○	○	○	○	○

1st torsional, $M_{eff}/M = 8/45$

●	○	●	○	○	○	○	○	○	○	○	○	○	○	○
○	●	●	○	○	○	○	○	○	○	○	●	○	●	○
●	●	○	○	○	○	○	○	○	○	○	○	○	○	○

Table 5.28 Reduced models for the modal analysis of a hinged-hinged rectangular cross-section beam. This table presents the reduced models required to detect the first bending and torsional modes of a hinged-hinged rectangular cross-section beam

1st bending z, $M_{eff}/M = 5/45$

△	△	△	△	△	▲	△	△	△	△	△	△	△	△	△
△	△	▲	△	△	△	△	△	△	△	△	△	△	△	△
▲	△	△	▲	△	▲	△	△	△	△	△	△	△	△	△

1st bending x, $M_{eff}/M = 4/45$

▲	△	△	▲	△	△	△	△	△	△	△	△	△	△	△
△	▲	△	△	△	△	△	△	△	△	△	△	△	△	△
△	△	△	△	▲	△	△	△	△	△	△	△	△	△	△

1st torsional, $M_{eff}/M = 6/45$

△	△	▲	△	△	△	△	△	△	△	△	△	△	△	△
△	△	△	△	▲	△	△	△	△	△	△	▲	△	▲	△
△	▲	△	△	△	△	△	△	▲	△	△	△	△	△	△

2. Clamped-clamped and hinged-hinged conditions instead require similar models and the simply-supported condition needs the most cumbersome 1D model.
3. In general, significant reductions in the number of generalized variables can be obtained for all the boundary conditions considered.

Table 5.29 Reduced models for the modal analysis of a simply-supported rectangular cross-section beam. This table presents the reduced models required to detect the first bending and torsional modes of a simply-supported rectangular cross-section beam

1st bending z, $M_{eff}/M = 13/45$

■	■	□	■	□	■	□	□	□	□	□	□	□	□	□
■	■	□	■	□	■	■	□	■	□	□	□	■	□	□
□	□	■	□	■	□	□	□	□	□	□	□	□	□	□

1st bending x, $M_{eff}/M = 8/45$

□	□	■	□	■	□	□	□	□	□	□	□	□	□	□
□	□	■	□	□	□	□	□	□	■	□	□	□	□	□
■	■	□	■	□	■	□	□	□	□	□	□	□	□	□

1st torsional, $M_{eff}/M = 9/45$

□	□	■	□	□	□	□	□	□	□	□	□	□	□	□
□	□	■	□	■	□	□	■	□	■	□	■	□	■	□
■	■	□	□	□	□	□	□	□	□	□	□	□	□	□

Table 5.30 Combined reduced models for the modal analysis of cantilever beams with various boundary conditions. This table presents the reduced models required to detect the first bending and torsional modes of cantilever beams with various boundary conditions

Clamped-clamped, $M_{eff}/M = 14/45$

●	○	●	●	○	●	○	○	○	○	○	○	○	○	○
○	●	●	○	○	○	○	○	○	●	○	●	○	●	○
●	●	○	●	●	●	○	○	○	○	○	○	○	○	○

Hinged-hinged, $M_{eff}/M = 15/45$

▲	△	▲	▲	△	▲	△	△	△	△	△	△	△	△	△
△	▲	▲	△	▲	△	△	△	△	△	△	▲	△	▲	△
▲	▲	△	▲	▲	▲	△	△	▲	△	△	△	△	△	△

Simply-supported, $M_{eff}/M = 25/45$

■	■	■	■	■	■	□	□	□	□	□	□	□	□	□
■	■	■	■	■	■	■	■	■	■	□	■	■	■	□
■	■	■	■	■	■	□	□	□	□	□	□	□	□	□

5.30 STATIC ANALYSIS OF PLATES AND SHELLS

A square plate, is analyzed in this section through the MAAA. The following displacement and stress variables have been considered to build reduced 2D models: u_z, σ_{xx}, σ_{yy} and σ_{zz} at $[a/2, b/2, 0]$, σ_{xz} at $[0, b/2, h/2]$ and σ_{yz} at $[a/2, 0, h/2]$. Four-node plate elements have been used with a uniform mesh of 15×15 elements.

The accuracy of the reduced models obtained through the MAAA is compared with results from other theories that can be obtained as particular cases of the present unified formulation. The already mentioned classical plate theory (CPT, by Kirchhoff) and the first-order shear deformation theory (FSDT, by Reissner and Mindlin) are considered, together with three other refined models. The first one, hereafter referred to as Pandya, was developed in Ref. [45],

$$\begin{aligned} u_x &= u_{x1} + z\,u_{x2} + z^2\,u_{x3} + z^3\,u_{x4} \\ u_y &= u_{y1} + z\,u_{y2} + z^2\,u_{y3} + z^3\,u_{y4} \\ u_z &= u_{z1} \end{aligned} \tag{5.219}$$

The second one, hereafter referred to as Kant-1, was developed in Ref. [44],

$$\begin{aligned} u_x &= u_{x1} + z\,u_{x2} + z^2\,u_{x3} + z^3\,u_{x4} \\ u_y &= u_{y1} + z\,u_{y2} + z^2\,u_{y3} + z^3\,u_{y4} \\ u_z &= u_{z1} + z\,u_{z2} + z^2\,u_{z3} + z^3\,u_{z4} \end{aligned} \tag{5.220}$$

The third one, hereafter referred to as Kant-2, was developed in Ref. [43],

$$\begin{aligned} u_x &= z\,u_{x2} + z^3\,u_{x4} \\ u_y &= z\,u_{y2} + z^3\,u_{y4} \\ u_z &= u_{z1} + z^2\,u_{z3} \end{aligned} \tag{5.221}$$

5.30.1 Influence of the boundary conditions

Reduced 2D models for various boundary conditions are given in this section.The following boundary conditions are considered:

1. *ssss*, Four simply-supported edges.
2. *cfcf*, Two clamped and two free edges.
3. *cccc*, Four clamped edges.

Table 5.31 shows the symbols that have been adopted to refer to each BCs set. A bi-sinusoidal transverse distributed load – $p_z = \bar{p}_z sin\left(\frac{x}{a}\right) cos\left(\frac{y}{b}\right)$ – is applied at the top surface. An $N = 4$ solution is used to check the effectiveness of each term. $N = 4$ can, in fact, provide 3D-like solutions for this class of problems.

Table 5.32 presents the reduced 2D models required to detect the 3D-like solution for a moderately thick plate ($a/h = 10$). Each row refers to a displacement or stress variable. Each column considers a different set of boundary conditions. M_{eff} indicates the number of terms (i.e. the number

Table 5.31 Symbols adopted to distinguish from various plate boundary conditions along the four edges. "c" indicates clamped, "s" simply-supported and "f" free

Active term	Inactive term	
▲	△	*ssss*
■	□	*cfcf*
▼	▽	*cccc*

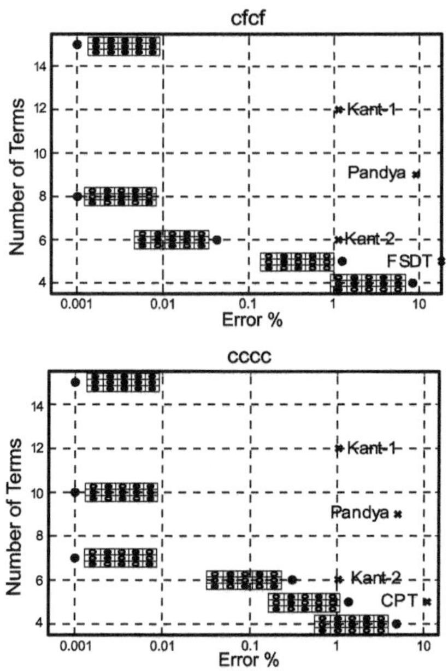

Figure 5.46 Number of terms vs error for different boundary conditions, σ_{xz}, moderately thick plate.

of DOFs per node) of the models that are equivalent to the fourth-order one. The last row shows the reduced models needed to detect all the considered outputs. The latter combined models are used to build Table 5.33, which shows a comparison with the accuracies obtained with CPT, FSDT and Kant-2 plate models.

It is possible to determine other theories that provide solutions with a certain degree of error, compared to the full fourth-order one. Figure 5.46 shows the number of terms needed to detect σ_{xz} with a given error. The corresponding plate models are also indicated. The analyses that have been carried out suggest:

Table 5.32 Reduced 2D models for a moderately thick plate with various boundary conditions. This table presents the reduced models required to detect displacement and stress variables for a plate under various boundary conditions

ssss					*cfcf*					*cccc*				
							u_z							
$M_{eff}/M=6/15$					$M_{eff}/M=7/15$					$M_{eff}/M=7/15$				
△	▲	△	▲	△	□	■	□	■	□	▽	▼	▽	▼	▽
△	▲	△	▲	△	□	■	□	■	□	▽	▼	▽	▼	▽
▲	△	▲	△	△	■	■	■	□	□	▼	▼	▼	▽	▽
							σ_{xx}							
$M_{eff}/M=10/15$					$M_{eff}/M=10/15$					$M_{eff}/M=10/15$				
▲	▲	△	▲	△	■	■	□	■	□	▼	▼	▽	▼	▽
▲	▲	△	▲	△	■	■	□	■	□	▼	▼	▽	▼	▽
▲	▲	▲	△	▲	■	■	■	□	■	▼	▼	▼	▽	▼
							σ_{yy}							
$M_{eff}/M=10/15$					$M_{eff}/M=10/15$					$M_{eff}/M=10/15$				
▲	▲	△	▲	△	■	■	□	■	□	▼	▼	▽	▼	▽
▲	▲	△	▲	△	■	■	□	■	□	▼	▼	▽	▼	▽
▲	▲	▲	△	▲	■	■	■	□	■	▼	▼	▼	▽	▼
							σ_{xz}							
$M_{eff}/M=6/15$					$M_{eff}/M=6/15$					$M_{eff}/M=7/15$				
△	▲	△	▲	△	□	■	□	■	□	▽	▼	▽	▼	▽
△	▲	△	▲	△	□	■	□	□	□	▽	▼	▽	▼	▽
▲	△	▲	△	△	■	□	■	□	■	▼	▽	▼	▽	▼
							σ_{yz}							
$M_{eff}/M=6/15$					$M_{eff}/M=7/15$					$M_{eff}/M=7/15$				
△	▲	△	▲	△	□	■	□	■	□	▽	▼	▽	▼	▽
△	▲	△	▲	△	□	■	□	■	□	▽	▼	▽	▼	▽
▲	△	▲	△	△	■	□	■	□	■	▼	▽	▼	▽	▼
							σ_{zz}							
$M_{eff}/M=11/15$					$M_{eff}/M=12/15$					$M_{eff}/M=13/15$				
▲	▲	▲	△	△	■	■	■	■	□	▼	▼	▼	▼	▽
▲	▲	▲	△	△	■	■	■	□	□	▼	▼	▼	▼	▽
▲	▲	▲	▲	▲	■	■	■	■	■	▼	▼	▼	▼	▼
							COMBINED							
$M_{eff}/M=13/15$					$M_{eff}/M=13/15$					$M_{eff}/M=13/15$				
▲	▲	▲	▲	△	■	■	■	■	□	▼	▼	▼	▼	▽
▲	▲	▲	▲	△	■	■	■	■	□	▼	▼	▼	▼	▽
▲	▲	▲	▲	▲	■	■	■	■	■	▼	▼	▼	▼	▼

Table 5.33 Accuracy of different models in detecting displacements and stresses, moderately thick plate. This table compares different plate models in terms of accuracies for given output variables, the model referred to as "COMBINED" can be found in Table 5.32

	δ_{u_z} %	$\delta_{\sigma_{xx}}$ %	$\delta_{\sigma_{yy}}$ %	$\delta_{\sigma_{xz}}$ %	$\delta_{\sigma_{yz}}$ %	$\delta_{\sigma_{zz}}$ %
ssss						
COMBINED	100.0	100.0	100.0	100.0	100.0	100.0
CPT	96.1	98.4	98.4	66.8	66.8	1974.2
FSDT	100.9	98.4	98.4	66.8	66.8	1974.2
Kant-2	99.9	100.2	100.2	100.0	100.0	79.4
cfcf						
COMBINED	100.0	100.0	100.0	100.0	100.0	100.0
CPT	91.2	98.3	94.9	239.5	−67.86	1171.3
FSDT	103.1	98.7	97.2	83.2	65.4	1183.7
Kant-2	99.9	100.0	101.0	98.8	99.7	80.4
cccc						
COMBINED	100.0	100.0	100.0	100.0	100.0	100.0
CPT	88.4	96.2	96.2	87.5	87.5	1074.0
FSDT	102.8	97.1	97.1	81.4	81.4	1084.8
Kant-2	99.8	99.9	99.9	98.9	98.9	80.0

1. Reduced plate models that are equivalent to a fourth-order theory vary significantly when different output variables are considered.
2. The influence of the boundary conditions is not very significant.
3. Classical models are unable to deal with shear stresses.
4. A significant computational cost reduction is only obtained when a limited number of output variables has to be detected.
5. The number of terms vs error diagram shows that all the theories derived from the present approach are able to satisfy a given error requirement, with lower computational costs than the open literature models that have been considered. This shows the strength of the present technique in detecting the possible best theories for a given structural problem.

5.30.2 Influence of the loading conditions

Reduced 2D models for various loading conditions are considered in this section for a simply-supported plate ($a/h = 10$). Three different loading conditions are taken into account (as shown in Fig. 5.47):

1. A bi-sinusoidal distributed load.

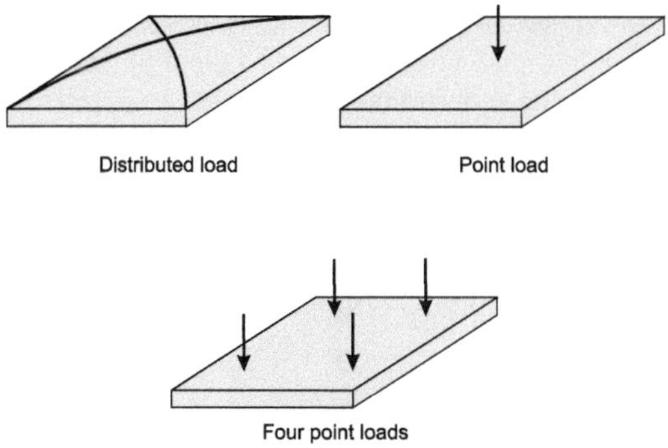

Figure 5.47 Loading conditions for the MAAA analysis of plates.

Table 5.34 Symbols adopted to distinguish from various plate loading conditions. This table shows the graphic notation adopted to indicate the different loading conditions for the MAAA analysis of a plate

Active term	Inactive term	
▲	△	*Distributed load*
■	□	*Point load*
▼	▽	*Four point load*

2. A point load.
3. Four point loads.

Table 5.34 shows the symbols adopted to distinguish each loading condition. The reduced models needed for different displacement and stress components are shown in Table 5.35. Each row refers to a different output variable, while each column considers a different loading condition. M_{eff} indicates the number of terms of the models that are equivalent to the fourth-order one. The last row shows the plate models that are needed to detect all the considered outputs. The latter combined models have been used to build Table 5.36, which shows a comparison of the accuracies given by the CPT, FSDT and Kant-2 plate models. Figure 5.48 shows the number of terms needed to compute σ_{xz} with a given error. The corresponding plate models are also indicated together with models retrieved from open literature. The following remarks can be made on the basis of the analyses:

1. The sets of effective displacement variables vary when distributed or concentrated loads are considered, whereas no differences arise if one or multiple point loads are introduced.

Table 5.35 Reduced 2D models for a simply-supported plate under various loading conditions. This table presents the reduced models required to detect displacement and stress variables for a plate under various loading conditions

Distributed load					**Point load**					**Four point loads**				
							u_z							
$M_{eff}/M = 6/15$					$M_{eff}/M = 6/15$					$M_{eff}/M = 6/15$				
△	▲	△	▲	△	□	■	□	■	□	▽	▼	▽	▼	▽
△	▲	△	▲	△	□	■	□	■	□	▽	▼	▽	▼	▽
▲	△	▲	△	△	■	□	■	□	□	▼	▽	▼	▽	▽
							σ_{xx}							
$M_{eff}/M = 10/15$					$M_{eff}/M = 7/15$					$M_{eff}/M = 7/15$				
▲	▲	△	▲	△	□	■	□	■	□	▽	▼	▽	▼	▽
▲	▲	△	▲	△	□	■	□	■	□	▽	▼	▽	▼	▽
▲	▲	▲	△	▲	■	□	■	□	■	▼	▽	▼	▽	▼
							σ_{yy}							
$M_{eff}/M = 10/15$					$M_{eff}/M = 7/15$					$M_{eff}/M = 7/15$				
▲	▲	△	▲	△	□	■	□	■	□	▽	▼	▽	▼	▽
▲	▲	△	▲	△	□	■	□	■	□	▽	▼	▽	▼	▽
▲	▲	▲	△	▲	■	□	■	□	■	▼	▽	▼	▽	▼
							σ_{xz}							
$M_{eff}/M = 6/15$					$M_{eff}/M = 7/15$					$M_{eff}/M = 7/15$				
△	▲	△	▲	△	□	■	□	■	□	▽	▼	▽	▼	▽
△	▲	△	▲	△	□	■	□	■	□	▽	▼	▽	▼	▽
▲	△	▲	△	△	■	□	■	□	■	▼	▽	▼	▽	▼
							σ_{yz}							
$M_{eff}/M = 6/15$					$M_{eff}/M = 7/15$					$M_{eff}/M = 7/15$				
△	▲	△	▲	△	□	■	□	■	□	▽	▼	▽	▼	▽
△	▲	△	▲	△	□	■	□	■	□	▽	▼	▽	▼	▽
▲	△	▲	△	△	■	□	■	□	■	▼	▽	▼	▽	▼
							σ_{zz}							
$M_{eff}/M = 11/15$					$M_{eff}/M = 7/15$					$M_{eff}/M = 7/15$				
▲	▲	▲	△	△	□	■	□	■	□	▽	▼	▽	▼	▽
▲	▲	▲	△	△	□	■	□	■	□	▽	▼	▽	▼	▽
▲	▲	▲	▲	▲	■	□	■	□	■	▼	▽	▼	▽	▼
							COMBINED							
$M_{eff}/M = 13/15$					$M_{eff}/M = 7/15$					$M_{eff}/M = 7/15$				
▲	▲	▲	▲	△	□	■	□	■	□	▽	▼	▽	▼	▽
▲	▲	▲	▲	△	□	■	□	■	□	▽	▼	▽	▼	▽
▲	▲	▲	▲	▲	■	□	■	□	■	▼	▽	▼	▽	▼

Table 5.36 Accuracy of different models in detecting displacements and stresses for various loading conditions. This table compares different plate models in terms of accuracies for given output variables, the model referred to as "COMBINED" can be found in Table 5.35

	δ_{u_z} %	$\delta_{\sigma_{xx}}$ %	$\delta_{\sigma_{yy}}$ %	$\delta_{\sigma_{xz}}$ %	$\delta_{\sigma_{xz}}$ %	$\delta_{\sigma_{zz}}$ %
Distributed						
COMBINED	100.0	100.0	100.0	100.0	100.0	100.0
CPT	96.1	98.4	98.4	66.8	66.8	1974.2
FSDT	100.9	98.4	98.4	66.8	66.8	1974.2
Kant-2	99.9	100.3	100.3	100.0	100.0	79.4
Point load						
COMBINED	100.0	100.0	100.0	100.0	100.0	100.0
CPT	91.4	83.7	83.7	67.0	67.0	540.1
FSDT	99.9	83.7	83.7	67.0	67.0	540.1
Kant-2	100.0	96.5	96.5	99.9	99.9	39.5
Four point load						
COMBINED	100.0	100.0	100.0	100.0	100.0	100.0
CPT	91.4	83.7	83.7	67.0	67.0	540.1
FSDT	99.9	83.7	83.7	67.0	67.0	540.1
Kant-2	100.0	96.5	96.5	99.9	99.9	39.5

2. The effect of the considered output variable is significant in the case of a distributed load, but it is almost negligible if point loads are considered.
3. The validity of the number of terms vs error diagram in providing guidelines for the construction of plate theories to achieve a given accuracy has been confirmed. As the error is fixed, the derived theories generally lie below classical and other refined models, that is, the proposed models are able to fulfil a certain accuracy demand but with lower computational costs.

5.30.3 Influence of the loading and thickness

The combined effect of loading and boundary conditions is considered in this section. The influence of the length-to-thickness ratio (a/h) is also investigated.The loading sets considered are the same as those of the previous analyses. Each output variable is associated with a symbol, as shown in Table 5.37.

First, a thick plate is considered assuming $a/h = 5$. Table 5.38 shows the sets of displacement variables that are needed to detect various out-

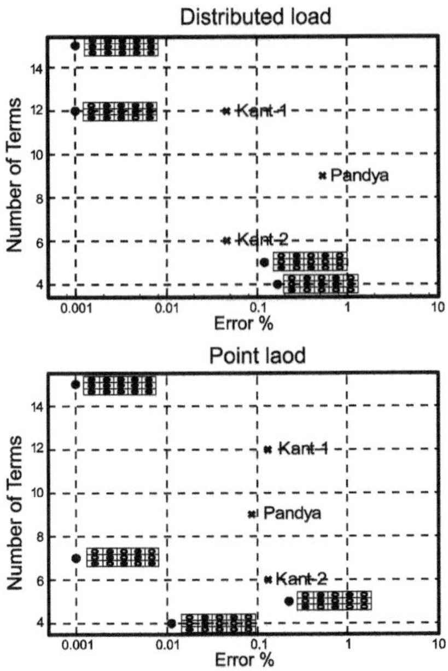

Figure 5.48 Number of terms vs error for different loading conditions, σ_{xz}, moderately thick plate.

Table 5.37 Symbols adopted to distinguish from various displacement and stress components. This table shows the graphic notation adopted to indicate various displacement and stress components for the MAAA analysis of a plate

	Active term	Inactive term
▲	△	u_z
■	□	σ_{xx}
▼	▽	σ_{xz}
◆	◇	σ_{zz}

put variables. Each row refers to a different loading condition; the plate is considered simply-supported. Table 5.39 shows the plate models that detect all the considered outputs for all the loading and boundary conditions, and for different thicknesses. Comparisons with other theories are given in Table 5.40. The analyses that have been carried out suggest the following comments:

1. The effective displacement variable sets depend on all of the three considered parameters: loadings, boundary conditions, and thickness.

Table 5.38 Reduced 2D models for a simply-supported thick plate for various displacement and stress components. This table presents the reduced models required to detect displacement and stress variables for a simply-supported thick plate

Distributed load					**Point load**					**Four point loads**				
$M_{eff}/M=9/15$					$M_{eff}/M=7/15$					$M_{eff}/M=7/15$				
▲	▲	△	▲	△	△	▲	△	▲	△	△	▲	△	▲	△
▲	▲	△	▲	△	△	▲	△	▲	△	△	▲	△	▲	△
▲	▲	▲	△	△	▲	△	▲	△	▲	▲	△	▲	△	▲
$M_{eff}/M=11/15$					$M_{eff}/M=7/15$					$M_{eff}/M=7/15$				
■	■	□	■	□	□	■	□	■	□	□	■	□	■	□
■	■	□	■	□	□	■	□	■	□	□	■	□	■	□
■	■	■	■	■	■	□	■	□	■	■	□	■	□	■
$M_{eff}/M=7/15$					$M_{eff}/M=7/15$					$M_{eff}/M=6/15$				
▽	▼	▽	▼	▽	▽	▼	▽	▼	▽	▽	▼	▽	▼	▽
▽	▼	▽	▼	▽	▽	▼	▽	▼	▽	▽	▼	▽	▼	▽
▼	▽	▼	▽	▼	▼	▽	▼	▽	▼	▼	▽	▼	▽	▽
$M_{eff}/M=11/15$					$M_{eff}/M=7/15$					$M_{eff}/M=7/15$				
◆	◆	◆	◇	◇	◇	◆	◇	◆	◇	◇	◆	◇	◆	◇
◆	◆	◆	◇	◇	◇	◆	◇	◆	◇	◇	◆	◇	◆	◇
◆	◆	◆	◆	◆	◆	◇	◆	◇	◆	◆	◇	◆	◇	◆

Table 5.39 Sets of effective terms for various thicknesses, boundary and loading conditions. This table shows the reduced 2D models needed to detect different displacement and stress components for different boundary and loading conditions

		$a/h=100$		
△□▽◇	▲■▼◆	△□▽◇	△■▼◆	△□▽◇
△□▽◇	▲■▼◆	△□▽◇	△■▼◆	△□▽◇
▲■▼◆	△□▽◆	▲■▼◆	△□▽◇	△□▼◆
		$a/h=10$		
△■▽◆	▲■▼◆	△□▽◆	▲■▼◆	△□▽◇
△■▽◆	▲■▼◆	△□▽◆	▲■▼◆	△□▽◇
▲■▼◆	▲■▽◆	▲■▼◆	△□▽◆	▲■▼◆
		$a/h=5$		
▲■▽◆	▲■▼◆	△□▽◆	▲■▼◆	△□▽◇
▲■▽◆	▲■▼◆	△□▽◆	▲■▼◆	△□▽◇
▲■▼◆	▲■▽◆	▲■▼◆	△■▽◆	▲■▼◆

2. The proper analysis of thick plates requires more sophisticated models, since the total number of expansion terms increases as a/h decreases. This result is analogous to that shown in [42].

Table 5.40 Accuracy of different models in detecting displacements and stresses for various thicknesses. This table compares different plate models in terms of accuracies for various output variables, BCs and thicknesses; the model referred to as "COMBINED" can be found in Table 5.39

	δ_{u_z} %	$\delta_{\sigma_{xx}}$ %	$\delta_{\sigma_{yy}}$ %	$\delta_{\sigma_{xz}}$ %	$\delta_{\sigma_{xz}}$ %	$\delta_{\sigma_{zz}}$ %
$a/h=100$						
COMBINED	100.0	100.0	100.0	100.0	100.0	100.0
CPT	102.7	98.5	98.0	158.6	710.2	1337.4
FSDT	102.9	98.5	98.0	80.0	114.3	1337.6
Kant-2	100.0	100.0	100.0	99.9	117.9	99.0
$a/h=10$						
COMBINED	100.0	100.0	100.0	100.0	100.0	100.0
CPT	86.2	82.5	79.2	202.9	−106.014	449.824
FSDT	101.6	82.7	80.24	83.6	72.7	453.2
Kant-2	100.0	96.2	95.7	98.7	97.7	39.8
$a/h=5$						
COMBINED	100.0	100.0	100.0	100.0	100.0	100.0
CPT	60.7	70.6	65.7	228.9	−90.6	189.9
FSDT	102.8	70.9	68.1	88.7	71.6	193.5
Kant-2	100.4	89.7	88.7	96.0	99.1	−23.1

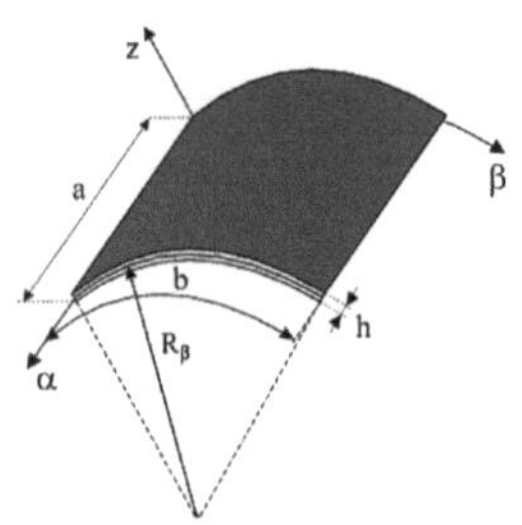

Figure 5.49 Geometry of shell for the MAAA analysis.

3. It has been confirmed that different output variables require different plate models in order to be detected properly.

5.30.4 Influence of the thickness ratio on shells

The MAAA is now applied to a shell structure, as in Fig. 5.49, under cylindrical bending (see also Ref. [40]). The shell is simply supported and a sinusoidal distribution of transverse pressure is applied at the top surface,

$$p_z = \bar{p}_z \, sin(\frac{\pi\beta}{b}) \tag{5.222}$$

where β is the curvilinear coordinate. The displacement u_z and the stresses $\sigma_{\beta\beta}$ and σ_{zz} are computed at $[a/2, b/2, h/2]$, while $\sigma_{\beta z}$ is computed at $[a/2, 0, 0]$.

Table 5.41 shows the reduced 2D models for different thickness ratios. The following remarks can be made:

1. As the thickness ratio decreases, the theories become more computationally expensive (M_{eff} increases).
2. $\sigma_{\beta\beta}$ and $\sigma_{\beta z}$ and all the terms in the u_z expansion are necessary for the exact evaluation of σ_{zz}.
3. In this particular case, the terms in the u_α expansion are not influential, because a cylindrical bending problem has been considered.
4. The constant term of the in plane displacement u_β is more important in a shell than in a plate because a shell undergoes a membranal deformation, even when it is very thin. This is due to curvature effects.

5.31 THE BEST THEORY DIAGRAM

The construction of reduced models through the MAAA allows one to obtain a diagram, for a given problem, which in terms of accuracy (input), answers the following fundamental questions:

- What is the "minimum" number of terms (N_{min}) that needs to be used in a finite element model?
- What terms need to be retained, that is, what generalized displacement variables need to be used as FE DOFs?

In other words, the MAAA is able to create plots, like the one shown in Fig. 5.50, which gives the number of terms vs the error. This plot can be defined as the Best Theory Diagram (BTD) since it allows one to edit an arbitrary given theory in order to have a lower number of terms for a given error (vertical shift, Δ_N) or, to increase the accuracy while keeping the computational costs constant (horizontal shift, Δ_{error}). The presented plot appears generally as a hyperbole. The UF makes the computation of such a plot possible. It should be noted that the diagram has the following properties:

- It changes as the problem changes (a/h, loadings, boundary conditions, etc.).

Table 5.41 Reduced 2D models for the cylindrical bending of isotropic shells with different thickness ratios. This table shows different 2D models to detect displacement and stress components of thin and thick shells

$R_\beta/h=100$					$R_\beta/h=10$					$R_\beta/h=4$				
							u_z							
$M_{eff}/M=4/15$					$M_{eff}/M=6/15$					$M_{eff}/M=6/15$				
△	△	△	△	△	△	△	△	△	△	△	△	△	△	△
▲	▲	△	△	△	▲	▲	△	▲	△	▲	▲	△	▲	△
▲	△	▲	△	△	▲	△	▲	▲	△	▲	△	▲	▲	△
							$\sigma_{\beta\beta}$							
$M_{eff}/M=6/15$					$M_{eff}/M=8/15$					$M_{eff}/M=8/15$				
△	△	△	△	△	△	△	△	△	△	△	△	△	△	△
▲	▲	△	△	△	▲	▲	△	▲	△	▲	▲	△	▲	△
▲	▲	▲	▲	△	▲	▲	▲	▲	▲	▲	▲	▲	▲	▲
							$\sigma_{\beta z}$							
$M_{eff}/M=5/15$					$M_{eff}/M=5/15$					$M_{eff}/M=8/15$				
△	△	△	△	△	△	△	△	△	△	△	△	△	△	△
▲	▲	△	▲	△	▲	▲	△	▲	△	▲	▲	△	▲	▲
▲	△	▲	△	△	▲	△	▲	△	△	▲	△	▲	▲	▲
							σ_{zz}							
$M_{eff}/M=9/15$					$M_{eff}/M=10/15$					$M_{eff}/M=10/15$				
△	△	△	△	△	△	△	△	△	△	△	△	△	△	△
▲	▲	▲	▲	△	▲	▲	▲	▲	▲	▲	▲	▲	▲	▲
▲	▲	▲	▲	▲	▲	▲	▲	▲	▲	▲	▲	▲	▲	▲
							COMBINED							
$M_{eff}/M=9/15$					$M_{eff}/M=10/15$					$M_{eff}/M=10/15$				
△	△	△	△	△	△	△	△	△	△	△	△	△	△	△
▲	▲	▲	▲	△	▲	▲	▲	▲	▲	▲	▲	▲	▲	▲
▲	▲	▲	▲	▲	▲	▲	▲	▲	▲	▲	▲	▲	▲	▲

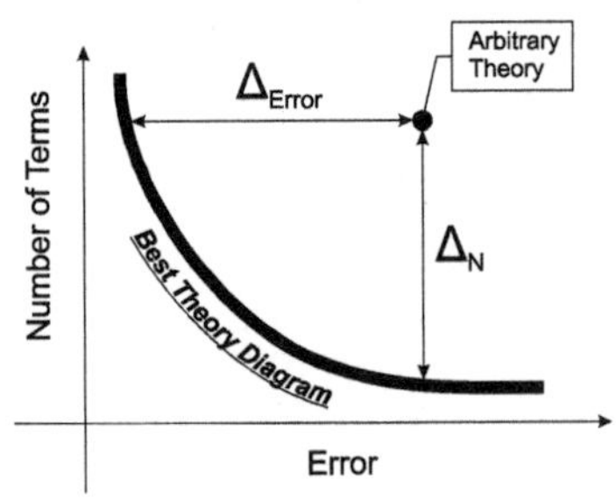

Figure 5.50 The best theory diagram.

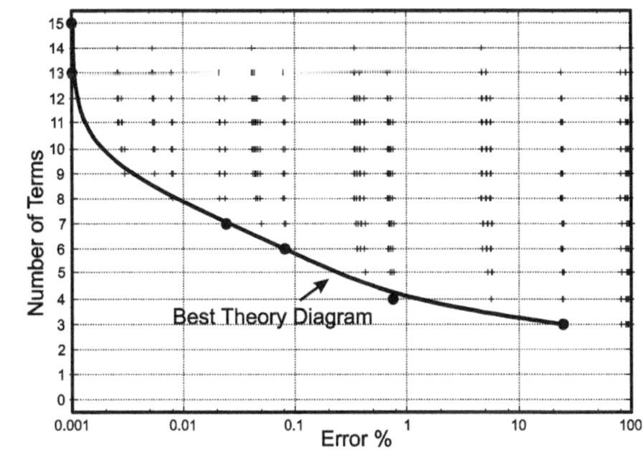

Figure 5.51 Accuracy of all the possible combinations of plate models in computing u_Z for the simply-supported plate under a distributed load (each "+" indicates a plate model).

- It changes as the output variable changes (displacement/stress components, or a combination of these).

The validity of the BTD has been tested by computing the accuracy of all the plate models obtainable as a combination of thc 15 terms of the fourth-order theory. The results are reported in Fig. 5.51 for the case of a simply-supported plate loaded by a distributed load where u_z is considered as the output variable. The BTD perfectly matches the lower boundaries of the region in which all the models lie. This confirms that the BTD can be considered as the best theory (i.e. the least cumbersome) for a given error. The BTD can be considered as a tool to evaluate any other structural theory.

REFERENCES

1. Oñate E. Structural analysis with the finite element method. Linear statics, vol. 1. Basis and solids. Springer; 2009.
2. Zienkiewicz OC, Taylor RL, Zhu JZ. The finite element method: its basis and fundamentals. Sixth edition. Elsevier; 2005.
3. Bathe KJ. Finite element procedures. Prentice-Hall; 1996.
4. Ambartsumian SA. On a theory of bending of anisotropic plates. Investiia Akad Nauk SSSR, Ot Tekh Nauk 1958;4.
5. Ambartsumian SA. Theory of anisotropic shells. Moskwa: Fizmatzig; 1961 [translated from Russian]. NASA TTF-118, 1964.
6. Ambartsumian SA. Theory of anisotropic plates. 1969 [translated from Russian by T. Cheron, edited by J.E. Ashton Tech. Pub. Co.].

7. Auricchio F, Sacco E. Partial-mixed formulation and refined models for the analysis of composites laminated within FSDT. Compos Struct 2001;46:103–13.
8. Carrera E. A class of two-dimensional theories for anisotropic multilayered plates analysis. Accad Sci Torino, Mem Sci Fis 1995–1996;19–20:1–39.
9. Carrera E. C_z^0 requirements–models for the two dimensional analysis of multilayered structures. Compos Struct 1997;37:373–84.
10. Carrera E. Developments, ideas and evaluations based upon Reissner's mixed variational theorem in the modeling of multilayered plates and shells. Appl Mech Rev 2001;54:301–29.
11. Carrera E. Historical review of zig-zag theories for multilayered plates and shell. Appl Mech Rev 2003;56(3):287–308.
12. Cheung YK, Wu CI. Free vibrations of thick, layered cylinders having finite length with various boundary conditions. J Sound Vib 1972;24:189–200.
13. Cho KN, Bert CW, Striz AG. Free vibrations of laminated rectangular plates analyzed by higher order individual-layer theory. J Sound Vib 1991;145:429–42.
14. Dennis ST, Palazotto AN. Laminated shell in cylindrical bending, two-dimensional approach vs exact. AIAA J 1991;29:647–50.
15. Gaudenzi P. A general formulation of higher order theories for the analysis of laminated plates. Compos Struct 1992;20:103–12.
16. Hsu T, Wang JT. A theory of laminated cylindrical shells consisting of layers of orthotropic laminae. AIAA J 1970;8:2141–6.
17. Lekhnitskii SG. Strength calculation of composite beams. Vestnik Inzhen i Tekhnikov 1935;9.
18. Librescu L, Schmidt R. Refined theories of elastic anisotropic shells accounting for small strains and moderate rotations. Int J Non-Linear Mech 1988;23:217–29.
19. Mindlin R. Influence of rotatory inertia and shear in flexural motions of isotropic elastic plates. J Appl Mech 1951;18:1031–6.
20. Murakami H. Laminated composite plate theory with improved in-plane responses. J Appl Mech 1986;53:661–6.
21. Nosier A, Kapania RK, Reddy JN. Free vibration analysis of laminated plates using a layer-wise theory. AIAA J 1993;31:2335–46.
22. Pagano NJ. Stress fields in composite laminates. Int J Solids Struct 1978;14:385–400.
23. Rabinovitch O, Frosting Y. Higher-order analysis of unidirectional sandwich panels with flat and generally curved faces and a "soft" core. Sandw Struct Mater 2001;3:89–116.
24. Reddy JN. A simple higher order theories for laminated composites plates. J Appl Mech 1984;52:745–52.
25. Reddy JN. Mechanics of laminated composite plates, theory and analysis. CRC Press; 1997.
26. Reddy JN, Phan ND. Stability and vibration of isotropic, orthotropic and laminated plates according to a higher order shear deformation theory. J Sound Vib 1985;98:157–70.
27. Reissner E. The effect of transverse shear deformation on the bending of elastic plates. J Appl Mech 1945;12:69–76.
28. Reissner E. On a certain mixed variational theory and a proposed application. Int J Numer Methods Eng 1984;20:1366–8.
29. Reissner E. On a certain mixed variational theorem and on laminated elastic shell theory. In: Proceedings of the Euromech–Colloquium, vol. 219. 1986. p. 17–27.

30. Reissner E. On a mixed variational theorem and on a shear deformable plate theory. Int J Numer Methods Eng 1986;23:193–8.
31. Reissner E, Stavsky Y. Bending and stretching of certain type of heterogeneous elastic plates. J Appl Mech 1961;9:402–8.
32. Robbins Jr DH, Reddy JN. Modeling of thick composites using a layer-wise theory. Int J Numer Methods Eng 1993;36:655–77.
33. Soldatos KP. Cylindrical bending of cross-ply laminated plates: refined 2D plate theories in comparison with the exact 3D elasticity solution. Tech Report No. 140. Greece: Dept. of Math., University of Ioannina; 1987.
34. Srinivas S. A refined analysis of composite laminates. J Sound Vib 1973;30:495–507.
35. Sun CT, Whitney JM. On the theories for the dynamic response of laminated plates. AIAA J 1973;11:372–98.
36. Touratier M. A refined theory for thick composite plates. Mech Res Commun 1988;15:229–36.
37. Touratier M. A refined theory of laminated shallow shells. Int J Solids Struct 1992;29:1401–15.
38. Vlasov BF. On the equations of bending of plates. Dokl Akad Nauk Azerbeijanskoi SSR 1957;3:955–79.
39. Yang PC, Norris CH, Stavsky Y. Elastic wave propagation in heterogeneous plates. Int J Solids Struct 1966;2:665–84.
40. Carrera E, Cinefra M, Petrolo M. A best theory diagram for metallic and laminated shells. In: Shell-like structures: non-classical theories and applications. 2011. p. 681–98.
41. Fazzolari FA, Carrera E. Accurate free vibration analysis of thermo-mechanically pre/post-buckled anisotropic multilayered plates based on a refined hierarchical trigonometric Ritz formulation. Compos Struct 2013;95:381–402.
42. Carrera E, Petrolo M. Guidelines and recommendations to construct theories for metallic and composite plates. AIAA J 2010;48(12):2852–66.
43. Kant T. Numerical analysis of thick plates. Comput Methods Appl Mech Eng 1982;31:1–18.
44. Kant T, Manjunatha B. An unsymmetric FRC laminate C^0 finite element model with 12 degrees of freedom per node. Eng Comput 1988;5(3):292–308.
45. Pandya B, Kant T. Finite element analysis of laminated composite plates using high-order displacement model. Compos Sci Technol 1988;32:137–55.

CHAPTER 6

Multilayered, anisotropic thermal stress structures

6.1 EQUATIONS OF ANISOTROPIC ELASTICITY

6.1.1 Reference system

The reference system used for the plate has the x and y axes in the same plane that correspond at the plate reference surface Ω and the z axis orthogonal at both. Such system is depicted in Fig. 6.1. In the same figure, the definition of the non-dimensional coordinate ζ^k is shown. It should be noted that at the *top* of the generic layer k, $\zeta^k = 1$, instead at the *bottom*, $\zeta^k = -1$.

6.1.2 Generalized Hooke's law

The kinematic relation and the mechanical and thermodynamic principles are applicable to any continuum irrespective of its physical constitution. The equations that characterize the individual material and its reaction to applied loads, are called *constitutive equations*. Material for which the constitutive behaviour is only a function of the current state of deformation are known as *elastic*. In the special case in which the work done by the stresses during a deformation is dependent only on the initial state and current configuration, the material is called *hyperelastic*. A material body is said to be *homogeneous* if the material properties are the same throughout the body. In the *heterogeneous* materials the material properties are a function of position. For instance, a structure composed from several uniform thickness layers of different materials stacked on top of each other and bounded to each other is heterogeneous through the thickness. An *anisotropic* body is one that has different value of material property in different direction at a point; i.e. material properties are direction dependent. An *isotropic* body is one for which every material property in all direction at a point is the same. An isotropic or anisotropic material can be non-homogeneous or homogeneous. A material body is said to be *ideally elastic* when, under isothermal condition, the body recovers its original form completely upon removal of the forces causing deformation, and there is one-to-one relationship between the state of

Thermal Stress Analysis of Composite Beams, Plates and Shells.
DOI: http://dx.doi.org/10.1016/B978-0-12-849892-7.00008-X

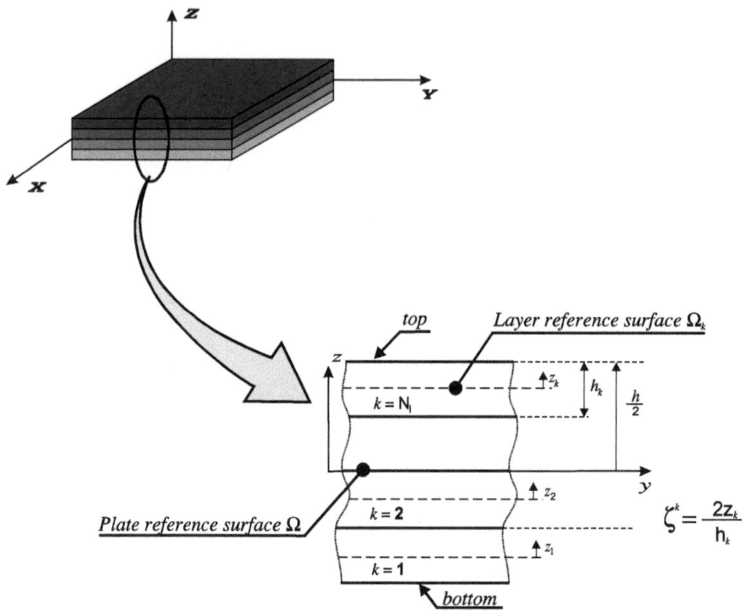

Figure 6.1 Geometry and notation.

stress and the state of strain in the current configuration. Thus, the material coefficients that specify the constitutive relationship between the stress and strain components are assumed to be constant during the deformation. The linear constitutive model for infinitesimal deformation is referred to as the *generalized Hooke's law*. Suppose that the reference configuration does not have a residual stress state σ_0. Then if the stress components are assumed to be linear functions of the components of strain, then the most general form of the constitutive equations for infinitesimal deformation is

$$\sigma_{ij} = C_{ijkl}\,\varepsilon_{kl} \quad \varepsilon_{kl} = \varepsilon_{lk} \tag{6.1}$$

where C_{ijkl} is the fourth-order tensor of material parameters and is termed *stiffness tensor*, there are, in general, $3^4 = 81$ scalar components for a fourth-order tensor. The number of the independent components of C_{ijkl} are considerably less because of the symmetry of σ_{ij}, symmetry of ε_{kl} and symmetry of C_{ijkl}. In absence of body couples the principle of conservation of angular momentum requires the stress tensor to be symmetric, $\sigma_{ij} = \sigma_{ji}$. Then it follows from Eq. (6.1) that C_{ijkl} is symmetric in the first two subscripts. Inasmuch the strain tensor is symmetric by its definition $\varepsilon_{kl} = \varepsilon_{lk}$, then C_{ijkl} must be symmetric in the last two subscripts as well, reducing the number of independent material stiffness components to $6 \times 6 = 36$. If we also assume that the material is hyperelastic, i.e., there exists an energy density function $\Phi(\varepsilon_{ij})$ such that:

$$\sigma_{ij} = \frac{\partial \Phi(\varepsilon_{ij})}{\partial \varepsilon_{ij}} = C_{ijkl}\,\varepsilon_{ij} \tag{6.2}$$

we have

$$\frac{\partial^2 \Phi(\varepsilon_{ij})}{\partial \varepsilon_{ij}\,\partial \varepsilon_{kl}} = C_{ijkl} \tag{6.3}$$

Since the order of differentiation is arbitrary

$$\frac{\partial^2 \Phi(\varepsilon_{ij})}{\partial \varepsilon_{ij}\,\partial \varepsilon_{kl}} = \frac{\partial^2 \Phi(\varepsilon_{ij})}{\partial \varepsilon_{kl}\,\partial \varepsilon_{ij}} \Longrightarrow C_{ijkl} = C_{klij} \tag{6.4}$$

This reduces the number of independent material stiffness components to 21. To show this we express Eq. (6.1) in an alternate form using single subscript notation for stress and strain and two subscript notation for the material stiffness coefficients:

$$\begin{gathered}
\sigma_1 = \sigma_{11}, \quad \sigma_2 = \sigma_{22}, \quad \sigma_3 = \sigma_{33}, \quad \sigma_4 = \sigma_{23}, \quad \sigma_5 = \sigma_{13}, \quad \sigma_6 = \sigma_{12} \\
\varepsilon_1 = \varepsilon_{11}, \quad \varepsilon_2 = \varepsilon_{22}, \quad \varepsilon_3 = \varepsilon_{33}, \quad \varepsilon_4 = 2\,\varepsilon_{23}, \quad \varepsilon_5 = 2\,\varepsilon_{13}, \quad \varepsilon_6 = 2\,\varepsilon_{12} \\
11 \Longrightarrow 1 \quad 22 \Longrightarrow 2 \quad 33 \Longrightarrow 3 \quad 23 \Longrightarrow 4 \quad 13 \Longrightarrow 5 \quad 12 \Longrightarrow 6
\end{gathered} \tag{6.5}$$

It should be borne in mind that the single subscript notation used for stresses and strains and the two-subscript components C_{ij} render them to non-tensor components (i.e., σ_i, ε_i, and C_{ij} do not transform like the components of a vector or tensor). The single subscript notation for stresses and strains is called the *engineering notation* or *Voigt-Kelvin* notation, Eq. (6.1) now takes form

$$\sigma_i = C_{ij}\,\varepsilon_j \tag{6.6}$$

where summation on repeated subscripts is implied, in matrix notation Eq. (6.6) can be written as:

$$\begin{bmatrix} \sigma_1 \\ \sigma_2 \\ \sigma_3 \\ \sigma_4 \\ \sigma_5 \\ \sigma_6 \end{bmatrix} = \begin{bmatrix} C_{11} & C_{12} & C_{13} & C_{14} & C_{15} & C_{16} \\ C_{21} & C_{22} & C_{23} & C_{24} & C_{25} & C_{26} \\ C_{31} & C_{32} & C_{33} & C_{34} & C_{35} & C_{36} \\ C_{41} & C_{42} & C_{43} & C_{44} & C_{45} & C_{46} \\ C_{51} & C_{52} & C_{53} & C_{54} & C_{55} & C_{56} \\ C_{61} & C_{62} & C_{63} & C_{64} & C_{65} & C_{66} \end{bmatrix} \begin{bmatrix} \varepsilon_1 \\ \varepsilon_2 \\ \varepsilon_3 \\ \varepsilon_4 \\ \varepsilon_5 \\ \varepsilon_6 \end{bmatrix} \tag{6.7}$$

Now the coefficients C_{ij} must be symmetric $\left(C_{ij} = C_{ji}\right)$ by virtue of the assumption that the material is hyperelastic. Hence, we have 21 independent stiffness coefficients for the most general elastic material. We assume that the stress-strain relation in Eq. (6.7) are invertible. Thus the components of strain are related to the components of stress by:

$$\varepsilon_i = S_{ij}\,\sigma_j \tag{6.8}$$

where S_{ij} are the compliance parameters (the compliance tensor is the inverse of the stiffness tensor $\mathbf{S} = \mathbf{C}^{-1}$). In the matrix form Eq. (6.8) becomes

$$\begin{bmatrix} \varepsilon_1 \\ \varepsilon_2 \\ \varepsilon_3 \\ \varepsilon_4 \\ \varepsilon_5 \\ \varepsilon_6 \end{bmatrix} = \begin{bmatrix} S_{11} & S_{12} & S_{13} & S_{14} & S_{15} & S_{16} \\ S_{21} & S_{22} & S_{23} & S_{24} & S_{25} & S_{26} \\ S_{31} & S_{32} & S_{33} & S_{34} & S_{35} & S_{36} \\ S_{41} & S_{42} & S_{43} & S_{44} & S_{45} & S_{46} \\ S_{51} & S_{52} & S_{53} & S_{54} & S_{55} & S_{56} \\ S_{61} & S_{62} & S_{63} & S_{64} & S_{65} & S_{66} \end{bmatrix} \begin{bmatrix} \sigma_1 \\ \sigma_2 \\ \sigma_3 \\ \sigma_4 \\ \sigma_5 \\ \sigma_6 \end{bmatrix} \tag{6.9}$$

Material symmetry

Further reduction in the number of independent stiffness parameters comes from the so-called material symmetry. Suppose that (x_1, x_2, x_3) denote the coordinate system with respect to which Eqs. (6.6) and (6.8) are defined. We shall call them *material coordinate system*. The coordinate system (x, y, z) used to write the equation of motion and strain-displacement equations will be called the *problem coordinate* to distinguish them from the material coordinate system. Both are fixed in the body, and the two systems are oriented with respect to each other. When elastic material parameters at a point have the same value for every pair of coordinate system that are mirror images of each other in a certain plane, that plane is called a *material plane of symmetry* (e.g., symmetry of internal structure due to crystallographic form, regular arrangement of fibres or molecules, etc.). It should be noted that the symmetry under discussion is a directional property and not a positional property. Thus a material may have certain elastic symmetry at every point of a material body the properties may vary from point to point. Positional dependence of material properties is what we called the inhomogeneity of material. In the following we discuss various planes of symmetry and forms of associated stress-strain relations. Note that the use of the tensor components of stress and strain is necessary to express the transformation law from a system to another. The fourth-order tensor, for example, transforms according to the formula

$$C'_{ijkl} = l_{ip}\, l_{jq}\, l_{kr}\, l_{ls}\, C_{pqrs} \tag{6.10}$$

where the l_{ij} are the direction cosines associated with the coordinate system (x_1, x_2, x_3) and $\left(x'_1, x'_2, x'_3\right)$, and C'_{ijkl} and C_{pqrs} are the components of the fourth-order tensor $\mathbf{C}$ in the primed and unprimed coordinate system respectively.

Isotropic materials

When there exist no preferred direction in the materials (i.e., the material has infinite number of planes of material symmetry), the number of independent elastic coefficients reduces to 2. Such materials are called isotropic. For isotropic material we have that the stress-strain relations take the following form:

$$\begin{bmatrix} \sigma_1 \\ \sigma_2 \\ \sigma_3 \\ \sigma_4 \\ \sigma_5 \\ \sigma_6 \end{bmatrix} = \lambda \begin{bmatrix} 1-\nu & \nu & \nu & 0 & 0 & C_{13} \\ \nu & 1-\nu & 0 & 0 & 0 & C_{23} \\ \nu & 0 & 1-\nu & 0 & 0 & 0 \\ 0 & 0 & 0 & \frac{1}{2}(1-2\nu) & 0 & 0 \\ 0 & 0 & 0 & 0 & \frac{1}{2}(1-2\nu) & 0 \\ C_{13} & C_{23} & 0 & 0 & 0 & \frac{1}{2}(1-2\nu) \end{bmatrix} \begin{bmatrix} \varepsilon_1 \\ \varepsilon_2 \\ \varepsilon_3 \\ \varepsilon_4 \\ \varepsilon_5 \\ \varepsilon_6 \end{bmatrix} \tag{6.11}$$

or in a more useful way for the theory that will follow throughout this book

$$\begin{bmatrix} \sigma_{xx} \\ \sigma_{yy} \\ \tau_{xy} \\ \tau_{xz} \\ \tau_{yz} \\ \sigma_{zz} \end{bmatrix} = \begin{bmatrix} C_{11} & C_{12} & 0 & 0 & 0 & C_{13} \\ C_{12} & C_{22} & 0 & 0 & 0 & C_{23} \\ 0 & 0 & C_{66} & 0 & 0 & 0 \\ 0 & 0 & 0 & C_{55} & 0 & 0 \\ 0 & 0 & 0 & 0 & C_{44} & 0 \\ C_{13} & C_{23} & 0 & 0 & 0 & C_{33} \end{bmatrix} \begin{bmatrix} \varepsilon_{xx} \\ \varepsilon_{yy} \\ \gamma_{xy} \\ \gamma_{xz} \\ \gamma_{yz} \\ \varepsilon_{zz} \end{bmatrix} \tag{6.12}$$

with

$$\begin{aligned} C_{11} &= C_{22} = C_{33} = \lambda + 2\mu \\ C_{12} &= C_{13} = C_{23} = \lambda \\ C_{44} &= C_{55} = C_{66} = \mu \end{aligned} \tag{6.13}$$

and

$$\begin{aligned} \mu &= G = \frac{E}{2(1+\nu)} \\ \lambda &= \frac{\nu E}{(1+\nu)(1-2\nu)} \end{aligned} \tag{6.14}$$

μ and λ are referred to as *Lamé constants*, E indicates the *Young's modulus*, G is the *transverse shear modulus* and ν the *Poisson's ratio*. Alternatively, the stress-strain relations can be also written in a more compact form using the fact that the fourth-order isotropic tensor can be expressed as

$$C_{ijkl} = \lambda\,\delta_{ij}\,\delta_{kl} + \mu\,\left(\delta_{ik}\,\delta_{jl} + \delta_{il}\,\delta_{jk}\right) \tag{6.15}$$

Therefore the stress-strain relation for the isotropic case takes the form

$$\sigma_{ij} = C_{ijkl}\,\varepsilon_{kl} = 2\,\mu\,\varepsilon_{ij} + \lambda\,\varepsilon_{kk}\delta_{ij}; \qquad \boldsymbol{\sigma} = 2\,\mu\,\boldsymbol{\varepsilon} + \lambda\,\mathrm{tr}\,(\boldsymbol{\varepsilon})\,\boldsymbol{I} \tag{6.16}$$

The strain-stress relations are

$$\varepsilon_{ij} = \frac{1}{2\,\mu}\left[\sigma_{ij} - \frac{\lambda}{2\,\mu + 3\,\lambda}\sigma_{kk}\,\delta_{ij}\right]; \qquad \boldsymbol{\varepsilon} = \frac{1}{2\,\mu}\left[\boldsymbol{\sigma} - \frac{\lambda}{2\,\mu + 3\,\lambda}\mathrm{tr}\,(\boldsymbol{\sigma})\,\boldsymbol{I}\right] \tag{6.17}$$

The relation between the Lamé constants λ and μ and the engineering constants E, ν and G for isotropic materials are

$$E = \frac{\mu\,3\,\lambda + 2\,\mu}{\lambda + \mu}, \qquad \nu = \frac{\lambda}{2\,(\mu + \lambda)}, \qquad G = \mu \tag{6.18}$$

Orthotropic materials

When three mutually orthogonal planes of material symmetry exist, the number of elastic coefficients is reduced to 9 and such materials are called *orthotropic*. The stress-strain relations for an orthotropic material takes the form

$$\begin{bmatrix} \sigma_1 \\ \sigma_2 \\ \sigma_3 \\ \sigma_4 \\ \sigma_5 \\ \sigma_6 \end{bmatrix} = \begin{bmatrix} C_{11} & C_{12} & C_{13} & 0 & 0 & 0 \\ C_{12} & C_{22} & C_{23} & 0 & 0 & 0 \\ C_{13} & C_{23} & C_{33} & 0 & 0 & 0 \\ 0 & 0 & 0 & C_{44} & 0 & 0 \\ 0 & 0 & 0 & 0 & C_{55} & 0 \\ 0 & 0 & 0 & 0 & 0 & C_{66} \end{bmatrix} \begin{bmatrix} \varepsilon_1 \\ \varepsilon_2 \\ \gamma_3 \\ \gamma_4 \\ \gamma_5 \\ \varepsilon_6 \end{bmatrix} \tag{6.19}$$

and as done for the isotropic case, the following arrangement is preferred

$$\begin{bmatrix} \sigma_{11} \\ \sigma_{22} \\ \sigma_{12} \\ \sigma_{13} \\ \sigma_{23} \\ \sigma_{33} \end{bmatrix} = \begin{bmatrix} C_{11} & C_{12} & 0 & 0 & 0 & C_{13} \\ C_{12} & C_{22} & 0 & 0 & 0 & C_{23} \\ 0 & 0 & C_{66} & 0 & 0 & 0 \\ 0 & 0 & 0 & C_{55} & 0 & 0 \\ 0 & 0 & 0 & 0 & C_{44} & 0 \\ C_{13} & C_{23} & 0 & 0 & 0 & C_{33} \end{bmatrix} \begin{bmatrix} \varepsilon_{11} \\ \varepsilon_{22} \\ \gamma_{12} \\ \gamma_{13} \\ \gamma_{23} \\ \varepsilon_{33} \end{bmatrix} \tag{6.20}$$

with:

$$
\begin{gathered}
C_{11}=E_1\frac{1-\nu_{23}\nu_{32}}{\Delta};\ C_{12}=E_1\frac{\nu_{21}-\nu_{31}\nu_{23}}{\Delta}=E_2\frac{\nu_{12}+\nu_{32}\nu_{13}}{\Delta}\\
C_{22}=E_2\frac{1-\nu_{13}\nu_{31}}{\Delta};\ C_{13}=E_1\frac{\nu_{31}-\nu_{21}\nu_{32}}{\Delta}=E_3\frac{\nu_{13}+\nu_{12}\nu_{23}}{\Delta}\\
C_{33}=E_3\frac{1-\nu_{12}\nu_{21}}{\Delta};\ C_{23}=E_1\frac{\nu_{32}-\nu_{12}\nu_{31}}{\Delta}=E_3\frac{\nu_{23}+\nu_{21}\nu_{13}}{\Delta}\\
C_{44}=G_{23};\ C_{55}=G_{13};\ C_{66}=G_{12}\\
\Delta=1-\nu_{12}\nu_{21}-\nu_{23}\nu_{32}-\nu_{31}\nu_{13}-2\nu_{12}\nu_{32}\nu_{13}
\end{gathered}
\tag{6.21}
$$

where E_1, E_2, E_3 are *Young's moduli* in 1, 2 and 3 material direction respectively, ν_{ij} is Poisson's ratio, defined as the ratio of transverse strain in the jth direction to the axial strain in the ith direction when stressed in the ith direction, and G_{23}, G_{13}, G_{12} are *shear moduli* in the 2–3, 1–3, and 1–2 planes, respectively. Since the compliance matrix **S** is the inverse of stiffness matrix **C** and the inverse of a symmetric matrix is symmetric, it follows that the compliance matrix **S** is also a symmetric matrix. This in turn implies that the following reciprocal relations hold:

$$
\frac{\nu_{21}}{E_2}=\frac{\nu_{12}}{E_1};\qquad \frac{\nu_{31}}{E_3}=\frac{\nu_{13}}{E_1};\qquad \frac{\nu_{32}}{E_3}=\frac{\nu_{23}}{E_2};
\tag{6.22}
$$

or, in short

$$
\frac{\nu_{ij}}{E_i}=\frac{\nu_{ji}}{E_j}\qquad (\text{no sum in } i,j)
\tag{6.23}
$$

for $i,j=1,2,3$. The nine independent material coefficients for an orthotropic material are

$$
E_1,\ E_2,\ E_3,\ G_{12},\ G_{13},\ G_{23},\ \nu_{12},\ \nu_{13},\ \nu_{23}
\tag{6.24}
$$

Further details on this topic can be found in Refs. [1–3].

6.1.3 Transformation of stress, strain components and material coefficients

The constitutive relation of Eq. (6.20) for an orthotropic material were written in terms of stress and strain components that are referred to a coordinate system that coincides with the principal material coordinate system, Fig. 6.2. The coordinate system used in the problem formulation, in general, does not coincide with the principal material coordinate system. Further, composite laminates have several layers, each with different orientation of their material coordinates with respect to the laminate coordinate.

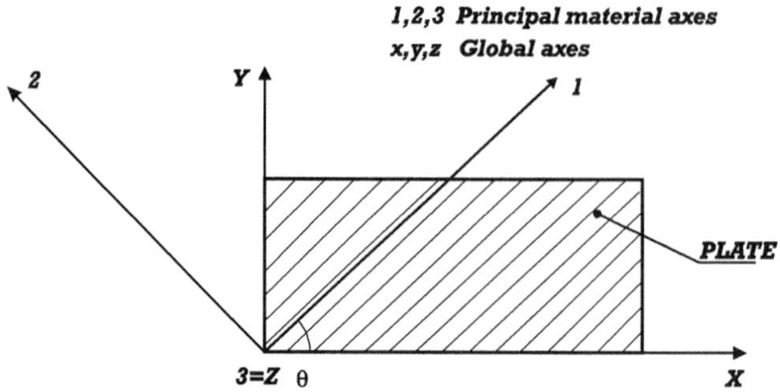

Figure 6.2 Reference systems.

Thus, there is a need to establish transformation relations among stresses and strains in one coordinate system to the corresponding quantities in other coordinate system. These relations can be used to transform constitutive equations from the material coordinates of each layer to the coordinate used in the problem description. Beginning from the stress and strain vector written in both coordinate systems

$$\begin{aligned}
\boldsymbol{\sigma}_m &= \begin{bmatrix} \sigma_{11} & \sigma_{22} & \sigma_{12} & \sigma_{13} & \sigma_{23} & \sigma_{33} \end{bmatrix}^T \\
\boldsymbol{\varepsilon}_m &= \begin{bmatrix} \varepsilon_{11} & \varepsilon_{22} & \gamma_{12} & \gamma_{13} & \gamma_{23} & \varepsilon_{33} \end{bmatrix}^T \\
\boldsymbol{\sigma} &= \begin{bmatrix} \sigma_{xx} & \sigma_{yy} & \sigma_{xy} & \sigma_{xz} & \sigma_{yz} & \sigma_{zz} \end{bmatrix}^T \\
\boldsymbol{\varepsilon} &= \begin{bmatrix} \varepsilon_{xx} & \varepsilon_{yy} & \gamma_{xy} & \gamma_{xz} & \gamma_{yz} & \varepsilon_{zz} \end{bmatrix}^T
\end{aligned} \tag{6.25}$$

the relation that links stress and strain components in the two different reference systems can be written as

$$\begin{aligned}
\boldsymbol{\sigma} &= \mathbf{T}\,\boldsymbol{\sigma}_m \\
\boldsymbol{\varepsilon}_m &= \mathbf{T}^T\,\boldsymbol{\varepsilon}
\end{aligned} \tag{6.26}$$

Where it has been placed

$$\mathbf{T} = \begin{bmatrix}
\cos^2\theta & \sin^2\theta & -\sin 2\theta & 0 & 0 & 0 \\
\sin^2\theta & \cos^2\theta & \sin 2\theta & 0 & 0 & 0 \\
\sin\theta\cos\theta & -\sin\theta\cos\theta & \cos^2\theta - \sin^2\theta & 0 & 0 & 0 \\
0 & 0 & 0 & \cos\theta & -\sin\theta & 0 \\
0 & 0 & 0 & \sin\theta & \cos\theta & 0 \\
0 & 0 & 0 & 0 & 0 & 1
\end{bmatrix} \tag{6.27}$$

Rewriting in a more compact way Eq. (6.20), it becomes

$$\boldsymbol{\sigma}_m = \mathbf{C}\boldsymbol{\varepsilon}_m \tag{6.28}$$

Substituting Eq. (6.25) in Eq. (6.27) and using Eq. (6.26), we obtain

$$\boldsymbol{\sigma} = \mathbf{T}\mathbf{C}\mathbf{T}^T\boldsymbol{\varepsilon} \tag{6.29}$$

Finally assuming

$$\tilde{\mathbf{C}} = \mathbf{T}\mathbf{C}\mathbf{T}^T = \begin{bmatrix} \tilde{C}_{11} & \tilde{C}_{12} & \tilde{C}_{16} & 0 & 0 & \tilde{C}_{13} \\ \tilde{C}_{12} & \tilde{C}_{22} & \tilde{C}_{26} & 0 & 0 & \tilde{C}_{23} \\ \tilde{C}_{16} & \tilde{C}_{26} & \tilde{C}_{66} & 0 & 0 & \tilde{C}_{36} \\ 0 & 0 & 0 & \tilde{C}_{55} & \tilde{C}_{45} & 0 \\ 0 & 0 & 0 & C_{45} & \tilde{C}_{44} & 0 \\ \tilde{C}_{13} & \tilde{C}_{23} & \tilde{C}_{36} & 0 & 0 & \tilde{C}_{33} \end{bmatrix} \tag{6.30}$$

Hooke's law becomes

$$\boldsymbol{\sigma} = \tilde{\mathbf{C}}\boldsymbol{\varepsilon} \tag{6.31}$$

6.1.4 Rearrangement of the generalized Hooke's law for the present analysis

For some reasons which will be clearer later on in this book, both stress and strain vectors, and Hooke's law are rearranged. In particular, the stress and strain vector components can be grouped as follows

$$\boldsymbol{\sigma}_p = \begin{bmatrix} \sigma_{xx} & \sigma_{yy} & \tau_{xy} \end{bmatrix}^T; \quad \boldsymbol{\sigma}_n = \begin{bmatrix} \tau_{xz} & \tau_{yz} & \sigma_{zz} \end{bmatrix}^T \tag{6.32}$$

$$\boldsymbol{\varepsilon}_p = \begin{bmatrix} \varepsilon_{xx} & \varepsilon_{yy} & \gamma_{xy} \end{bmatrix}^T; \quad \boldsymbol{\varepsilon}_n = \begin{bmatrix} \gamma_{xz} & \gamma_{yz} & \varepsilon_{zz} \end{bmatrix}^T \tag{6.33}$$

Moreover, the constitutive matrix $\tilde{\mathbf{C}}$ is split in the following four parts

$$\begin{aligned} \tilde{\mathbf{C}}_{pp} &= \begin{bmatrix} \tilde{C}_{11} & \tilde{C}_{12} & \tilde{C}_{16} \\ \tilde{C}_{12} & \tilde{C}_{22} & \tilde{C}_{26} \\ \tilde{C}_{16} & \tilde{C}_{26} & \tilde{C}_{66} \end{bmatrix}; \quad \tilde{\mathbf{C}}_{pn} = \begin{bmatrix} 0 & 0 & \tilde{C}_{13} \\ 0 & 0 & \tilde{C}_{23} \\ 0 & 0 & \tilde{C}_{36} \end{bmatrix} \\ \tilde{\mathbf{C}}_{np} &= \begin{bmatrix} 0 & 0 & 0 \\ 0 & 0 & 0 \\ \tilde{C}_{13} & \tilde{C}_{23} & \tilde{C}_{36} \end{bmatrix}; \quad \tilde{\mathbf{C}}_{nn} = \begin{bmatrix} \tilde{C}_{55} & \tilde{C}_{45} & 0 \\ \tilde{C}_{45} & \tilde{C}_{44} & 0 \\ 0 & 0 & \tilde{C}_{33} \end{bmatrix} \end{aligned} \tag{6.34}$$

Using Eqs. (6.34), (6.33) and (6.32), Eq. (6.31) becomes

$$\boldsymbol{\sigma}_p = \tilde{\mathbf{C}}_{pp}\boldsymbol{\varepsilon}_p + \tilde{\mathbf{C}}_{pn}\boldsymbol{\varepsilon}_n \tag{6.35}$$

$$\boldsymbol{\sigma}_n = \tilde{\mathbf{C}}_{np}\boldsymbol{\varepsilon}_p + \tilde{\mathbf{C}}_{nn}\boldsymbol{\varepsilon}_n \tag{6.36}$$

where the subscripts, p and n, state for in-plane and out-of-plane components.

6.2 FUNCTIONALLY GRADED MATERIALS CONSTITUTIVE LAW

In the case of functionally graded materials (see Refs. [4,5]) the 3D constitutive equations according to Hooke's law are given as

$$\boldsymbol{\sigma}^k = \mathbf{C}^k(z)\,\boldsymbol{\varepsilon}^k \tag{6.37}$$

Splitting Eq. (6.37) in in-plane and out-of-plane components, it assumes the following form

$$\begin{aligned} \boldsymbol{\sigma}^k_{pH} &= \tilde{\mathbf{C}}^k_{pp}(z)\,\boldsymbol{\varepsilon}^k_{pG} + \tilde{\mathbf{C}}^k_{pn}(z)\,\boldsymbol{\varepsilon}^k_{nG} \\ \boldsymbol{\sigma}^k_{nH} &= \tilde{\mathbf{C}}^k_{np}(z)\,\boldsymbol{\varepsilon}^k_{pG} + \tilde{\mathbf{C}}^k_{nn}(z)\,\boldsymbol{\varepsilon}^k_{nG} \end{aligned} \tag{6.38}$$

where matrices $\tilde{\mathbf{C}}^k_{pp}(z)$, $\tilde{\mathbf{C}}^k_{nn}(z)$, $\tilde{\mathbf{C}}^k_{pn}(z)$ and $\tilde{\mathbf{C}}^k_{np}(z)$ are

$$\tilde{\mathbf{C}}^k_{pp}(z) = \begin{bmatrix} \tilde{C}_{11}(z) & \tilde{C}_{12}(z) & 0 \\ \tilde{C}_{12}(z) & \tilde{C}_{22}(z) & 0 \\ 0 & 0 & \tilde{C}_{66}(z) \end{bmatrix}^k,$$

$$\tilde{\mathbf{C}}^k_{nn}(z) = \begin{bmatrix} \tilde{C}_{55}(z) & 0 & 0 \\ 0 & \tilde{C}_{44}(z) & 0 \\ 0 & 0 & \tilde{C}_{33}(z) \end{bmatrix}^k,$$

$$\tilde{\mathbf{C}}^k_{pn}(z) = \begin{bmatrix} 0 & 0 & \tilde{C}_{13}(z) \\ 0 & 0 & \tilde{C}_{23}(z) \\ 0 & 0 & 0 \end{bmatrix}^k, \quad \tilde{\mathbf{C}}^k_{np}(z) = \begin{bmatrix} 0 & 0 & 0 \\ 0 & 0 & 0 \\ \tilde{C}_{13}(z) & \tilde{C}_{23}(z) & 0 \end{bmatrix}^k \tag{6.39}$$

The computation of the effective FG material properties $\tilde{C}_{ij}$ in Eq. (6.39) is independent of the FGM structures considered and it follows three steps:

1. Computation of volume fraction of ceramic and metal phases.
2. Computation of elastic properties, Young's modulus E^k and Poisson's coefficient ν^k.
3. Computation of the effective FG material properties $\tilde{C}_{ij}$.

Only the evaluation of the volume fraction depends on the analyzed FGM structure. In the present book the following cases are examined:

1. FGM isotropic plates, the bottom skin is metallic and the top skin is ceramic (see Fig. 6.3(a)), the volume fraction of the ceramic phase is

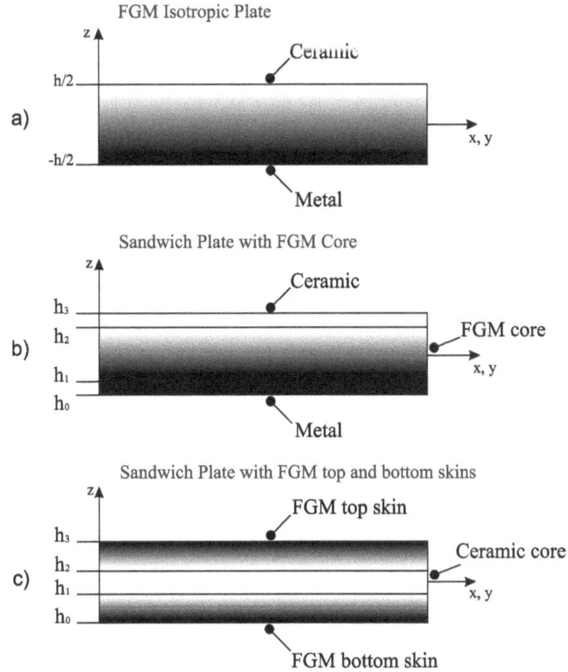

Figure 6.3 FGM isotropic and sandwich plates.

defined according to the following power-law:

$$V_c(z) = \left(\frac{z}{h} + \frac{1}{2}\right)^p \quad z \in [-h/2, h/2] \tag{6.40}$$

and its trend against the dimensionless thickness coordinate z/h is shown in Fig. 6.4(a).

2. FGM sandwich plates, the bottom skin is ceramic, the top skin is metal and the core is considered FGM (see Fig. 6.3(b)). The volume fraction of the ceramic phase is defined according to the following power-law:

$$\begin{aligned} V_c^1(z) &= 0 && z \in [h_0, h_1] \\ V_c^2(z) &= \left(\tfrac{z-h_1}{h_2-h_1}\right)^p && z \in [h_1, h_2] \\ V_c^3(z) &= 1 && z \in [h_2, h_3] \end{aligned} \tag{6.41}$$

As in the previous case its trend against the dimensionless thickness coordinate z/h is shown in Fig. 6.4(b).

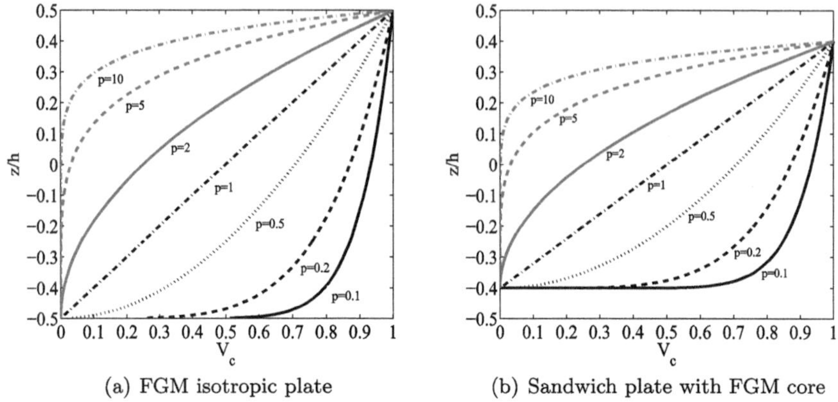

(a) FGM isotropic plate

(b) Sandwich plate with FGM core

Figure 6.4 Volume fraction V_C distribution along the thickness plate direction for different values of the volume fraction index p.

3. FGM sandwich plate, the core is fully ceramic and the top and bottom skins are FGM across the thickness direction (see Fig. 6.3(c)). The volume fraction of the ceramic phase is defined according to the following power-law:

$$\begin{aligned} V_c^1(z) &= \left(\frac{z-h_0}{h_1-h_0}\right)^p \quad & z \in [h_0,\, h_1] \\ V_c^2(z) &= 1 & z \in [h_1,\, h_2] \\ V_c^3(z) &= \left(\frac{h_3-z}{h_3-h_2}\right)^p & z \in [h_2,\, h_3] \end{aligned} \tag{6.42}$$

and the trends, against the dimensionless thickness coordinate z/h, of several sandwich plate configurations are shown in Fig. 6.5.

4. The volume fraction of the FGM can be assumed to obey a sigmoid power-law function along the thickness. The volume fraction using sigmoid power-law functions, which ensure a smooth distribution of stresses, is defined in Ref. [6].

$$\begin{aligned} V_c(z)^1 &= \frac{1}{2}\left(\frac{h/2-z}{h/2}\right)^p & z \in \left[-\frac{h}{2},\, 0\right] \\ V_c(z)^2 &= 1-\frac{1}{2}\left(\frac{h/2-z}{h/2}\right)^p \quad & z \in \left[0,\, \frac{h}{2}\right] \end{aligned} \tag{6.43}$$

where p is the power-law index, which indicates the material variation profile through the thickness. The material properties of the S-FGM can be expressed by using the rule of mixture, as:

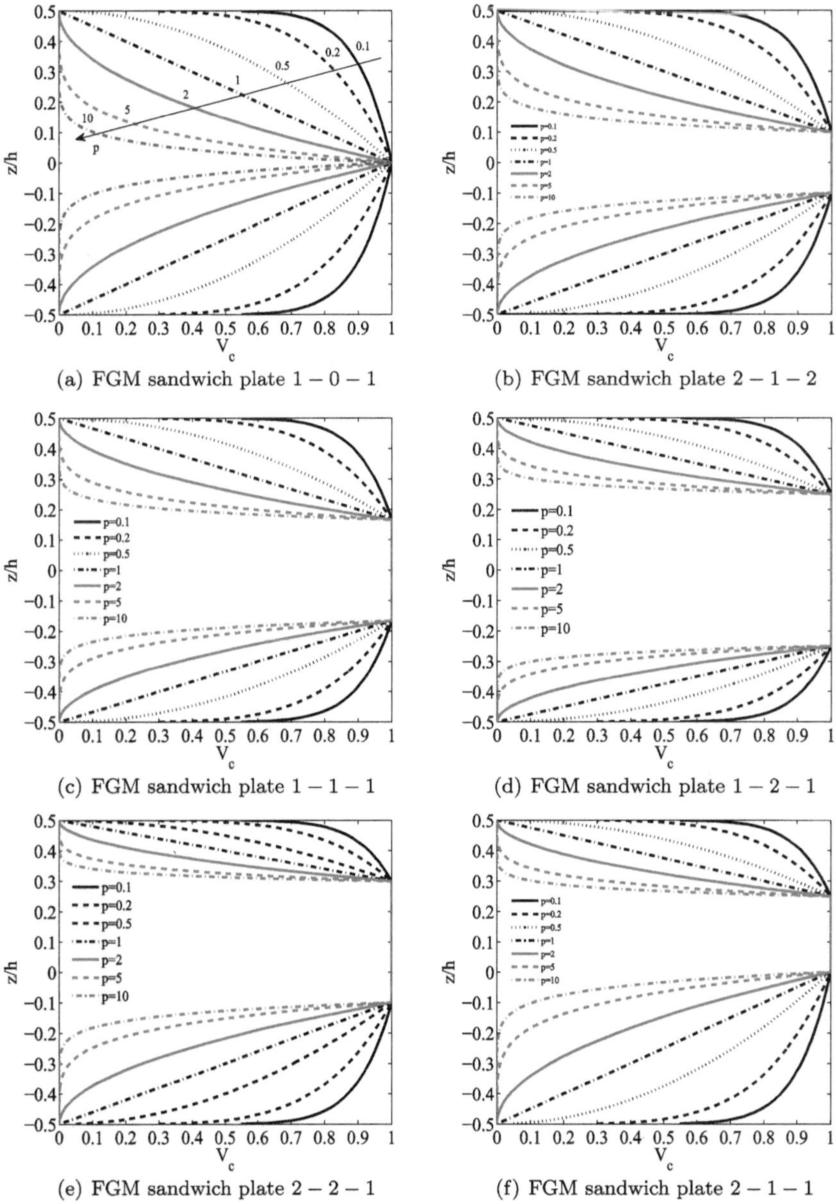

Figure 6.5 Volume fraction V_C distribution along the thickness plate direction for different values of the volume fraction index p and several sandwich plate configurations with FGM top and bottom skins.

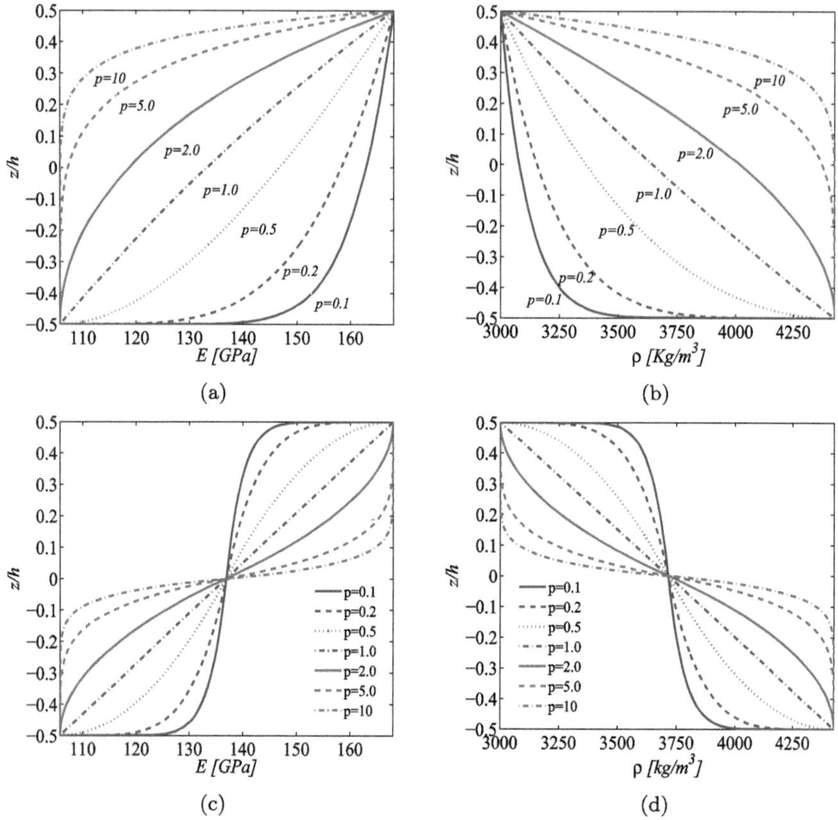

Figure 6.6 Young's modulus E and density ρ distributions along the thickness-plate-direction for different values of the volume fraction index p at room temperature ($T = 300\ K$): (a) and (b) P-FGM, (c) and (d) S-FGM.

$$\begin{aligned} \mathcal{A} &= V_c^1 \mathcal{A}_c + \left(1 - V_c^1\right) \mathcal{A}_m \quad z \in \left[-\tfrac{h}{2}, 0\right] \\ \mathcal{A} &= V_c^2 \mathcal{A}_c + \left(1 - V_c^2\right) \mathcal{A}_m \quad z \in \left[0, \tfrac{h}{2}\right] \end{aligned} \tag{6.44}$$

and the trends, against the dimensionless thickness coordinate z/h, of several sandwich plate configurations are shown in Fig. 6.6 and compared to those obtained in the case of power-law FGM for the Young's modulus E and the density ρ.

The thickness of the plate is h and the exponent p is the volume fraction index indicating the material variation through-the-thickness direction. The volume fraction of the metal phase is given as $V_m^k(z) = 1 - V_c^k(z)$. Poisson's coefficient ν^k, Young's modulus $E^k(z)$ and the coefficient of thermal expansion $\alpha^k(z)$ are computed by the following law-of-mixtures:

$$\left\{\begin{array}{l} E^k(z) = (E_c - E_m)\ V_c^k(z) + E_m \\ \alpha^k(z) = (\alpha_c - \alpha_m)\ V_c^k(z) + \alpha_m \\ \nu^k(z) = (\nu_c - \nu_m)\ V_c^k(z) + \nu_m \end{array}\right. \tag{6.45}$$

Finally the elastic coefficients $\tilde{C}_{ij}$ are given as:

$$\left\{\begin{array}{l} \tilde{C}_{11}^k(z) = \tilde{C}_{22}^k(z) = \tilde{C}_{33}^k(z) = \dfrac{E^k(z)\left[1-\left(\nu(z)^k\right)^2\right]}{1-3\left(\nu(z)^k\right)^2-2\left(\nu(z)^k\right)^3} \\ \tilde{C}_{12}^k(z) = \tilde{C}_{13}^k(z) = \tilde{C}_{23}^k(z) = \dfrac{E^k(z)\,\nu_0^k\left[1-\nu(z)^k\right]}{1-3\left(\nu(z)^k\right)^2-2\left(\nu(z)^k\right)^3} \\ \tilde{C}_{44}^k(z) = \tilde{C}_{55}^k(z) = \tilde{C}_{66}^k(z) = \dfrac{E^k(z)}{2\left(1+\nu(z)^k\right)} \end{array}\right. \tag{6.46}$$

The thermo-mechanical coupling coefficients are expressed as:

$$\begin{array}{l} \boldsymbol{\lambda}_p^k(z) = \tilde{\mathbf{C}}_{pp}^k(z)\,\tilde{\boldsymbol{\alpha}}_p^k(z) + \tilde{\mathbf{C}}_{pn}^k(z)\,\tilde{\boldsymbol{\alpha}}_n^k(z) \\ \boldsymbol{\lambda}_n^k(z) = \tilde{\mathbf{C}}_{np}^k(z)\,\tilde{\boldsymbol{\alpha}}_p^k(z) + \tilde{\mathbf{C}}_{nn}^k(z)\,\tilde{\boldsymbol{\alpha}}_n^k(z) \end{array} \tag{6.47}$$

where

$$\tilde{\boldsymbol{\alpha}}_p^k(z) = \begin{bmatrix} \alpha(z) \\ \alpha(z) \\ 0 \end{bmatrix}^k, \quad \tilde{\boldsymbol{\alpha}}_n^k(z) = \begin{bmatrix} 0 \\ 0 \\ \alpha(z) \end{bmatrix}^k \tag{6.48}$$

are the thermal expansion coefficients split in in-plane and out-of-plane components. In the explicit vectorial form they can be written as:

$$\boldsymbol{\lambda}_p^k(z) = \begin{bmatrix} \lambda(z) \\ \lambda(z) \\ 0 \end{bmatrix}^k, \quad \boldsymbol{\lambda}_n^k(z) = \begin{bmatrix} 0 \\ 0 \\ \lambda(z) \end{bmatrix}^k \tag{6.49}$$

6.2.1 Temperature-dependent functionally graded materials

More generally, the constituents of FGM may have temperature-dependent properties (for instance see Ref. [6]). Therefore, the properties of the common structural ceramics and metals are expressed as functions of tem-

Table 6.1 Temperature-dependent coefficients for Si_3N_4, SUS304, ZrO_2 and Ti-6Al-4V (Ref. [7])

Material	Properties	$\mathcal{P}_{-1}$	$\mathcal{P}_0$	$\mathcal{P}_1$	$\mathcal{P}_2$	$\mathcal{P}_3$
Si_3N_4	E_c [GPa]	0.0	348.43	-3.070×10^{-4}	2.160×10^{-7}	-8.946×10^{-11}
	ν_c	0.0	0.2400	0.0	0.0	0.0
	α_c [1/K]	0.0	5.8723×10^{-6}	9.095×10^{-6}	0.0	0.0
	ρ_c [Kg/m^3]	0.0	2370	0.0	0.0	0.0
SUS304	E_m [GPa]	0.0	201.04	3.079×10^{-4}	-6.534×10^{-7}	0.0
	ν_m	0.0	0.3262	-2.002×10^{-4}	3.797×10^{-7}	0.0
	α_m [1/K]	0.0	12.330×10^{-6}	8.086×10^{-6}	0.0	0.0
	ρ_m [Kg/m^3]	0.0	8166	0.0	0.0	0.0
ZrO_2	E_c [GPa]	0.0	244.27	-1.371×10^{-3}	1.214×10^{-6}	-3.681×10^{-10}
	ν_c	0.0	0.2882	1.133×10^{-4}	0.0	0.0
	α_c [1/K]	0.0	12.766×10^{-6}	-1.491×10^{-3}	1.006×10^{-5}	-6.778×10^{-11}
	ρ_c [Kg/m^3]	0.0	3000	0.0	0.0	0.0
Ti-6Al-4V	E_m [GPa]	0.0	122.56	-4.586×10^{-4}	0.0	0.0
	ν_m	0.0	0.2884	1.121×10^{-4}	0.0	0.0
	α_m [1/K]	0.0	7.5788×10^{-6}	6.638×10^{-4}	-3.147×10^{-6}	0.0
	ρ_m [Kg/m^3]	0.0	4429	0.0	0.0	0.0

perature, and Eq. (6.46) becomes

$$\begin{cases} \tilde{C}_{11}(z,T)=\tilde{C}_{22}(z,T)=\tilde{C}_{33}(z,T)=\dfrac{E(z,T)\left[1-(\nu(z,T))^2\right]}{1-3(\nu(z,T))^2-2(\nu(z,T))^3} \\ \tilde{C}_{12}(z,T)=\tilde{C}_{13}(z,T)=\tilde{C}_{23}(z,T)=\dfrac{E(z,T)\,\nu(z,T)\left[1-\nu(z,T)\right]}{1-3(\nu(z,T))^2-2(\nu(z,T))^3} \\ \tilde{C}_{44}(z,T)=\tilde{C}_{55}(z,T)=\tilde{C}_{66}(z,T)=\dfrac{E(z,T)}{2(1+\nu(z,T))} \end{cases} \tag{6.50}$$

The data are obtained from experiments reported in the literature (see Ref. [7]). All material properties (P) are expressed in the form

$$\mathcal{P}=\mathcal{P}_0\left(\mathcal{P}_{-1}T^{-1}+1+\mathcal{P}T+\mathcal{P}_2T^2+\mathcal{P}_3T^3\right) \tag{6.51}$$

where $\mathcal{P}_0$, $\mathcal{P}^{-1}$, $\mathcal{P}$, $\mathcal{P}^2$ and $\mathcal{P}^3$ are constants in the cubic fit of the material property. With the above equation the higher-order effect of the temperature on the material is captured. The constant values for some materials, namely, Silicon Nitride (Si_3N_4), Stainless Still (SUS304), Zirconia (ZrO_2) and Ti-6Al-4V are given in Table 6.1 The effect of the temperature increase on the top of the FGM plate with temperature-dependent materials, as expected, generates a degradation of the mechanical properties. The latter phenomenon is shown in Fig. 6.7 where the classical power-law (P-FGM)

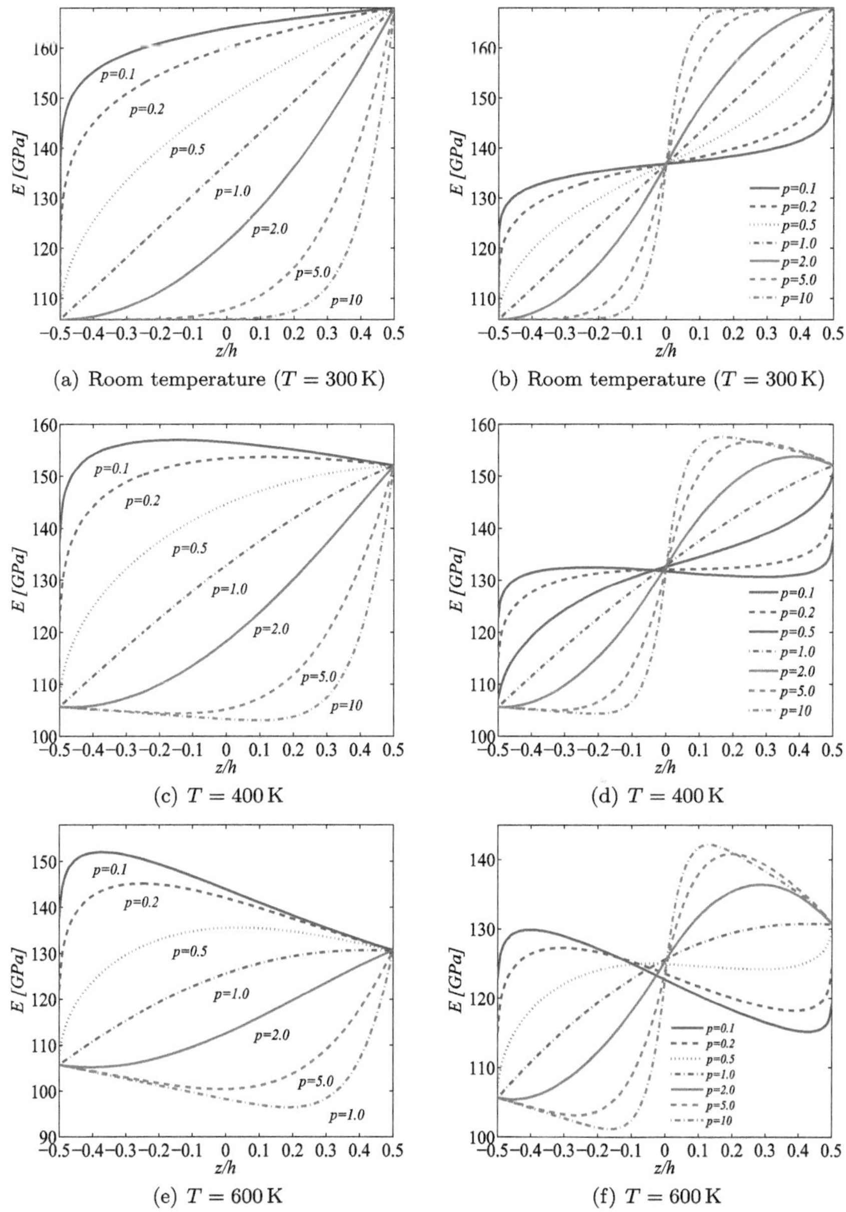

Figure 6.7 Young's modulus *E* degradation when increasing the temperature.

and the sigmoid-law (S-FGM) of the Young's modulus are depicted for different temperature values.

6.3 RMVT CONSTITUTIVE LAW

In the case of the application of RMVT the stress-strain relationships assume the following form (see Ref. [8]):

$$\begin{aligned}\boldsymbol{\sigma}^k_{pH} &= \hat{\mathbf{C}}^k_{pp}\,\boldsymbol{\varepsilon}^k_{pG} + \hat{\mathbf{C}}^k_{pn}\,\boldsymbol{\sigma}^k_{nM}\\ \boldsymbol{\sigma}^k_{nH} &= \hat{\mathbf{C}}^k_{np}\,\boldsymbol{\varepsilon}^k_{pG} + \hat{\mathbf{C}}^k_{nn}\,\boldsymbol{\sigma}^k_{nM}\end{aligned} \tag{6.52}$$

The subscript M in Eq. (6.52) states that the transverse shear and normal stresses are those of the assumed model. Consequently, by exploiting the use of Eq. (6.52), the relation between the elastic coefficients of the mixed model and the classical ones is immediately found:

$$\begin{aligned}\hat{\mathbf{C}}^k_{pp} &= \tilde{\mathbf{C}}^k_{pp} - \tilde{\mathbf{C}}^k_{pn}\left(\tilde{\mathbf{C}}^k_{nn}\right)^{-1}\tilde{\mathbf{C}}^k_{np}, & \hat{\mathbf{C}}^k_{pn} &= \tilde{\mathbf{C}}^k_{pn}\left(\tilde{\mathbf{C}}^k_{nn}\right)^{-1}\\ \hat{\mathbf{C}}^k_{np} &= -\left(\tilde{\mathbf{C}}^k_{nn}\right)^{-1}\tilde{\mathbf{C}}^k_{np}, & \hat{\mathbf{C}}^k_{nn} &= \left(\tilde{\mathbf{C}}^k_{nn}\right)^{-1}\end{aligned} \tag{6.53}$$

Superscript -1 denotes an inversion of the array.

6.4 CONSTITUTIVE EQUATIONS FOR THERMOELASTIC PROBLEMS

The constitutive equations for the thermoelastic problems are proposed according to that derived in Ref. [9]. The coupling between the mechanical and thermal fields can be determined by using thermodynamical principles and Maxwell's relations [10–14]. For this purpose, it is necessary to define a *Gibbs free-energy function* G and a *thermomechanical enthalpy density* H:

$$\begin{aligned}G\left(\varepsilon_{ij},\theta\right) &= \sigma_{ij}\,\varepsilon_{ij} - \eta\,\theta\\ H\left(\varepsilon_{ij},\theta,\vartheta_i\right) &= G\left(\varepsilon_{ij},\theta\right) - F\left(\vartheta_i\right)\end{aligned} \tag{6.54}$$

where σ_{ij} and ε_{ij} are the stress and strain components, η is the entropy per unit of volume, and θ the temperature considered with respect to the reference temperature T_0. The function F is the dissipation function, it depends on the spatial temperature gradient:

$$F\left(\vartheta_i\right) = \frac{1}{2}\kappa_{ij}\,\vartheta_i\,\vartheta_j - \tau_0\,\dot{h}_i \tag{6.55}$$

where κ_{ij} is the symmetric, positive semi-definite conductivity tensor. In the second term, τ_0 is a thermal relaxation parameter which multiplies the temporal derivative of the heat flux $\dot{h}_i$. The thermal relaxation parameter is omitted in the present work. The thermo-mechanical enthalpy density H

can be expanded in order to obtain a quadratic form for a linear interaction:

$$H\left(\varepsilon_{ij},\theta,\vartheta_i\right)=\frac{1}{2}C_{ijkl}\,\varepsilon_{ij}\,\varepsilon_{kl}-\lambda_{ij}\,\varepsilon_{ij}\,\theta-\frac{1}{2}\chi\,\theta^2-\frac{1}{2}\kappa_{ij}\,\vartheta_i\,\vartheta_j \tag{6.56}$$

where C_{ijkl} is the elastic coefficients tensor considered for an orthotropic material, λ_{ij} are the thermo-mechanical coupling coefficients, $\chi=\frac{\rho}{C_v\,T_0}$ where ρ is the material density, C_v is the specific heat per unit mass and T_0 is the reference temperature. The constitutive equations are obtained by considering the following relations:

$$\sigma_{ij}=\frac{\partial H}{\partial\varepsilon_{ij}},\qquad \eta=-\frac{\partial H}{\partial\theta},\qquad h_i=-\frac{\partial H}{\partial\vartheta_i} \tag{6.57}$$

By combining Eqs. (6.56) and (6.57), the constitutive equations for the thermo-mechanical problem are obtained

$$\begin{aligned}\sigma_{ij}&=C_{ijkl}\,\varepsilon_{ij}-\lambda_{ij}\,\theta\\ \eta&=\lambda_{ij}\,\varepsilon_{ij}+\chi\,\theta\\ h_i&=\kappa_{ij}\,\vartheta_i\end{aligned} \tag{6.58}$$

The above equations can be written in matrix form introducing the usual bold scripture. Considering a generic multilayer structure Eqs. (6.58) are written for a generic k-layer as

$$\begin{aligned}\boldsymbol{\sigma}^k&=\tilde{\mathbf{C}}^k\,\boldsymbol{\varepsilon}^k-\tilde{\boldsymbol{\lambda}}^k\,\theta^k\\ \eta^k&=\tilde{\boldsymbol{\lambda}}^k\,\boldsymbol{\varepsilon}+\chi^k\,\theta^k\\ \mathbf{h}^k&=\tilde{\boldsymbol{\kappa}}^k\,\boldsymbol{\vartheta}^k\end{aligned} \tag{6.59}$$

the symbol $(\tilde{\ })$ indicates that the equations refer to the problem reference system (x,y,z) for plates or (α,β,z) for shells. In order to use the relations given in Eqs. (6.59) in the proposed variational statements, that will be presented in the next section, it is convenient to split them in in-plane components (subscript p) and out-of-plane components (subscript n) as follows:

$$\begin{aligned}\boldsymbol{\sigma}^k_{pC}&=\tilde{\mathbf{C}}_{pp}\,\varepsilon^k_{pG}+\tilde{\mathbf{C}}_{pn}\,\varepsilon^k_{nG}-\tilde{\boldsymbol{\lambda}}^k_p\,\theta^k\\ \boldsymbol{\sigma}^k_{nC}&=\tilde{\mathbf{C}}_{np}\,\varepsilon^k_{pG}+\tilde{\mathbf{C}}_{nn}\,\varepsilon^k_{nG}-\tilde{\boldsymbol{\lambda}}^k_n\,\theta^k\\ \eta^k_C&=\boldsymbol{\lambda}^{k\,T}_p\,\varepsilon^k_{pG}+\boldsymbol{\lambda}^{k\,T}_n\,\varepsilon^k_{nG}+\chi^k\,\theta^k\\ \boldsymbol{h}^k_p&=\tilde{\boldsymbol{\kappa}}^k_{pp}\,\boldsymbol{\vartheta}^k_{pG}+\tilde{\boldsymbol{\kappa}}^k_{pn}\,\boldsymbol{\vartheta}^k_{nG}\\ \boldsymbol{h}^k_n&=\tilde{\boldsymbol{\kappa}}^k_{np}\,\boldsymbol{\vartheta}^k_{pG}+\tilde{\boldsymbol{\kappa}}^k_{nn}\,\boldsymbol{\vartheta}^k_{nG}\end{aligned} \tag{6.60}$$

Other two new subscripts are introduced: the subscript C for those variables, in the variational statements, which need the substitution of constitutive equations; the subscript G for those variables, in constitutive equations, which need the substitution of geometrical relations. The explicit forms of the split matrices in Eqs. (6.60) are

- Thermo-mechanical coupling coefficients

$$\tilde{\boldsymbol{\lambda}}_p^k = \begin{bmatrix} \lambda_x \\ \lambda_y \\ \lambda_{xy} \end{bmatrix}^k, \quad \tilde{\boldsymbol{\lambda}}_n^k = \begin{bmatrix} 0 \\ 0 \\ \lambda_z \end{bmatrix}^k \tag{6.61}$$

- Heat flux and spatial gradient of the temperature

$$\mathbf{h}_{pC}^k = \begin{bmatrix} h_x \\ h_y \end{bmatrix}^k, \quad \mathbf{h}_{nC}^k = \begin{bmatrix} h_z \end{bmatrix}^k, \quad \boldsymbol{\vartheta}_{pG}^k = \begin{bmatrix} \vartheta_x \\ \vartheta_y \end{bmatrix}^k, \quad \boldsymbol{\vartheta}_{nG}^k = \begin{bmatrix} \vartheta_z \end{bmatrix}^k \tag{6.62}$$

- Conductivity coefficients

$$\tilde{\boldsymbol{\kappa}}_{pp}^k = \begin{bmatrix} \kappa_x & \kappa_{xy} \\ \kappa_{xy} & \kappa_y \end{bmatrix}^k, \quad \tilde{\boldsymbol{\kappa}}_{nn}^k = \begin{bmatrix} \kappa_z \end{bmatrix}^k, \quad \tilde{\boldsymbol{\kappa}}_{pn}^k = \begin{bmatrix} 0 \\ 0 \end{bmatrix}^k, \quad \tilde{\boldsymbol{\kappa}}_{np} = \begin{bmatrix} 0 & 0 \end{bmatrix}^k \tag{6.63}$$

REFERENCES

1. Reddy JN. Energy principles and variational methods in applied mechanics. 2nd edition. New Jersey: John Wiley & Sons; 2002.
2. Reddy JN. Variational methods in theoretical mechanics. 2nd edition. Berlin: Springer-Verlag; 1982.
3. Reddy JN. Mechanics of laminated composite plates and shells – theory and analysis. 2nd edition. CRC Press; 2004.
4. Fazzolari FA. Natural frequencies and critical temperatures of functionally graded sandwich plates subjected to uniform and non-uniform temperature distributions. Compos Struct 2015;121:197–210.
5. Fazzolari FA. Stability analysis of FGM sandwich plates by using variable-kinematics Ritz models. Mech Adv Mat Struct 2016;23(9):1–27.
6. Fazzolari FA. Modal characteristics of P- and S-FGM plates with temperature-dependent materials in thermal environment. J Therm Stresses 2016;39(7):854–73.
7. Reddy JN, Chin CD. Thermomechanical analysis of functionally graded cylinders and plates. J Therm Stresses 1998;21:593–626.
8. Fazzolari FA. Reissner's mixed variational theorem and variable kinematics in the modelling of laminated composite and FGM doubly-curved shells. Composites, Part B, Eng 2016;89:408–23.

9. Fazzolari FA, Carrera E. Coupled thermoelastic effect in free vibration analysis of anisotropic multilayered plates and FGM plates by using a variable-kinematics Ritz formulation. Eur J Mech A, Solids 2014;44:157–74.
10. Altay GA, Dokmeci M. Some variational principles for linear coupled thermoelasticity. Int J Solids Struct 1996;33(26):3937–48.
11. Altay GA, Dokmeci M. Fundamental variational equations of discontinuous thermopiezoelectric fields. Int J Eng Sci 1996;34(7):769–82.
12. Cannarozzi AA, Umbertini FA. A mixed variational method for linear coupled thermoelastic analysis. Int J Solids Struct 2001;38(4):717–39.
13. Brischetto S, Carrera E. Coupled thermo-mechanical analysis of one-layered and multilayered plates. Compos Struct 2010;92:1793–812.
14. Brischetto S, Carrera E. Thermomechanical effect in vibration analysis of one-layered and two-layered plates. Int J Appl Mech 2011;3(1):165–85.

CHAPTER 7

Computational methods for thermal stress analysis

7.1 APPROXIMATE SOLUTION METHODS

When dealing with boundary value problems, the coupled differential governing equations and related variationally consistent boundary conditions, which come from the variational statements proposed in chapter 3, can be solved in a closed-form manner, in very few cases. In particular, the latter take into account particular geometries, boundary conditions and lamination schemes. In the most general case approximate solution methods such as Ritz, Galerkin, Generalized Galerkin and/or other meshless methods as well as finite element method (FEM) are essential to have a solution of various structural problems. In the derivation of what follows, Ritz, Galerkin and Generalized Galerkin formulations are fully developed for beam, plate and shell structures.

7.2 RITZ METHOD

The fundamentals of the Ritz method are briefly introduced using the notation for the 2D case. In the Ritz method the functional degrees of freedom, $\mathbf{u}_\tau^k$ are expressed in series expansion as follows

$$\mathbf{u}_\tau^k\left(x, y, t\right) = \mathbf{U}_{\tau i}^k\left(t\right)\,\boldsymbol{\Psi}_i\left(x, y\right) \quad \text{where} \quad i = 1, \ldots, \mathcal{N} \quad \tau = \tau_{u_x}, \tau_{u_y}, \tau_{u_z} \tag{7.1}$$

$\mathcal{N}$ indicates the order of expansion in the approximation. Consequently the displacement field, in compact way, assumes the following form

$$\mathbf{u}^k\left(x, y, z, t\right) = \mathbf{F}_\tau\left(z\right)\,\mathbf{U}_{\tau i}^k\left(t\right)\,\boldsymbol{\Psi}_i\left(x, y\right) \tag{7.2}$$

where

$$\boldsymbol{\Psi}_i\left(x, y\right) = \begin{bmatrix} \psi_{x_i}\left(x, y\right) & 0 & 0 \\ 0 & \psi_{y_i}\left(x, y\right) & 0 \\ 0 & 0 & \psi_{z_i}\left(x, y\right) \end{bmatrix} \tag{7.3}$$

The vectors $\mathbf{U}_{\tau i}^k$ and the matrix $\mathbf{F}_\tau$ have already been defined in chapter 5, section 5.25, Eq. (5.196). The Ritz functions ψ_{x_i}, ψ_{z_i}, ψ_{z_i} are chosen

Thermal Stress Analysis of Composite Beams, Plates and Shells.
DOI: http://dx.doi.org/10.1016/B978-0-12-849892-7.00009-1

according to the type of problem addressed. The results are strongly dependent on this choice. Convergence to the exact solution is guaranteed if the Ritz functions are admissible functions, i.e., they satisfy the following three conditions [1–5] :

- be continuous as required in the variational statement (i.e., should be such that it has a non-zero contribution to the virtual work statement);
- satisfy the homogeneous form of the specified geometric boundary condition;
- the set is linearly independent and complete.

While applying the Ritz method, it is useful to rewrite the PVD as

$$\delta L^k_{int} - \delta L^k_{ext} + \delta L^k_{ine} = 0 \tag{7.4}$$

where L_{int} is the virtual internal work, L_{ext} is the virtual external work and L_{ine} is virtual inertial work. In particular, the latter is related to the kinetic energy by the following relation

$$\delta T^k = -\delta L^k_{F_{ine}} \tag{7.5}$$

The mathematical proof has been briefly recalled in chapter 3. Moreover, for elastic systems subjected to conservative forces the following equivalences hold

$$\delta L^k_{int} = \delta \Phi^k_{se}, \quad \delta L^k_{ext} = -\delta \Phi^k_{ef} \tag{7.6}$$

where δU is the virtual potential strain energy and δV is the virtual potential energy related to the external forces. Being

$$\Pi^k = \Phi^k_{se} + \Phi^k_{ef} \tag{7.7}$$

the total potential energy functional, then Eq. (7.4) corresponds to a minimization of the functional related to the energy system under investigation

$$\delta\left(T^k - \Pi^k\right) = 0 \tag{7.8}$$

The minimization is respect to the unknown coefficients of linear combination that derived to the approximation solution in Eq. (7.2). In particular, Π^k is a function of $U^k_{x\tau_{u_x}i}$, $U^k_{y\tau_{u_y}i}$, $U^k_{z\tau_{u_z}i}$ and the condition given in

Eq. (7.8) can be alternatively written in the following form

$$
\begin{aligned}
&\frac{\partial\left(T^k-\Pi^k\right)}{\partial U^k_{x\tau_{u_x}i}}=0\\
&\quad\text{with}\quad i=1,\cdots,\mathcal{N};\quad \tau_{u_x}=b_{u_x},r_{u_x},t_{u_x}\quad r_{u_x}=2,3,\cdots,N_{u_x}-1\\
&\frac{\partial\left(T^k-\Pi^k\right)}{\partial U^k_{y\tau_{u_y}i}}=0\\
&\quad\text{with}\quad i=1,\cdots,\mathcal{N};\quad \tau_{u_y}=b_{u_y},r_{u_y},t_{u_y}\quad r_{u_y}=2,3,\cdots,N_{u_y}-1\\
&\frac{\partial\left(T^k-\Pi^k\right)}{\partial U^k_{z\tau_{u_z}i}}=0\\
&\quad\text{with}\quad i=1,\cdots,\mathcal{N};\quad \tau_{u_z}=b_{u_z},b_{u_z},t_{u_z}\quad r_{u_z}=2,3,\cdots,N_{u_z}-1
\end{aligned}
\tag{7.9}
$$

7.2.1 Beams

Stiffness nucleus K

In the derivation of what follows the Principle of Virtual Displacements (PVD) is employed to derive the Hierarchical Ritz Formulation (HRF). The PVD variational statement for beam structures used in the present analysis at multilayer level can be written as

$$
\begin{aligned}
&\sum_{k=1}^{N_l}\int_{\Omega^k}\int_l\left(\delta\boldsymbol{\varepsilon}^{k\,T}_{pG}\,\boldsymbol{\sigma}^k_{pC}+\delta\boldsymbol{\varepsilon}^{k\,T}_{nG}\,\boldsymbol{\sigma}^k_{nC}\right)\mathrm{d}\Omega^k\,\mathrm{d}z=-\sum_{k=1}^{N_l}\int_{\Omega^k}\int_l\rho^k\,\delta\mathbf{u}^k\,\ddot{\mathbf{u}}^k\,\mathrm{d}\Omega^k\,\mathrm{d}z\\
&\sum_{k=1}^{N_l}\delta L^k_{ext}
\end{aligned}
\tag{7.10}
$$

where Ω_k and l are the area of the beam cross-section and the beam length, respectively. As shown at the beginning of the section 7.2 for the 2D case, similarly for the 1D case the displacement amplitude vector components $u_{x_{\tau_{u_x}}}$, $u_{y_{\tau_{u_y}}}$ and $u_{z_{\tau_{u_z}}}$ are expressed in series expansion as follows

$$
\begin{aligned}
u^k_{x_{\tau_{u_x}}}(z,t)&=\sum_i^{\mathcal{N}}U^k_{x\tau_{u_x}i}\,\psi_{x_i}(z)\,e^{\imath\,\omega_{ij}\,t}\\
u^k_{y_{\tau_{u_y}}}(z,t)&=\sum_i^{\mathcal{N}}U^k_{y\tau_{u_y}i}\,\psi_{y_i}(z)\,e^{\imath\,\omega_{ij}\,t}\\
u^k_{z_{\tau_{u_z}}}(z,t)&=\sum_i^{\mathcal{N}}U^k_{z\tau_{u_z}i}\,\psi_{z_i}(z)\,e^{\imath\,\omega_{ij}\,t}
\end{aligned}
\tag{7.11}
$$

where $\iota = \sqrt{-1}$, t is the time and ω_{ij} the circular frequency; $\mathcal{N}$ indicates the order of expansion in the approximation; $U_{x\tau_{ux}i}$, $U_{y\tau_{uy}i}$, $U_{z\tau_{uz}i}$ are the unknown coefficients and ψ_{x_i}, ψ_{y_i}, ψ_{z_i} are the Ritz functions appropriately selected making reference to the features of the analyzed problem. The latter have to fulfil the conditions briefly recalled in section 7.2. The displacement field is then given as

$$
\begin{aligned}
u_x^k(x, y, z, t) &= \sum_i^{\mathcal{N}} U_{x\tau_{ux}i}^k F_{\tau_{ux}}(x, y)\, \psi_{x_i}(z)\, e^{\iota\, \omega_{ij} t} \\
u_y^k(x, y, z, t) &= \sum_i^{\mathcal{N}} U_{y\tau_{uy}i}^k F_{\tau_{uy}}(x, y)\, \psi_{y_i}(z)\, e^{\iota\, \omega_{ij} t} \\
u_z^k(x, y, z, t) &= \sum_i^{\mathcal{N}} U_{z\tau_{uz}i}^k F_{\tau_{uz}}(x, y)\, \psi_{z_i}(z)\, e^{\iota\, \omega_{ij} t}
\end{aligned} \tag{7.12}
$$

and in compact form

$$
\mathbf{u}^k = \mathbf{F}_\tau\, \mathbf{U}_{\tau i}^k\, \mathbf{\Psi}_i \tag{7.13}
$$

where

$$
\mathbf{U}_{\tau i}^k(t) = \begin{Bmatrix} U_{x\tau_{ux}i}^k e^{\iota\, \omega_{ij} t} \\ U_{y\tau_{uy}i}^k e^{\iota\, \omega_{ij} t} \\ U_{z\tau_{uz}i}^k e^{\iota\, \omega_{ij} t} \end{Bmatrix}, \quad \mathbf{\Psi}_i(z) = \begin{bmatrix} \psi_{x_i}(z) & 0 & 0 \\ 0 & \psi_{y_i}(z) & 0 \\ 0 & 0 & \psi_{z_i(z)} \end{bmatrix} \tag{7.14}
$$

By introducing Eq. (7.13) in the geometrical relations the following expression is found

$$
\begin{aligned}
\boldsymbol{\varepsilon}_{pG}^k &= \mathbf{D}_p (\mathbf{F}_\tau\, \mathbf{\Psi}_i)\, \mathbf{U}_{\tau i}^k \\
\boldsymbol{\varepsilon}_{nG}^k &= \mathbf{D}_{np} (\mathbf{F}_\tau\, \mathbf{\Psi}_i)\, \mathbf{U}_{\tau i}^k + \mathbf{D}_{nz} (\mathbf{F}_\tau\, \mathbf{\Psi}_i)\, \mathbf{U}_{\tau i}^k
\end{aligned} \tag{7.15}
$$

By substituting the abode equation in Eq. (7.10) the explicit expressions of the internal work, split in its fourth contributions, and the inertial work at layer level are obtained

$$
\begin{aligned}
\delta\left(L_{int}^k\right)_{pp} &= \int_{\Omega^k} \int_l \delta\boldsymbol{\varepsilon}_{pG}^{k\,T}\, \tilde{\mathbf{C}}_{pp}^k\, \boldsymbol{\varepsilon}_{pG}^k\, d\Omega^k\, dz \\
&= \int_{\Omega^k} \int_l \delta\mathbf{U}_{\tau i}^{kT} \left(\left[\mathbf{D}_p (\mathbf{F}_\tau\, \mathbf{\Psi}_i)\right]^T \tilde{\mathbf{C}}_{pp}^k \mathbf{D}_p (\mathbf{F}_s\, \mathbf{\Psi}_j)\right) \mathbf{U}_{sj}^k\, \mathrm{d}\Omega\, \mathrm{d}z \\
\delta\left(L_{int}^k\right)_{pn} &= \int_{\Omega^k} \int_l \delta\boldsymbol{\varepsilon}_{pG}^{k\,T}\, \tilde{\mathbf{C}}_{pn}^k\, \boldsymbol{\varepsilon}_{nG}^k\, d\Omega^k\, dz \\
&= \int_{\Omega^k} \int_l \delta\mathbf{U}_{\tau i}^{kT} \left(\left[\mathbf{D}_p (\mathbf{F}_\tau\, \mathbf{\Psi}_i)\right]^T \tilde{\mathbf{C}}_{pn}^k \mathbf{D}_{np} (\mathbf{F}_s\, \mathbf{\Psi}_j) + \right. \\
&\qquad \left. \left[\mathbf{D}_p (\mathbf{F}_\tau\, \mathbf{\Psi}_i)\right]^T \tilde{\mathbf{C}}_{pn}^k \mathbf{D}_{nz} (\mathbf{F}_s\, \mathbf{\Psi}_j)\right) \mathbf{U}_{sj}^k\, \mathrm{d}\Omega\, \mathrm{d}z
\end{aligned}
$$

$$
\begin{aligned}
\delta\left(L_{int}^{k}\right)_{np} &= \int_{\Omega^k}\int_{l} \delta\boldsymbol{\varepsilon}_{nG}^{k\ T}\,\tilde{\mathbf{C}}_{pn}^{k}\,\boldsymbol{\varepsilon}_{pG}^{k}\,d\Omega^k\,dz\\
&= \int_{\Omega^k}\int_{l} \delta\mathbf{U}_{\tau i}^{kT}\left(\left[\mathbf{D}_{np}\left(\mathbf{F}_\tau\,\boldsymbol{\Psi}_i\right)\right]^T\tilde{\mathbf{C}}_{np}^{k}\mathbf{D}_p\left(\mathbf{F}_s\,\boldsymbol{\Psi}_j\right)+\right.\\
&\qquad\left.\left[\mathbf{D}_{nz}\left(\mathbf{F}_\tau\,\boldsymbol{\Psi}_i\right)\right]^T\tilde{\mathbf{C}}_{np}^{k}\mathbf{D}_p\left(\mathbf{F}_s\,\boldsymbol{\Psi}_j\right)\right)\mathbf{U}_{sj}^{k}\,\mathrm{d}\Omega\,\mathrm{d}z\\
\delta\left(L_{int}^{k}\right)_{nn} &= \int_{\Omega^k}\int_{l} \delta\boldsymbol{\varepsilon}_{nG}^{k\ T}\,\tilde{\mathbf{C}}_{nn}^{k}\,\boldsymbol{\varepsilon}_{nG}^{k}\,d\Omega^k\,dz\\
&= \int_{\Omega^k}\int_{l} \delta\mathbf{U}_{\tau i}^{kT}\left(\left[\mathbf{D}_{np}\left(\mathbf{F}_\tau\,\boldsymbol{\Psi}_i\right)\right]^T\tilde{\mathbf{C}}_{nn}^{k}\mathbf{D}_{np}\left(\mathbf{F}_s\,\boldsymbol{\Psi}_j\right)+\right.\\
&\qquad\left[\mathbf{D}_{np}\left(\mathbf{F}_\tau\,\boldsymbol{\Psi}_i\right)\right]^T\tilde{\mathbf{C}}_{nn}^{k}\mathbf{D}_{nz}\left(\mathbf{F}_s\,\boldsymbol{\Psi}_j\right)+\\
&\qquad\left[\mathbf{D}_{nz}\left(\mathbf{F}_\tau\,\boldsymbol{\Psi}_i\right)\right]^T\tilde{\mathbf{C}}_{nn}^{k}\mathbf{D}_{np}\left(\mathbf{F}_s\,\boldsymbol{\Psi}_j\right)+\\
&\qquad\left.\left[\mathbf{D}_{nz}\left(\mathbf{F}_\tau\,\boldsymbol{\Psi}_i\right)\right]^T\tilde{\mathbf{C}}_{nn}^{k}\mathbf{D}_{nz}\left(\mathbf{F}_s\,\boldsymbol{\Psi}_j\right)\right)\mathbf{U}_{sj}^{k}\,\mathrm{d}\Omega\,\mathrm{d}z
\end{aligned}
\tag{7.16}
$$

The overall internal work can then be obtained by the summation of the in-plane and out-of-plane contributions

$$
\begin{aligned}
\delta L_{int}^{k} &= \delta\left(L_{int}^{k}\right)_{pp}+\delta\left(L_{int}^{k}\right)_{pn}+\delta\left(L_{int}^{k}\right)_{np}+\delta\left(L_{int}^{k}\right)_{nn}\\
&= \delta\mathbf{U}_{\tau i}^{kT}\left(\mathbf{K}_{pp}^{k\tau sij}+\mathbf{K}_{pn}^{k\tau sij}+\mathbf{K}_{np}^{k\tau sij}+\mathbf{K}_{nn}^{k\tau sij}\right)\mathbf{U}_{sj}^{k}\\
&= \delta\mathbf{U}_{\tau i}^{kT}\,\mathbf{K}^{k\tau sij}\,\mathbf{U}_{sj}^{k}
\end{aligned}
\tag{7.17}
$$

where

$$
\mathbf{K}^{k\tau sij}=\mathbf{K}_{pp}^{k\tau sij}+\mathbf{K}_{pn}^{k\tau sij}+\mathbf{K}_{np}^{k\tau sij}+\mathbf{K}_{nn}^{k\tau sij}
\tag{7.18}
$$

and its explicit final form is given as follows

$$
\begin{aligned}
\mathbf{K}^{k\tau sij} = &\int_{\Omega^k}\int_{l}\left(\left[\mathbf{D}_p\left(\mathbf{F}_\tau\,\boldsymbol{\Psi}_i\right)\right]^T\right.\\
&\left[\tilde{C}_{pp}^{k}\mathbf{D}_p\left(\mathbf{F}_s\,\boldsymbol{\Psi}_j\right)+\tilde{C}_{pn}^{k}\mathbf{D}_{np}\left(\mathbf{F}_s\,\boldsymbol{\Psi}_j\right)+\tilde{C}_{pn}^{k}\mathbf{D}_{nz}\left(\mathbf{F}_s\,\boldsymbol{\Psi}_j\right)\right]\\
&+\left[\mathbf{D}_{np}\left(\mathbf{F}_\tau\,\boldsymbol{\Psi}_i\right)\right]^T\\
&\left[\tilde{C}_{np}^{k}\mathbf{D}_p\left(\mathbf{F}_s\,\boldsymbol{\Psi}_j\right)+\tilde{C}_{nn}^{k}\mathbf{D}_{np}\left(\mathbf{F}_s\,\boldsymbol{\Psi}_j\right)+\tilde{C}_{nn}^{k}\mathbf{D}_{nz}\left(\mathbf{F}_s\,\boldsymbol{\Psi}_j\right)\right]\\
&+\left[\mathbf{D}_{nz}\left(\mathbf{F}_\tau\,\boldsymbol{\Psi}_i\right)\right]^T\\
&\left.\left[\tilde{C}_{np}^{k}\mathbf{D}_p\left(\mathbf{F}_s\,\boldsymbol{\Psi}_j\right)+\tilde{C}_{nn}^{k}\mathbf{D}_{np}\left(\mathbf{F}_s\,\boldsymbol{\Psi}_j\right)+\tilde{C}_{nn}^{k}\mathbf{D}_{nz}\left(\mathbf{F}_s\,\boldsymbol{\Psi}_j\right)\right]\right)\mathrm{d}\Omega^k\,\mathrm{d}z
\end{aligned}
\tag{7.19}
$$

After performing the matrix calculus in Eq. (7.19) the nine secondary stiffness nuclei are obtained

$$K_{u_x u_x}^{k\,\tau_{u_x} s_{u_x}} = \tilde{C}_{11}^k \int_{\Omega^k} F_{\tau_{u_{x,x}}} F_{s_{u_{x,x}}} \mathrm{d}\Omega \int_l \psi_{x_i} \psi_{x_j} \mathrm{d}z + \tilde{C}_{16}^k \int_{\Omega^k} F_{\tau_{u_{x,y}}} F_{s_{u_{x,x}}} \mathrm{d}\Omega \int_l \psi_{x_i} \psi_{x_j} \mathrm{d}z + \tilde{C}_{11}^k \int_{\Omega^k} F_{\tau_{u_{x,x}}} F_{s_{u_{x,y}}} \mathrm{d}\Omega \int_l \psi_{x_i} \psi_{x_j} \mathrm{d}z + \tilde{C}_{66}^k \int_{\Omega^k} F_{\tau_{u_{x,y}}} F_{s_{u_{x,y}}} \mathrm{d}\Omega \int_l \psi_{x_i} \psi_{x_j} \mathrm{d}z + \tilde{C}_{55}^k \int_{\Omega^k} F_{\tau_{u_x}} F_{s_{u_x}} \mathrm{d}\Omega \int_l \psi_{x_{i,z}} \psi_{x_{j,z}} \mathrm{d}z$$

$$K_{u_x u_y}^{k\,\tau_{u_x} s_{u_y}} = \tilde{C}_{12}^k \int_{\Omega^k} F_{\tau_{u_{x,x}}} F_{s_{u_{y,y}}} \mathrm{d}\Omega \int_l \psi_{x_i} \psi_{y_j} \mathrm{d}z + \tilde{C}_{26}^k \int_{\Omega^k} F_{\tau_{u_{x,y}}} F_{s_{u_{y,y}}} \mathrm{d}\Omega \int_l \psi_{x_i} \psi_{y_j} \mathrm{d}z + \tilde{C}_{16}^k \int_{\Omega^k} F_{\tau_{u_{x,x}}} F_{s_{u_{y,x}}} \mathrm{d}\Omega \int_l \psi_{x_i} \psi_{y_j} \mathrm{d}z + \tilde{C}_{66}^k \int_{\Omega^k} F_{\tau_{u_{x,y}}} F_{s_{u_{y,x}}} \mathrm{d}\Omega \int_l \psi_{x_i} \psi_{y_j} \mathrm{d}z + \tilde{C}_{45}^k \int_{\Omega^k} F_{\tau_{u_x}} F_{s_{u_y}} \mathrm{d}\Omega \int_l \psi_{x_{i,z}} \psi_{y_{j,z}} \mathrm{d}z$$

$$K_{u_x u_z}^{k\,\tau_{u_x} s_{u_z}} = \tilde{C}_{13}^k \int_{\Omega^k} F_{\tau_{u_{x,x}}} F_{s_{u_z}} \mathrm{d}\Omega \int_l \psi_{x_i} \psi_{z_{j,z}} \mathrm{d}z + \tilde{C}_{36}^k \int_{\Omega^k} F_{\tau_{u_{x,y}}} F_{s_{u_z}} \mathrm{d}\Omega \int_l \psi_{x_i} \psi_{z_{j,z}} \mathrm{d}z + \tilde{C}_{55}^k \int_{\Omega^k} F_{\tau_{u_x}} F_{s_{u_{z,x}}} \mathrm{d}\Omega \int_l \psi_{x_{i,z}} \psi_{z_j} \mathrm{d}z + \tilde{C}_{45}^k \int_{\Omega^k} F_{\tau_{u_x}} F_{s_{u_{z,y}}} \mathrm{d}\Omega \int_l \psi_{x_{i,z}} \psi_{z_j} \mathrm{d}z$$

$$K_{u_y u_x}^{k\,\tau_{u_y} s_{u_x}} = \tilde{C}_{12}^k \int_{\Omega^k} F_{\tau_{u_{y,y}}} F_{s_{u_{x,x}}} \mathrm{d}\Omega \int_l \psi_{y_i} \psi_{x_j} \mathrm{d}z + \tilde{C}_{16}^k \int_{\Omega^k} F_{\tau_{u_{y,x}}} F_{s_{u_{x,x}}} \mathrm{d}\Omega \int_l \psi_{y_i} \psi_{x_j} \mathrm{d}z + \tilde{C}_{26}^k \int_{\Omega^k} F_{\tau_{u_{y,y}}} F_{s_{u_{x,y}}} \mathrm{d}\Omega \int_l \psi_{y_i} \psi_{x_j} \mathrm{d}z + \tilde{C}_{66}^k \int_{\Omega^k} F_{\tau_{u_{y,x}}} F_{s_{u_{x,y}}} \mathrm{d}\Omega \int_l \psi_{y_i} \psi_{x_j} \mathrm{d}z + \tilde{C}_{45}^k \int_{\Omega^k} F_{\tau_{u_y}} F_{s_{u_x}} \mathrm{d}\Omega \int_l \psi_{y_{i,z}} \psi_{x_{j,z}} \mathrm{d}z$$

$$K_{u_y u_y}^{k\,\tau_{u_y} s_{u_y}} = \tilde{C}_{22}^k \int_{\Omega^k} F_{\tau_{u_{y,y}}} F_{s_{u_{y,y}}} \mathrm{d}\Omega \int_l \psi_{y_i} \psi_{y_j} \mathrm{d}z + \tilde{C}_{26}^k \int_{\Omega^k} F_{\tau_{u_{y,x}}} F_{s_{u_{y,y}}} \mathrm{d}\Omega \int_l \psi_{y_i} \psi_{y_j} \mathrm{d}z + \tilde{C}_{26}^k \int_{\Omega^k} F_{\tau_{u_{y,y}}} F_{s_{u_{y,x}}} \mathrm{d}\Omega \int_l \psi_{y_i} \psi_{y_j} \mathrm{d}z + \tilde{C}_{66}^k \int_{\Omega^k} F_{\tau_{u_{y,x}}} F_{s_{u_{y,x}}} \mathrm{d}\Omega \int_l \psi_{y_i} \psi_{y_j} \mathrm{d}z + \tilde{C}_{44}^k \int_{\Omega^k} F_{\tau_{u_y}} F_{s_{u_y}} \mathrm{d}\Omega \int_l \psi_{y_{i,z}} \psi_{y_{j,z}} \mathrm{d}z$$

$$K_{u_y u_z}^{k\,\tau_{u_y} s_{u_z}} = \tilde{C}_{23}^k \int_{\Omega^k} F_{\tau_{u_{y,y}}} F_{s_{u_z}} \mathrm{d}\Omega \int_l \psi_{y_i} \psi_{z_{j,z}} \mathrm{d}z + \tilde{C}_{36}^k \int_{\Omega^k} F_{\tau_{u_{y,x}}} F_{s_{u_z}} \mathrm{d}\Omega \int_l \psi_{y_i} \psi_{z_{j,z}} \mathrm{d}z + \tilde{C}_{45}^k \int_{\Omega^k} F_{\tau_{u_y}} F_{s_{u_{z,x}}} \mathrm{d}\Omega \int_l \psi_{y_{i,z}} \psi_{z_j} \mathrm{d}z + \tilde{C}_{44}^k \int_{\Omega^k} F_{\tau_{u_y}} F_{s_{u_{z,y}}} \mathrm{d}\Omega \int_l \psi_{y_{i,z}} \psi_{z_j} \mathrm{d}z$$

$$
\begin{aligned}
&K_{u_z u_x}^{k\,\tau_{u_z} s_{u_x}} \\
&= \tilde{C}_{55}^k \int_{\Omega^k} F_{\tau_{u_z,x}} F_{s_{u_x}} \mathrm{d}\Omega \int_l \psi_{z_i} \psi_{x_{j,z}} \mathrm{d}z + \tilde{C}_{45}^k \int_{\Omega^k} F_{\tau_{u_z,y}} F_{s_{u_x}} \mathrm{d}\Omega \int_l \psi_{z_i} \psi_{x_{j,z}} \mathrm{d}z \\
&+ \tilde{C}_{13}^k \int_{\Omega^k} F_{\tau_{u_z}} F_{s_{u_x,x}} \mathrm{d}\Omega \int_l \psi_{z_{i,z}} \psi_{x_j} \mathrm{d}z + \tilde{C}_{36}^k \int_{\Omega^k} F_{\tau_{u_z}} F_{s_{u_x,y}} \mathrm{d}\Omega \int_l \psi_{z_{i,z}} \psi_{x_j} \mathrm{d}z \\
&K_{u_z u_y}^{k\,\tau_{u_z} s_{u_y}} \\
&= \tilde{C}_{45}^k \int_{\Omega^k} F_{\tau_{u_z,x}} F_{s_{u_y}} \mathrm{d}\Omega \int_l \psi_{z_i} \psi_{y_{j,z}} \mathrm{d}z + \tilde{C}_{44}^k \int_{\Omega^k} F_{\tau_{u_z,y}} F_{s_{u_y}} \mathrm{d}\Omega \int_l \psi_{z_i} \psi_{y_{j,z}} \mathrm{d}z \\
&+ \tilde{C}_{23}^k \int_{\Omega^k} F_{\tau_{u_z}} F_{s_{u_y,y}} \mathrm{d}\Omega \int_l \psi_{z_{i,z}} \psi_{y_j} \mathrm{d}z + \tilde{C}_{36}^k \int_{\Omega^k} F_{\tau_{u_z}} F_{s_{u_y,x}} \mathrm{d}\Omega \int_l \psi_{z_{i,z}} \psi_{y_j} \mathrm{d}z \\
&K_{u_z u_z}^{k\,\tau_{u_z} s_{u_z}} \\
&= \tilde{C}_{55}^k \int_{\Omega^k} F_{\tau_{u_z,x}} F_{s_{u_z,x}} \mathrm{d}\Omega \int_l \psi_{z_i} \psi_{z_j} \mathrm{d}z + \tilde{C}_{45}^k \int_{\Omega^k} F_{\tau_{u_z,y}} F_{s_{u_z,x}} \mathrm{d}\Omega \int_l \psi_{z_i} \psi_{z_j} \mathrm{d}z \\
&+ \tilde{C}_{45}^k \int_{\Omega^k} F_{\tau_{u_z,x}} F_{s_{u_z,y}} \mathrm{d}\Omega \int_l \psi_{z_i} \psi_{z_j} \mathrm{d}z + \tilde{C}_{44}^k \int_{\Omega^k} F_{\tau_{u_z,y}} F_{s_{u_z,y}} \mathrm{d}\Omega \int_l \psi_{z_i} \psi_{z_j} \mathrm{d}z \\
&+ \tilde{C}_{33}^k \int_{\Omega^k} F_{\tau_{u_z}} F_{s_{u_z}} \mathrm{d}\Omega \int_l \psi_{z_{i,z}} \psi_{z_{j,z}} \mathrm{d}z
\end{aligned}
\tag{7.20}
$$

In order to simplify the derived expressions a compact notation, for both line and surface integrals, is introduced. In particular, as regard the surface integral the following contraction is used

$$
J^{k\,\tau_{,\zeta} s_{,\eta}} = \int_{\Omega} F_{\tau_{,\zeta}} F_{s_{,\eta}} \mathrm{d}\Omega \qquad \tau = \tau_{u_x}, \tau_{u_y}, \tau_{u_z}; \qquad s = s_{u_x}, s_{u_y}, s_{u_z}; \tag{7.21}
$$

where $,\zeta$ and $,\eta$ represent differentiation with respect to that variable. The line integrals are instead contracted as follows

$$
I_{pq}^{\chi\xi} = \int_0^l \frac{\mathrm{d}^{\chi} \psi_{p_i}(z)}{\mathrm{d}z^{\xi}} \frac{\mathrm{d}^{\xi} \psi_{q_j}(z)}{\mathrm{d}z^{\chi}} \mathrm{d}z \quad i = 1, \cdots, \mathcal{M}; \quad J = 1, \cdots, \mathcal{N}; \tag{7.22}
$$

where $p, q = u_x, u_y, u_z$; χ and ξ indicate differentiation orders; and $\mathcal{M}$ and $\mathcal{N}$ indicate the Ritz expansion orders. Substituting Eqs. (7.21) and (7.22) in Eq. (7.20) the compact form of the nucleus components is then obtained

$$
\begin{aligned}
K_{u_x u_x}^{k\,\tau_{u_x} s_{u_x}} &= \tilde{C}_{11}^k J^{k\tau_{u_x,x} s_{u_x,x}} I_{u_x u_x}^{00} + \tilde{C}_{16}^k J^{k\tau_{u_x,y} s_{u_x,x}} I_{u_x u_x}^{00} + \tilde{C}_{11}^k J^{k\tau_{u_x,x} s_{u_x,y}} I_{u_x u_x}^{00} \\
&+ \tilde{C}_{66}^k J^{k\tau_{u_x,y} s_{u_x,y}} I_{u_x u_x}^{00} + \tilde{C}_{55}^k J^{k\tau_{u_x} s_{u_x}} I_{u_x u_x}^{11} \\
K_{u_x u_y}^{k\,\tau_{u_x} s_{u_y}} &= \tilde{C}_{12}^k J^{k\tau_{u_x,x} s_{u_y,y}} I_{u_x u_y}^{00} + \tilde{C}_{26}^k J^{k\tau_{u_x,y} s_{u_y,y}} I_{u_x u_y}^{00} + \tilde{C}_{16}^k J^{k\tau_{u_x,x} s_{u_y,x}} I_{u_x u_y}^{00} \\
&+ \tilde{C}_{66}^k J^{k\tau_{u_x,y} s_{u_y,x}} I_{u_x u_y}^{00} + \tilde{C}_{45}^k J^{k\tau_{u_x} s_{u_y}} I_{u_x u_y}^{11}
\end{aligned}
$$

$$
\begin{aligned}
K^{k\,\tau_{u_x} s_{u_z}}_{u_x u_z} &= \tilde{C}^k_{13} J^{k\tau_{u_x,x}\, s_{u_z}} I^{01}_{u_x u_z} + \tilde{C}^k_{36} J^{k\tau_{u_x,y}\, s_{u_z}} I^{01}_{u_x u_z} + \tilde{C}^k_{55} J^{k\tau_{u_x} s_{u_z,x}} I^{10}_{u_x u_z} \\
&+ \tilde{C}^k_{45} J^{k\tau_{u_x} s_{u_z,y}} I^{10}_{u_x u_z} \\
K^{k\,\tau_{u_y} s_{u_x}}_{u_y u_x} &= \tilde{C}^k_{12} J^{k\tau_{u_y,y}\, s_{u_x,x}} I^{00}_{u_y u_x} + \tilde{C}^k_{16} J^{k\tau_{u_y,x}\, s_{u_x,x}} I^{00}_{u_y u_x} + \tilde{C}^k_{26} J^{k\tau_{u_y,y}\, s_{u_x,y}} I^{00}_{u_y u_x} \\
&+ \tilde{C}^k_{66} J^{k\tau_{u_y,x}\, s_{u_x,y}} I^{00}_{u_y u_x} + \tilde{C}^k_{45} J^{k\tau_{u_y} s_{u_x}} I^{11}_{u_y u_x} \\
K^{k\,\tau_{u_y} s_{u_y}}_{u_y u_y} &= \tilde{C}^k_{22} J^{k\tau_{u_y,y}\, s_{u_y,y}} I^{00}_{u_y u_y} + \tilde{C}^k_{26} J^{k\tau_{u_y,x}\, s_{u_y,y}} I^{00}_{u_y u_y} + \tilde{C}^k_{26} J^{k\tau_{u_y,y}\, s_{u_y,x}} I^{00}_{u_y u_y} \\
&+ \tilde{C}^k_{66} J^{k\tau_{u_y,x}\, s_{u_y,x}} I^{00}_{u_y u_y} + \tilde{C}^k_{44} J^{k\tau_{u_y} s_{u_y}} I^{11}_{u_y u_y} \\
K^{k\,\tau_{u_y} s_{u_z}}_{u_y u_z} &= \tilde{C}^k_{23} J^{k\tau_{u_y,y}\, s_{u_z}} I^{01}_{u_y u_z} + \tilde{C}^k_{36} J^{k\tau_{u_y,x}\, s_{u_z}} I^{01}_{u_y u_z} + \tilde{C}^k_{45} J^{k\tau_{u_y} s_{u_z,x}} I^{10}_{u_y u_z} \\
&+ \tilde{C}^k_{44} J^{k\tau_{u_y} s_{u_z,y}} I^{10}_{u_y u_z} \\
K^{k\,\tau_{u_z} s_{u_x}}_{u_z u_x} &= \tilde{C}^k_{55} J^{k\tau_{u_z,x}\, s_{u_x}} I^{01}_{u_z u_x} + \tilde{C}^k_{45} J^{k\tau_{u_z,y}\, s_{u_x}} I^{01}_{u_z u_x} + \tilde{C}^k_{13} J^{k\tau_{u_z} s_{u_x,x}} I^{10}_{u_z u_x} \\
&+ \tilde{C}^k_{36} J^{k\tau_{u_z} s_{u_x,y}} I^{10}_{u_z u_x} \\
K^{k\,\tau_{u_z} s_{u_y}}_{u_z u_y} &= \tilde{C}^k_{45} J^{k\tau_{u_z,x}\, s_{u_y}} I^{01}_{u_z u_y} + \tilde{C}^k_{44} J^{k\tau_{u_z,y}\, s_{u_y}} I^{01}_{u_z u_y} + \tilde{C}^k_{23} J^{k\tau_{u_z} s_{u_y,y}} I^{10}_{u_z u_y} \\
&+ \tilde{C}^k_{36} J^{k\tau_{u_z} s_{u_y,x}} I^{10}_{u_z u_y} \\
K^{k\,\tau_{u_z} s_{u_z}}_{u_z u_z} &= \tilde{C}^k_{55} J^{k\tau_{u_z,x}\, s_{u_z,x}} I^{00}_{u_z u_z} + \tilde{C}^k_{45} J^{k\tau_{u_z,y}\, s_{u_z,x}} I^{00}_{u_z u_z} + \tilde{C}^k_{45} J^{k\tau_{u_z,x}\, s_{u_z,y}} I^{00}_{u_z u_z} \\
&+ \tilde{C}^k_{44} J^{k\tau_{u_z,y}\, s_{u_z,y}} I^{00}_{u_z u_z} + \tilde{C}^k_{33} J^{k\tau_{u_z} s_{u_z}} I^{11}_{u_z u_z}
\end{aligned}
\tag{7.23}
$$

Mass nucleus $\mathbf{M}$

The inertial work at layer level assumes the following form

$$
\delta L^k_{F_{in}} = \int_{\Omega^k} \int_l \delta \mathbf{U}^{k^T}_{\tau i} \left[\rho^k \left(\mathbf{F}_\tau \mathbf{\Psi}_i\right)^T \left(\mathbf{F}_s \mathbf{\Psi}_j\right)\right] \ddot{\mathbf{U}}_{sj}\, d\Omega^k\, dz \tag{7.24}
$$

Equation (7.11) can be written in a compact vectorial notation as follows

$$
\delta L^k_{F_{in}} = \delta \mathbf{U}^{k^T}_{\tau i}\, \mathbf{M}^{k\tau sij}\, \ddot{\mathbf{U}}^k_{sj} \tag{7.25}
$$

where

$$
\mathbf{M}^{k\tau sij} = \int_{\Omega^k} \int_l \left(\rho^k \left[(\mathbf{F}_\tau \mathbf{\Psi}_i)^T (\mathbf{F}_s \mathbf{\Psi}_j)\right]\right) \mathrm{d}\Omega^k\, \mathrm{d}z \tag{7.26}
$$

and the three non-zero components of the mass nucleus are

$$
\begin{aligned}
M^{k\,\tau_{u_x} s_{u_x}}_{u_x u_x} &= \rho^k J^{k\,\tau_{u_x}\, s_{u_x}} I^{00}_{u_x u_x} \\
M^{k\,\tau_{u_y} s_{u_y}}_{u_y u_y} &= \rho^k J^{k\,\tau_{u_y}\, s_{u_y}} I^{00}_{u_y u_y} \\
M^{k\,\tau_{u_z} s_{u_z}}_{u_z u_z} &= \rho^k J^{k\,\tau_{u_z}\, s_{u_z}} I^{00}_{u_z u_z}
\end{aligned}
\tag{7.27}
$$

Initial stress nucleus $\mathbf{K}_\sigma$

The initial stress matrix is constructed from the work done by the virtual non-linear strains with the actual initial stresses

$$\delta L_{ext}^k = \int_{\Omega^k} \int_l \left(\delta \varepsilon_{zz_{nl}}^k \sigma_{zz0}^k \right) \mathrm{d}\Omega^k \, \mathrm{d}z \tag{7.28}$$

where σ_{zz0}^k is the actual initial stress, and $\varepsilon_{zz_{nl}}$ represents the geometric non-linearity. In particular, it is considered in the axial strain component in a Green–Lagrange sense:

$$\varepsilon_{zz_{nl}} = \frac{1}{2}\left(u_{x,z}^2 + u_{y,z}^2 + u_{z,z}^2 \right) \tag{7.29}$$

The buckling load can be then defined via a scalar load factor λ as the load $\sigma = \lambda \, \sigma_0$ for which an equilibrium configuration exists such that

$$\delta \mathbf{U}_{\tau i}^{k^T} : \qquad \left[\mathbf{K}^{k\tau sij} + \lambda_{ij} \, \mathbf{K}_\sigma^{k\tau sij} \left(\sigma_0 \right) \right] \mathbf{U}_{sj}^k = 0 \tag{7.30}$$

$\mathbf{K}^{k\tau sij}$ is the usual linear stiffness matrix and $\mathbf{K}_\sigma^{k\tau sij}$ is the initial stress primary nucleus whose non-zero terms assume the following form

$$\begin{aligned} K_{\sigma_{u_x u_x}}^{\tau_{u_x} s_{u_x}} &= \sigma_{zz0} J^{k\tau_{u_x} s_{u_x}} I_{u_x u_x}^{11} \\ K_{\sigma_{u_y u_y}}^{\tau_{u_y} s_{u_y}} &= \sigma_{zz0} J^{k\tau_{u_y} s_{u_y}} I_{u_y u_y}^{11} \\ K_{\sigma_{u_z u_z}}^{\tau_{u_z} s_{u_z}} &= \sigma_{zz0} J^{k\tau_{u_z} s_{u_z}} I_{u_z u_z}^{11} \end{aligned} \tag{7.31}$$

7.2.2 Plates

Stiffness nucleus $\mathbf{K}$

The weak form of the governing equation can be obtained using the PVD (see chapter 3) [5]. For this purpose, the strain vectors can be written by introducing Eq. (7.2) in the geometrical relations as follows

$$\begin{aligned} \boldsymbol{\varepsilon}_{pG}^k &= \mathbf{D}_p \left(\mathbf{F}_\tau \, \boldsymbol{\Psi}_i \right) \mathbf{U}_{\tau i}^k \\ \boldsymbol{\varepsilon}_{nG}^k &= \mathbf{D}_{np} \left(\mathbf{F}_\tau \, \boldsymbol{\Psi}_i \right) \mathbf{U}_{\tau i}^k + \mathbf{D}_{nz} \left(\mathbf{F}_\tau \, \boldsymbol{\Psi}_i \right) \mathbf{U}_{\tau i}^k \end{aligned} \tag{7.32}$$

By substituting the previous expression in the PVD variational statement, the internal work becomes

$$
\begin{aligned}
\delta L_{int}^{k} = & \int_{\Omega^k}\int_{A^k} \delta \mathbf{U}_{\tau i}^{k^T} \left[\mathbf{D}_p \left(\mathbf{F}_\tau \boldsymbol{\Psi}_i\right)\right]^T \tilde{\mathbf{C}}_{pp}^{k} \mathbf{D}_p \left(\mathbf{F}_s \boldsymbol{\Psi}_j\right) \mathbf{U}_{sj}^{k}\, d\Omega^k\, dz + \\
& \int_{\Omega^k}\int_{A^k} \delta \mathbf{U}_{\tau i}^{k^T} \left[\mathbf{D}_p \left(\mathbf{F}_\tau \boldsymbol{\Psi}_i\right)\right]^T \tilde{\mathbf{C}}_{pn}^{k} \mathbf{D}_{np} \left(\mathbf{F}_s \boldsymbol{\Psi}_j\right) \mathbf{U}_{sj}^{k}\, d\Omega^k\, dz + \\
& \int_{\Omega^k}\int_{A^k} \delta \mathbf{U}_{\tau i}^{k^T} \left[\mathbf{D}_p \left(\mathbf{F}_\tau \boldsymbol{\Psi}_i\right)\right]^T \tilde{\mathbf{C}}_{pn}^{k} \mathbf{D}_{nz} \left(\mathbf{F}_s \boldsymbol{\Psi}_j\right) \mathbf{U}_{sj}^{k}\, d\Omega^k\, dz + \\
& \int_{\Omega^k}\int_{A^k} \delta \mathbf{U}_{\tau i}^{k^T} \left[\mathbf{D}_{np} \left(\mathbf{F}_\tau \boldsymbol{\Psi}_i\right)\right]^T \tilde{\mathbf{C}}_{np}^{k} \mathbf{D}_p \left(\mathbf{F}_s \boldsymbol{\Psi}_j\right) \mathbf{U}_{sj}^{k}\, d\Omega^k\, dz + \\
& \int_{\Omega^k}\int_{A^k} \delta \mathbf{U}_{\tau i}^{k^T} \left[\mathbf{D}_{np} \left(\mathbf{F}_\tau \boldsymbol{\Psi}_i\right)\right]^T \tilde{\mathbf{C}}_{nn}^{k} \mathbf{D}_{np} \left(\mathbf{F}_s \boldsymbol{\Psi}_j\right) \mathbf{U}_{sj}^{k}\, d\Omega^k\, dz + \\
& \int_{\Omega^k}\int_{A^k} \delta \mathbf{U}_{\tau i}^{k^T} \left[\mathbf{D}_{np} \left(\mathbf{F}_\tau \boldsymbol{\Psi}_i\right)\right]^T \tilde{\mathbf{C}}_{nn}^{k} \mathbf{D}_{nz} \left(\mathbf{F}_s \boldsymbol{\Psi}_j\right) \mathbf{U}_{sj}^{k}\, d\Omega^k\, dz + \\
& \int_{\Omega^k}\int_{A^k} \delta \mathbf{U}_{\tau i}^{k^T} \left[\mathbf{D}_{nz} \left(\mathbf{F}_\tau \boldsymbol{\Psi}_i\right)\right]^T \tilde{\mathbf{C}}_{np}^{k} \mathbf{D}_p \left(\mathbf{F}_s \boldsymbol{\Psi}_j\right) \mathbf{U}_{sj}^{k}\, d\Omega^k\, dz + \\
& \int_{\Omega^k}\int_{A^k} \delta \mathbf{U}_{\tau i}^{k^T} \left[\mathbf{D}_{nz} \left(\mathbf{F}_\tau \boldsymbol{\Psi}_i\right)\right]^T \tilde{\mathbf{C}}_{nn}^{k} \mathbf{D}_{np} \left(\mathbf{F}_s \boldsymbol{\Psi}_j\right) \mathbf{U}_{sj}^{k}\, d\Omega^k\, dz + \\
& \int_{\Omega^k}\int_{A^k} \delta \mathbf{U}_{\tau i}^{k^T} \left[\mathbf{D}_{nz} \left(\mathbf{F}_\tau \boldsymbol{\Psi}_i\right)\right]^T \tilde{\mathbf{C}}_{nn}^{k} \mathbf{D}_{nz} \left(\mathbf{F}_s \boldsymbol{\Psi}_j\right) \mathbf{U}_{sj}^{k}\, \mathrm{d}\Omega^k\, \mathrm{d}z
\end{aligned}
\tag{7.33}
$$

Recall the general expression of the virtual work

$$
\delta L_{int}^{k} = \delta \mathbf{U}_{\tau i}^{k^T} \mathbf{K}^{k\tau sij} \mathbf{U}_{sj}^{k} \tag{7.34}
$$

By comparison with Eq. (7.33)

$$
\begin{aligned}
\boldsymbol{K}^{k\tau sij} = & \int_{\Omega^k}\int_{A^k} \Big(\left[\mathbf{D}_p \left(\mathbf{F}_\tau \boldsymbol{\Psi}_i\right)\right]^T \times \\
& \left[\tilde{\mathbf{C}}_{pp}^{k} \mathbf{D}_p \left(\mathbf{F}_s \boldsymbol{\Psi}_j\right) + \tilde{\mathbf{C}}_{pn}^{k} \mathbf{D}_{np} \left(\mathbf{F}_s \boldsymbol{\Psi}_j\right) + \tilde{\mathbf{C}}_{pn}^{k} \mathbf{D}_{nz} \left(\mathbf{F}_s \boldsymbol{\Psi}_j\right)\right] + \\
& \left[\mathbf{D}_{np} \left(\mathbf{F}_\tau \boldsymbol{\Psi}_i\right)\right]^T \times \\
& \left[\tilde{\mathbf{C}}_{np}^{k} \mathbf{D}_p \left(\mathbf{F}_s \boldsymbol{\Psi}_j\right) + \tilde{\mathbf{C}}_{nn}^{k} \mathbf{D}_{np} \left(\mathbf{F}_s \boldsymbol{\Psi}_j\right) + \tilde{\mathbf{C}}_{nn}^{k} \mathbf{D}_{nz} \left(\mathbf{F}_s \boldsymbol{\Psi}_j\right)\right] + \\
& \left[\mathbf{D}_{nz} \left(\mathbf{F}_\tau \boldsymbol{\Psi}_i\right)\right]^T \times \\
& \left[\tilde{\mathbf{C}}_{np}^{k} \mathbf{D}_p \left(\mathbf{F}_s \boldsymbol{\Psi}_j\right) + \tilde{\mathbf{C}}_{nn}^{k} \mathbf{D}_{np} \left(\mathbf{F}_s \boldsymbol{\Psi}_j\right) + \tilde{\mathbf{C}}_{nn}^{k} \mathbf{D}_{nz} \left(\mathbf{F}_s \boldsymbol{\Psi}_j\right)\right] \Big)\, \mathrm{d}\Omega^k\, \mathrm{d}z
\end{aligned}
\tag{7.35}
$$

Equation (7.18) leads to a matrix composed of nine *secondary invariant fundamental nuclei*, for the sake of accuracy they are listed below

$$K_{u_x u_x}^{k\,\tau_{u_x} s_{u_x}} = \tilde{C}_{11}^k \int_{A^k} F_{\tau_{u_x}} F_{s_{u_x}} \mathrm{d}z \int_{\Omega^k} \psi_{x_i,x} \psi_{x_j,x} d\Omega^k + \tilde{C}_{16}^k \int_{A^k} F_{\tau_{u_x}} F_{s_{u_x}} \mathrm{d}z \int_{\Omega^k} \psi_{x_i,x} \psi_{x_j,y} d\Omega^k$$
$$+ \tilde{C}_{16}^k \int_{A^k} F_{\tau_{u_x}} F_{s_{u_x}} \mathrm{d}z \int_{\Omega^k} \psi_{x_i,y} \psi_{x_j,x} d\Omega^k + \tilde{C}_{66}^k \int_{A^k} F_{\tau_{u_x}} F_{s_{u_x}} \mathrm{d}z \int_{\Omega^k} \psi_{x_i,y} \psi_{x_j,y} d\Omega^k$$
$$+ \tilde{C}_{55}^k \int_{A^k} F_{\tau_{u_x,z}} F_{s_{u_x,z}} \mathrm{d}z \int_{\Omega^k} \psi_{x_i} \psi_{x_j} d\Omega^k$$

$$K_{u_x u_y}^{k\,\tau_{u_x} s_{u_y}} = \tilde{C}_{16}^k \int_{A^k} F_{\tau_{u_x}} F_{s_{u_y}} \mathrm{d}z \int_{\Omega^k} \psi_{x_i,x} \psi_{y_j,x} d\Omega^k + \tilde{C}_{12}^k \int_{A^k} F_{\tau_{u_x}} F_{s_{u_y}} \mathrm{d}z \int_{\Omega^k} \psi_{x_i,x} \psi_{y_j,y} d\Omega^k$$
$$+ \tilde{C}_{66}^k \int_{A^k} F_{\tau_{u_x}} F_{s_{u_y}} \mathrm{d}z \int_{\Omega^k} \psi_{x_i,y} \psi_{y_j,x} d\Omega^k + \tilde{C}_{26}^k \int_{A^k} F_{\tau_{u_x}} F_{s_{u_y}} \mathrm{d}z \int_{\Omega^k} \psi_{x_i,y} \psi_{y_j,y} d\Omega^k$$
$$+ \tilde{C}_{45}^k \int_{A^k} F_{\tau_{u_x,z}} F_{s_{u_y,z}} \mathrm{d}z \int_{\Omega^k} \psi_{x_i} \psi_{y_j} d\Omega^k$$

$$K_{u_x u_z}^{k\,\tau_{u_x} s_{u_z}} = \tilde{C}_{55}^k \int_{A^k} F_{\tau_{u_x,z}} F_{s_{u_z}} \mathrm{d}z \int_{\Omega^k} \psi_{x_i} \psi_{z_j,x} d\Omega^k + \tilde{C}_{45}^k \int_{A^k} F_{\tau_{u_x,z}} F_{s_{u_z}} \mathrm{d}z \int_{\Omega^k} \psi_{x_i} \psi_{z_j,y} d\Omega^k$$
$$+ \tilde{C}_{13}^k \int_{A^k} F_{\tau_{u_x}} F_{s_{u_z,z}} \mathrm{d}z \int_{\Omega^k} \psi_{x_i,x} \psi_{z_j} d\Omega^k + \tilde{C}_{36}^k \int_{A^k} F_{\tau_{u_x}} F_{s_{u_z,z}} \mathrm{d}z \int_{\Omega^k} \psi_{x_i,y} \psi_{z_j} d\Omega^k$$

$$K_{u_y u_x}^{k\,\tau_{u_y} s_{u_x}} = \tilde{C}_{16}^k \int_{A^k} F_{\tau_{u_y}} F_{s_{u_x}} \mathrm{d}z \int_{\Omega^k} \psi_{x_i} \psi_{y_j,xx} \mathrm{d}\Omega^k + \tilde{C}_{12}^k \int_{A^k} F_{\tau_{u_y}} F_{s_{u_x}} \mathrm{d}z \int_{\Omega^k} \psi_{x_i} \psi_{y_j,xy} \mathrm{d}\Omega^k$$
$$+ \tilde{C}_{66}^k \int_{A^k} F_{\tau_{u_y}} F_{s_{u_x}} \mathrm{d}z \int_{\Omega^k} \psi_{x_i} \psi_{y_j,yx} \mathrm{d}\Omega^k + \tilde{C}_{26}^k \int_{A^k} F_{\tau_{u_y}} F_{s_{u_x}} \mathrm{d}z \int_{\Omega^k} \psi_{x_i} \psi_{y_j,yy} \mathrm{d}\Omega^k$$
$$+ \tilde{C}_{45}^k \int_{A^k} F_{\tau_{u_y,z}} F_{s_{u_x,z}} \mathrm{d}z \int_{\Omega^k} \psi_{x_i} \psi_{y_j} \mathrm{d}\Omega^k$$

$$K_{u_y u_y}^{k\,\tau_{u_y} s_{u_y}} = \tilde{C}_{26}^k \int_{A^k} F_{\tau_{u_y}} F_{s_{u_y}} \mathrm{d}z \int_{\Omega^k} \psi_{y_i,y} \psi_{y_j,x} \mathrm{d}\Omega^k + \tilde{C}_{22}^k \int_{A^k} F_{\tau_{u_y}} F_{s_{u_y}} \mathrm{d}z \int_{\Omega^k} \psi_{y_i,y} \psi_{y_j,y} \mathrm{d}\Omega^k$$
$$+ \tilde{C}_{66}^k \int_{A^k} F_{\tau_{u_y}} F_{s_{u_y}} \mathrm{d}z \int_{\Omega^k} \psi_{y_i,y} \psi_{y_j,x} \mathrm{d}\Omega^k + \tilde{C}_{26}^k \int_{A^k} F_{\tau_{u_y}} F_{s_{u_y}} \mathrm{d}z \int_{\Omega^k} \psi_{y_i,y} \psi_{y_j,y} \mathrm{d}\Omega^k$$
$$+ \tilde{C}_{44}^k \int_{A^k} F_{\tau_{u_y,z}} F_{s_{u_y,z}} \mathrm{d}z \int_{\Omega^k} \psi_{y_i} \psi_{y_j} \mathrm{d}\Omega^k$$

$$K_{u_y u_z}^{k\,\tau_{u_y} s_{u_z}} = \tilde{C}_{45}^k \int_{A^k} F_{\tau_{u_y,z}} F_{s_{u_z}} \mathrm{d}z \int_{\Omega^k} \psi_{y_i} \psi_{z_j,x} \mathrm{d}\Omega^k + \tilde{C}_{44}^k \int_{A^k} F_{\tau_{u_y,z}} F_{s_{u_z}} \mathrm{d}z \int_{\Omega^k} \psi_{y_i} \psi_{z_j,y} \mathrm{d}\Omega^k$$
$$+ \tilde{C}_{36}^k \int_{A^k} F_{\tau_{u_y}} F_{s_{u_z,z}} \mathrm{d}z \int_{\Omega^k} \psi_{y_i,x} \psi_{z_j} \mathrm{d}\Omega^k + \tilde{C}_{23}^k \int_{A^k} F_{\tau_{u_y}} F_{s_{u_z,z}} \mathrm{d}z \int_{\Omega^k} \psi_{y_i,y} \psi_{z_j} \mathrm{d}\Omega^k$$

$$
\begin{aligned}
&K^{k\,\tau_{u_z} s_{u_x}}_{u_z u_x}\\
&= \tilde{C}^k_{55} \int_{A^k} F_{\tau_{u_z}} F_{s_{u_x,z}} \mathrm{d}z \int_{\Omega^k} \psi_{x_i} \psi_{z_j,x} \mathrm{d}\Omega^k + \tilde{C}^k_{45} \int_{A^k} F_{\tau_{u_z}} F_{s_{u_x,z}} \mathrm{d}z \int_{\Omega^k} \psi_{x_i} \psi_{z_j,y} \mathrm{d}\Omega^k\\
&+ \tilde{C}^k_{13} \int_{A^k} F_{\tau_{u_z,z}} F_{s_{u_x}} \mathrm{d}z \int_{\Omega^k} \psi_{x_i,x} \psi_{z_j} \mathrm{d}\Omega^k + \tilde{C}^k_{36} \int_{A^k} F_{\tau_{u_z,z}} F_{s_{u_x}} \mathrm{d}z \int_{\Omega^k} \psi_{x_i,y} \psi_{z_j} \mathrm{d}\Omega^k\\
&K^{k\,\tau_{u_z} s_{u_y}}_{u_z u_y}\\
&= \tilde{C}^k_{45} \int_{A^k} F_{\tau_{u_z}} F_{s_{u_y,z}} \mathrm{d}z \int_{\Omega^k} \psi_{y_i} \psi_{z_j,x} \mathrm{d}\Omega^k + \tilde{C}^k_{44} \int_{A^k} F_{\tau_{u_z}} F_{s_{u_y,z}} \mathrm{d}z \int_{\Omega^k} \psi_{y_i} \psi_{z_j,y} \mathrm{d}\Omega^k\\
&+ \tilde{C}^k_{36} \int_{A^k} F_{\tau_{u_z,z}} F_{s_{u_y}} \mathrm{d}z \int_{\Omega^k} \psi_{y_i,x} \psi_{z_j} \mathrm{d}\Omega^k + \tilde{C}^k_{23} \int_{A^k} F_{\tau_{u_z,z}} F_{s_{u_y}} \mathrm{d}z \int_{\Omega^k} \psi_{y_i,y} \psi_{z_j} \mathrm{d}\Omega^k\\
&K^{k\,\tau_{u_z} s_{u_z}}_{u_z u_z}\\
&= \tilde{C}^k_{45} \int_{A^k} F_{\tau_{u_z}} F_{s_{u_z}} \mathrm{d}z \int_{\Omega^k} \psi_{z_i,y} \psi_{z_j,x} \mathrm{d}\Omega^k + \tilde{C}^k_{44} \int_{A^k} F_{\tau_{u_z}} F_{s_{u_z}} \mathrm{d}z \int_{\Omega^k} \psi_{z_i,y} \psi_{z_j,y} \mathrm{d}\Omega^k\\
&+ \tilde{C}^k_{55} \int_{A^k} F_{\tau_{u_z}} F_{s_{u_z}} \mathrm{d}z \int_{\Omega^k} \psi_{z_i,x} \psi_{z_j,x} \mathrm{d}\Omega^k + \tilde{C}^k_{45} \int_{A^k} F_{\tau_{u_z}} F_{s_{u_z}} \mathrm{d}z \int_{\Omega^k} \psi_{z_i,x} \psi_{z_j,y} \mathrm{d}\Omega^k\\
&+ \tilde{C}^k_{33} \int_{A^k} F_{\tau_{u_z,z}} F_{s_{u_z,z}} \mathrm{d}z \int_{\Omega^k} \psi_{z_i} \psi_{z_j} \mathrm{d}\Omega^k
\end{aligned} \tag{7.36}
$$

As previously done for beam structures, at this stage it is convenient to introduce a concise and compact notation for both thickness and surface integrals. The former assume the following form

$$
\begin{aligned}
J^{k\,\tau s} &= \int_{A^k} F_\tau F_s \mathrm{d}z, & J^{k\,\tau_{,z} s} &= \int_{A^k} \frac{\partial F_\tau}{\partial z} F_s \mathrm{d}z,\\
J^{k\,\tau s_{,z}} &= \int_{A^k} F_\tau \frac{\partial F_s}{\partial z} \mathrm{d}z, & J^{k\,\tau_{,z} s_{,z}} &= \int_{A^k} \frac{\partial F_\tau}{\partial z} \frac{\partial F_s}{\partial z} \mathrm{d}z
\end{aligned} \tag{7.37}
$$

where $\tau = \tau_{u_x}, \tau_{u_y}, \tau_{u_z}$ and $s = s_{u_x}, s_{u_y}, s_{u_z}$. Concerning the in-plane integrals it is convenient to rewrite the trial functions ψ_{x_i}, ψ_{y_i}, ψ_{z_i} as

$$
\begin{aligned}
\psi_{x_i}(x, y) &= \sum_m \sum_n \phi^{u_x}_m(x)\, \phi^{u_x}_n(y)\\
\psi_{y_i}(x, y) &= \sum_m \sum_n \phi^{u_y}_m(x)\, \phi^{u_y}_n(y)\\
\psi_{z_i}(x, y) &= \sum_m \sum_n \phi^{u_z}_m(x)\, \phi^{u_z}_n(y)
\end{aligned} \tag{7.38}
$$

by exploiting the use of Eq. (7.38) the general in-plane integrals can be written as

$$
{}^i_j I^{\xi\zeta}_{mp} = \int_0^a \frac{\mathrm{d}^\xi \phi^i_m(x)}{\mathrm{d}x^\xi} \frac{\mathrm{d}^\zeta \phi^j_p(x)}{\mathrm{d}x^\zeta} \mathrm{d}x \qquad m = \cdots, M, \quad p = 1, \cdots, P
$$

$$
{}^{i}_{j}I^{\xi\zeta}_{nq} = \int_0^b \frac{\mathrm{d}^{\xi}\phi^{i}_{n}(y)}{\mathrm{d}y^{\xi}} \frac{\mathrm{d}^{\zeta}\phi^{j}_{q}(y)}{\mathrm{d}y^{\zeta}} \mathrm{d}y \qquad n=\cdots, N, \quad q=1,\cdots, Q \tag{7.39}
$$

where $i, j = u_x, u_y, u_z$ and $\xi,\ \zeta$ are differentiation orders. By substituting Eqs. (7.37) and (7.39) into Eq. (7.36), the explicit forms of the Ritz fundamental secondary nuclei are derived as follows

$$
\begin{aligned}
K^{k\tau_{u_x}s_{u_x}}_{u_xu_x} &= \tilde{C}^{k}_{11}J^{k\tau_{u_x}s_{u_x}}\,{}^{u_x}_{u_x}I^{11}_{mp}\,{}^{u_x}_{u_x}I^{00}_{nq} + \tilde{C}^{k}_{16}J^{k\tau_{u_x}s_{u_x}}\,{}^{u_x}_{u_x}I^{10}_{mp}\,{}^{u_x}_{u_x}I^{01}_{nq} + \tilde{C}^{k}_{16}J^{k\tau_{u_x}s_{u_x}}\,{}^{u_x}_{u_x}I^{01}_{mp}\,{}^{u_x}_{u_x}I^{10}_{nq} \\
&\quad + \tilde{C}^{k}_{66}J^{k\tau_{u_x}s_{u_x}}\,{}^{u_x}_{u_x}I^{00}_{mp}\,{}^{u_x}_{u_x}I^{11}_{nq} + \tilde{C}^{k}_{55}J^{k\tau_{u_x,z}s_{u_x,z}}\,{}^{u_x}_{u_x}I^{00}_{mp}\,{}^{u_x}_{u_x}I^{00}_{nq} \\
K^{k\tau_{u_x}s_{u_y}}_{u_xu_y} &= \tilde{C}^{k}_{16}J^{k\tau_{u_x}s_{u_y}}\,{}^{u_x}_{u_y}I^{11}_{mp}\,{}^{u_x}_{u_y}I^{00}_{nq} + \tilde{C}^{k}_{12}J^{k\tau_{u_x}s_{u_y}}\,{}^{u_x}_{u_y}I^{10}_{mp}\,{}^{u_x}_{u_y}I^{01}_{nq} + \tilde{C}^{k}_{66}J^{k\tau_{u_x}s_{u_y}}\,{}^{u_x}_{u_y}I^{01}_{mp}\,{}^{u_x}_{u_y}I^{10}_{nq} \\
&\quad + \tilde{C}^{k}_{26}J^{k\tau_{u_x}s_{u_y}}\,{}^{u_x}_{u_y}I^{00}_{mp}\,{}^{u_x}_{u_y}I^{11}_{nq} + \tilde{C}^{k}_{45}J^{k\tau_{u_x,z}s_{u_y,z}}\,{}^{u_x}_{u_y}I^{00}_{mp}\,{}^{u_x}_{u_y}I^{00}_{nq} \\
K^{k\tau_{u_x}s_{u_z}}_{u_xu_z} &= \tilde{C}^{k}_{55}J^{k\tau_{u_x,z}s_{u_z}}\,{}^{u_x}_{u_z}I^{01}_{mp}\,{}^{u_x}_{u_z}I^{10}_{nq} + \tilde{C}^{k}_{45}J^{k\tau_{u_x,z}s_{u_z}}\,{}^{u_x}_{u_z}I^{00}_{mp}\,{}^{u_x}_{u_z}I^{01}_{nq} \\
&\quad + \tilde{C}^{k}_{13}J^{k\tau_{u_x}s_{u_z,z}}\,{}^{u_x}_{u_z}I^{10}_{mp}\,{}^{u_x}_{u_z}I^{00}_{nq} + \tilde{C}^{k}_{36}J^{k\tau_{u_x}s_{u_z,z}}\,{}^{u_x}_{u_z}I^{00}_{mp}\,{}^{u_x}_{u_z}I^{10}_{nq} \\
K^{k\tau_{u_y}s_{u_x}}_{u_yu_x} &= \tilde{C}^{k}_{16}J^{k\tau_{u_y}s_{u_x}}\,{}^{u_y}_{u_x}I^{11}_{mp}\,{}^{u_y}_{u_x}I^{00}_{nq} + \tilde{C}^{k}_{12}J^{k\tau_{u_y}s_{u_x}}\,{}^{u_y}_{u_x}I^{01}_{mp}\,{}^{u_y}_{u_x}I^{10}_{nq} + \tilde{C}^{k}_{66}J^{k\tau_{u_y}s_{u_x}}\,{}^{u_y}_{u_x}I^{10}_{mp}\,{}^{u_y}_{u_x}I^{01}_{nq} \\
&\quad + \tilde{C}^{k}_{26}J^{k\tau_{u_y}s_{u_x}}\,{}^{u_y}_{u_x}I^{00}_{mp}\,{}^{u_y}_{u_x}I^{11}_{nq} + \tilde{C}^{k}_{45}J^{k\tau_{u_y,z}s_{u_x,z}}\,{}^{u_y}_{u_x}I^{00}_{mp}\,{}^{u_y}_{u_x}I^{00}_{nq} \\
K^{k\tau_{u_y}s_{u_y}}_{u_yu_y} &= \tilde{C}^{k}_{26}J^{k\tau_{u_y}s_{u_y}}\,{}^{u_y}_{u_y}I^{01}_{mp}\,{}^{u_y}_{u_y}I^{10}_{nq} + \tilde{C}^{k}_{22}J^{k\tau_{u_y}s_{u_y}}\,{}^{u_y}_{u_y}I^{00}_{mp}\,{}^{u_y}_{u_y}I^{11}_{nq} + \tilde{C}^{k}_{66}J^{k\tau_{u_y}s_{u_y}}\,{}^{u_y}_{u_y}I^{01}_{mp}\,{}^{u_y}_{u_y}I^{10}_{nq} \\
&\quad + \tilde{C}^{k}_{26}J^{k\tau_{u_y}s_{u_y}}\,{}^{u_y}_{u_y}I^{00}_{mp}\,{}^{u_y}_{u_y}I^{11}_{nq} + \tilde{C}^{k}_{44}J^{k\tau_{u_y,z}s_{u_y,z}}\,{}^{u_y}_{u_y}I^{00}_{mp}\,{}^{u_y}_{u_y}I^{00}_{nq} \\
K^{k\tau_{u_y}s_{u_z}}_{u_yu_z} &= \tilde{C}^{k}_{45}J^{k\tau_{u_y,z}s_{u_z}}\,{}^{u_y}_{u_z}I^{01}_{mp}\,{}^{u_y}_{u_z}I^{00}_{nq} + \tilde{C}^{k}_{44}J^{k\tau_{u_y,z}s_{u_z}}\,{}^{u_y}_{u_z}I^{00}_{mp}\,{}^{u_y}_{u_z}I^{01}_{nq} \\
&\quad + \tilde{C}^{k}_{36}J^{k\tau_{u_y}s_{u_z,z}}\,{}^{u_y}_{u_z}I^{10}_{mp}\,{}^{u_y}_{u_z}I^{00}_{nq} + \tilde{C}^{k}_{23}J^{k\tau_{u_y}s_{u_z,z}}\,{}^{u_y}_{u_z}I^{00}_{mp}\,{}^{u_y}_{u_z}I^{10}_{nq} \\
K^{k\tau_{u_z}s_{u_x}}_{u_zu_x} &= \tilde{C}^{k}_{55}J^{k\tau_{u_z}s_{u_x,z}}\,{}^{u_z}_{u_x}I^{01}_{mp}\,{}^{u_z}_{u_x}I^{00}_{nq} + \tilde{C}^{k}_{45}J^{k\tau_{u_z}s_{u_x,z}}\,{}^{u_z}_{u_x}I^{00}_{mp}\,{}^{u_z}_{u_x}I^{01}_{nq} \\
&\quad + \tilde{C}^{k}_{13}J^{k\tau_{u_z,z}s_{u_x}}\,{}^{u_z}_{u_x}I^{10}_{mp}\,{}^{u_z}_{u_x}I^{00}_{nq} + \tilde{C}^{k}_{36}J^{k\tau_{u_z,z}s_{u_x}}\,{}^{u_z}_{u_x}I^{00}_{mp}\,{}^{u_z}_{u_x}I^{10}_{nq} \\
K^{k\tau_{u_z}s_{u_y}}_{u_zu_y} &= \tilde{C}^{k}_{45}J^{k\tau_{u_z}s_{u_y,z}}\,{}^{u_z}_{u_y}I^{01}_{mp}\,{}^{u_z}_{u_y}I^{00}_{nq} + \tilde{C}^{k}_{44}J^{k\tau_{u_z}s_{u_y,z}}\,{}^{u_z}_{u_y}I^{00}_{mp}\,{}^{u_z}_{u_y}I^{01}_{nq} \\
&\quad + \tilde{C}^{k}_{36}J^{k\tau_{u_z,z}s_{u_y}}\,{}^{u_z}_{u_y}I^{10}_{mp}\,{}^{u_z}_{u_y}I^{00}_{nq} + \tilde{C}^{k}_{23}J^{k\tau_{u_z,z}s_{u_y}}\,{}^{u_z}_{u_y}I^{00}_{mp}\,{}^{u_z}_{u_y}I^{10}_{nq} \\
K^{k\tau_{u_z}s_{u_z}}_{u_zu_z} &= \tilde{C}^{k}_{45}J^{k\tau_{u_z}s_{u_z}}\,{}^{u_z}_{u_z}I^{01}_{mp}\,{}^{u_z}_{u_z}I^{10}_{nq} + \tilde{C}^{k}_{44}J^{k\tau_{u_z}s_{u_z}}\,{}^{u_z}_{u_z}I^{00}_{mp}\,{}^{u_z}_{u_z}I^{11}_{nq} + \tilde{C}^{k}_{55}J^{k\tau_{u_z}s_{u_z}}\,{}^{u_z}_{u_z}I^{11}_{mp}\,{}^{u_z}_{u_z}I^{00}_{nq} \\
&\quad + \tilde{C}^{k}_{45}J^{k\tau_{u_z}s_{u_z}}\,{}^{u_z}_{u_z}I^{10}_{mp}\,{}^{u_z}_{u_z}I^{01}_{nq} + \tilde{C}^{k}_{33}J^{k\tau_{u_z,z}s_{u_z,z}}\,{}^{u_z}_{u_z}I^{00}_{mp}\,{}^{u_z}_{u_z}I^{00}_{nq}
\end{aligned} \tag{7.40}
$$

Each *nucleus* or *kernel* of the UF, has to be expanded individually according to the expansion order chosen for the displacement components. When the expansions have been performed then the 9 secondary invariant nuclei generate the main fundamental nucleus related to the particular used theory and the expansion order of the Ritz functions. The secondary nuclei are then formally and mathematically invariant.

Mass nucleus M

From the work done by the inertial forces

$$\delta L^k_{F_{in}} = \int_{\Omega^k}\int_{A^k} \rho^k\, \delta \mathbf{u}^{k^T}\, \ddot{\mathbf{u}}^k\, \mathrm{d}\Omega^k\, \mathrm{d}z \tag{7.41}$$

using Eq. (7.2):

$$\delta L^k_{F_{in}} = \int_{\Omega^k}\int_{A^k} \left(\rho^k \delta \mathbf{U}^{k^T}_{\tau i} \left[(\mathbf{F}_\tau \boldsymbol{\Psi}_i)^T (\mathbf{F}_s \boldsymbol{\Psi}_j)\right] \ddot{\mathbf{U}}_{sj}\right) \mathrm{d}\Omega^k\, \mathrm{d}z \tag{7.42}$$

Being

$$\delta L^k_{F_{in}} = \delta \mathbf{U}^{k^T}_{\tau i}\, \mathbf{M}^{k\tau s ij}\, \ddot{\mathbf{U}}^k_{sj} \tag{7.43}$$

it follows

$$\mathbf{M}^{k\tau s ij} = \int_{\Omega^k}\int_{A^k} \left(\rho^k \left[(\mathbf{F}_\tau \boldsymbol{\Psi}_i)^T (\mathbf{F}_s \boldsymbol{\Psi}_j)\right]\right) \mathrm{d}\Omega^k\, \mathrm{d}z \tag{7.44}$$

Eq. (7.44) leads to a matrix composed by the three *secondary invariant fundamental nuclei*, which take the following form:

$$\begin{aligned} M^{\tau_{u_x} s_{u_x}}_{u_x u_x} &= \rho^k J^{k\, \tau_{u_x} s_{u_x}}\; {}^{u_x}_{u_x}I^{00}_{mp}\; {}^{u_x}_{u_x}I^{00}_{nq} \\ M^{\tau_{u_y} s_{u_y}}_{u_y u_y} &= \rho^k J^{k\, \tau_{u_y} s_{u_y}}\; {}^{u_y}_{u_y}I^{00}_{mp}\; {}^{u_y}_{u_y}I^{00}_{nq} \\ M^{\tau_{u_z} s_{u_z}}_{u_z u_z} &= \rho^k J^{k\, \tau_{u_z} s_{u_z}}\; {}^{u_z}_{u_z}I^{00}_{mp}\; {}^{u_z}_{u_z}I^{00}_{nq} \end{aligned} \tag{7.45}$$

Initial stress nucleus $\mathbf{K}_\sigma$

The buckling analysis is addressed via Euler's method of adjacent equilibrium states [5,6]. It consists in a linearized stability analysis of an undeformed equilibrium configuration, whose critical condition is defined by a proportionally scaled load in combination with a *geometric* or *initial stress* stiffness built up from the exact non-linear strain-displacement relations or von Kàrmàn's approximation. The linearized buckling analysis requires the fulfilment of the following condition:

- The pre-buckling deformation can be neglected;
- The initial stress σ_0 remains constant and varies neither in magnitude nor in direction during buckling;
- At bifurcation the equilibrium states are infinitesimally adjacent so that a linearization is possible.

The initial stress matrix is constructed from the work done by the virtual non-linear strains with the actual initial stresses

$$\boldsymbol{\varepsilon}^k_{p_{nl}} = \left[\varepsilon^k_{xx_{nl}}, \varepsilon^k_{yy_{nl}}, \gamma^k_{xy_{nl}}\right]^T \qquad \boldsymbol{\sigma}^k_{p0} = \left[\sigma^k_{xx_0}, \sigma^k_{yy_0}, \sigma^k_{xy_0}\right]^T \tag{7.46}$$

$$\delta L^k_{ext} = \int_{\Omega^k}\int_{A^k} \left(\delta\varepsilon^k_{xx_{nl}}\sigma^k_{xx_0} + \delta\varepsilon^k_{yy_{nl}}\sigma^k_{yy_0} + \delta\gamma^k_{xy_{nl}}\sigma^k_{xy_0}\right) \mathrm{d}\Omega^k\, \mathrm{d}z \tag{7.47}$$

The buckling load can be then defined via a scalar load factor λ as the load $\sigma = \lambda\, \sigma_0$ for which an equilibrium configuration exists such that

$$\delta \mathbf{U}_{\tau i}^{k^T}: \qquad \left[\mathbf{K}^{k\tau sij} + \lambda_{ij}\, \mathbf{K}_{\sigma}^{k\tau sij}\right] \mathbf{U}_{sj}^{k} = 0 \tag{7.48}$$

$\mathbf{K}^{k\tau sij}$ is the usual linear stiffness matrix and $\mathbf{K}_{\sigma}^{k\tau sij}$ is the initial stress secondary nuclei assuming the following form

$$\begin{aligned}
K_{\sigma_{u_x u_x}}^{\tau_{u_\alpha} s_{u_\alpha}} &= \sigma_{xx_0} J^{k\tau_{u_x} s_{u_x}}\, {}^{u_x}_{u_x} I^{11}_{mp}\, {}^{u_x}_{u_x} I^{00}_{nq} + \sigma_{yy_0} J^{k\tau_{u_x} s_{u_x}}\, {}^{u_x}_{u_x} I^{00}_{mp}\, {}^{u_x}_{u_x} I^{11}_{nq} \\
&+ \tau_{xy_0} J^{k\tau_{u_z} s_{u_z}} \left({}^{u_x}_{u_x} I^{10}_{mp}\, {}^{u_x}_{u_x} I^{01}_{nq} + {}^{u_x}_{u_x} I^{01}_{mp}\, {}^{u_x}_{u_x} I^{10}_{nq} \right) \\
K_{\sigma_{u_y u_y}}^{\tau_{u_\beta} s_{u_\beta}} &= \sigma_{xx_0} J^{k\tau_{u_y} s_{u_y}}\, {}^{u_y}_{u_y} I^{11}_{mp}\, {}^{u_y}_{u_y} I^{00}_{nq} + \sigma_{yy_0} J^{k\tau_{u_y} s_{u_y}}\, {}^{u_y}_{u_y} I^{00}_{mp}\, {}^{u_y}_{u_y} I^{11}_{nq} \\
&+ \tau_{xy_0} J^{k\tau_{u_z} s_{u_z}} \left({}^{u_y}_{u_y} I^{10}_{mp}\, {}^{u_y}_{u_y} I^{01}_{nq} + {}^{u_y}_{u_y} I^{01}_{mp}\, {}^{u_y}_{u_y} I^{10}_{nq} \right) \\
K_{\sigma_{u_z u_z}}^{\tau_{u_z} s_{u_z}} &= \sigma_{xx_0} J^{k\tau_{u_z} s_{u_z}}\, {}^{u_z}_{u_z} I^{11}_{mp}\, {}^{u_z}_{u_z} I^{00}_{nq} + \sigma_{yy_0} J^{k\tau_{u_z} s_{u_z}}\, {}^{u_z}_{u_z} I^{00}_{mp}\, {}^{u_z}_{u_z} I^{11}_{nq} \\
&+ \tau_{xy_0} J^{k\tau_{u_z} s_{u_z}} \left({}^{u_z}_{u_z} I^{10}_{mp}\, {}^{u_z}_{u_z} I^{01}_{nq} + {}^{u_z}_{u_z} I^{01}_{mp}\, {}^{u_z}_{u_z} I^{10}_{nq} \right)
\end{aligned} \tag{7.49}$$

Furthermore in both cases, the tracer δ_{vK} is used to introduce von Kàrmàn's approximation.

7.2.3 Shells

Stiffness nucleus K

By substituting the first of Eq. (7.2) within the strain-displacement relationships and upon substitution in the PVD the explicit expressions of the internal virtual work and the virtual work done by the inertial forces in terms of Ritz functions and unknown coefficients are obtained [7],

$$\begin{aligned}
\delta L_{int}^{k} = &\int_{\Omega_k}\int_{A^k} \delta \mathbf{U}_{\tau i}^{k^T} \left[\mathbf{D}_p (\mathbf{F}_\tau \boldsymbol{\Psi}_i)\right]^T \tilde{\mathbf{C}}_{pp}^{k} \mathbf{D}_p \left(\mathbf{F}_s \boldsymbol{\Psi}_j\right) \mathbf{U}_{sj}^{k}\, \mathrm{d}\Omega_k\, \mathrm{d}z + \\
&\int_{\Omega_k}\int_{A^k} \delta \mathbf{U}_{\tau i}^{k^T} \left[\mathbf{D}_p (\mathbf{F}_\tau \boldsymbol{\Psi}_i)\right]^T \tilde{\mathbf{C}}_{pp}^{k} \mathbf{A}_p \left(\mathbf{F}_s \boldsymbol{\Psi}_j\right) \mathbf{U}_{sj}^{k}\, \mathrm{d}\Omega_k\, \mathrm{d}z + \\
&\int_{\Omega_k}\int_{A^k} \delta \mathbf{U}_{\tau i}^{k^T} \left[\mathbf{D}_p (\mathbf{F}_\tau \boldsymbol{\Psi}_i)\right]^T \tilde{\mathbf{C}}_{pn}^{k} \mathbf{D}_{np} \left(\mathbf{F}_s \boldsymbol{\Psi}_j\right) \mathbf{U}_{sj}^{k}\, \mathrm{d}\Omega_k\, \mathrm{d}z + \\
&\int_{\Omega_k}\int_{A^k} \delta \mathbf{U}_{\tau i}^{k^T} \left[\mathbf{D}_p (\mathbf{F}_\tau \boldsymbol{\Psi}_i)\right]^T \tilde{\mathbf{C}}_{pn}^{k}\, \delta_D\, \mathbf{A}_n \left(\mathbf{F}_s \boldsymbol{\Psi}_j\right) \mathbf{U}_{sj}^{k}\, \mathrm{d}\Omega_k\, \mathrm{d}z + \\
&\int_{\Omega_k}\int_{A^k} \delta \mathbf{U}_{\tau i}^{k^T} \left[\mathbf{D}_p (\mathbf{F}_\tau \boldsymbol{\Psi}_i)\right]^T \tilde{\mathbf{C}}_{pn}^{k} \mathbf{D}_{nz} \left(\mathbf{F}_s \boldsymbol{\Psi}_j\right) \mathbf{U}_{sj}^{k}\, \mathrm{d}\Omega_k\, \mathrm{d}z + \\
&\int_{\Omega_k}\int_{A^k} \delta \mathbf{U}_{\tau i}^{k^T} \left[\mathbf{A}_p (\mathbf{F}_\tau \boldsymbol{\Psi}_i)\right]^T \tilde{\mathbf{C}}_{pp}^{k} \mathbf{D}_p \left(\mathbf{F}_s \boldsymbol{\Psi}_j\right) \mathbf{U}_{sj}^{k}\, \mathrm{d}\Omega_k\, \mathrm{d}z + \\
&\int_{\Omega_k}\int_{A^k} \delta \mathbf{U}_{\tau i}^{k^T} \left[\mathbf{A}_p (\mathbf{F}_\tau \boldsymbol{\Psi}_i)\right]^T \tilde{\mathbf{C}}_{pp}^{k} \mathbf{A}_p \left(\mathbf{F}_s \boldsymbol{\Psi}_j\right) \mathbf{U}_{sj}^{k}\, \mathrm{d}\Omega_k\, \mathrm{d}z +
\end{aligned}$$

$$
\begin{aligned}
&\int_{\Omega_k}\int_{A^k} \delta\mathbf{U}_{\tau i}^{kT}\left[\mathbf{A}_p\left(\mathbf{F}_\tau\,\mathbf{\Psi}_i\right)\right]^T\tilde{C}_{pn}^k\mathbf{D}_{np}\left(\mathbf{F}_s\,\mathbf{\Psi}_j\right)\mathbf{U}_{sj}^k\,\mathrm{d}\Omega_k\,\mathrm{d}z+\\
&\int_{\Omega_k}\int_{A^k} \delta\mathbf{U}_{\tau i}^{kT}\left[\mathbf{A}_p\left(\mathbf{F}_\tau\,\mathbf{\Psi}_i\right)\right]^T\tilde{C}_{pn}^k\,\delta_D\,\mathbf{A}_n\left(\mathbf{F}_s\,\mathbf{\Psi}_j\right)\mathbf{U}_{sj}^k\,\mathrm{d}\Omega_k\,\mathrm{d}z+\\
&\int_{\Omega_k}\int_{A^k} \delta\mathbf{U}_{\tau i}^{kT}\left[\mathbf{A}_p\left(\mathbf{F}_\tau\,\mathbf{\Psi}_i\right)\right]^T\tilde{C}_{pn}^k\mathbf{D}_{nz}\left(\mathbf{F}_s\,\mathbf{\Psi}_j\right)\mathbf{U}_{sj}^k\,\mathrm{d}\Omega_k\,\mathrm{d}z+\\
&\int_{\Omega_k}\int_{A^k} \delta\mathbf{U}_{\tau i}^{kT}\left[\mathbf{D}_{np}\left(\mathbf{F}_\tau\,\mathbf{\Psi}_i\right)\right]^T\tilde{C}_{np}^k\mathbf{D}_{p}\left(\mathbf{F}_s\,\mathbf{\Psi}_j\right)\mathbf{U}_{sj}^k\,\mathrm{d}\Omega_k\,\mathrm{d}z+\\
&\int_{\Omega_k}\int_{A^k} \delta\mathbf{U}_{\tau i}^{kT}\left[\mathbf{D}_{np}\left(\mathbf{F}_\tau\,\mathbf{\Psi}_i\right)\right]^T\tilde{C}_{np}^k\mathbf{A}_{p}\left(\mathbf{F}_s\,\mathbf{\Psi}_j\right)\mathbf{U}_{sj}^k\,\mathrm{d}\Omega_k\,\mathrm{d}z+\\
&\int_{\Omega_k}\int_{A^k} \delta\mathbf{U}_{\tau i}^{kT}\left[\mathbf{D}_{np}\left(\mathbf{F}_\tau\,\mathbf{\Psi}_i\right)\right]^T\tilde{C}_{nn}^k\mathbf{D}_{np}\left(\mathbf{F}_s\,\mathbf{\Psi}_j\right)\mathbf{U}_{sj}^k\,\mathrm{d}\Omega_k\,\mathrm{d}z+\\
&\int_{\Omega_k}\int_{A^k} \delta\mathbf{U}_{\tau i}^{kT}\left[\mathbf{D}_{np}\left(\mathbf{F}_\tau\,\mathbf{\Psi}_i\right)\right]^T\tilde{C}_{nn}^k\,\delta_D\,\mathbf{A}_{n}\left(\mathbf{F}_s\,\mathbf{\Psi}_j\right)\mathbf{U}_{sj}^k\,\mathrm{d}\Omega_k\,\mathrm{d}z+\\
&\int_{\Omega_k}\int_{A^k} \delta\mathbf{U}_{\tau i}^{kT}\left[\mathbf{D}_{np}\left(\mathbf{F}_\tau\,\mathbf{\Psi}_i\right)\right]^T\tilde{C}_{nn}^k\mathbf{D}_{nz}\left(\mathbf{F}_s\,\mathbf{\Psi}_j\right)\mathbf{U}_{sj}^k\,\mathrm{d}\Omega_k\,\mathrm{d}z+\\
&\int_{\Omega_k}\int_{A^k} \delta\mathbf{U}_{\tau i}^{kT}\left[\delta_D\,\mathbf{A}_{n}\left(\mathbf{F}_\tau\,\mathbf{\Psi}_i\right)\right]^T\tilde{C}_{np}^k\mathbf{D}_{p}\left(\mathbf{F}_s\,\mathbf{\Psi}_j\right)\mathbf{U}_{sj}^k\,\mathrm{d}\Omega_k\,\mathrm{d}z+\\
&\int_{\Omega_k}\int_{A^k} \delta\mathbf{U}_{\tau i}^{kT}\left[\delta_D\,\mathbf{A}_{n}\left(\mathbf{F}_\tau\,\mathbf{\Psi}_i\right)\right]^T\tilde{C}_{np}^k\mathbf{A}_{p}\left(\mathbf{F}_s\,\mathbf{\Psi}_j\right)\mathbf{U}_{sj}^k\,\mathrm{d}\Omega_k\,\mathrm{d}z+\\
&\int_{\Omega_k}\int_{A^k} \delta\mathbf{U}_{\tau i}^{kT}\left[\delta_D\,\mathbf{A}_{n}\left(\mathbf{F}_\tau\,\mathbf{\Psi}_i\right)\right]^T\tilde{C}_{nn}^k\mathbf{D}_{np}\left(\mathbf{F}_s\,\mathbf{\Psi}_j\right)\mathbf{U}_{sj}^k\,\mathrm{d}\Omega_k\,\mathrm{d}z+\\
&\int_{\Omega_k}\int_{A^k} \delta\mathbf{U}_{\tau i}^{kT}\left[\delta_D\,\mathbf{A}_{n}\left(\mathbf{F}_\tau\,\mathbf{\Psi}_i\right)\right]^T\tilde{C}_{nn}^k\,\delta_D\,\mathbf{A}_{n}\left(\mathbf{F}_s\,\mathbf{\Psi}_j\right)\mathbf{U}_{sj}^k\,\mathrm{d}\Omega_k\,\mathrm{d}z+\\
&\int_{\Omega_k}\int_{A^k} \delta\mathbf{U}_{\tau i}^{kT}\left[\delta_D\,\mathbf{A}_{n}\left(\mathbf{F}_\tau\,\mathbf{\Psi}_i\right)\right]^T\tilde{C}_{nn}^k\mathbf{D}_{nz}\left(\mathbf{F}_s\,\mathbf{\Psi}_j\right)\mathbf{U}_{sj}^k\,\mathrm{d}\Omega_k\,\mathrm{d}z+\\
&\int_{\Omega_k}\int_{A^k} \delta\mathbf{U}_{\tau i}^{kT}\left[\mathbf{D}_{nz}\left(\mathbf{F}_\tau\,\mathbf{\Psi}_i\right)\right]^T\tilde{C}_{np}^k\mathbf{D}_{p}\left(\mathbf{F}_s\,\mathbf{\Psi}_j\right)\mathbf{U}_{sj}^k\,\mathrm{d}\Omega_k\,\mathrm{d}z+\\
&\int_{\Omega_k}\int_{A^k} \delta\mathbf{U}_{\tau i}^{kT}\left[\mathbf{D}_{nz}\left(\mathbf{F}_\tau\,\mathbf{\Psi}_i\right)\right]^T\tilde{C}_{np}^k\mathbf{A}_{p}\left(\mathbf{F}_s\,\mathbf{\Psi}_j\right)\mathbf{U}_{sj}^k\,\mathrm{d}\Omega_k\,\mathrm{d}z+\\
&\int_{\Omega_k}\int_{A^k} \delta\mathbf{U}_{\tau i}^{kT}\left[\mathbf{D}_{nz}\left(\mathbf{F}_\tau\,\mathbf{\Psi}_i\right)\right]^T\tilde{C}_{nn}^k\mathbf{D}_{np}\left(\mathbf{F}_s\,\mathbf{\Psi}_j\right)\mathbf{U}_{sj}^k\,\mathrm{d}\Omega_k\,\mathrm{d}z+\\
&\int_{\Omega_k}\int_{A^k} \delta\mathbf{U}_{\tau i}^{kT}\left[\mathbf{D}_{nz}\left(\mathbf{F}_\tau\,\mathbf{\Psi}_i\right)\right]^T\tilde{C}_{nn}^k\,\delta_D\,\mathbf{A}_{n}\left(\mathbf{F}_s\,\mathbf{\Psi}_j\right)\mathbf{U}_{sj}^k\,\mathrm{d}\Omega_k\,\mathrm{d}z+\\
&\int_{\Omega_k}\int_{A^k} \delta\mathbf{U}_{\tau i}^{kT}\left[\mathbf{D}_{nz}\left(\mathbf{F}_\tau\,\mathbf{\Psi}_i\right)\right]^T\tilde{C}_{nn}^k\mathbf{D}_{nz}\left(\mathbf{F}_s\,\mathbf{\Psi}_j\right)\mathbf{U}_{sj}^k\,\mathrm{d}\Omega_k\,\mathrm{d}z
\end{aligned}
\tag{7.50}
$$

The general quadratic forms of the internal virtual work and the virtual work done by the inertial forces can be written as

$$\delta L_{int}^{k} = \delta \mathbf{U}_{\tau i}^{k^T} \mathbf{K}^{k\tau sij} \mathbf{U}_{sj}^{k} \tag{7.51}$$

Consequently, the Ritz fundamental primary nuclei are easily obtained comparing Eq. (7.50) with Eq. (7.51):

$$\begin{aligned}
\mathbf{K}^{k\tau sij} = \int_{\Omega_k}\int_{A^k} \Big(& \left[\left(\mathbf{D}_p\left(\mathbf{F}_\tau \boldsymbol{\Psi}_i\right)\right)^T + \left(\mathbf{A}_p\left(\mathbf{F}_\tau \boldsymbol{\Psi}_i\right)\right)^T \right] \times \\
& \left[\tilde{\mathbf{C}}_{pp}^{k} \mathbf{D}_p\left(\mathbf{F}_s \boldsymbol{\Psi}_j\right) + \tilde{\mathbf{C}}_{pp}^{k} \mathbf{A}_p\left(\mathbf{F}_s \boldsymbol{\Psi}_j\right) + \tilde{\mathbf{C}}_{pn}^{k} \mathbf{D}_{np}\left(\mathbf{F}_s \boldsymbol{\Psi}_j\right) + \right. \\
& \left. \tilde{\mathbf{C}}_{pn}^{k}\, \delta_D\, \mathbf{A}_n\left(\mathbf{F}_s \boldsymbol{\Psi}_j\right) + \tilde{\mathbf{C}}_{pn}^{k} \mathbf{D}_{nz}\left(\mathbf{F}_s \boldsymbol{\Psi}_j\right) \right] + \\
& \left[\left[\mathbf{D}_{np}\left(\mathbf{F}_\tau \boldsymbol{\Psi}_i\right)\right]^T + \left[\delta_D\, \mathbf{A}_n\left(\mathbf{F}_\tau \boldsymbol{\Psi}_i\right)\right]^T + \left[\mathbf{D}_{nz}\left(\mathbf{F}_\tau \boldsymbol{\Psi}_i\right)\right]^T \right] \times \\
& \left[\tilde{\mathbf{C}}_{np}^{k} \mathbf{D}_p\left(\mathbf{F}_s \boldsymbol{\Psi}_j\right) + \tilde{\mathbf{C}}_{np}^{k} \mathbf{A}_p\left(\mathbf{F}_s \boldsymbol{\Psi}_j\right) + \tilde{\mathbf{C}}_{nn}^{k} \mathbf{D}_{np}\left(\mathbf{F}_s \boldsymbol{\Psi}_j\right) + \right. \\
& \left. \tilde{\mathbf{C}}_{nn}^{k}\, \delta_D\, \mathbf{A}_n\left(\mathbf{F}_s \boldsymbol{\Psi}_j\right) + \tilde{\mathbf{C}}_{nn}^{k} \mathbf{D}_{nz}\left(\mathbf{F}_s \boldsymbol{\Psi}_j\right) \right] \Big)\, \mathrm{d}\Omega_k\, \mathrm{d}z
\end{aligned} \tag{7.52}$$

the explicit forms of the Ritz fundamental secondary stiffness nuclei are following reported:

$$\begin{aligned}
K_{u_\alpha u_\alpha}^{\tau_{u_\alpha} s_{u_\alpha}} = {} & \tilde{C}_{11}^{k} \int_{A^k} \left(F_{\tau_{u_\alpha}} F_{s_{u_\alpha}} \frac{H_\beta^k}{H_\alpha^k} \right) \mathrm{d}z \int_{\Omega^k} \left(\psi_{\alpha_i,\alpha}\, \psi_{\alpha_j,\alpha} \right) \mathrm{d}\alpha_k\, \mathrm{d}\beta_k \\
& + \tilde{C}_{16}^{k} \int_{A^k} \left(F_{\tau_{u_\alpha}} F_{s_{u_\alpha}} \right) \mathrm{d}z \int_{\Omega^k} \left(\psi_{\alpha_i,\alpha}\, \psi_{\alpha_j,\beta} \right) \mathrm{d}\alpha_k\, \mathrm{d}\beta_k \\
& + \tilde{C}_{16}^{k} \int_{A^k} \left(F_{\tau_{u_\alpha}} F_{s_{u_\alpha}} \right) \mathrm{d}z \int_{\Omega^k} \left(\psi_{\alpha_i,\beta}\, \psi_{\alpha_j,\alpha} \right) \mathrm{d}\alpha_k\, \mathrm{d}\beta_k \\
& + \tilde{C}_{66}^{k} \int_{A^k} \left(F_{\tau_{u_\alpha}} F_{s_{u_\alpha}} \frac{H_\alpha^k}{H_\beta^k} \right) \mathrm{d}z \int_{\Omega^k} \left(\psi_{\alpha_i,\beta}\, \psi_{\alpha_j,\beta} \right) \mathrm{d}\alpha_k\, \mathrm{d}\beta_k \\
& + \tilde{C}_{55}^{k} \int_{A^k} \left(F_{\tau_{u_\alpha,z}} F_{s_{u_\alpha,z}} \frac{H_\alpha^k}{H_\beta^k} \right) \mathrm{d}z \int_{\Omega^k} \left(\psi_{\alpha_i}\, \psi_{\alpha_j} \right) \mathrm{d}\alpha_k\, \mathrm{d}\beta_k \\
& + \delta_D \times \left[-\tilde{C}_{55}^{k} \frac{1}{R_\alpha^k} \int_{A^k} \left(F_{\tau_{u_\alpha,z}} F_{s_{u_\alpha}} H_\beta^k \right) \mathrm{d}z \int_{\Omega^k} \left(\psi_{\alpha_i}\, \psi_{\alpha_j} \right) \mathrm{d}\alpha_k\, \mathrm{d}\beta_k \right. \\
& - \tilde{C}_{55}^{k} \frac{1}{R_\alpha^k} \int_{A^k} \left(F_{\tau_{u_\alpha}} F_{s_{u_\alpha,z}} H_\beta^k \right) \mathrm{d}z \int_{\Omega^k} \left(\psi_{\alpha_i}\, \psi_{\alpha_j} \right) \mathrm{d}\alpha_k\, \mathrm{d}\beta_k \\
& \left. + \tilde{C}_{55}^{k} \frac{1}{\left(R_\alpha^k\right)^2} \int_{A^k} \left(F_{\tau_{u_\alpha}} F_{s_{u_\alpha}} \frac{H_\beta^k}{H_\alpha^k} \right) \mathrm{d}z \int_{\Omega^k} \left(\psi_{\alpha_i}\, \psi_{\alpha_j} \right) \mathrm{d}\alpha_k\, \mathrm{d}\beta_k \right]
\end{aligned}$$

$$
\begin{aligned}
K_{u_\alpha u_\beta}^{\tau_{u\alpha} s_{u\beta}} &= \tilde{C}_{16}^k \int_{A^k} \left(F_{\tau_{u\alpha}} F_{s_{u\beta}} \frac{H_\beta^k}{H_\alpha^k} \right) \mathrm{d}z \int_{\Omega^k} \left(\psi_{\alpha_i,\alpha} \psi_{\beta_j,\alpha} \right) \mathrm{d}\alpha_k \, \mathrm{d}\beta_k \\
&+ \tilde{C}_{12}^k \int_{A^k} \left(F_{\tau_{u\alpha}} F_{s_{u\beta}} \right) \mathrm{d}z \int_{\Omega^k} \left(\psi_{\alpha_i,\alpha} \psi_{\beta_j,\beta} \right) \mathrm{d}\alpha_k \, \mathrm{d}\beta_k \\
&+ \tilde{C}_{66}^k \int_{A^k} \left(F_{\tau_{u\alpha}} F_{s_{u\beta}} \right) \mathrm{d}z \int_{\Omega^k} \left(\psi_{\alpha_i,\beta} \psi_{\beta_j,\alpha} \right) \mathrm{d}\alpha_k \, \mathrm{d}\beta_k \\
&+ \tilde{C}_{26}^k \int_{A^k} \left(F_{\tau_{u\alpha}} F_{s_{u\beta}} \frac{H_\alpha^k}{H_\beta^k} \right) \mathrm{d}z \int_{\Omega^k} \left(\psi_{\alpha_i,\beta} \psi_{\beta_j,\beta} \right) \mathrm{d}\alpha_k \, \mathrm{d}\beta_k \\
&+ \tilde{C}_{45}^k \int_{A^k} \left(F_{\tau_{u\alpha}} F_{s_{u\beta}} \frac{H_\alpha^k}{H_\beta^k} \right) \mathrm{d}z \int_{\Omega^k} \left(\psi_{\alpha_i} \psi_{\beta_j} \right) \mathrm{d}\alpha_k \, \mathrm{d}\beta_k \\
&+ \delta_D \times \left[-\tilde{C}_{45}^k \frac{1}{R_\beta^k} \int_{A^k} \left(F_{\tau_{u\alpha,z}} F_{s_{u\beta}} H_\alpha^k \right) \mathrm{d}z \int_{\Omega^k} \left(\psi_{\alpha_i} \psi_{\beta_j} \right) \mathrm{d}\alpha_k \, \mathrm{d}\beta_k \right. \\
&- \tilde{C}_{45}^k \frac{1}{R_\alpha^k} \int_{A^k} \left(F_{\tau_{u\alpha}} F_{s_{u\beta,z}} H_\beta^k \right) \mathrm{d}z \int_{\Omega^k} \left(\psi_{\alpha_i} \psi_{\beta_j} \right) \mathrm{d}\alpha_k \, \mathrm{d}\beta_k \\
&\left. + \tilde{C}_{45}^k \frac{1}{R_\alpha^k R_\beta^k} \int_{A^k} \left(F_{\tau_{u\alpha,z}} F_{s_{u\beta,z}} \right) \mathrm{d}z \int_{\Omega^k} \left(\psi_{\alpha_i} \psi_{\beta_j} \right) \mathrm{d}\alpha_k \, \mathrm{d}\beta_k \right] \\
K_{u_\alpha u_z}^{\tau_{u\alpha} s_{uz}} &= \tilde{C}_{13}^k \int_{A^k} \left(F_{\tau_{u\alpha}} F_{s_{uz,z}} H_\beta^k \right) \mathrm{d}z \int_{\Omega^k} \left(\psi_{\alpha_i,\alpha} \psi_{z_j} \right) \mathrm{d}\alpha_k \, \mathrm{d}\beta_k \\
&+ \tilde{C}_{55}^k \int_{A^k} \left(F_{\tau_{u\alpha,z}} F_{s_{uz}} H_\beta^k \right) \mathrm{d}z \int_{\Omega^k} \left(\psi_{\alpha_i} \psi_{z_j,\alpha} \right) \mathrm{d}\alpha_k \, \mathrm{d}\beta_k \\
&+ \tilde{C}_{45}^k \int_{A^k} \left(F_{\tau_{u\alpha,z}} F_{s_{uz}} H_\alpha^k \right) \mathrm{d}z \int_{\Omega^k} \left(\psi_{\alpha_i} \psi_{z_j,\beta} \right) \mathrm{d}\alpha_k \, \mathrm{d}\beta_k \\
&+ \tilde{C}_{36}^k \int_{A^k} \left(F_{\tau_{u\alpha,z}} F_{s_{uz}} H_\alpha^k \right) \mathrm{d}z \int_{\Omega^k} \left(\psi_{\alpha_i} \psi_{z_j,\beta} \right) \mathrm{d}\alpha_k \, \mathrm{d}\beta_k \\
&+ \tilde{C}_{11}^k \frac{1}{R_\alpha^k} \int_{A^k} \left(F_{\tau_{u\alpha}} F_{s_{uz}} \frac{H_\beta^k}{H_\alpha^k} \right) \mathrm{d}z \int_{\Omega^k} \left(\psi_{\alpha_i,\alpha} \psi_{z_j} \right) \mathrm{d}\alpha_k \, \mathrm{d}\beta_k \\
&+ \tilde{C}_{12}^k \frac{1}{R_\beta^k} \int_{A^k} \left(F_{\tau_{u\alpha}} F_{s_{uz}} \right) \mathrm{d}z \int_{\Omega^k} \left(\psi_{\alpha_i,\alpha} \psi_{z_j} \right) \mathrm{d}\alpha_k \, \mathrm{d}\beta_k \\
&+ \tilde{C}_{16}^k \frac{1}{R_\alpha^k} \int_{A^k} \left(F_{\tau_{u\alpha}} F_{s_{uz}} \right) \mathrm{d}z \int_{\Omega^k} \left(\psi_{\alpha_i,\beta} \psi_{z_j} \right) \mathrm{d}\alpha_k \, \mathrm{d}\beta_k \\
&+ \tilde{C}_{26}^k \frac{1}{R_\beta^k} \int_{A^k} \left(F_{\tau_{u\alpha}} F_{s_{uz}} \frac{H_\alpha^k}{H_\beta^k} \right) \mathrm{d}z \int_{\Omega^k} \left(\psi_{\alpha_i,\beta} \psi_{z_j} \right) \mathrm{d}\alpha_k \, \mathrm{d}\beta_k \\
&- \delta_D \times \left[\tilde{C}_{55}^k \frac{1}{R_\alpha^k} \int_{A^k} \left(F_{\tau_{u\alpha}} F_{s_{uz}} \frac{H_\beta^k}{H_\alpha^k} \right) \mathrm{d}z \int_{\Omega^k} \left(\psi_{\alpha_i} \psi_{z_j,\alpha} \right) \mathrm{d}\alpha_k \, \mathrm{d}\beta_k \right. \\
&\left. + \tilde{C}_{45}^k \frac{1}{R_\alpha^k} \int_{A^k} \left(F_{\tau_{u\alpha}} F_{s_{uz}} \right) \mathrm{d}z \int_{\Omega^k} \left(\psi_{\alpha_i} \psi_{z_j,\beta} \right) \mathrm{d}\alpha_k \, \mathrm{d}\beta_k \right]
\end{aligned}
$$

$$
\begin{aligned}
K_{u_\beta u_\alpha}^{\tau_{u_\beta} s_{u_\alpha}} - & \tilde{C}_{16}^k \int_{A^k} \left(F_{\tau_{u_\beta}} F_{s_{u_\alpha}} \frac{H_\beta^k}{H_\alpha^k} \right) \mathrm{d}z \int_{\Omega^k} \left(\psi_{\alpha_i,\alpha} \, \psi_{\beta_j,\alpha} \right) \mathrm{d}\alpha_k \, \mathrm{d}\beta_k \\
& + \tilde{C}_{12}^k \int_{A^k} \left(F_{\tau_{u_\beta}} F_{s_{u_\alpha}} \right) \mathrm{d}z \int_{\Omega^k} \left(\psi_{\alpha_i,\beta} \, \psi_{\beta_j,\alpha} \right) \mathrm{d}\alpha_k \, \mathrm{d}\beta_k \\
& + \tilde{C}_{66}^k \int_{A^k} \left(F_{\tau_{u_\beta}} F_{s_{u_\alpha}} \right) \mathrm{d}z \int_{\Omega^k} \left(\psi_{\alpha_i,\alpha} \, \psi_{\beta_j,\beta} \right) \mathrm{d}\alpha_k \, \mathrm{d}\beta_k \\
& + \tilde{C}_{26}^k \int_{A^k} \left(F_{\tau_{u_\beta}} F_{s_{u_\alpha}} \frac{H_\alpha^k}{H_\beta^k} \right) \mathrm{d}z \int_{\Omega^k} \left(\psi_{\alpha_i,\beta} \, \psi_{\beta_j,\beta} \right) \mathrm{d}\alpha_k \, \mathrm{d}\beta_k \\
& + \tilde{C}_{45}^k \int_{A^k} \left(F_{\tau_{u_\beta}} F_{s_{u_\alpha}} H_\alpha^k H_\beta^k \right) \mathrm{d}z \int_{\Omega^k} \left(\psi_{\alpha_i} \, \psi_{\beta_j} \right) \mathrm{d}\alpha_k \, \mathrm{d}\beta_k \\
& + \delta_D \times \Bigg[- \tilde{C}_{45}^k \frac{1}{R_\alpha^k} \int_{A^k} \left(F_{\tau_{u_{\beta,z}}} F_{s_{u_\alpha}} H_\beta^k \right) \mathrm{d}z \int_{\Omega^k} \left(\psi_{\alpha_i} \, \psi_{\beta_j} \right) \mathrm{d}\alpha_k \, \mathrm{d}\beta_k \\
& - \tilde{C}_{45}^k \frac{1}{R_\beta^k} \int_{A^k} \left(F_{\tau_{u_\beta}} F_{s_{u_{\alpha,z}}} H_\alpha^k \right) \mathrm{d}z \int_{\Omega^k} \left(\psi_{\alpha_i} \, \psi_{\beta_j} \right) \mathrm{d}\alpha_k \, \mathrm{d}\beta_k \\
& + \tilde{C}_{45}^k \frac{1}{R_\alpha^k R_\beta^k} \int_{A^k} \left(F_{\tau_{u_{\beta,z}}} F_{s_{u_{\alpha,z}}} \right) \mathrm{d}z \int_{\Omega^k} \left(\psi_{\alpha_i} \, \psi_{\beta_j} \right) \mathrm{d}\alpha_k \, \mathrm{d}\beta_k \Bigg] \\
K_{u_\beta u_\beta}^{\tau_{u_\beta} s_{u_\beta}} = & \tilde{C}_{66}^k \int_{A^k} \left(F_{\tau_{u_\beta}} F_{s_{u_\beta}} \frac{H_\beta^k}{H_\alpha^k} \right) \mathrm{d}z \int_{\Omega^k} \left(\psi_{\alpha_i,\alpha} \, \psi_{\alpha_j,\alpha} \right) \mathrm{d}\alpha_k \, \mathrm{d}\beta_k \\
& + \tilde{C}_{26}^k \int_{A^k} \left(F_{\tau_{u_\beta}} F_{s_{u_\beta}} \right) \mathrm{d}z \int_{\Omega^k} \left(\psi_{\alpha_i,\alpha} \, \psi_{\alpha_j,\beta} \right) \mathrm{d}\alpha_k \, \mathrm{d}\beta_k \\
& + \tilde{C}_{26}^k \int_{A^k} \left(F_{\tau_{u_\beta}} F_{s_{u_\beta}} \right) \mathrm{d}z \int_{\Omega^k} \left(\psi_{\alpha_i,\beta} \, \psi_{\alpha_j,\alpha} \right) \mathrm{d}\alpha_k \, \mathrm{d}\beta_k \\
& + \tilde{C}_{22}^k \int_{A^k} \left(F_{\tau_{u_\beta}} F_{s_{u_\beta}} \frac{H_\alpha^k}{H_\beta^k} \right) \mathrm{d}z \int_{\Omega^k} \left(\psi_{\alpha_i,\beta} \, \psi_{\alpha_j,\beta} \right) \mathrm{d}\alpha_k \, \mathrm{d}\beta_k \\
& + \tilde{C}_{44}^k \int_{A^k} \left(F_{\tau_{u_{\beta,z}}} F_{s_{u_{\beta,z}}} \frac{H_\alpha^k}{H_\beta^k} \right) \mathrm{d}z \int_{\Omega^k} \left(\psi_{\alpha_i} \, \psi_{\alpha_j} \right) \mathrm{d}\alpha_k \, \mathrm{d}\beta_k \\
& + \delta_D \times \Bigg[- \tilde{C}_{44}^k \frac{1}{R_\beta^k} \int_{A^k} \left(F_{\tau_{u_{\beta,z}}} F_{s_{u_\beta}} H_\alpha^k \right) \mathrm{d}z \int_{\Omega^k} \left(\psi_{\alpha_i} \, \psi_{\alpha_j} \right) \mathrm{d}\alpha_k \, \mathrm{d}\beta_k \\
& - \tilde{C}_{44}^k \frac{1}{R_\beta^k} \int_{A^k} \left(F_{\tau_{u_\beta}} F_{s_{u_{\beta,z}}} H_\alpha^k \right) \mathrm{d}z \int_{\Omega^k} \left(\psi_{\alpha_i} \, \psi_{\alpha_j} \right) \mathrm{d}\alpha_k \, \mathrm{d}\beta_k \\
& + \tilde{C}_{44}^k \frac{1}{\left(R_\beta^k\right)^2} \int_{A^k} \left(F_{\tau_{u_\beta}} F_{s_{u_\beta}} \frac{H_\alpha^k}{H_\beta^k} \right) \mathrm{d}z \int_{\Omega^k} \left(\psi_{\alpha_i} \, \psi_{\alpha_j} \right) \mathrm{d}\alpha_k \, \mathrm{d}\beta_k \Bigg] \\
K_{u_\beta u_z}^{\tau_{u_\beta} s_{u_z}} = & \tilde{C}_{45}^k \int_{A^k} \left(F_{\tau_{u_{\beta,z}}} F_{s_{u_z}} H_\beta^k \right) \mathrm{d}z \int_{\Omega^k} \left(\psi_{\alpha_i} \, \psi_{\alpha_j,\alpha} \right) \mathrm{d}\alpha_k \, \mathrm{d}\beta_k \\
& + \tilde{C}_{44}^k \int_{A^k} \left(F_{\tau_{u_{\beta,z}}} F_{s_{u_z}} H_\alpha^k \right) \mathrm{d}z \int_{\Omega^k} \left(\psi_{\alpha_i} \, \psi_{\alpha_j,\beta} \right) \mathrm{d}\alpha_k \, \mathrm{d}\beta_k
\end{aligned}
$$

$$
\begin{aligned}
&+ \tilde{C}_{23}^k \int_{A^k} \left(F_{\tau_{u_{\beta,z}}} F_{s_{u_z}} H_\alpha^k \right) \mathrm{d}z \int_{\Omega^k} \left(\psi_{\alpha_i,\beta} \psi_{\alpha_j} \right) \mathrm{d}\alpha_k \, \mathrm{d}\beta_k \\
&+ \tilde{C}_{12}^k \frac{1}{R_\alpha^k} \int_{A^k} \left(F_{\tau_{u_\beta}} F_{s_{u_z}} \right) \mathrm{d}z \int_{\Omega^k} \left(\psi_{\alpha_i,\beta} \psi_{\alpha_j} \right) \mathrm{d}\alpha_k \, \mathrm{d}\beta_k \\
&+ \tilde{C}_{22}^k \frac{1}{R_\beta^k} \int_{A^k} \left(F_{\tau_{u_\beta}} F_{s_{u_z}} \frac{H_\alpha^k}{H_\beta^k} \right) \mathrm{d}z \int_{\Omega^k} \left(\psi_{\alpha_i,\beta} \psi_{\alpha_j} \right) \mathrm{d}\alpha_k \, \mathrm{d}\beta_k \\
&+ \tilde{C}_{36}^k \int_{A^k} \left(F_{\tau_{u_\beta}} F_{s_{u_{z,z}}} H_\beta \right) \mathrm{d}z \int_{\Omega^k} \left(\psi_{\alpha_i,\alpha} \psi_{\alpha_j} \right) \mathrm{d}\alpha_k \, \mathrm{d}\beta_k \\
&+ \tilde{C}_{16}^k \frac{1}{R_\alpha^k} \int_{A^k} \left(F_{\tau_{u_\beta}} F_{s_{u_z}} \frac{H_\beta^k}{H_\alpha^k} \right) \mathrm{d}z \int_{\Omega^k} \left(\psi_{\alpha_i,\beta} \psi_{\alpha_j} \right) \mathrm{d}\alpha_k \, \mathrm{d}\beta_k \\
&+ \tilde{C}_{26}^k \frac{1}{R_\beta^k} \int_{A^k} \left(F_{\tau_{u_\beta}} F_{s_{u_z}} \right) \mathrm{d}z \int_{\Omega^k} \left(\psi_{\alpha_i,\beta} \psi_{\alpha_j} \right) \mathrm{d}\alpha_k \, \mathrm{d}\beta_k \\
&- \delta_D \times \left[\tilde{C}_{45}^k \frac{1}{R_\beta^k} \int_{A^k} \left(F_{\tau_{u_\beta}} F_{s_{u_z}} \right) \mathrm{d}z \int_{\Omega^k} \left(\psi_{\alpha_i} \psi_{\alpha_j,\alpha} \right) \mathrm{d}\alpha_k \, \mathrm{d}\beta_k \right. \\
&\left. + \tilde{C}_{44}^k \frac{1}{R_\beta^k} \int_{A^k} \left(F_{\tau_{u_\beta}} F_{s_{u_{z,z}}} \frac{H_\alpha^k}{H_\beta^k} \right) \mathrm{d}z \int_{\Omega^k} \left(\psi_{\alpha_i} \psi_{\alpha_j,\beta} \right) \mathrm{d}\alpha_k \, \mathrm{d}\beta_k \right]
\end{aligned}
$$

$$
\begin{aligned}
K_{u_z u_\alpha}^{\tau_{u_z} s_{u_\alpha}} &= \tilde{C}_{13}^k \int_{A^k} \left(F_{\tau_{u_{z,z}}} F_{s_{u_\alpha}} H_\beta^k \right) \mathrm{d}z \int_{\Omega^k} \left(\psi_{\alpha_i} \psi_{\alpha_j,\alpha} \right) \mathrm{d}\alpha_k \, \mathrm{d}\beta_k \\
&+ \tilde{C}_{36}^k \int_{A^k} \left(F_{\tau_{u_{z,z}}} F_{s_{u_\alpha}} H_\alpha^k \right) \mathrm{d}z \int_{\Omega^k} \left(\psi_{\alpha_i} \psi_{\alpha_j,\beta} \right) \mathrm{d}\alpha_k \, \mathrm{d}\beta_k \\
&+ \tilde{C}_{45}^k \int_{A^k} \left(F_{\tau_{u_z}} F_{s_{u_{\alpha,z}}} H_\alpha^k \right) \mathrm{d}z \int_{\Omega^k} \left(\psi_{\alpha_i,\beta} \psi_{\alpha_j} \right) \mathrm{d}\alpha_k \, \mathrm{d}\beta_k \\
&+ \tilde{C}_{16}^k \frac{1}{R_\alpha^k} \int_{A^k} \left(F_{\tau_{u_z}} F_{s_{u_\alpha}} \right) \mathrm{d}z \int_{\Omega^k} \left(\psi_{\alpha_i,\beta} \psi_{\alpha_j} \right) \mathrm{d}\alpha_k \, \mathrm{d}\beta_k \\
&+ \tilde{C}_{26}^k \frac{1}{R_\beta^k} \int_{A^k} \left(F_{\tau_{u_z}} F_{s_{u_\alpha}} \frac{H_\alpha^k}{H_\beta^k} \right) \mathrm{d}z \int_{\Omega^k} \left(\psi_{\alpha_i,\beta} \psi_{\alpha_j} \right) \mathrm{d}\alpha_k \, \mathrm{d}\beta_k \\
&+ \tilde{C}_{55}^k \int_{A^k} \left(F_{\tau_{u_z}} F_{s_{u_{\alpha,z}}} H_\beta^k \right) \mathrm{d}z \int_{\Omega^k} \left(\psi_{\alpha_i,\alpha} \psi_{\alpha_j} \right) \mathrm{d}\alpha_k \, \mathrm{d}\beta_k \\
&+ \tilde{C}_{11}^k \frac{1}{R_\alpha^k} \int_{A^k} \left(F_{\tau_{u_z}} F_{s_{u_\alpha}} \frac{H_\alpha^k}{H_\beta^k} \right) \mathrm{d}z \int_{\Omega^k} \left(\psi_{\alpha_i,\beta} \psi_{\alpha_j} \right) \mathrm{d}\alpha_k \, \mathrm{d}\beta_k \\
&+ \tilde{C}_{12}^k \frac{1}{R_\beta^k} \int_{A^k} \left(F_{\tau_{u_z}} F_{s_{u_\alpha}} \right) \mathrm{d}z \int_{\Omega^k} \left(\psi_{\alpha_i,\beta} \psi_{\alpha_j} \right) \mathrm{d}\alpha_k \, \mathrm{d}\beta_k \\
&- \delta_D \times \left[\tilde{C}_{55}^k \frac{1}{R_\alpha^k} \int_{A^k} \left(F_{\tau_{u_z}} F_{s_{u_\alpha}} \frac{H_\beta^k}{H_\alpha^k} \right) \mathrm{d}z \int_{\Omega^k} \left(\psi_{\alpha_i} \psi_{\alpha_j,\alpha} \right) \mathrm{d}\alpha_k \, \mathrm{d}\beta_k \right. \\
&\left. + \tilde{C}_{45}^k \frac{1}{R_\alpha^k} \int_{A^k} \left(F_{\tau_{u_z}} F_{s_{u_{\alpha,z}}} \right) \mathrm{d}z \int_{\Omega^k} \left(\psi_{\alpha_i} \psi_{\alpha_j,\beta} \right) \mathrm{d}\alpha_k \, \mathrm{d}\beta_k \right]
\end{aligned}
$$

$$
\begin{aligned}
K_{u_z u_\beta}^{\tau_{u_z} s_{u_\beta}} &= \tilde{C}_{36}^k \int_{A^k} \left(F_{\tau_{u_{z,z}}} F_{s_{u_\beta}} H_\beta^k\right) \mathrm{d}z \int_{\Omega^k} \left(\psi_{\alpha_i} \psi_{\alpha_j,\alpha}\right) \mathrm{d}\alpha_k \, \mathrm{d}\beta_k \\
&+ \tilde{C}_{23}^k \int_{A^k} \left(F_{\tau_{u_{z,z}}} F_{s_{u_\beta}} H_\alpha^k\right) \mathrm{d}z \int_{\Omega^k} \left(\psi_{\alpha_i} \psi_{\alpha_j,\beta}\right) \mathrm{d}\alpha_k \, \mathrm{d}\beta_k \\
&+ \tilde{C}_{44}^k \int_{A^k} \left(F_{\tau_{u_z}} F_{s_{u_{\beta,z}}} H_\beta^k\right) \mathrm{d}z \int_{\Omega^k} \left(\psi_{\alpha_i,\beta} \psi_{\alpha_j}\right) \mathrm{d}\alpha_k \, \mathrm{d}\beta_k \\
&+ \tilde{C}_{12}^k \frac{1}{R_\alpha^k} \int_{A^k} \left(F_{\tau_{u_z}} F_{s_{u_\beta}}\right) \mathrm{d}z \int_{\Omega^k} \left(\psi_{\alpha_i,\beta} \psi_{\alpha_j}\right) \mathrm{d}\alpha_k \, \mathrm{d}\beta_k \\
&+ \tilde{C}_{22}^k \frac{1}{R_\beta^k} \int_{A^k} \left(F_{\tau_{u_z}} F_{s_{u_\beta}} \frac{H_\alpha^k}{H_\beta^k}\right) \mathrm{d}z \int_{\Omega^k} \left(\psi_{\alpha_i,\beta} \psi_{\alpha_j}\right) \mathrm{d}\alpha_k \, \mathrm{d}\beta_k \\
&+ \tilde{C}_{45}^k \int_{A^k} \left(F_{\tau_{u_z}} F_{s_{u_\beta}} H_\beta^k\right) \mathrm{d}z \int_{\Omega^k} \left(\psi_{\alpha_i,\alpha} \psi_{\alpha_j}\right) \mathrm{d}\alpha_k \, \mathrm{d}\beta_k \\
&+ \tilde{C}_{16}^k \frac{1}{R_\alpha^k} \int_{A^k} \left(F_{\tau_{u_z}} F_{s_{u_\beta}} \frac{H_\beta^k}{H_\alpha^k}\right) \mathrm{d}z \int_{\Omega^k} \left(\psi_{\alpha_i,\beta} \psi_{\alpha_j}\right) \mathrm{d}\alpha_k \, \mathrm{d}\beta_k \\
&+ \tilde{C}_{26}^k \frac{1}{R_\beta^k} \int_{A^k} \left(F_{\tau_{u_z}} F_{s_{u_\beta}}\right) \mathrm{d}z \int_{\Omega^k} \left(\psi_{\alpha_i,\beta} \psi_{\alpha_j}\right) \mathrm{d}\alpha_k \, \mathrm{d}\beta_k \\
&- \delta_D \times \left[\tilde{C}_{45}^k \frac{1}{R_\alpha^k} \int_{A^k} \left(F_{\tau_{u_z}} F_{s_{u_\beta}} \frac{H_\beta^k}{H_\alpha^k}\right) \mathrm{d}z \int_{\Omega^k} \left(\psi_{\alpha_i} \psi_{\alpha_j,\alpha}\right) \mathrm{d}\alpha_k \, \mathrm{d}\beta_k \right. \\
&\left. + \tilde{C}_{44}^k \frac{1}{R_\alpha^k} \int_{A^k} \left(F_{\tau_{u_z}} F_{s_{u_{\beta,z}}}\right) \mathrm{d}z \int_{\Omega^k} \left(\psi_{\alpha_i} \psi_{\alpha_j,\beta}\right) \mathrm{d}\alpha_k \, \mathrm{d}\beta_k \right] \\
K_{u_z u_z}^{\tau_{u_z} s_{u_z}} &= \tilde{C}_{55}^k \int_{A^k} \left(F_{\tau_{u_z}} F_{s_{u_z}} \frac{H_\beta^k}{H_\alpha^k}\right) \mathrm{d}z \int_{\Omega^k} \left(\psi_{\alpha_i,\alpha} \psi_{\alpha_j,\alpha}\right) \mathrm{d}\alpha_k \, \mathrm{d}\beta_k \\
&+ \tilde{C}_{45}^k \int_{A^k} \left(F_{\tau_{u_z}} F_{s_{u_z}}\right) \mathrm{d}z \int_{\Omega^k} \left(\psi_{\alpha_i,\alpha} \psi_{\alpha_j,\beta}\right) \mathrm{d}\alpha_k \, \mathrm{d}\beta_k \\
&+ \tilde{C}_{45}^k \int_{A^k} \left(F_{\tau_{u_z}} F_{s_{u_z}}\right) \mathrm{d}z \int_{\Omega^k} \left(\psi_{\alpha_i,\beta} \psi_{\alpha_j,\alpha}\right) \mathrm{d}\alpha_k \, \mathrm{d}\beta_k \\
&+ \tilde{C}_{44}^k \int_{A^k} \left(F_{\tau_{u_z}} F_{s_{u_z}} \frac{H_\alpha^k}{H_\beta^k}\right) \mathrm{d}z \int_{\Omega^k} \left(\psi_{\alpha_i,\beta} \psi_{\alpha_j,\beta}\right) \mathrm{d}\alpha_k \, \mathrm{d}\beta_k \\
&+ \tilde{C}_{33}^k \int_{A^k} \left(F_{\tau_{u_{z,z}}} F_{s_{u_{z,z}}} H_\beta^k H_\alpha^k\right) \mathrm{d}z \int_{\Omega^k} \left(\psi_{\alpha_i} \psi_{\alpha_j}\right) \mathrm{d}\alpha_k \, \mathrm{d}\beta_k \\
&+ \tilde{C}_{13}^k \frac{1}{R_\alpha^k} \int_{A^k} \left(F_{\tau_{u_{z,z}}} F_{s_{u_z}} H_\beta^k\right) \mathrm{d}z \int_{\Omega^k} \left(\psi_{\alpha_i} \psi_{\alpha_j}\right) \mathrm{d}\alpha_k \, \mathrm{d}\beta_k \\
&+ \tilde{C}_{23}^k \frac{1}{R_\beta^k} \int_{A^k} \left(F_{\tau_{u_{z,z}}} F_{s_{u_z}} H_\alpha^k\right) \mathrm{d}z \int_{\Omega^k} \left(\psi_{\alpha_i} \psi_{\alpha_j}\right) \mathrm{d}\alpha_k \, \mathrm{d}\beta_k \\
&+ \tilde{C}_{13}^k \frac{1}{R_\alpha^k} \int_{A^k} \left(F_{\tau_{u_z}} F_{s_{u_{z,z}}} H_\beta^k\right) \mathrm{d}z \int_{\Omega^k} \left(\psi_{\alpha_i} \psi_{\alpha_j}\right) \mathrm{d}\alpha_k \, \mathrm{d}\beta_k
\end{aligned}
$$

$$
\begin{aligned}
&+ \tilde{C}^{k}_{23} \frac{1}{R^{k}_{\beta}} \int_{A^{k}} \left(F_{\tau_{u_z}} F_{s_{u_z,z}} H^{k}_{\alpha} \right) \mathrm{d}z \int_{\Omega^{k}} \left(\psi_{\alpha_i} \psi_{\alpha_j} \right) \mathrm{d}\alpha_k \, \mathrm{d}\beta_k \\
&+ \tilde{C}^{k}_{11} \frac{1}{\left(R^{k}_{\alpha}\right)^2} \int_{A^{k}} \left(F_{\tau_{u_z}} F_{s_{u_z}} \frac{H^{k}_{\beta}}{H^{k}_{\alpha}} \right) \mathrm{d}z \int_{\Omega^{k}} \left(\psi_{\alpha_i} \psi_{\alpha_j} \right) \mathrm{d}\alpha_k \, \mathrm{d}\beta_k \\
&+ \tilde{C}^{k}_{12} \frac{1}{\left(R^{k}_{\alpha} R^{k}_{\beta}\right)} \int_{A^{k}} \left(F_{\tau_{u_z}} F_{s_{u_z}} \right) \mathrm{d}z \int_{\Omega^{k}} \left(\psi_{\alpha_i} \psi_{\alpha_j} \right) \mathrm{d}\alpha_k \, \mathrm{d}\beta_k \\
&+ \tilde{C}^{k}_{12} \frac{1}{\left(R^{k}_{\alpha} R^{k}_{\beta}\right)} \int_{A^{k}} \left(F_{\tau_{u_z}} F_{s_{u_z}} \right) \mathrm{d}z \int_{\Omega^{k}} \left(\psi_{\alpha_i} \psi_{\alpha_j} \right) \mathrm{d}\alpha_k \, \mathrm{d}\beta_k \\
&+ \tilde{C}^{k}_{22} \frac{1}{\left(R^{k}_{\beta}\right)^2} \int_{A^{k}} \left(F_{\tau_{u_z}} F_{s_{u_z}} \frac{H^{k}_{\alpha}}{H^{k}_{\beta}} \right) \mathrm{d}z \int_{\Omega^{k}} \left(\psi_{\alpha_i} \psi_{\alpha_j} \right) \mathrm{d}\alpha_k \, \mathrm{d}\beta_k
\end{aligned}
\tag{7.53}
$$

As previously done for beams and plates, at this stage it is useful to introduce the following thickness integrals in order to write in a concise manner the explicit forms of the Ritz fundamental secondary nuclei

$$
\left(J^{k\tau_u s_u}_{pq}, J^{k\hat{\tau}_u s_u}_{pq}, J^{k\tau_u \hat{s}_u}_{pq}, J^{k\hat{\tau}_u \hat{s}_u}_{pq} \right) = \int_{A^{k}} \left(F_{\tau_u} F_{s_u}, \, \partial z F_{\tau_u} F_{s_u}, \, F_{\tau_u} \partial z F_{s_u}, \, \partial z F_{\tau_u} \partial z F_{s_u} \right) \times \frac{H^{k}_{\alpha} H^{k}_{\beta}}{\left(H^{k}_{\alpha}\right)^{p} \left(H^{k}_{\beta}\right)^{q}} \mathrm{d}z
\tag{7.54}
$$

where

$$
\tau_u = \tau_{u_\alpha}, \tau_{u_\beta}, \tau_{u_z}; \quad s_u = s_{u_\alpha}, s_{u_\beta}, s_{u_z}; \quad p = 0, 1, 2; \quad q = 0, 1, 2;
$$

The above expressions include all the possible combinations of product between displacement thickness functions. To the same aim, it is convenient to rewrite the Ritz functions involved in the in-plane integrals, namely ψ_{α_i}, ψ_{β_i}, ψ_{z_i} for displacements and $\chi_{\sigma_{\alpha z_i}}$, $\chi_{\sigma_{\beta z_i}}$, $\chi_{\sigma_{zz_i}}$ for transverse normal and shear stresses, as:

$$
\begin{aligned}
\psi_{\alpha_i}(\alpha, \beta) &= \sum_{m}^{M} \sum_{n}^{N} \phi^{u_\alpha}_{m}(\alpha) \, \phi^{u_\alpha}_{n}(\beta); \\
\psi_{\beta_i}(\alpha, \beta) &= \sum_{m}^{M} \sum_{n}^{N} \phi^{u_\beta}_{m}(\alpha) \, \phi^{u_\beta}_{n}(\beta); \\
\psi_{z_i}(\alpha, \beta) &= \sum_{m}^{M} \sum_{n}^{N} \phi^{u_z}_{m}(\alpha) \, \phi^{u_z}_{n}(\beta)
\end{aligned}
\tag{7.55}
$$

Then, by using Eq. (7.55) the general expressions of the in-plane integrals can be written as

$$
{}^{i}_{j}I^{\xi\zeta}_{mp} = \int_0^a \frac{\mathrm{d}^{\xi}\phi^{i}_{m}(\alpha)}{\mathrm{d}\alpha^{\xi}}\frac{\mathrm{d}^{\zeta}\phi^{j}_{p}(\alpha)}{\mathrm{d}\alpha^{\zeta}}\mathrm{d}\alpha\,; \quad {}^{i}_{j}I^{\xi\zeta}_{nq} = \int_0^b \frac{\mathrm{d}^{\xi}\phi^{i}_{n}(\beta)}{\mathrm{d}\beta^{\xi}}\frac{\mathrm{d}^{\zeta}\phi^{j}_{q}(\beta)}{\mathrm{d}\beta^{\zeta}}\mathrm{d}\beta \qquad (7.56)
$$
$$
m=1,\cdots,M;\quad p=1,\cdots,P;\quad n=1,\cdots,N;\quad q=1,\cdots,Q;
$$

where $i,j=u_\alpha,u_\beta,u_z$ and ξ, ζ are differentiation orders. Therefore, considering Eqs. (7.54) and (7.56), the explicit forms of the Ritz fundamental secondary stiffness nuclei are following reported

$$
\begin{aligned}
K^{\tau_{u_\alpha}s_{u_\alpha}}_{u_\alpha u_\alpha} &= \tilde{C}^{k}_{11}J^{k\tau_{u_\alpha}s_{u_\alpha}}_{20}\,{}^{u_\alpha}_{u_\alpha}I^{11}_{mp}\,{}^{u_\alpha}_{u_\alpha}I^{00}_{nq} + \tilde{C}^{k}_{16}J^{k\tau_{u_\alpha}s_{u_\alpha}}_{11}\,{}^{u_\alpha}_{u_\alpha}I^{10}_{mp}\,{}^{u_\alpha}_{u_\alpha}I^{01}_{nq}\\
&+ \tilde{C}^{k}_{16}J^{k\tau_{u_\alpha}s_{u_\alpha}}_{11}\,{}^{u_\alpha}_{u_\alpha}I^{01}_{mp}\,{}^{u_\alpha}_{u_\alpha}I^{10}_{nq} + \tilde{C}^{k}_{66}J^{k\tau_{u_\alpha}s_{u_\alpha}}_{02}\,I^{00}_{mp}\,I^{11}_{nq}\\
&+ \tilde{C}^{k}_{55}J^{k\hat{\tau}_{u_\alpha}\hat{s}_{u_\alpha}}_{00}\,{}^{u_\alpha}_{u_\alpha}I^{00}_{mp}\,{}^{u_\alpha}_{u_\alpha}I^{00}_{nq} + \delta_D\Big(-\tilde{C}^{k}_{55}\frac{1}{R^{k}_{\alpha}}J^{k\hat{\tau}_{u_\alpha}s_{u_\alpha}}_{10}\,{}^{u_\alpha}_{u_\alpha}I^{00}_{mp}\,{}^{u_\alpha}_{u_\alpha}I^{00}_{nq}\\
&- \tilde{C}^{k}_{55}\frac{1}{R^{k}_{\alpha}}J^{k\tau_{u_\alpha}\hat{s}_{u_\alpha}}_{10}\,{}^{u_\alpha}_{u_\alpha}I^{00}_{mp}\,{}^{u_\alpha}_{u_\alpha}I^{00}_{nq} + \tilde{C}^{k}_{55}\frac{1}{\left(R^{k}_{\alpha}\right)^{2}}J^{k\tau_{u_\alpha}s_{u_\alpha}}_{20}\,{}^{u_\alpha}_{u_\alpha}I^{00}_{mp}\,{}^{u_\alpha}_{u_\alpha}I^{00}_{nq}\Big)\\
K^{\tau_{u_\alpha}s_{u_\beta}}_{u_\alpha u_\beta} &= \tilde{C}^{k}_{16}J^{k\tau_{u_\alpha}s_{u_\beta}}_{20}\,{}^{u_\alpha}_{u_\beta}I^{11}_{mp}\,{}^{u_\alpha}_{u_\beta}I^{00}_{nq} + \tilde{C}^{k}_{12}J^{k\tau_{u_\alpha}s_{u_\beta}}_{11}\,{}^{u_\alpha}_{u_\beta}I^{10}_{mp}\,{}^{u_\alpha}_{u_\beta}I^{01}_{nq}\\
&+ \tilde{C}^{k}_{66}J^{k\tau_{u_\alpha}s_{u_\beta}}_{11}\,{}^{u_\alpha}_{u_\beta}I^{01}_{mp}\,{}^{u_\alpha}_{u_\beta}I^{10}_{nq} + \tilde{C}^{k}_{26}J^{k\tau_{u_\alpha}s_{u_\beta}}_{02}\,I^{00}_{mp}\,I^{11}_{nq}\\
&+ \tilde{C}^{k}_{45}J^{k\tau_{u_\alpha}s_{u_\beta}}_{00}\,{}^{u_\alpha}_{u_\beta}I^{00}_{mp}\,{}^{u_\alpha}_{u_\beta}I^{00}_{nq} + \delta_D\Big(-\tilde{C}^{k}_{45}\frac{1}{R^{k}_{\beta}}J^{k\hat{\tau}_{u_\alpha}s_{u_\beta}}_{01}\,{}^{u_\alpha}_{u_\beta}I^{00}_{mp}\,{}^{u_\alpha}_{u_\beta}I^{00}_{nq}\\
&- \tilde{C}^{k}_{45}\frac{1}{R^{k}_{\alpha}}J^{k\tau_{u_\alpha}\hat{s}_{u_\beta}}_{10}\,{}^{u_\alpha}_{u_\beta}I^{00}_{mp}\,{}^{u_\alpha}_{u_\beta}I^{00}_{nq} + \tilde{C}^{k}_{45}\frac{1}{R^{k}_{\alpha}R^{k}_{\beta}}J^{k\hat{\tau}_{u_\alpha}\hat{s}_{u_\beta}}_{11}\,{}^{u_\alpha}_{u_\beta}I^{00}_{mp}\,{}^{u_\alpha}_{u_\beta}I^{00}_{nq}\Big)\\
K^{\tau_{u_\alpha}s_{u_z}}_{u_\alpha u_z} &= \tilde{C}^{k}_{13}J^{k\tau_{u_\alpha}s_{u_z}}_{10}\,{}^{u_\alpha}_{u_z}I^{10}_{mp}\,{}^{u_\alpha}_{u_z}I^{00}_{nq} + \tilde{C}^{k}_{55}J^{k\hat{\tau}_{u_\alpha}s_{u_z}}_{10}\,{}^{u_\alpha}_{u_z}I^{01}_{mp}\,{}^{u_\alpha}_{u_z}I^{00}_{nq}\\
&+ \tilde{C}^{k}_{45}J^{k\hat{\tau}_{u_\alpha}s_{u_z}}_{01}\,{}^{u_\alpha}_{u_z}I^{00}_{mp}\,{}^{u_\alpha}_{u_z}I^{01}_{nq} + \tilde{C}^{k}_{36}J^{k\hat{\tau}_{u_\alpha}s_{u_z}}_{01}\,{}^{u_\alpha}_{u_z}I^{00}_{mp}\,{}^{u_\alpha}_{u_z}I^{01}_{nq}\\
&+ \tilde{C}^{k}_{11}\frac{1}{R^{k}_{\alpha}}J^{k\tau_{u_\alpha}s_{u_z}}_{20}\,{}^{u_\alpha}_{u_z}I^{10}_{mp}\,{}^{u_\alpha}_{u_z}I^{00}_{nq} + \tilde{C}^{k}_{12}\frac{1}{R^{k}_{\beta}}J^{k\tau_{u_\alpha}s_{u_z}}_{11}\,{}^{u_\alpha}_{u_z}I^{10}_{mp}\,{}^{u_\alpha}_{u_z}I^{00}_{nq}\\
&+ \tilde{C}^{k}_{16}\frac{1}{R^{k}_{\alpha}}J^{k\tau_{u_\alpha}s_{u_z}}_{11}\,{}^{u_\alpha}_{u_z}I^{01}_{mp}\,{}^{u_\alpha}_{u_z}I^{00}_{nq} + \tilde{C}^{k}_{26}\frac{1}{R^{k}_{\beta}}J^{k\tau_{u_\alpha}s_{u_z}}_{02}\,{}^{u_\alpha}_{u_z}I^{01}_{mp}\,{}^{u_\alpha}_{u_z}I^{00}_{nq}\\
&- \delta_D\Big(\tilde{C}^{k}_{55}\frac{1}{R^{k}_{\alpha}}J^{k\tau_{u_\alpha}s_{u_z}}_{20}\,{}^{u_\alpha}_{u_z}I^{00}_{mp}\,{}^{u_\alpha}_{u_z}I^{10}_{nq} + \tilde{C}^{k}_{45}\frac{1}{R^{k}_{\alpha}}J^{k\tau_{u_\alpha}s_{u_z}}_{11}\,{}^{u_\alpha}_{u_z}I^{00}_{mp}\,{}^{u_\alpha}_{u_z}I^{01}_{nq}\Big)\\
K^{\tau_{u_\beta}s_{u_\alpha}}_{u_\beta u_\alpha} &= \tilde{C}^{k}_{16}J^{k\tau_{u_\beta}s_{u_\alpha}}_{20}\,{}^{u_\beta}_{u_\alpha}I^{11}_{mp}\,{}^{u_\beta}_{u_\alpha}I^{00}_{nq} + \tilde{C}^{k}_{12}J^{k\tau_{u_\beta}s_{u_\alpha}}_{11}\,{}^{u_\beta}_{u_\alpha}I^{01}_{mp}\,{}^{u_\beta}_{u_\alpha}I^{10}_{nq}\\
&+ \tilde{C}^{k}_{66}J^{k\tau_{u_\beta}s_{u_\alpha}}_{11}\,{}^{u_\beta}_{u_\alpha}I^{10}_{mp}\,{}^{u_\beta}_{u_\alpha}I^{01}_{nq} + \tilde{C}^{k}_{26}J^{k\tau_{u_\beta}s_{u_\alpha}}_{02}\,{}^{u_\beta}_{u_\alpha}I^{00}_{mp}\,{}^{u_\beta}_{u_\alpha}I^{11}_{nq}\\
&+ \tilde{C}^{k}_{45}J^{k\tau_{u_\beta}s_{u_\alpha}}_{00}\,{}^{u_\beta}_{u_\alpha}I^{00}_{mp}\,{}^{u_\beta}_{u_\alpha}I^{00}_{nq} + \delta_D\Big(-\tilde{C}^{k}_{45}\frac{1}{R^{k}_{\alpha}}J^{k\hat{\tau}_{u_\alpha}s_{u_\alpha}}_{10}\,{}^{u_\beta}_{u_\alpha}I^{00}_{mp}\,{}^{u_\beta}_{u_\alpha}I^{00}_{nq}\\
&- \tilde{C}^{k}_{45}\frac{1}{R^{k}_{\beta}}J^{k\tau_{u_\beta}\hat{s}_{u_\beta}}_{01}\,{}^{u_\beta}_{u_\alpha}I^{00}_{mp}\,{}^{u_\beta}_{u_\alpha}I^{00}_{nq} + \tilde{C}^{k}_{45}\frac{1}{R^{k}_{\alpha}R^{k}_{\beta}}J^{k\hat{\tau}_{u_\alpha}\hat{s}_{u_\beta}}_{11}\,{}^{u_\beta}_{u_\alpha}I^{00}_{mp}\,{}^{u_\beta}_{u_\alpha}I^{00}_{nq}\Big)
\end{aligned}
$$

$$
\begin{aligned}
K^{\tau_{u_\beta} s_{u_\beta}}_{u_\beta u_\beta} = {} & \tilde{C}^k_{66} J^{k\,\tau_{u_\beta}\,s_{u_\beta}}_{20}\, {}^{u_\beta}_{u_\beta}I^{11}_{mp}\, {}^{u_\beta}_{u_\beta}I^{00}_{nq} + \tilde{C}^k_{26} J^{k\,\tau_{u_\beta}\,s_{u_\beta}}_{11}\, {}^{u_\beta}_{u_\beta}I^{01}_{mp}\, {}^{u_\beta}_{u_\beta}I^{10}_{nq} \\
& + \tilde{C}^k_{26} J^{k\,\tau_{u_\beta}\,s_{u_\beta}}_{11}\, {}^{u_\beta}_{u_\beta}I^{10}_{mp}\, {}^{u_\beta}_{u_\beta}I^{01}_{nq} + \tilde{C}^k_{22} J^{k\,\tau_{u_\beta}\,s_{u_\beta}}_{02}\, {}^{u_\beta}_{u_\beta}I^{00}_{mp}\, {}^{u_\beta}_{u_\beta}I^{11}_{nq} \\
& + \tilde{C}^k_{44} J^{k\,\hat{\tau}_{u_\alpha}\,\hat{s}_{u_\alpha}}_{00}\, {}^{u_\beta}_{u_\beta}I^{00}_{mp}\, {}^{u_\beta}_{u_\beta}I^{00}_{nq} + \delta_D \Big(-\tilde{C}^k_{44} \frac{1}{R^k_\beta} J^{k\,\hat{\tau}_{u_\alpha}\,s_{u_\beta}}_{01}\, {}^{u_\beta}_{u_\beta}I^{00}_{mp}\, {}^{u_\beta}_{u_\beta}I^{00}_{nq} \\
& - \tilde{C}^k_{44} \frac{1}{R^k_\beta} J^{k\,\tau_{u_\beta}\,\hat{s}_{u_\alpha}}_{01}\, {}^{u_\beta}_{u_\beta}I^{00}_{mp}\, {}^{u_\beta}_{u_\beta}I^{00}_{nq} + \tilde{C}^k_{44} \frac{1}{\left(R^k_\beta\right)^2} J^{k\,\tau_{u_\beta}\,s_{u_\beta}}_{02}\, {}^{u_\beta}_{u_\beta}I^{00}_{mp}\, {}^{u_\beta}_{u_\beta}I^{00}_{nq} \Big)
\end{aligned}
$$

$$
\begin{aligned}
K^{\tau_{u_\beta} s_{u_z}}_{u_\beta u_z} = {} & \tilde{C}^k_{45} J^{k\,\hat{\tau}_{u_\beta}\,s_{u_z}}_{10}\, {}^{u_\beta}_{u_z}I^{01}_{mp}\, {}^{u_\beta}_{u_z}I^{00}_{nq} + \tilde{C}^k_{44} J^{k\,\hat{\tau}_{u_\beta}\,s_{u_z}}_{01}\, {}^{u_\beta}_{u_z}I^{00}_{mp}\, {}^{u_\beta}_{u_z}I^{01}_{nq} \\
& + \tilde{C}^k_{23} J^{k\,\tau_{u_\beta}\,\hat{s}_{u_z}}_{01}\, {}^{u_\beta}_{u_z}I^{00}_{mp}\, {}^{u_\beta}_{u_z}I^{10}_{nq} + \tilde{C}^k_{12} \frac{1}{R^k_\alpha} J^{k\,\tau_{u_\beta}\,s_{u_z}}_{11}\, {}^{u_\beta}_{u_z}I^{00}_{mp}\, {}^{u_\beta}_{u_z}I^{10}_{nq} \\
& + \tilde{C}^k_{22} \frac{1}{R^k_\beta} J^{k\,\tau_{u_\alpha}\,s_{u_\alpha}}_{02}\, {}^{u_\beta}_{u_z}I^{00}_{mp}\, {}^{u_\beta}_{u_z}I^{10}_{nq} + \tilde{C}^k_{36} J^{k\,\tau_{u_\beta}\,\hat{s}_{u_z}}_{10}\, {}^{u_\beta}_{u_z}I^{10}_{mp}\, {}^{u_\beta}_{u_z}I^{00}_{nq} \\
& + \tilde{C}^k_{16} \frac{1}{R^k_\alpha} J^{k\,\tau_{u_\beta}\,s_{u_z}}_{20}\, {}^{u_\beta}_{u_z}I^{00}_{mp}\, {}^{u_\beta}_{u_z}I^{10}_{nq} + \tilde{C}^k_{26} \frac{1}{R^k_\beta} J^{k\,\tau_{u_\alpha}\,s_{u_\alpha}}_{11}\, {}^{u_\beta}_{u_z}I^{00}_{mp}\, {}^{u_\beta}_{u_z}I^{10}_{nq} \\
& - \delta_D \Big(\tilde{C}^k_{45} \frac{1}{R^k_\beta} J^{k\,\tau_{u_\alpha}\,s_{u_z}}_{11}\, {}^{u_\beta}_{u_z}I^{01}_{mp}\, {}^{u_\beta}_{u_z}I^{00}_{nq} + \tilde{C}^k_{44} \frac{1}{R^k_\beta} J^{k\,\tau_{u_\beta}\,\hat{s}_{u_\alpha}}_{02}\, {}^{u_\beta}_{u_z}I^{00}_{mp}\, {}^{u_\beta}_{u_z}I^{01}_{nq} \Big)
\end{aligned}
$$

$$
\begin{aligned}
K^{\tau_{u_z} s_{u_\alpha}}_{u_z u_\alpha} = {} & \tilde{C}^k_{13} J^{k\,\hat{\tau}_{u_\beta}\,s_{u_\alpha}}_{10}\, {}^{u_z}_{u_\alpha}I^{01}_{mp}\, {}^{u_z}_{u_\alpha}I^{00}_{nq} + \tilde{C}^k_{36} J^{k\,\hat{\tau}_{u_\beta}\,s_{u_\alpha}}_{01}\, {}^{u_z}_{u_\alpha}I^{00}_{mp}\, {}^{u_z}_{u_\alpha}I^{01}_{nq} \\
& + \tilde{C}^k_{45} J^{k\,\tau_{u_z}\,\hat{s}_{u_z}}_{01}\, {}^{u_z}_{u_\alpha}I^{00}_{mp}\, {}^{u_z}_{u_\alpha}I^{10}_{nq} + \tilde{C}^k_{16} \frac{1}{R^k_\alpha} J^{k\,\tau_{u_z}\,s_{u_\alpha}}_{11}\, {}^{u_z}_{u_\alpha}I^{00}_{mp}\, {}^{u_z}_{u_\alpha}I^{10}_{nq} \\
& + \tilde{C}^k_{26} \frac{1}{R^k_\beta} J^{k\,\tau_{u_\alpha}\,s_{u_\alpha}}_{02}\, {}^{u_z}_{u_\alpha}I^{00}_{mp}\, {}^{u_z}_{u_\alpha}I^{10}_{nq} + \tilde{C}^k_{55} J^{k\,\tau_{u_z}\,\hat{s}_{u_z}}_{10}\, {}^{u_z}_{u_\alpha}I^{10}_{mp}\, {}^{u_z}_{u_\alpha}I^{00}_{nq} \\
& + \tilde{C}^k_{11} \frac{1}{R^k_\alpha} J^{k\,\tau_{u_z}\,s_{u_\alpha}}_{20}\, {}^{u_z}_{u_\alpha}I^{00}_{mp}\, {}^{u_z}_{u_\alpha}I^{10}_{nq} + \tilde{C}^k_{12} \frac{1}{R^k_\beta} J^{k\,\tau_{u_\alpha}\,s_{u_\alpha}}_{11}\, {}^{u_z}_{u_\alpha}I^{00}_{mp}\, {}^{u_z}_{u_\alpha}I^{10}_{nq} \\
& - \delta_D \Big(\tilde{C}^k_{55} \frac{1}{R^k_\alpha} J^{k\,\tau_{u_\alpha}\,s_{u_\alpha}}_{20}\, {}^{u_z}_{u_\alpha}I^{01}_{mp}\, {}^{u_z}_{u_\alpha}I^{00}_{nq} + \tilde{C}^k_{45} \frac{1}{R^k_\alpha} J^{k\,\tau_{u_z}\,\hat{s}_{u_\alpha}}_{11}\, {}^{u_z}_{u_\alpha}I^{00}_{mp}\, {}^{u_z}_{u_\alpha}I^{01}_{nq} \Big)
\end{aligned}
$$

$$
\begin{aligned}
K^{\tau_{u_z} s_{u_\beta}}_{u_z u_\beta} = {} & \tilde{C}^k_{36} J^{k\,\hat{\tau}_{u_\beta}\,s_{u_\beta}}_{10}\, {}^{u_z}_{u_\beta}I^{01}_{mp}\, {}^{u_z}_{u_\beta}I^{00}_{nq} + \tilde{C}^k_{23} J^{k\,\hat{\tau}_{u_\beta}\,s_{u_\beta}}_{01}\, {}^{u_z}_{u_\beta}I^{00}_{mp}\, {}^{u_z}_{u_\beta}I^{01}_{nq} \\
& + \tilde{C}^k_{44} J^{k\,\tau_{u_z}\,\hat{s}_{u_z}}_{01}\, {}^{u_z}_{u_\beta}I^{00}_{mp}\, {}^{u_z}_{u_\beta}I^{10}_{nq} + \tilde{C}^k_{12} \frac{1}{R^k_\alpha} J^{k\,\tau_{u_z}\,s_{u_\beta}}_{11}\, {}^{u_z}_{u_\beta}I^{00}_{mp}\, {}^{u_z}_{u_\beta}I^{10}_{nq} \\
& + \tilde{C}^k_{22} \frac{1}{R^k_\beta} J^{k\,\tau_{u_\alpha}\,s_{u_\beta}}_{02}\, {}^{u_z}_{u_\beta}I^{00}_{mp}\, {}^{u_z}_{u_\beta}I^{10}_{nq} + \tilde{C}^k_{45} J^{k\,\tau_{u_z}\,\hat{s}_{u_z}}_{10}\, {}^{u_z}_{u_\beta}I^{10}_{mp}\, {}^{u_z}_{u_\beta}I^{00}_{nq} \\
& + \tilde{C}^k_{16} \frac{1}{R^k_\alpha} J^{k\,\tau_{u_z}\,s_{u_\beta}}_{20}\, {}^{u_z}_{u_\beta}I^{00}_{mp}\, {}^{u_z}_{u_\beta}I^{10}_{nq} + \tilde{C}^k_{26} \frac{1}{R^k_\beta} J^{k\,\tau_{u_\alpha}\,s_{u_\beta}}_{11}\, {}^{u_z}_{u_\beta}I^{00}_{mp}\, {}^{u_z}_{u_\beta}I^{10}_{nq} \\
& - \delta_D \Big(\tilde{C}^k_{45} \frac{1}{R^k_\alpha} J^{k\,\tau_{u_\alpha}\,s_{u_\beta}}_{20}\, {}^{u_z}_{u_\beta}I^{01}_{mp}\, {}^{u_z}_{u_\beta}I^{00}_{nq} + \tilde{C}^k_{44} \frac{1}{R^k_\alpha} J^{k\,\tau_{u_z}\,\hat{s}_{u_\alpha}}_{11}\, {}^{u_z}_{u_\beta}I^{00}_{mp}\, {}^{u_z}_{u_\beta}I^{01}_{nq} \Big)
\end{aligned}
$$

$$K^{\tau_{u_z} s_{u_z}}_{u_z u_z} = \tilde{C}^k_{55} J^{k\,\tau_{u_z}\,s_{u_z}}_{20}\; {}^{u_z}_{u_z}I^{11}_{mp}\; {}^{u_z}_{u_z}I^{00}_{nq} + \tilde{C}^k_{45} J^{k\,\tau_{u_z}\,s_{u_z}}_{11}\; {}^{u_z}_{u_z}I^{10}_{mp}\; {}^{u_z}_{u_z}I^{01}_{nq}$$
$$+ \tilde{C}^k_{45} J^{k\,\tau_{u_z}\,s_{u_z}}_{11}\; {}^{u_z}_{u_z}I^{01}_{mp}\; {}^{u_z}_{u_z}I^{10}_{nq} + \tilde{C}^k_{44} J^{k\,\tau_{u_z}\,s_{u_z}}_{02}\; {}^{u_z}_{u_z}I^{00}_{mp}\; {}^{u_z}_{u_z}I^{11}_{nq}$$
$$+ \tilde{C}^k_{33} J^{k\,\hat{\tau}_{u_\alpha}\,\hat{s}_{u_\alpha}}_{00}\; {}^{u_z}_{u_z}I^{00}_{mp}\; {}^{u_z}_{u_z}I^{00}_{nq} + \tilde{C}^k_{13} \frac{1}{R^k_\alpha} J^{k\,\hat{\tau}_{u_z}\,s_{u_z}}_{10}\; {}^{u_z}_{u_z}I^{00}_{mp}\; {}^{u_z}_{u_z}I^{00}_{nq}$$
$$+ \tilde{C}^k_{23} \frac{1}{R^k_\beta} J^{k\,\hat{\tau}_{u_z}\,s_{u_z}}_{01}\; {}^{u_z}_{u_z}I^{00}_{mp}\; {}^{u_z}_{u_z}I^{00}_{nq} + \tilde{C}^k_{13} \frac{1}{R^k_\alpha} J^{k\,\tau_{u_z}\,\hat{s}_{u_z}}_{10}\; {}^{u_z}_{u_z}I^{00}_{mp}\; {}^{u_z}_{u_z}I^{00}_{nq}$$
$$+ \tilde{C}^k_{23} \frac{1}{R^k_\beta} J^{k\,\tau_{u_z}\,\hat{s}_{u_z}}_{01}\; {}^{u_z}_{u_z}I^{00}_{mp}\; {}^{u_z}_{u_z}I^{00}_{nq} + \tilde{C}^k_{11} \frac{1}{\left(R^k_\alpha\right)^2} J^{k\,\tau_{u_z}\,s_{u_z}}_{20}\; {}^{u_z}_{u_z}I^{00}_{mp}\; {}^{u_z}_{u_z}I^{00}_{nq}$$
$$+ \tilde{C}^k_{12} \frac{1}{\left(R^k_\alpha R^k_\beta\right)} J^{k\,\tau_{u_z}\,s_{u_z}}_{11}\; {}^{u_z}_{u_z}I^{00}_{mp}\; {}^{u_z}_{u_z}I^{00}_{nq} + \tilde{C}^k_{12} \frac{1}{\left(R^k_\alpha R^k_\beta\right)} J^{k\,\tau_{u_z}\,s_{u_z}}_{11}\; {}^{u_z}_{u_z}I^{00}_{mp}\; {}^{u_z}_{u_z}I^{00}_{nq}$$
$$+ \tilde{C}^k_{22} \frac{1}{\left(R^k_\beta\right)^2} J^{k\,\tau_{u_z}\,s_{u_z}}_{02}\; {}^{u_z}_{u_z}I^{00}_{mp}\; {}^{u_z}_{u_z}I^{00}_{nq} \tag{7.57}$$

Mass nucleus M

The mass nucleus is obtained following the same procedure carried out for plate and beam structures, and in the present case leads to

$$\delta L^k_{F_{in}} = \int_{\Omega_k}\int_{A^k} \delta \mathbf{U}^{k^T}_{\tau i} \left[\rho^k \left(\mathbf{F}_\tau \mathbf{\Psi}_i\right)^T \left(\mathbf{F}_s \mathbf{\Psi}_j\right)\right] \ddot{\mathbf{U}}_{sj}\, \mathrm{d}\Omega_k\, \mathrm{d}z \tag{7.58}$$

$$\mathbf{M}^{k\tau sij} = \int_{\Omega_k}\int_{A^k} \left(\rho^k \left[\left(\mathbf{F}_\tau \mathbf{\Psi}_i\right)^T \left(\mathbf{F}_s \mathbf{\Psi}_j\right)\right]\right) \mathrm{d}\Omega_k\, \mathrm{d}z \tag{7.59}$$

and the Ritz fundamental secondary mass nuclei are given as:

$$\begin{aligned} M^{\tau_{u_\alpha} s_{u_\alpha}}_{u_\alpha u_\alpha} &= \rho^k J^{k\,\tau_{u_\alpha}\,s_{u_\alpha}}_{11}\; {}^{u_\alpha}_{u_\alpha}I^{00}_{mp}\; {}^{u_\alpha}_{u_\alpha}I^{00}_{nq} \\ M^{\tau_{u_\beta} s_{u_\beta}}_{u_\beta u_\beta} &= \rho^k J^{k\,\tau_{u_\beta}\,s_{u_\beta}}_{11}\; {}^{u_\beta}_{u_\beta}I^{00}_{mp}\; {}^{u_\beta}_{u_\beta}I^{00}_{nq} \\ M^{\tau_{u_z} s_{u_z}}_{u_z u_z} &= \rho^k J^{k\,\tau_{u_z}\,s_{u_z}}_{11}\; {}^{u_z}_{u_z}I^{00}_{mp}\; {}^{u_z}_{u_z}I^{00}_{nq} \end{aligned} \tag{7.60}$$

7.3 RITZ METHOD AND REISSNER'S MIXED VARIATIONAL THEOREM

The weak-form of the governing equations for shell structures based on the RMVT is explicitly derived below [8,9]. An advanced Ritz formulation is developed by virtue of the RMVT (see Chapter 3 for more details). After the series expansions of the displacement amplitude vector $\mathbf{u}^k_\tau$ and the transverse stress amplitude vector $\boldsymbol{\sigma}^k_{n_{\tau\sigma}}$, the displacement field is given in Eq. (7.2) and the transverse stress fields assume the following forms

$$
\begin{aligned}
\sigma_{\alpha z_{\tau_{\sigma\alpha z}}}(\alpha,\beta,z,t) &= \sum_i^{\mathcal{N}} F_{\tau_{\sigma\alpha z}}(z)\, S^k_{\alpha\tau_{\sigma\alpha z}i}\, \chi_{\alpha_i}(\alpha,\beta)\, e^{\imath\,\omega_{ij}\,t} \\
\sigma_{\beta z_{\tau_{\sigma\beta z}}}(\alpha,\beta,z,t) &= \sum_i^{\mathcal{N}} F_{\tau_{\sigma\beta z}}(z)\, S^k_{\beta\tau_{\sigma\beta z}i}\, \chi_{\beta_i}(\alpha,\beta)\, e^{\imath\,\omega_{ij}\,t} \\
\sigma_{zz_{\tau_{\sigma zz}}}(\alpha,\beta,z,t) &= \sum_i^{\mathcal{N}} F_{\tau_{\sigma zz}}(z)\, S^k_{z\tau_{\sigma zz}i}\, \chi_{z_i}(\alpha,\beta)\, e^{\imath\,\omega_{ij}\,t}
\end{aligned} \tag{7.61}
$$

where $\mathcal{N}$ indicates the expansion order in the Ritz approximation. Equations (7.1) and (7.61) can be written in compact form as

$$
\boldsymbol{\sigma}^k_n = \mathbf{F}_{\tau_\sigma}\,\mathbf{S}^k_{\tau_\sigma i}\,\mathbf{X}_i \qquad \tau_\sigma = \tau_{\sigma\alpha z}, \tau_{\sigma\beta z}, \tau_{\sigma zz}; \quad i = 1, \ldots, \mathcal{N} \tag{7.62}
$$

where

$$
\mathbf{S}^k_{\tau_\sigma i} = \begin{Bmatrix} S^k_{\alpha z\tau_{\sigma\alpha z}i}\, e^{\imath\,\omega_{ij}\,t} \\ S^k_{\beta z\tau_{\sigma\beta z}i}\, e^{\imath\,\omega_{ij}\,t} \\ S^k_{zz\tau_{\sigma zz}i}\, e^{\imath\,\omega_{ij}\,t} \end{Bmatrix}, \quad \mathbf{X}_i = \begin{bmatrix} \chi_{\sigma_{\alpha z_i}} & 0 & 0 \\ 0 & \chi_{\sigma_{\beta z_i}} & 0 \\ 0 & 0 & \chi_{\sigma_{zz_i}} \end{bmatrix} \tag{7.63}
$$

and $\imath = \sqrt{-1}$, t is the time and ω_{ij} the circular frequency, $U^k_{\alpha\tau_{u\alpha}i}$, $U^k_{\beta\tau_{u\beta}i}$, $U^k_{z\tau_{uz}i}$ and $S^k_{\alpha z\tau_{\sigma\alpha z}i}$, $S^k_{\beta z\tau_{\sigma\beta z}i}$, $S^k_{zz\tau_{\sigma zz}i}$ are the displacement and transverse stress unknown coefficients, respectively. Finally, ψ_{α_i}, ψ_{β_i}, ψ_{z_i} and $\chi_{\sigma_{\alpha z_i}}$, $\chi_{\sigma_{\beta z_i}}$, $\chi_{\sigma_{zz_i}}$ are the problem-dependent Ritz functions. At this stage it is useful to introduce the following thickness integrals in order to write in a concise manner the explicit forms of the Ritz fundamental secondary nuclei

$$
\begin{aligned}
&\left(J^{k\tau_u s_\sigma}_{pq}, J^{k\hat{\tau}_u s_\sigma}_{pq}, J^{k\tau_u \hat{s}_\sigma}_{pq}, J^{k\hat{\tau}_u \hat{s}_\sigma}_{pq}\right) \\
&= \int_{A^k} \left(F_{\tau_u} F_{s_\sigma}, \partial z F_{\tau_u} F_{s_\sigma}, F_{\tau_u}\partial z F_{s_\sigma}, \partial z F_{\tau_u}\partial z F_{s_\sigma}\right) \frac{H^k_\alpha H^k_\beta}{(H^k_\alpha)^p\,(H^k_\beta)^q}\,\mathrm{d}z \\
&\left(J^{k\tau_\sigma s_u}_{pq}, J^{k\hat{\tau}_\sigma s_u}_{pq}, J^{k\tau_\sigma \hat{s}_u}_{pq}, J^{k\hat{\tau}_\sigma \hat{s}_u}_{pq}\right) \\
&= \int_{A^k} \left(F_{\tau_\sigma} F_{s_u}, \partial z F_{\tau_\sigma} F_{s_u}, F_{\tau_\sigma}\partial z F_{s_u}, \partial z F_{\tau_\sigma}\partial z F_{s_u}\right) \frac{H^k_\alpha H^k_\beta}{(H^k_\alpha)^p\,(H^k_\beta)^q}\,\mathrm{d}z \\
&\left(J^{k\tau_\sigma s_\sigma}_{pq}, J^{k\hat{\tau}_\sigma s_\sigma}_{pq}, J^{k\tau_\sigma \hat{s}_\sigma}_{pq}, J^{k\hat{\tau}_\sigma \hat{s}_u}_{pq}\right) \\
&= \int_{A^k} \left(F_{\tau_\sigma} F_{s_\sigma}, \partial z F_{\tau_\sigma} F_{s_\sigma}, F_{\tau_\sigma}\partial z F_{s_\sigma}, \partial z F_{\tau_\sigma}\partial z F_{s_\sigma}\right) \frac{H^k_\alpha H^k_\beta}{(H^k_\alpha)^p\,(H^k_\beta)^q}\,\mathrm{d}z
\end{aligned} \tag{7.64}
$$

The above expressions include all the possible combinations of product between displacement and/or transverse stress thickness functions.

To the same aim, it is convenient to rewrite the Ritz functions involved in the in-plane integrals, namely ψ_{α_i}, ψ_{β_i}, ψ_{z_i} for displacements and $\chi_{\sigma_{\alpha z_i}}$, $\chi_{\sigma_{\beta z_i}}$, $\chi_{\sigma_{zz_i}}$ for transverse normal and shear stresses, as

$$
\begin{aligned}
\chi_{\sigma_{\alpha z_i}}(\alpha,\beta) &= \sum_m^M \sum_n^N \varphi_m^{\sigma_{\alpha z_i}}(\alpha)\,\varphi_n^{\sigma_{\alpha z_i}}(\beta)\\
\chi_{\sigma_{\beta z_i}}(\alpha,\beta) &= \sum_m^M \sum_n^N \varphi_m^{\sigma_{\beta z_i}}(\alpha)\,\varphi_n^{\sigma_{\beta z_i}}(\beta)\\
\chi_{\sigma_{zz_i}}(\alpha,\beta) &= \sum_m^M \sum_n^N \varphi_m^{\sigma_{zz_i}}(\alpha)\,\varphi_n^{\sigma_{zz_i}}(\beta)
\end{aligned}
\tag{7.65}
$$

Then, by using Eq. (7.65) the general expressions of the in-plane integrals can be written as

$$
\begin{aligned}
{}_k^iI_{mp}^{\xi\zeta} &= \int_0^a \frac{d^\xi\phi_m^i(\alpha)}{d\alpha^\xi}\frac{d^\zeta\varphi_p^k(\alpha)}{d\alpha^\zeta}d\alpha\,; \quad {}_k^iI_{nq}^{\xi\zeta} = \int_0^b \frac{d^\xi\phi_n^i(\beta)}{d\beta^\xi}\frac{d^\zeta\varphi_q^k(\beta)}{d\beta^\zeta}d\beta\\
{}_i^lI_{mp}^{\xi\zeta} &= \int_0^a \frac{d^\xi\varphi_m^l(\alpha)}{d\alpha^\xi}\frac{d^\zeta\phi_p^i(\alpha)}{d\alpha^\zeta}d\alpha\,; \quad {}_i^lI_{nq}^{\xi\zeta} = \int_0^b \frac{d^\xi\varphi_n^l(\beta)}{d\beta^\xi}\frac{d^\zeta\phi_q^i(\beta)}{d\beta^\zeta}d\beta\\
{}_k^lI_{mp}^{\xi\zeta} &= \int_0^a \frac{d^\xi\varphi_m^l(\alpha)}{d\alpha^\xi}\frac{d^\zeta\varphi_p^k(\alpha)}{d\alpha^\zeta}d\alpha\,; \quad {}_k^lI_{nq}^{\xi\zeta} = \int_0^b \frac{d^\xi\varphi_n^l(\beta)}{d\beta^\xi}\frac{d^\zeta\varphi_q^k(\beta)}{d\beta^\zeta}d\beta\\
m &= 1,\cdots,M; \quad p=1,\cdots,P; \quad n=1,\cdots,N; \quad q=1,\cdots,Q;
\end{aligned}
\tag{7.66}
$$

where $i,j=u_\alpha,u_\beta,u_z$; $l,k=\sigma_{\alpha z},\sigma_{\beta z},\sigma_{zz}$ and ξ, ζ are differentiation orders. The explicit expression of the internal and inertial virtual works is given as

$$
\begin{aligned}
\delta L_{int}^k = \int_{\Omega_k}\int_{A^k} & \delta\mathbf{U}_{\tau i}^{k^T}\left(\left[\left[\mathbf{D}_p\left(\mathbf{F}_{\tau_u}\boldsymbol{\Psi}_i\right)\right]^T+\left[\mathbf{A}_p\left(\mathbf{F}_{\tau_u}\boldsymbol{\Psi}_i\right)\right]^T\right]\times\right.\\
&\left.\left[\hat{\mathbf{C}}_{pp}^k\mathbf{D}_p\left(\mathbf{F}_{s_u}\boldsymbol{\Psi}_j\right)+\hat{\mathbf{C}}_{pp}^k\mathbf{A}_p\left(\mathbf{F}_{s_u}\boldsymbol{\Psi}_j\right)\right]\right)\mathbf{U}_{sj}^k+\\
&\delta\mathbf{U}_{\tau i}^{k^T}\left(\left[\left[\mathbf{D}_p\left(\mathbf{F}_{\tau_u}\boldsymbol{\Psi}_i\right)\right]^T+\left[\mathbf{A}_p\left(\mathbf{F}_{\tau_u}\boldsymbol{\Psi}_i\right)\right]^T\right]\left[\hat{\mathbf{C}}_{pn}^k\left(\mathbf{F}_{s_\sigma}\mathbf{X}_j\right)\right]+\right.\\
&\left[\left[\mathbf{D}_{np}\left(\mathbf{F}_{\tau_u}\boldsymbol{\Psi}_i\right)\right]^T+\left[\delta_D\mathbf{A}_n\left(\mathbf{F}_{\tau_u}\boldsymbol{\Psi}_i\right)\right]^T+\right.\\
&\left.\left.\left[\mathbf{D}_{nz}\left(\mathbf{F}_{\tau_u}\boldsymbol{\Psi}_i\right)\right]^T\right]\left[\left(\mathbf{F}_{s_\sigma}\mathbf{X}_j\right)\right]\right)\mathbf{S}_{sj}^k+\\
&\delta\mathbf{S}_{\tau i}^{k^T}\left(\left[\hat{\mathbf{C}}_{pn}^k\left(\mathbf{F}_{\tau_\sigma}\mathbf{X}_i\right)\right]^T\left[\left[\mathbf{D}_p\left(\mathbf{F}_{s_u}\boldsymbol{\Psi}_j\right)\right]^T+\left[\mathbf{A}_p\left(\mathbf{F}_{s_u}\boldsymbol{\Psi}_j\right)\right]^T\right]+\right.\\
&\left(\mathbf{F}_{\tau_\sigma}\mathbf{X}_i\right)^T\left[\left[\mathbf{D}_{np}\left(\mathbf{F}_{s_u}\boldsymbol{\Psi}_j\right)\right]^T+\left[\delta_D\mathbf{A}_n\left(\mathbf{F}_{s_u}\boldsymbol{\Psi}_j\right)\right]^T\right.\\
&\left.\left.+\left[\mathbf{D}_{nz}\left(\mathbf{F}_{s_u}\boldsymbol{\Psi}_j\right)\right]^T\right]\right)\mathbf{U}_{sj}^k+
\end{aligned}
$$

$$\delta \mathbf{S}_{\tau i}^{k^T} \left(\left[\hat{C}_{nn}^{k} \left(\mathbf{F}_{\tau_\sigma} \mathbf{X}_i \right)^T \left(\mathbf{F}_{s_\sigma} \mathbf{X}_j \right) \right] \right) \mathbf{S}_{sj}^{k} \, d\Omega_k \, dz \tag{7.67}$$

They can be equivalently written as

$$\delta L_{int}^{k} = \delta \mathbf{U}_{\tau i}^{k^T} \mathbf{K}_{u\,u}^{k\tau sij} \mathbf{U}_{sj}^{k} + \delta \mathbf{U}_{\tau i}^{k^T} \mathbf{K}_{u\,\sigma}^{k\tau sij} \mathbf{S}_{sj}^{k} + \delta \mathbf{S}_{\tau i}^{k^T} \mathbf{K}_{\sigma\,u}^{k\tau sij} \mathbf{U}_{sj}^{k} + \delta \mathbf{S}_{\tau i}^{k^T} \mathbf{K}_{\sigma\,\sigma}^{k\tau sij} \mathbf{S}_{sj}^{k} \tag{7.68}$$

It follows immediately that the mixed Ritz fundamental primary nuclei take the following form

$$\begin{aligned}
\mathbf{K}_{u\,u}^{k\tau sij} &= \int_{\Omega_k} \int_{A^k} \left(\left[\left[\mathbf{D}_p \left(\mathbf{F}_{\tau_u} \boldsymbol{\Psi}_i \right) \right]^T + \left[\mathbf{A}_p \left(\mathbf{F}_{\tau_u} \boldsymbol{\Psi}_i \right) \right]^T \right] \right. \\
&\qquad \left. \left[\hat{C}_{pp}^{k} \mathbf{D}_p \left(\mathbf{F}_{s_u} \boldsymbol{\Psi}_j \right) + \hat{C}_{pp}^{k} \mathbf{A}_p \left(\mathbf{F}_{s_u} \boldsymbol{\Psi}_j \right) \right] \right) d\Omega_k \, dz \\
\mathbf{K}_{u\,\sigma}^{k\tau sij} &= \int_{\Omega_k} \int_{A^k} \left(\left[\left[\mathbf{D}_p \left(\mathbf{F}_{\tau_u} \boldsymbol{\Psi}_i \right) \right]^T + \left[\mathbf{A}_p \left(\mathbf{F}_{\tau_u} \boldsymbol{\Psi}_i \right) \right]^T \right] \left[\hat{C}_{pn}^{k} \left(\mathbf{F}_{s_\sigma} \mathbf{X}_j \right) \right] + \right. \\
&\qquad \left[\left[\mathbf{D}_{np} \left(\mathbf{F}_{\tau_u} \boldsymbol{\Psi}_i \right) \right]^T + \left[\delta_D \mathbf{A}_n \left(\mathbf{F}_{\tau_u} \boldsymbol{\Psi}_i \right) \right]^T + \left[\mathbf{D}_{nz} \left(\mathbf{F}_{\tau_u} \boldsymbol{\Psi}_i \right) \right]^T \right] \\
&\qquad \left. \left[\left(\mathbf{F}_{s_\sigma} \mathbf{X}_j \right) \right] \right) d\Omega_k \, dz \\
\mathbf{K}_{\sigma\,u}^{k\tau sij} &= \int_{\Omega_k} \int_{A^k} \left(\left[\hat{C}_{pn}^{k} \left(\mathbf{F}_{\tau_\sigma} \mathbf{X}_i \right) \right]^T \left[\left[\mathbf{D}_p \left(\mathbf{F}_{s_u} \boldsymbol{\Psi}_j \right) \right]^T + \left[\mathbf{A}_p \left(\mathbf{F}_{s_u} \boldsymbol{\Psi}_j \right) \right]^T \right] + \right. \\
&\qquad \left(\mathbf{F}_{\tau_\sigma} \mathbf{X}_i \right)^T \left[\left[\mathbf{D}_{np} \left(\mathbf{F}_{s_u} \boldsymbol{\Psi}_j \right) \right]^T + \left[\delta_D \mathbf{A}_n \left(\mathbf{F}_{s_u} \boldsymbol{\Psi}_j \right) \right]^T + \right. \\
&\qquad \left. \left. \left[\mathbf{D}_{nz} \left(\mathbf{F}_{s_u} \boldsymbol{\Psi}_j \right) \right]^T \right] \right) d\Omega_k \, dz \\
\mathbf{K}_{\sigma\,\sigma}^{k\tau sij} &= \int_{\Omega_k} \int_{A^k} \left[\hat{C}_{nn}^{k} \left(\mathbf{F}_{\tau_\sigma} \mathbf{X}_i \right)^T \left(\mathbf{F}_{s_\sigma} \mathbf{X}_j \right) \right] d\Omega_k \, dz
\end{aligned} \tag{7.69}$$

After performing the matrix calculus in Eq. (7.69) and using the concise form of the in-plane and thickness integrals given in Eqs. (7.64) and (7.56), the explicit forms of the primary nucleus $\mathbf{K}_{u\,u}^{k\tau sij}$ components are obtained

$$\begin{aligned}
K_{u_\alpha u_\alpha}^{\tau_{u_\alpha} s_{u_\alpha}} &= \hat{C}_{11}^{k} J_{20}^{k\,\tau_{u_\alpha}\,s_{u_\alpha}} \, {}_{u_\alpha}^{u_\alpha} I_{mp}^{11} \, {}_{u_\alpha}^{u_\alpha} I_{nq}^{00} + \hat{C}_{16}^{k} J_{11}^{k\,\tau_{u_\alpha}\,s_{u_\alpha}} \, {}_{u_\alpha}^{u_\alpha} I_{mp}^{10} \, {}_{u_\alpha}^{u_\alpha} I_{nq}^{01} + \\
&\quad \hat{C}_{16}^{k} J_{11}^{k\,\tau_{u_\alpha}\,s_{u_\alpha}} \, {}_{u_\alpha}^{u_\alpha} I_{mp}^{01} \, {}_{u_\alpha}^{u_\alpha} I_{nq}^{10} + \hat{C}_{66}^{k} J_{02}^{k\,\tau_{u_\alpha}\,s_{u_\alpha}} \, I_{mp}^{00} \, I_{nq}^{11} \\
K_{u_\alpha u_\beta}^{\tau_{u_\alpha} s_{u_\beta}} &= \hat{C}_{16}^{k} J_{20}^{k\,\tau_{u_\alpha}\,s_{u_\beta}} \, {}_{u_\beta}^{u_\alpha} I_{mp}^{11} \, {}_{u_\beta}^{u_\alpha} I_{nq}^{00} + \hat{C}_{12}^{k} J_{11}^{k\,\tau_{u_\alpha}\,s_{u_\beta}} \, {}_{u_\beta}^{u_\alpha} I_{mp}^{10} \, {}_{u_\beta}^{u_\alpha} I_{nq}^{01} + \\
&\quad \hat{C}_{66}^{k} J_{11}^{k\,\tau_{u_\alpha}\,s_{u_\beta}} \, {}_{u_\beta}^{u_\alpha} I_{mp}^{01} \, {}_{u_\beta}^{u_\alpha} I_{nq}^{10} + \hat{C}_{26}^{k} J_{02}^{k\,\tau_{u_\alpha}\,s_{u_\beta}} \, I_{mp}^{00} \, I_{nq}^{11} \\
K_{u_\alpha u_z}^{\tau_{u_\alpha} s_{u_z}} &= \hat{C}_{11}^{k} \frac{1}{R_\alpha^k} J_{20}^{k\tau_{u_\alpha} s_{u_z}} \, {}_{u_z}^{u_\alpha} I_{mp}^{10} \, {}_{u_z}^{u_\alpha} I_{nq}^{00} + \hat{C}_{12}^{k} \frac{1}{R_\beta^k} J_{11}^{k\tau_{u_\alpha} s_{u_z}} \, {}_{u_z}^{u_\alpha} I_{mp}^{10} \, {}_{u_z}^{u_\alpha} I_{nq}^{00} +
\end{aligned}$$

$$
\begin{aligned}
& \hat{C}^k_{16}\,\frac{1}{R^k_\alpha}J^{k\tau_{u_\alpha}s_{u_z}}_{11}\;{}^{u_\alpha}_{u_z}I^{01}_{mp}\;{}^{u_\alpha}_{u_z}I^{00}_{nq} + \hat{C}^k_{26}\,\frac{1}{R^k_\beta}J^{k\tau_{u_\alpha}s_{u_z}}_{02}\;{}^{u_\alpha}_{u_z}I^{01}_{mp}\;{}^{u_\alpha}_{u_z}I^{00}_{nq}\\
K^{\tau_{u_\beta}s_{u_\alpha}}_{u_\beta u_\alpha} &= \hat{C}^k_{16}J^{k\,\tau_{u_\beta}\,s_{u_\alpha}}_{20}\;{}^{u_\beta}_{u_\alpha}I^{11}_{mp}\;{}^{u_\beta}_{u_\alpha}I^{00}_{nq} + \hat{C}^k_{12}J^{k\,\tau_{u_\beta}\,s_{u_\alpha}}_{11}\;{}^{u_\beta}_{u_\alpha}I^{01}_{mp}\;{}^{u_\beta}_{u_\alpha}I^{10}_{nq} +\\
& \hat{C}^k_{66}J^{k\,\tau_{u_\beta}\,s_{u_\alpha}}_{11}\;{}^{u_\beta}_{u_\alpha}I^{10}_{mp}\;{}^{u_\beta}_{u_\alpha}I^{01}_{nq} + \hat{C}^k_{26}J^{k\,\tau_{u_\beta}\,s_{u_\alpha}}_{02}\;{}^{u_\beta}_{u_\alpha}I^{00}_{mp}\;{}^{u_\beta}_{u_\alpha}I^{11}_{nq}\\
K^{\tau_{u_\beta}s_{u_\beta}}_{u_\beta u_\beta} &= \hat{C}^k_{66}J^{k\,\tau_{u_\beta}\,s_{u_\beta}}_{20}\;{}^{u_\beta}_{u_\beta}I^{11}_{mp}\;{}^{u_\beta}_{u_\beta}I^{00}_{nq} + \hat{C}^k_{26}J^{k\,\tau_{u_\beta}\,s_{u_\beta}}_{11}\;{}^{u_\beta}_{u_\beta}I^{01}_{mp}\;{}^{u_\beta}_{u_\beta}I^{10}_{nq} +\\
& \hat{C}^k_{26}J^{k\,\tau_{u_\beta}\,s_{u_\beta}}_{11}\;{}^{u_\beta}_{u_\beta}I^{10}_{mp}\;{}^{u_\beta}_{u_\beta}I^{01}_{nq} + \hat{C}^k_{22}J^{k\,\tau_{u_\beta}\,s_{u_\beta}}_{02}\;{}^{u_\beta}_{u_\beta}I^{00}_{mp}\;{}^{u_\beta}_{u_\beta}I^{11}_{nq}\\
K^{\tau_{u_\beta}s_{u_z}}_{u_\beta u_z} &= \hat{C}^k_{12}\,\frac{1}{R^k_\alpha}J^{k\,\tau_{u_\beta}\,s_{u_z}}_{11}\;{}^{u_\beta}_{u_z}I^{00}_{mp}\;{}^{u_\beta}_{u_z}I^{10}_{nq} + \hat{C}^k_{22}\,\frac{1}{R^k_\beta}J^{k\,\tau_{u_\alpha}\,s_{u_\alpha}}_{02}\;{}^{u_\beta}_{u_z}I^{00}_{mp}\;{}^{u_\beta}_{u_z}I^{10}_{nq} +\\
& \hat{C}^k_{16}\,\frac{1}{R^k_\alpha}J^{k\,\tau_{u_\beta}\,s_{u_z}}_{20}\;{}^{u_\beta}_{u_z}I^{00}_{mp}\;{}^{u_\beta}_{u_z}I^{10}_{nq} + \hat{C}^k_{26}\,\frac{1}{R^k_\beta}J^{k\,\tau_{u_\alpha}\,s_{u_\alpha}}_{11}\;{}^{u_\beta}_{u_z}I^{00}_{mp}\;{}^{u_\beta}_{u_z}I^{10}_{nq}\\
K^{\tau_{u_z}s_{u_\alpha}}_{u_z u_\alpha} &= \hat{C}^k_{16}\,\frac{1}{R^k_\alpha}J^{k\,\tau_{u_z}\,s_{u_\alpha}}_{11}\;{}^{u_z}_{u_\alpha}I^{00}_{mp}\;{}^{u_z}_{u_\alpha}I^{10}_{nq} + \hat{C}^k_{26}\,\frac{1}{R^k_\beta}J^{k\,\tau_{u_\alpha}\,s_{u_\alpha}}_{02}\;{}^{u_z}_{u_\alpha}I^{00}_{mp}\;{}^{u_z}_{u_\alpha}I^{10}_{nq} +\\
& \hat{C}^k_{11}\,\frac{1}{R^k_\alpha}J^{k\,\tau_{u_z}\,s_{u_\alpha}}_{20}\;{}^{u_z}_{u_\alpha}I^{00}_{mp}\;{}^{u_z}_{u_\alpha}I^{10}_{nq} + \hat{C}^k_{12}\,\frac{1}{R^k_\beta}J^{k\,\tau_{u_\alpha}\,s_{u_\alpha}}_{11}\;{}^{u_z}_{u_\alpha}I^{00}_{mp}\;{}^{u_z}_{u_\alpha}I^{10}_{nq}\\
K^{\tau_{u_z}s_{u_\beta}}_{u_z u_\beta} &= \hat{C}^k_{12}\,\frac{1}{R^k_\alpha}J^{k\,\tau_{u_z}\,s_{u_\beta}}_{11}\;{}^{u_z}_{u_\beta}I^{00}_{mp}\;{}^{u_z}_{u_\beta}I^{10}_{nq} + \hat{C}^k_{22}\,\frac{1}{R^k_\beta}J^{k\,\tau_{u_\alpha}\,s_{u_\beta}}_{02}\;{}^{u_z}_{u_\beta}I^{00}_{mp}\;{}^{u_z}_{u_\beta}I^{10}_{nq} +\\
& \hat{C}^k_{16}\,\frac{1}{R^k_\alpha}J^{k\,\tau_{u_z}\,s_{u_\beta}}_{20}\;{}^{u_z}_{u_\beta}I^{00}_{mp}\;{}^{u_z}_{u_\beta}I^{10}_{nq} + \hat{C}^k_{26}\,\frac{1}{R^k_\beta}J^{k\,\tau_{u_\alpha}\,s_{u_\beta}}_{11}\;{}^{u_z}_{u_\beta}I^{00}_{mp}\;{}^{u_z}_{u_\beta}I^{10}_{nq}\\
K^{\tau_{u_z}s_{u_z}}_{u_z u_z} &= \hat{C}^k_{11}\,\frac{1}{(R^k_\alpha)^2}J^{k\,\tau_{u_z}\,s_{u_z}}_{20}\;{}^{u_z}_{u_z}I^{00}_{mp}\;{}^{u_z}_{u_z}I^{00}_{nq} + \hat{C}^k_{12}\,\frac{1}{(R^k_\alpha R^k_\beta)}J^{k\,\tau_{u_z}\,s_{u_z}}_{11}\;{}^{u_z}_{u_z}I^{00}_{mp}\;{}^{u_z}_{u_z}I^{00}_{nq} +\\
& \hat{C}^k_{12}\,\frac{1}{(R^k_\alpha R^k_\beta)}J^{k\,\tau_{u_z}\,s_{u_z}}_{11}\;{}^{u_z}_{u_z}I^{00}_{mp}\;{}^{u_z}_{u_z}I^{00}_{nq} + \hat{C}^k_{22}\,\frac{1}{(R^k_\beta)^2}J^{k\,\tau_{u_z}\,s_{u_z}}_{02}\;{}^{u_z}_{u_z}I^{00}_{mp}\;{}^{u_z}_{u_z}I^{00}_{nq}
\end{aligned}
\tag{7.70}
$$

For the mixed Ritz fundamental primary stiffness nucleus $\mathbf{K}^{k\tau sij}_{u\sigma}$ the components are

$$
\begin{aligned}
K^{\tau_{u_\alpha}s_{\sigma_{\alpha z}}}_{u_\alpha\sigma_{\alpha z}} &= J^{k\,\hat{\tau}_{u_\alpha}\,s_{\sigma_{\alpha z}}}_{00}\;{}^{\sigma_{\alpha z}}_{u_\alpha}I^{00}_{ml}\;{}^{\sigma_{\alpha z}}_{u_\alpha}I^{00}_{nk} - J^{k\,\tau_{u_\alpha}\,s_{\sigma_{\alpha z}}}_{01}\,\delta_D\,\frac{1}{R^k_\alpha}\;{}^{\sigma_{\alpha z}}_{u_\alpha}I^{00}_{ml}\;{}^{\sigma_{\alpha z}}_{u_\alpha}I^{00}_{nk}\\
K^{\tau_{u_\alpha}s_{\sigma_{\beta z}}}_{u_\alpha\sigma_{\beta z}} &= 0\\
K^{\tau_{u_\alpha}s_{\sigma_{zz}}}_{u_\alpha\sigma_{zz}} &= \hat{C}^k_{13}J^{k\tau_{u_\alpha}s_{\sigma_{zz}}}_{10}\;{}^{\sigma_{zz}}_{u_\alpha}I^{10}_{mp}\;{}^{\sigma_{zz}}_{u_\alpha}I^{00}_{nq} + \hat{C}^k_{36}J^{k\tau_{u_\alpha}s_{\sigma_{zz}}}_{11}\;{}^{\sigma_{\alpha z}}_{u_\alpha}I^{00}_{mp}\;{}^{\sigma_{\alpha z}}_{u_\alpha}I^{01}_{nq}\\
K^{\tau_{u_\beta}s_{\sigma_{\alpha z}}}_{u_\beta\sigma_{\alpha z}} &= 0\\
K^{\tau_{u_\beta}s_{\sigma_{\beta z}}}_{u_\beta\sigma_{\beta z}} &= J^{k\,\hat{\tau}_{u_\beta}\,s_{\sigma_{\beta z}}}_{00}\;{}^{\sigma_{\beta z}}_{u_\beta}I^{00}_{ml}\;{}^{\sigma_{\beta z}}_{u_\beta}I^{00}_{nk} - J^{k\,\tau_{u_\beta}\,s_{\sigma_{\beta z}}}_{10}\,\delta_D\,\frac{1}{R^k_\beta}\;{}^{\sigma_{\beta z}}_{u_\beta}I^{00}_{ml}\;{}^{\sigma_{\beta z}}_{u_\beta}I^{00}_{nk}\\
K^{\tau_{u_\beta}s_{\sigma_{zz}}}_{u_\beta\sigma_{zz}} &= \hat{C}^k_{36}J^{k\tau_{u_\beta}s_{\sigma_{zz}}}_{10}\;{}^{\sigma_{zz}}_{u_\beta}I^{10}_{mp}\;{}^{\sigma_{zz}}_{u_\beta}I^{00}_{nq} + \hat{C}^k_{23}J^{k\tau_{u_\beta}s_{\sigma_{zz}}}_{11}\;{}^{\sigma_{zz}}_{u_\beta}I^{00}_{mp}\;{}^{\sigma_{zz}}_{u_\beta}I^{01}_{nq}
\end{aligned}
$$

$$
\begin{aligned}
K^{\tau_{u_z} s_{\sigma_{\alpha z}}}_{u_z \sigma_{\alpha z}} &= J^{k\,\tau_{u_z}\, s_{\sigma_{\alpha z}}}_{10}\; {}^{\sigma_{\alpha z}}_{u_z}I^{10}_{mp}\; {}^{\sigma_{\alpha z}}_{u_z}I^{00}_{nq}\\
K^{\tau_{u_z} s_{\sigma_{\beta z}}}_{u_z \sigma_{\beta z}} &= J^{k\,\tau_{u_z}\, s_{\sigma_{\beta z}}}_{01}\; {}^{\sigma_{\beta z}}_{u_z}I^{01}_{mp}\; {}^{\sigma_{\beta z}}_{u_z}I^{00}_{nq}\\
K^{\tau_{u_z} s_{\sigma_{zz}}}_{u_z \sigma_{zz}} &= J^{k\,\hat{\tau}_{u_z}\, s_{\sigma_{zz}}}_{00}\; {}^{\sigma_{zz}}_{u_z}I^{00}_{ml}\; {}^{\sigma_{zz}}_{u_z}I^{00}_{nk} + \hat{C}^k_{13}\frac{1}{R^k_\alpha} J^{k\tau_{u_z} s_{\sigma_{zz}}}_{10}\; {}^{\sigma_{zz}}_{u_z}I^{10}_{mp}\; {}^{\sigma_{zz}}_{u_z}I^{00}_{nq} +\\
&\quad \hat{C}^k_{23}\frac{1}{R^k_\beta} J^{k\tau_{u_z} s_{\sigma_{zz}}}_{11}\; {}^{\sigma_{zz}}_{u_z}I^{00}_{mp}\; {}^{\sigma_{zz}}_{u_z}I^{01}_{nq}
\end{aligned}
\tag{7.71}
$$

In the case of the mixed Ritz fundamental primary stiffness nucleus $\mathbf{K}^{k\tau sij}_{\sigma\, u}$ the secondary stiffness nuclei are

$$
\begin{aligned}
K^{\tau_{\sigma_{\alpha z}} s_{u_\alpha}}_{\sigma_{\alpha z} u_\alpha} &= J^{k\,\tau_{\sigma_{\alpha z}}\, \hat{s}_{u_\alpha}}_{00}\; {}^{u_\alpha}_{\sigma_{\alpha z}}I^{00}_{ml}\; {}^{u_\alpha}_{\sigma_{\alpha z}}I^{00}_{nk} - J^{k\,\tau_{\sigma_{\alpha z}}\, s_{u_\alpha}}_{01}\,\delta_D \frac{1}{R^k_\alpha}\; {}^{u_\alpha}_{\sigma_{\alpha z}}I^{00}_{ml}\; {}^{u_\alpha}_{\sigma_{\alpha z}}I^{00}_{nk}\\
K^{\tau_{\sigma_{\alpha z}} s_{u_\beta}}_{\sigma_{\alpha z} u_\beta} &= 0\\
K^{\tau_{u_\alpha} s_{u_z}}_{\sigma_{\alpha z} u_z} &= J^{k\,\tau_{\sigma_{zz}}\, s_{u_\alpha}}_{10}\; {}^{u_\alpha}_{\sigma_{zz}}I^{10}_{mp}\; {}^{u_\alpha}_{\sigma_{zz}}I^{00}_{nq}\\
K^{\tau_{\sigma_{\beta z}} s_{u_\alpha}}_{\sigma_{\beta z} u_\alpha} &= 0\\
K^{\tau_{\sigma_{\beta z}} s_{u_\beta}}_{\sigma_{\beta z} u_\beta} &= J^{k\,\tau_{\sigma_{\beta z}}\, \hat{s}_{u_\beta}}_{00}\; {}^{u_\beta}_{\sigma_{\beta z}}I^{00}_{ml}\; {}^{u_\beta}_{\sigma_{\beta z}}I^{00}_{nk} - J^{k\,\tau_{\sigma_{\beta z}}\, s_{u_\beta}}_{10}\,\delta_D \frac{1}{R^k_\beta}\; {}^{u_\beta}_{\sigma_{\beta z}}I^{00}_{ml}\; {}^{u_\beta}_{\sigma_{\beta z}}I^{00}_{nk}\\
K^{\tau_{\sigma_{\beta z}} s_{u_z}}_{\sigma_{\beta z} u_z} &= J^{k\,\tau_{\sigma_{zz}}\, s_{u_\beta}}_{01}\; {}^{u_\beta}_{\sigma_{zz}}I^{01}_{mp}\; {}^{u_\beta}_{\sigma_{zz}}I^{00}_{nq}\\
K^{\tau_{\sigma_{zz}} s_{u_\alpha}}_{\sigma_{zz} u_\alpha} &= -\hat{C}^k_{13} J^{k\tau_{\sigma_{\alpha z}} s_{u_z}}_{10}\; {}^{u_z}_{\sigma_{\alpha z}}I^{10}_{mp}\; {}^{u_z}_{\sigma_{\alpha z}}I^{00}_{nq} - \hat{C}^k_{36} J^{k\tau_{\sigma_{\alpha z}} s_{u_z}}_{11}\; {}^{u_z}_{\sigma_{\alpha z}}I^{00}_{mp}\; {}^{u_z}_{\sigma_{\alpha z}}I^{01}_{nq}\\
K^{\tau_{\sigma_{zz}} s_{\sigma_\beta}}_{\sigma_{zz} u_\beta} &= -\hat{C}^k_{36} J^{k\tau_{\sigma_{\beta z}} s_{u_z}}_{10}\; {}^{u_\beta}_{\sigma_{\beta z}}I^{10}_{mp}\; {}^{u_\beta}_{\sigma_{\beta z}}I^{00}_{nq} - \hat{C}^k_{23} J^{k\tau_{\sigma_{\beta z}} s_{u_z}}_{11}\; {}^{\sigma_{zz}}_{u_\beta}I^{00}_{mp}\; {}^{\sigma_{zz}}_{u_\beta}I^{01}_{nq}\\
K^{\tau_{\sigma_{zz}} s_{u_z}}_{\sigma_{zz} u_z} &= J^{k\,\tau_{\sigma_{\alpha z}}\, \hat{s}_{u_\alpha}}_{00}\; {}^{u_z}_{\sigma_{zz}}I^{00}_{ml}\; {}^{\sigma_{\alpha z}}_{u_\alpha}I^{00}_{nk} - \hat{C}^k_{13}\frac{1}{R^k_\alpha} J^{k\tau_{u_\alpha} s_{\sigma_{zz}}}_{10}\; {}^{u_z}_{\sigma_{zz}}I^{10}_{mp}\; {}^{u_\alpha}_{\sigma_{zz}}I^{00}_{nq} -\\
&\quad \hat{C}^k_{23}\frac{1}{R^k_\beta} J^{k\tau_{\sigma_{zz}} s_{u_z}}_{11}\; {}^{u_z}_{\sigma_{zz}}I^{00}_{mp}\; {}^{u_z}_{\sigma_{zz}}I^{01}_{nq}
\end{aligned}
\tag{7.72}
$$

Finally, for the mixed Ritz fundamental primary stiffness nucleus $\mathbf{K}^{k\tau sij}_{\sigma\,\sigma}$ the components are given as follows

$$
\begin{aligned}
K^{\tau_{\sigma_{\alpha z}} s_{\sigma_{\alpha z}}}_{\sigma_{\alpha z}\sigma_{\alpha z}} &= -\hat{C}^k_{55} J^{k\,\tau_{\sigma_{\alpha z}}\, s_{\sigma_{\alpha z}}}_{00}\; {}^{\sigma_{\alpha z}}_{\sigma_{\alpha z}}I^{00}_{lk}\\
K^{\tau_{\sigma_{\alpha z}} s_{\sigma_{\beta z}}}_{\sigma_{\alpha z}\sigma_{\beta z}} &= -\hat{C}^k_{45} J^{k\,\tau_{\sigma_{\alpha z}}\, s_{\sigma_{\beta z}}}_{00}\; {}^{\sigma_{\beta z}}_{\sigma_{\alpha z}}I^{00}_{lk}\\
K^{\tau_{\sigma_{\alpha z}} s_{\sigma_{zz}}}_{\sigma_{\alpha z}\sigma_{zz}} &= 0\\
K^{\tau_{\sigma_{\beta z}} s_{\sigma_{\alpha z}}}_{\sigma_{\beta z}\sigma_{\alpha z}} &= -\hat{C}^k_{45} J^{k\,\tau_{\sigma_{\beta z}}\, s_{\sigma_{\alpha z}}}_{00}\; {}^{\sigma_{\alpha z}}_{\sigma_{\beta z}}I^{00}_{lk}\\
K^{\tau_{\sigma_{\beta z}} s_{\sigma_{\beta z}}}_{\sigma_{\beta z}\sigma_{\beta z}} &= -\hat{C}^k_{44} J^{k\,\tau_{\sigma_{\beta z}}\, s_{\sigma_{\alpha z}}}_{00}\; {}^{\sigma_{\beta z}}_{\sigma_{\beta z}}I^{00}_{lk}
\end{aligned}
\tag{7.73}
$$

$$
\begin{aligned}
K_{\sigma_{\beta z}\sigma_{zz}}^{\tau_{\sigma_{\beta z}}s_{\sigma_{zz}}} &= 0 \\
K_{\sigma_{zz}\sigma_{\alpha z}}^{\tau_{\sigma_{zz}}s_{\sigma_{\alpha z}}} &= 0 \\
K_{\sigma_{zz}\sigma_{\beta z}}^{\tau_{\sigma_{zz}}s_{\sigma_{\beta z}}} &= 0 \\
K_{\sigma_{zz}\sigma_{zz}}^{\tau_{\sigma_{zz}}s_{\sigma_{zz}}} &= -\hat{C}_{33}^{k} J_{00}^{k\,\tau_{\sigma_{zz}}s_{\sigma_{zz}}}\; {}_{\sigma_{zz}}^{\sigma_{zz}} I_{lk}^{00}
\end{aligned}
$$

7.4 GALERKIN AND GENERALIZED GALERKIN METHODS

In this section Galerkin and Generalized Galerkin methods are developed [5]. The displacement approach is formulated by variationally imposing the equilibrium via the principle of virtual displacements. Then, PDV statement is used to derive strong form of governing differential equations and related boundary conditions. In order to derive the governing differential equations and natural boundary conditions the Gauss theorem is applied:

$$
\begin{aligned}
\int_{\Omega_k} \left((\mathbf{D}_p)\, \delta \mathbf{a}^k \right)^T \mathbf{a}^k \mathbf{d}\Omega_k &= -\int_{\Omega_k} \delta \mathbf{a}^k \left((\mathbf{D}_p)^T \delta \mathbf{a}^k \right) \mathbf{d}\Omega_k \\
&\quad + \int_{\Gamma^k} \delta \mathbf{a}^k \left((\mathbf{I}_p)^T \delta \mathbf{a}^k \right) \mathbf{d}\Gamma^k \\
\int_{\Omega_k} \left((\mathbf{D}_\Omega)\, \delta \mathbf{a}^k \right)^T \mathbf{a}^k \mathbf{d}\Omega_k &= -\int_{\Omega_k} \delta \mathbf{a}^k \left((\mathbf{D}_\Omega)^T \delta \mathbf{a}^k \right) \mathbf{d}\Omega_k \\
&\quad + \int_{\Gamma^k} \delta \mathbf{a}^k \left((\mathbf{I}_\Omega)^T \delta \mathbf{a}^k \right) \mathbf{d}\Gamma^k
\end{aligned}
\tag{7.74}
$$

where $\mathbf{a}$ can be displacement or stress variables and the introduced $\mathbf{I}_p$ and $\mathbf{I}_\Omega$ arrays are

$$
\mathbf{I}_p = \begin{bmatrix} n_\alpha & 0 & 0 \\ 0 & n_\beta & 0 \\ n_\beta & n_\alpha & 0 \end{bmatrix}; \qquad \mathbf{I}_\Omega = \begin{bmatrix} 0 & 0 & n_\alpha \\ 0 & 0 & n_\beta \\ 0 & 0 & 0 \end{bmatrix} \tag{7.75}
$$

The normal to the boundary of domain Ω_k is:

$$
\hat{\mathbf{n}} = \begin{bmatrix} n_\alpha \\ n_\beta \end{bmatrix} = \begin{bmatrix} \cos(\varphi_\alpha) \\ \cos(\varphi_\beta) \end{bmatrix} \tag{7.76}
$$

where φ_α and φ_β are the direction cosines, namely, the angles between the normal $\hat{\mathbf{n}}$ and the directions α and β, respectively.

In the derivation of what follows Galerkin and generalized Galerkin methods have been restricted, for the sake of conciseness, to the analysis of plate structures.

7.4.1 Plates

Stiffness and boundary nuclei

The functional related to the Galerkin method can be derived integrating by parts Eq. (7.10), which leads to the new expression of the internal virtual work

$$
\begin{aligned}
\delta L_{int}^{k} = & -\int_{\Omega^k}\int_{A^k} \delta \mathbf{U}_{\tau i}^{k^T} \left(\mathbf{F}_\tau \boldsymbol{\Psi}_i\right)^T \left\{(\mathbf{D}_p)^T \left[\tilde{\mathbf{C}}_{pp}^{k} \mathbf{D}_p \left(\mathbf{F}_s \boldsymbol{\Psi}_j\right)\right]\right\} \mathbf{U}_{sj}^{k}\, d\Omega^k\, dz \\
& -\int_{\Omega^k}\int_{A^k} \delta \mathbf{U}_{\tau i}^{k^T} \left(\mathbf{F}_\tau \boldsymbol{\Psi}_i\right)^T \left\{(\mathbf{D}_p)^T \left[\tilde{\mathbf{C}}_{pn}^{k} \mathbf{D}_{np} \left(\mathbf{F}_s \boldsymbol{\Psi}_j\right)\right]\right\} \mathbf{U}_{sj}^{k}\, d\Omega^k\, dz \\
& -\int_{\Omega^k}\int_{A^k} \delta \mathbf{U}_{\tau i}^{k^T} \left(\mathbf{F}_\tau \boldsymbol{\Psi}_i\right)^T \left\{(\mathbf{D}_p)^T \left[\tilde{\mathbf{C}}_{pn}^{k} \mathbf{D}_{nz} \left(\mathbf{F}_s \boldsymbol{\Psi}_j\right)\right]\right\} \mathbf{U}_{sj}^{k}\, d\Omega^k\, dz \\
& -\int_{\Omega^k}\int_{A^k} \delta \mathbf{U}_{\tau i}^{k^T} \left(\mathbf{F}_\tau \boldsymbol{\Psi}_i\right)^T \left\{(\mathbf{D}_{np})^T \left[\tilde{\mathbf{C}}_{np}^{k} \mathbf{D}_p \left(\mathbf{F}_s \boldsymbol{\Psi}_j\right)\right]\right\} \mathbf{U}_{sj}^{k}\, d\Omega^k\, dz \\
& -\int_{\Omega^k}\int_{A^k} \delta \mathbf{U}_{\tau i}^{k^T} \left(\mathbf{F}_\tau \boldsymbol{\Psi}_i\right)^T \left\{(\mathbf{D}_{np})^T \left[\tilde{\mathbf{C}}_{nn}^{k} \mathbf{D}_{np} \left(\mathbf{F}_s \boldsymbol{\Psi}_j\right)\right]\right\} \mathbf{U}_{sj}^{k}\, d\Omega^k\, dz \\
& -\int_{\Omega^k}\int_{A^k} \delta \mathbf{U}_{\tau i}^{k^T} \left(\mathbf{F}_\tau \boldsymbol{\Psi}_i\right)^T \left\{(\mathbf{D}_{np})^T \left[\tilde{\mathbf{C}}_{nn}^{k} \mathbf{D}_{nz} \left(\mathbf{F}_s \boldsymbol{\Psi}_j\right)\right]\right\} \mathbf{U}_{sj}^{k}\, d\Omega^k\, dz \\
& -\int_{\Omega^k}\int_{A^k} \delta \mathbf{U}_{\tau i}^{k^T} \left(\mathbf{F}_\tau \boldsymbol{\Psi}_i\right)^T \left\{(\mathbf{D}_{nz})^T \left[\tilde{\mathbf{C}}_{np}^{k} \mathbf{D}_p \left(\mathbf{F}_s \boldsymbol{\Psi}_j\right)\right]\right\} \mathbf{U}_{sj}^{k}\, d\Omega^k\, dz \\
& -\int_{\Omega^k}\int_{A^k} \delta \mathbf{U}_{\tau i}^{k^T} \left(\mathbf{F}_\tau \boldsymbol{\Psi}_i\right)^T \left\{(\mathbf{D}_{nz})^T \left[\tilde{\mathbf{C}}_{nn}^{k} \mathbf{D}_{np} \left(\mathbf{F}_s \boldsymbol{\Psi}_j\right)\right]\right\} \mathbf{U}_{sj}^{k}\, d\Omega^k\, dz \\
& -\int_{\Omega^k}\int_{A^k} \delta \mathbf{U}_{\tau i}^{k^T} \left(\mathbf{F}_\tau \boldsymbol{\Psi}_i\right)^T \left\{(\mathbf{D}_{nz})^T \left[\tilde{\mathbf{C}}_{nn}^{k} \mathbf{D}_{nz} \left(\mathbf{F}_s \boldsymbol{\Psi}_j\right)\right]\right\} \mathbf{U}_{sj}^{k}\, d\Omega^k\, dz \\
& +\int_{\Gamma^k}\int_{A^k} \delta \mathbf{U}_{\tau i}^{k^T} \left(\mathbf{F}_\tau \boldsymbol{\Psi}_i\right)^T \left\{(\mathbf{I}_p)^T \left[\tilde{\mathbf{C}}_{pp}^{k} \mathbf{D}_p \left(\mathbf{F}_s \boldsymbol{\Psi}_j\right)\right]\right\} \mathbf{U}_{sj}^{k}\, d\Gamma^k\, dz + \\
& +\int_{\Gamma^k}\int_{A^k} \delta \mathbf{U}_{\tau i}^{k^T} \left(\mathbf{F}_\tau \boldsymbol{\Psi}_i\right)^T \left\{(\mathbf{I}_p)^T \left[\tilde{\mathbf{C}}_{pn}^{k} \mathbf{D}_{np} \left(\mathbf{F}_s \boldsymbol{\Psi}_j\right)\right]\right\} \mathbf{U}_{sj}^{k}\, d\Gamma^k\, dz \\
& +\int_{\Gamma^k}\int_{A^k} \delta \mathbf{U}_{\tau i}^{k^T} \left(\mathbf{F}_\tau \boldsymbol{\Psi}_i\right)^T \left\{(\mathbf{I}_p)^T \left[\tilde{\mathbf{C}}_{pn}^{k} \mathbf{D}_{nz} \left(\mathbf{F}_s \boldsymbol{\Psi}_j\right)\right]\right\} \mathbf{U}_{sj}^{k}\, d\Gamma^k\, dz \\
& +\int_{\Gamma^k}\int_{A^k} \delta \mathbf{U}_{\tau i}^{k^T} \left(\mathbf{F}_\tau \boldsymbol{\Psi}_i\right)^T \left\{(\mathbf{I}_{np})^T \left[\tilde{\mathbf{C}}_{np}^{k} \mathbf{D}_p \left(\mathbf{F}_s \boldsymbol{\Psi}_j\right)\right]\right\} \mathbf{U}_{sj}^{k}\, d\Gamma^k\, dz \\
& +\int_{\Gamma^k}\int_{A^k} \delta \mathbf{U}_{\tau i}^{k^T} \left(\mathbf{F}_\tau \boldsymbol{\Psi}_i\right)^T \left\{(\mathbf{I}_{np})^T \left[\tilde{\mathbf{C}}_{nn}^{k} \mathbf{D}_{np} \left(\mathbf{F}_s \boldsymbol{\Psi}_j\right)\right]\right\} \mathbf{U}_{sj}^{k}\, d\Gamma^k\, dz \\
& +\int_{\Gamma^k}\int_{A^k} \delta \mathbf{U}_{\tau i}^{k^T} \left(\mathbf{F}_\tau \boldsymbol{\Psi}_i\right)^T \left\{(\mathbf{I}_{np})^T \left[\tilde{\mathbf{C}}_{nn}^{k} \mathbf{D}_{nz} \left(\mathbf{F}_s \boldsymbol{\Psi}_j\right)\right]\right\} \mathbf{U}_{sj}^{k}\, \mathrm{d}\Gamma^k\, \mathrm{d}z
\end{aligned}
\tag{7.77}
$$

where

$$\mathbf{I}_p = \begin{bmatrix} 1 & 0 & 0 \\ 0 & 1 & 0 \\ 1 & 1 & 0 \end{bmatrix}; \qquad \mathbf{I}_{np} = \begin{bmatrix} 0 & 0 & 1 \\ 0 & 0 & 1 \\ 0 & 0 & 0 \end{bmatrix} \tag{7.78}$$

The functional in Eq. (7.77) encompasses both the stiffness and the boundary nuclei. If, only the stiffness nucleus is taken into account then the classical Galerkin method is obtained, otherwise the contribution provided by the boundary nucleus leads to the so called Generalized Galerkin method. It should be noted that if the functional in Eq. (7.77) is used, then a higher differentiability grade, for the basis functions $\boldsymbol{\psi}_i$, is required.

This is the main and substantial difference between Ritz and Galerkin methods. When for particular boundary conditions or lamination schemes the boundary terms are not zero then the use of the Generalized Galerkin method is suggested. The functional in Eq. (7.77) should be used both to obtain the Galerkin stiffness nucleus, with related boundary nucleus, and to obtain the governing differential equations with variationally consistent boundary conditions.

The Galerkin stiffness and boundary nuclei assume the following form

$$\begin{aligned}
\mathbf{K}^{k\tau sij} = \int_{\Omega^k}\int_{A^k} \Big(& -(\mathbf{F}_\tau \boldsymbol{\Psi}_i)^T \\
& \Big\{(\mathbf{D}_p)^T\Big[\tilde{\mathbf{C}}^k_{pp}\mathbf{D}_p(\mathbf{F}_s\boldsymbol{\Psi}_j) + \tilde{\mathbf{C}}^k_{pn}\mathbf{D}_{np}(\mathbf{F}_s\boldsymbol{\Psi}_j) + \tilde{\mathbf{C}}^k_{pn}\mathbf{D}_{nz}(\mathbf{F}_s\boldsymbol{\Psi}_j)\Big]\Big\} + \\
& -(\mathbf{F}_\tau \boldsymbol{\Psi}_i)^T \\
& \Big\{(\mathbf{D}_{np})^T\Big[\tilde{\mathbf{C}}^k_{np}\mathbf{D}_p(\mathbf{F}_s\boldsymbol{\Psi}_j) + \tilde{\mathbf{C}}^k_{nn}\mathbf{D}_{np}(\mathbf{F}_s\boldsymbol{\Psi}_j) + \tilde{\mathbf{C}}^k_{nn}\mathbf{D}_{nz}(\mathbf{F}_s\boldsymbol{\Psi}_j)\Big]\Big\} + \\
& +(\mathbf{F}_\tau \boldsymbol{\Psi}_i)^T \\
& \Big\{(\mathbf{D}_{nz})^T\Big[\tilde{\mathbf{C}}^k_{np}\mathbf{D}_p(\mathbf{F}_s\boldsymbol{\Psi}_j) + \tilde{\mathbf{C}}^k_{nn}\mathbf{D}_{np}(\mathbf{F}_s\boldsymbol{\Psi}_j) + \tilde{\mathbf{C}}^k_{nn}\mathbf{D}_{nz}(\mathbf{F}_s\boldsymbol{\Psi}_j)\Big]\Big\}\Big)\, \mathrm{d}\Omega^k\, \mathrm{d}z \\
\mathbf{\Pi}^{k\tau sij} = \int_{\Gamma^k}\int_{A^k} \Big(& (\mathbf{F}_\tau \boldsymbol{\Psi}_i)^T \\
& \Big\{(\mathbf{I}_p)^T\Big[\tilde{\mathbf{C}}^k_{pp}\mathbf{D}_p(\mathbf{F}_s\boldsymbol{\Psi}_j) + \tilde{\mathbf{C}}^k_{pn}\mathbf{D}_{np}(\mathbf{F}_\tau\boldsymbol{\Psi}_j) + \tilde{\mathbf{C}}^k_{pn}\mathbf{D}_{nz}(\mathbf{F}_\tau\boldsymbol{\Psi}_j)\Big]\Big\} + \\
& +(\mathbf{F}_\tau \boldsymbol{\Psi}_i)^T \\
& \Big\{(\mathbf{I}_{np})^T\Big[\tilde{\mathbf{C}}^k_{np}\mathbf{D}_p(\mathbf{F}_s\boldsymbol{\Psi}_j) + \tilde{\mathbf{C}}^k_{nn}\mathbf{D}_{np}(\mathbf{F}_s\boldsymbol{\Psi}_j) + \tilde{\mathbf{C}}^k_{nn}\mathbf{D}_{nz}(\mathbf{F}_s\boldsymbol{\Psi}_j)\Big]\Big\}\Big)\, \mathrm{d}\Gamma^k\, \mathrm{d}z
\end{aligned} \tag{7.79}$$

performing the matrix calculus in Eq. (7.79), the 9 stiffness secondary nuclei become

$$K^{\tau_{u_x} s_{u_x}}_{u_x u_x} = \tilde{C}^k_{11} J^{k\tau_{u_x} s_{u_x}}\, {}^{u_x}_{u_x}I^{11}_{mp}\, {}^{u_x}_{u_x}I^{00}_{nq} + \tilde{C}^k_{16} J^{k\tau_{u_x} s_{u_x}}\, {}^{u_x}_{u_x}I^{10}_{mp}\, {}^{u_x}_{u_x}I^{01}_{nq}$$

$$
\begin{aligned}
&+\tilde{C}^k_{16}J^{k\tau_{u_x}s_{u_x}}\,{}^{u_x}_{u_x}I^{01}_{mp}\,{}^{u_x}_{u_x}I^{10}_{nq}+\tilde{C}^k_{66}J^{k\tau_{u_x}s_{u_x}}\,{}^{u_x}_{u_x}I^{00}_{mp}\,{}^{u_x}_{u_x}I^{11}_{nq}\\
&+\tilde{C}^k_{55}J^{k\tau_{u_x,z}s_{u_x,z}}\,{}^{u_x}_{u_x}I^{00}_{mp}\,{}^{u_x}_{u_x}I^{00}_{nq}\\
K^{\tau_{u_x}s_{u_y}}_{u_xu_y}&=\tilde{C}^k_{16}J^{k\tau_{u_x}s_{u_y}}\,{}^{u_x}_{u_y}I^{11}_{mp}\,{}^{u_x}_{u_y}I^{00}_{nq}+\tilde{C}^k_{12}J^{k\tau_{u_x}s_{u_y}}\,{}^{u_x}_{u_y}I^{10}_{mp}\,{}^{u_x}_{u_y}I^{01}_{nq}\\
&+\tilde{C}^k_{66}J^{k\tau_{u_x}s_{u_y}}\,{}^{u_x}_{u_y}I^{01}_{mp}\,{}^{u_x}_{u_y}I^{10}_{nq}+\tilde{C}^k_{26}J^{k\tau_{u_x}s_{u_y}}\,{}^{u_x}_{u_y}I^{00}_{mp}\,{}^{u_x}_{u_y}I^{11}_{nq}\\
&+\tilde{C}^k_{45}J^{k\tau_{u_x,z}s_{u_y,z}}\,{}^{u_x}_{u_y}I^{00}_{mp}\,{}^{u_x}_{u_y}I^{00}_{nq}\\
K^{\tau_{u_x}s_{u_z}}_{u_xu_z}&=\tilde{C}^k_{55}J^{k\tau_{u_x,z}s_{u_z}}\,{}^{u_x}_{u_z}I^{01}_{mp}\,{}^{u_x}_{u_z}I^{10}_{nq}+\tilde{C}^k_{45}J^{k\tau_{u_x,z}s_{u_z}}\,{}^{u_x}_{u_z}I^{00}_{mp}\,{}^{u_x}_{u_z}I^{01}_{nq}\\
&+\tilde{C}^k_{13}J^{k\tau_{u_x}s_{u_z,z}}\,{}^{u_x}_{u_z}I^{10}_{mp}\,{}^{u_x}_{u_z}I^{00}_{nq}+\tilde{C}^k_{36}J^{k\tau_{u_x}s_{u_z,z}}\,{}^{u_x}_{u_z}I^{00}_{mp}\,{}^{u_x}_{u_z}I^{10}_{nq}\\
K^{\tau_{u_y}s_{u_x}}_{u_yu_x}&=\tilde{C}^k_{16}J^{k\tau_{u_y}s_{u_x}}\,{}^{u_y}_{u_x}I^{11}_{mp}\,{}^{u_y}_{u_x}I^{00}_{nq}+\tilde{C}^k_{12}J^{k\tau_{u_y}s_{u_x}}\,{}^{u_y}_{u_x}I^{01}_{mp}\,{}^{u_y}_{u_x}I^{10}_{nq}\\
&+\tilde{C}^k_{66}J^{k\tau_{u_y}s_{u_x}}\,{}^{u_y}_{u_x}I^{10}_{mp}\,{}^{u_y}_{u_x}I^{01}_{nq}+\tilde{C}^k_{26}J^{k\tau_{u_y}s_{u_x}}\,{}^{u_y}_{u_x}I^{00}_{mp}\,{}^{u_y}_{u_x}I^{11}_{nq}\\
&+\tilde{C}^k_{45}J^{k\tau_{u_y,z}s_{u_x,z}}\,{}^{u_y}_{u_x}I^{00}_{mp}\,{}^{u_y}_{u_x}I^{00}_{nq}\\
K^{\tau_{u_y}s_{u_y}}_{u_yu_y}&=\tilde{C}^k_{26}J^{k\tau_{u_y}s_{u_y}}\,{}^{u_y}_{u_y}I^{01}_{mp}\,{}^{u_y}_{u_y}I^{10}_{nq}+\tilde{C}^k_{22}J^{k\tau_{u_y}s_{u_y}}\,{}^{u_y}_{u_y}I^{00}_{mp}\,{}^{u_y}_{u_y}I^{11}_{nq}\\
&+\tilde{C}^k_{66}J^{k\tau_{u_y}s_{u_y}}\,{}^{u_y}_{u_y}I^{01}_{mp}\,{}^{u_y}_{u_y}I^{10}_{nq}+\tilde{C}^k_{26}J^{k\tau_{u_y}s_{u_y}}\,{}^{u_y}_{u_y}I^{00}_{mp}\,{}^{u_y}_{u_y}I^{11}_{nq}\\
&+\tilde{C}^k_{44}J^{k\tau_{u_y,z}s_{u_y,z}}\,{}^{u_y}_{u_y}I^{00}_{mp}\,{}^{u_y}_{u_y}I^{00}_{nq}\\
K^{\tau_{u_y}s_{u_z}}_{u_yu_z}&=\tilde{C}^k_{45}J^{k\tau_{u_y,z}s_{u_z}}\,{}^{u_y}_{u_z}I^{01}_{mp}\,{}^{u_y}_{u_z}I^{00}_{nq}+\tilde{C}^k_{44}J^{k\tau_{u_y,z}s_{u_z}}\,{}^{u_y}_{u_z}I^{00}_{mp}\,{}^{u_y}_{u_z}I^{01}_{nq}\\
&+\tilde{C}^k_{36}J^{k\tau_{u_y}s_{u_z,z}}\,{}^{u_y}_{u_z}I^{10}_{mp}\,{}^{u_y}_{u_z}I^{00}_{nq}+\tilde{C}^k_{23}J^{k\tau_{u_y}s_{u_z,z}}\,{}^{u_y}_{u_z}I^{00}_{mp}\,{}^{u_y}_{u_z}I^{10}_{nq}\\
K^{\tau_{u_z}s_{u_x}}_{u_zu_x}&=\tilde{C}^k_{55}J^{k\tau_{u_z}s_{u_x,z}}\,{}^{u_z}_{u_x}I^{01}_{mp}\,{}^{u_z}_{u_x}I^{00}_{nq}+\tilde{C}^k_{45}J^{k\tau_{u_z}s_{u_x,z}}\,{}^{u_z}_{u_x}I^{00}_{mp}\,{}^{u_z}_{u_x}I^{01}_{nq}\\
&+\tilde{C}^k_{13}J^{k\tau_{u_z,z}s_{u_x}}\,{}^{u_z}_{u_x}I^{10}_{mp}\,{}^{u_z}_{u_x}I^{00}_{nq}+\tilde{C}^k_{36}J^{k\tau_{u_z,z}s_{u_x}}\,{}^{u_z}_{u_x}I^{00}_{mp}\,{}^{u_z}_{u_x}I^{10}_{nq}\\
K^{\tau_{u_z}s_{u_y}}_{u_zu_y}&=\tilde{C}^k_{45}J^{k\tau_{u_z}s_{u_y,z}}\,{}^{u_z}_{u_y}I^{01}_{mp}\,{}^{u_z}_{u_y}I^{00}_{nq}+\tilde{C}^k_{44}J^{k\tau_{u_z}s_{u_y,z}}\,{}^{u_z}_{u_y}I^{00}_{mp}\,{}^{u_z}_{u_y}I^{01}_{nq}\\
&+\tilde{C}^k_{36}J^{k\tau_{u_z,z}s_{u_y}}\,{}^{u_z}_{u_y}I^{10}_{mp}\,{}^{u_z}_{u_y}I^{00}_{nq}+\tilde{C}^k_{23}J^{k\tau_{u_z,z}s_{u_y}}\,{}^{u_z}_{u_y}I^{00}_{mp}\,{}^{u_z}_{u_y}I^{10}_{nq}\\
K^{\tau_{u_z}s_{u_z}}_{u_zu_z}&=\tilde{C}^k_{45}J^{k\tau_{u_z}s_{u_z}}\,{}^{u_z}_{u_z}I^{01}_{mp}\,{}^{u_z}_{u_z}I^{10}_{nq}+\tilde{C}^k_{44}J^{k\tau_{u_z}s_{u_z}}\,{}^{u_z}_{u_z}I^{00}_{mp}\,{}^{u_z}_{u_z}I^{11}_{nq}\\
&+\tilde{C}^k_{55}J^{k\tau_{u_z}s_{u_z}}\,{}^{u_z}_{u_z}I^{11}_{mp}\,{}^{u_z}_{u_z}I^{00}_{nq}+\tilde{C}^k_{45}J^{k\tau_{u_z}s_{u_z}}\,{}^{u_z}_{u_z}I^{10}_{mp}\,{}^{u_z}_{u_z}I^{01}_{nq}\\
&+\tilde{C}^k_{33}J^{k\tau_{u_z,z}s_{u_z,z}}\,{}^{u_z}_{u_z}I^{00}_{mp}\,{}^{u_z}_{u_z}I^{00}_{nq}
\end{aligned}
\tag{7.80}
$$

and the 9 boundary secondary nuclei are

$$
\begin{aligned}
\Pi^{\tau_{u_x}s_{u_x}}_{u_xu_x}&=\tilde{C}^k_{11}J^{k\tau_{u_x}s_{u_x}}\,{}^{u_x}_{u_x}I^{11}_{mp}\,{}^{u_x}_{u_x}I^{00}_{nq}+\tilde{C}^k_{16}J^{k\tau_{u_x}s_{u_x}}\,{}^{u_x}_{u_x}I^{10}_{mp}\,{}^{u_x}_{u_x}I^{01}_{nq}\\
&+\tilde{C}^k_{16}J^{k\tau_{u_x}s_{u_x}}\,{}^{u_x}_{u_x}I^{01}_{mp}\,{}^{u_x}_{u_x}I^{10}_{nq}+\tilde{C}^k_{66}J^{k\tau_{u_x}s_{u_x}}\,{}^{u_x}_{u_x}I^{00}_{mp}\,{}^{u_x}_{u_x}I^{11}_{nq}\\
\Pi^{\tau_{u_x}s_{u_y}}_{u_xu_y}&=\tilde{C}^k_{16}J^{k\tau_{u_x}s_{u_y}}\,{}^{u_x}_{u_y}I^{11}_{mp}\,{}^{u_x}_{u_y}I^{00}_{nq}+\tilde{C}^k_{12}J^{k\tau_{u_x}s_{u_y}}\,{}^{u_x}_{u_y}I^{10}_{mp}\,{}^{u_x}_{u_y}I^{01}_{nq}\\
&+\tilde{C}^k_{66}J^{k\tau_{u_x}s_{u_y}}\,{}^{u_x}_{u_y}I^{01}_{mp}\,{}^{u_x}_{u_y}I^{10}_{nq}+\tilde{C}^k_{26}J^{k\tau_{u_x}s_{u_y}}\,{}^{u_x}_{u_y}I^{00}_{mp}\,{}^{u_x}_{u_y}I^{11}_{nq}
\end{aligned}
$$

$$
\begin{aligned}
\Pi^{\tau_{u_x}s_{u_z}}_{u_x u_z} &= \tilde{C}^k_{13} J^{k\tau_{u_x}s_{u_z,z}} \, {}^{u_x}_{u_z}I^{10}_{mp} \, {}^{u_x}_{u_z}I^{00}_{nq} + \tilde{C}^k_{36} J^{k\tau_{u_x}s_{u_z,z}} \, {}^{u_x}_{u_z}I^{00}_{mp} \, {}^{u_x}_{u_z}I^{10}_{nq} \\
\Pi^{\tau_{u_y}s_{u_x}}_{u_y u_x} &= \tilde{C}^k_{16} J^{k\tau_{u_y}s_{u_x}} \, {}^{u_y}_{u_x}I^{11}_{mp} \, {}^{u_y}_{u_x}I^{00}_{nq} + \tilde{C}^k_{12} J^{k\tau_{u_y}s_{u_x}} \, {}^{u_y}_{u_x}I^{01}_{mp} \, {}^{u_y}_{u_x}I^{10}_{nq} \\
&\quad + \tilde{C}^k_{66} J^{k\tau_{u_y}s_{u_x}} \, {}^{u_y}_{u_x}I^{10}_{mp} \, {}^{u_y}_{u_x}I^{01}_{nq} + \tilde{C}^k_{26} J^{k\tau_{u_y}s_{u_x}} \, {}^{u_y}_{u_x}I^{00}_{mp} \, {}^{u_y}_{u_x}I^{11}_{nq} \\
\Pi^{\tau_{u_y}s_{u_y}}_{u_y u_y} &= \tilde{C}^k_{26} J^{k\tau_{u_y}s_{u_y}} \, {}^{u_y}_{u_y}I^{01}_{mp} \, {}^{u_y}_{u_y}I^{10}_{nq} + \tilde{C}^k_{22} J^{k\tau_{u_y}s_{u_y}} \, {}^{u_y}_{u_y}I^{00}_{mp} \, {}^{u_y}_{u_y}I^{11}_{nq} \\
&\quad + \tilde{C}^k_{66} J^{k\tau_{u_y}s_{u_y}} \, {}^{u_y}_{u_y}I^{01}_{mp} \, {}^{u_y}_{u_y}I^{10}_{nq} + \tilde{C}^k_{26} J^{k\tau_{u_y}s_{u_y}} \, {}^{u_y}_{u_y}I^{00}_{mp} \, {}^{u_y}_{u_y}I^{11}_{nq} \\
\Pi^{\tau_{u_y}s_{u_z}}_{u_y u_z} &= \tilde{C}^k_{36} J^{k\tau_{u_y}s_{u_z,z}} \, {}^{u_y}_{u_z}I^{10}_{mp} \, {}^{u_y}_{u_z}I^{00}_{nq} + \tilde{C}^k_{23} J^{k\tau_{u_y}s_{u_z,z}} \, {}^{u_y}_{u_z}I^{00}_{mp} \, {}^{u_y}_{u_z}I^{10}_{nq} \\
\Pi^{\tau_{u_z}s_{u_x}}_{u_z u_x} &= \tilde{C}^k_{55} J^{k\tau_{u_z}s_{u_x,z}} \, {}^{u_z}_{u_x}I^{01}_{mp} \, {}^{u_z}_{u_x}I^{00}_{nq} + \tilde{C}^k_{45} J^{k\tau_{u_z}s_{u_x,z}} \, {}^{u_z}_{u_x}I^{00}_{mp} \, {}^{u_z}_{u_x}I^{01}_{nq} \\
\Pi^{\tau_{u_z}s_{u_y}}_{u_z u_y} &= \tilde{C}^k_{45} J^{k\tau_{u_z}s_{u_y,z}} \, {}^{u_z}_{u_y}I^{01}_{mp} \, {}^{u_z}_{u_y}I^{00}_{nq} + \tilde{C}^k_{44} J^{k\tau_{u_z}s_{u_y,z}} \, {}^{u_z}_{u_y}I^{00}_{mp} \, {}^{u_z}_{u_y}I^{01}_{nq} \\
\Pi^{\tau_{u_z}s_{u_z}}_{u_z u_z} &= \tilde{C}^k_{45} J^{k\tau_{u_z}s_{u_z}} \, {}^{u_z}_{u_z}I^{01}_{mp} \, {}^{u_z}_{u_z}I^{10}_{nq} + \tilde{C}^k_{44} J^{k\tau_{u_z}s_{u_z}} \, {}^{u_z}_{u_z}I^{00}_{mp} \, {}^{u_z}_{u_z}I^{11}_{nq} \\
&\quad + \tilde{C}^k_{55} J^{k\tau_{u_z}s_{u_z}} \, {}^{u_z}_{u_z}I^{11}_{mp} \, {}^{u_z}_{u_z}I^{00}_{nq} + \tilde{C}^k_{45} J^{k\tau_{u_z}s_{u_z}} \, {}^{u_z}_{u_z}I^{10}_{mp} \, {}^{u_z}_{u_z}I^{01}_{nq}
\end{aligned}
\tag{7.81}
$$

If the Galerkin method is applied to the total potential energy functional in Eq. (7.7) then the results perfectly coincide with those obtained by using the Ritz method. It should be borne in mind that if the boundary terms are zero (condition satisfied under particular assumptions) then the three different methodologies lead to the same results. The advantage to use Galerkin method is that only the governing differential equations are required and the boundary conditions are negligible. Nevertheless, when the boundary terms are not zero, the stiffness matrix can lose some important properties like symmetry and positivity-definite. This is extremely important from a computation point of view, above all when increasing the value of the half-waves m and n due to convergence reasons. Indeed in these cases some algorithms for the calculation of the eigenvalues which are computationally more expensive have to be used.

7.5 GOVERNING DIFFERENTIAL EQUATIONS

In this section the governing differential equations are derived for anisotropic composite beam, plate and shell structures. In particular, for beams and plates, they will be provided in a compact vectorial form, in the case of shells, the equations will be given in an unabridged way, namely, the explicit form of the secondary stiffness and mass nuclei will be proposed. In the particular case of shell structures the equation will be also derived in the case of Reissner's Mixed Variational Theorems (RMVT).

7.5.1 Beams and plates

According to the constitutive and geometrical relationships provided in the analysis of beam structures, the strong form of the governing differential equations in a compact matrix form can be written as follows

$$
\begin{aligned}
&\sum_{k=1}^{N_l}\int_l \delta\mathbf{u}_\tau^{k^T}\int_{\Omega^k}\Big\{\Big(\Big[(\mathbf{F}_\tau)^T(\mathbf{D}_p)^T\Big]\times\\
&\qquad\Big[\tilde{\mathbf{C}}_{pp}^k\mathbf{D}_p(\mathbf{F}_s)+\tilde{\mathbf{C}}_{pn}^k\mathbf{D}_{np}(\mathbf{F}_s)+\tilde{\mathbf{C}}_{pn}^k\mathbf{D}_{nz}(\mathbf{F}_s)\Big]\Big)+\\
&\qquad\Big(\Big[(\mathbf{F}_\tau)^T(\mathbf{D}_{np})^T\Big]\times\\
&\qquad\Big[\tilde{\mathbf{C}}_{np}^k\mathbf{D}_p(\mathbf{F}_s)+\tilde{\mathbf{C}}_{nn}^k\mathbf{D}_{np}(\mathbf{F}_s)+\tilde{\mathbf{C}}_{nn}^k\mathbf{D}_{nz}(\mathbf{F}_s)\Big]\Big)+\\
&\qquad\Big(-\Big[(\mathbf{F}_\tau)^T(\mathbf{D}_{nz})^T\Big]\times\\
&\qquad\Big[\tilde{\mathbf{C}}_{np}^k\mathbf{D}_p(\mathbf{F}_s)+\tilde{\mathbf{C}}_{nn}^k\mathbf{D}_{np}(\mathbf{F}_s)+\tilde{\mathbf{C}}_{nn}^k\mathbf{D}_{nz}(\mathbf{F}_s)\Big]\Big)\Big\}\mathrm{d}\Omega^k\,\mathbf{u}_s^k\,\mathrm{d}z\\
&+\sum_{k=1}^{N_l}\delta\mathbf{u}_\tau^{k^T}\int_{\Omega^k}\Big\{\Big[(\mathbf{F}_\tau)^T(\mathbf{I}_{np})^T\Big]\times\\
&\qquad\Big[\tilde{\mathbf{C}}_{np}^k\mathbf{D}_p(\mathbf{F}_s)+\tilde{\mathbf{C}}_{nn}^k\mathbf{D}_{np}(\mathbf{F}_s)+\tilde{\mathbf{C}}_{nn}^k\mathbf{D}_{nz}(\mathbf{F}_s)\Big]\Big\}\,\mathrm{d}\Omega^k\mathbf{u}_\tau^k\Big|_{z=0}^{l}=\\
&-\sum_{k=1}^{N_l}\int_l\delta\mathbf{u}_\tau^{k^T}\int_{\Omega^k}\Big[\rho(\mathbf{F}_\tau)^T(\mathbf{F}_s)\Big]\mathrm{d}\Omega^k\ddot{\mathbf{u}}_\tau^k\mathrm{d}z
\end{aligned}\tag{7.82}
$$

Similarly for plates the governing differential equations in vectorial form can be given as follows

$$
\begin{aligned}
&\sum_{k=1}^{N_l}\int_{\Omega_k}\int_{A^k}\delta\mathbf{u}_\tau^{k^T}\Big\{\Big(-\Big[(\mathbf{F}_\tau)^T(\mathbf{D}_p)^T\Big]\times\\
&\qquad\Big[\tilde{\mathbf{C}}_{pp}^k\mathbf{D}_p(\mathbf{F}_s)+\tilde{\mathbf{C}}_{pn}^k\mathbf{D}_{np}(\mathbf{F}_s)+\tilde{\mathbf{C}}_{pn}^k\mathbf{D}_{nz}(\mathbf{F}_s)\Big]\Big)+\\
&\qquad\Big(-\Big[(\mathbf{F}_\tau)^T(\mathbf{D}_{np})^T\Big]\times\\
&\qquad\Big[\tilde{\mathbf{C}}_{np}^k\mathbf{D}_p(\mathbf{F}_s)+\tilde{\mathbf{C}}_{nn}^k\mathbf{D}_{np}(\mathbf{F}_s)+\tilde{\mathbf{C}}_{nn}^k\mathbf{D}_{nz}(\mathbf{F}_s)\Big]\Big)+
\end{aligned}
$$

$$
\left(\left[(\mathbf{F}_\tau)^T(\mathbf{D}_{nz})^T\right]\times\right.
$$

$$
\left.\left.\left[\tilde{\mathbf{C}}^k_{np}\mathbf{D}_p(\mathbf{F}_s)+\tilde{\mathbf{C}}^k_{nn}\mathbf{D}_{np}(\mathbf{F}_s)+\tilde{\mathbf{C}}^k_{nn}\mathbf{D}_{nz}(\mathbf{F}_s)\right]\right)\right\}\mathrm{d}\Omega^k\,\mathbf{u}^k_s\,\mathrm{d}z
$$

$$
+\sum_{k=1}^{N_l}\int_{\Gamma_k}\int_{A^k}\delta\mathbf{u}_\tau\left\{\left[(\mathbf{F}_\tau)^T(\boldsymbol{I}_p)^T\right]\times\right.
$$

$$
\left[\tilde{\mathbf{C}}^k_{pp}\mathbf{D}_p(\mathbf{F}_s)+\tilde{\mathbf{C}}^k_{pn}\mathbf{D}_{np}(\mathbf{F}_s)+\tilde{\mathbf{C}}^k_{pn}\mathbf{D}_{nz}(\mathbf{F}_s)\right]+
$$

$$
\left[(\mathbf{F}_\tau)^T(\boldsymbol{I}_{np})^T\right]\times
$$

$$
\left.\left[\tilde{\mathbf{C}}^k_{np}\mathbf{D}_p(\mathbf{F}_s)+\tilde{\mathbf{C}}^k_{nn}\mathbf{D}_{np}(\mathbf{F}_s)+\tilde{\mathbf{C}}^k_{nn}\mathbf{D}_{nz}(\mathbf{F}_s)\right]\right\}\mathbf{u}_s\,d\Omega_k dz=
$$

$$
-\sum_{k=1}^{N_l}\int_{\Omega_k}\int_{A^k}\delta\mathbf{u}^{k^T}_\tau\left[\rho(\mathbf{F}_\tau)^T(\mathbf{F}_s)\right]\mathrm{d}\Omega^k\ddot{\mathbf{u}}^k_s\mathrm{d}z \qquad (7.83)
$$

It must be noted the extraordinary similarity between the two sets of equations. The fact that formally there are no substantial changes to carry out when writing beam and plate equations, is one of the strong point of the unified formulation. And as will be seen in the next subsection, the same similarity can be noted with the governing differential equations of shell structures. However, it must be borne in mind that, of course, the functions involved in the formulations assume different forms. For instance the typical function F_τ must be interpreted as thickness function, namely, $F_\tau(z)$, in the case of plates and shells, but it becomes a cross-section function, $F_\tau(x,y)$, when involved in the beam equations.

7.5.2 Shells

The governing differential equations and natural boundary conditions (Neumann-type) on Γ^m_k, for a doubly-curved anisotropic composite shell at multilayer level can be written as

$$
\sum_{k=1}^{N_l}\int_{\Omega_k}\int_{A^k}\delta\mathbf{u}_\tau\left\{-\left[(\mathbf{F}_\tau)^T(\mathbf{D}_p)^T+(\mathbf{F}_\tau)^T(\mathbf{A}_p)^T\right]\times\right.
$$

$$
\left[\tilde{\mathbf{C}}^k_{pp}\mathbf{D}_p(\mathbf{F}_\tau)+\tilde{\mathbf{C}}^k_{pp}\mathbf{A}_p(\mathbf{F}_\tau)+\tilde{\mathbf{C}}^k_{pn}\mathbf{D}_{np}(\mathbf{F}_\tau)+\right.
$$

$$
\left.\delta_D\tilde{\mathbf{C}}^k_{pn}\tilde{\mathbf{A}}^k_n\mathbf{D}_{np}(\mathbf{F}_\tau)+\tilde{\mathbf{C}}^k_{pn}\mathbf{D}_{nz}(\mathbf{F}_\tau)\right]-
$$

$$
\left[(\mathbf{F}_\tau)^T(\mathbf{D}_{np})^T+(\mathbf{F}_\tau)^T(\delta_D\mathbf{A}_n)^T+(\mathbf{F}_\tau)^T(\mathbf{D}_{nz})^T\right]\times
$$

$$\left[\tilde{\mathbf{C}}^k_{np}\mathbf{D}_p(\mathbf{F}_\tau)+\tilde{\mathbf{C}}^k_{np}\mathbf{A}_p(\mathbf{F}_\tau)+\tilde{\mathbf{C}}^k_{nn}\mathbf{D}_{np}(\mathbf{F}_\tau)+\right.$$

$$\left.\left.\delta_D\tilde{\mathbf{C}}^k_{nn}\tilde{\mathbf{A}}^k_n\mathbf{D}_{np}(\mathbf{F}_\tau)+\tilde{\mathbf{C}}^k_{nn}\mathbf{D}_{nz}(\mathbf{F}_\tau)\right]\right\}\mathbf{u}_s\,d\Omega_k dz$$

$$+\sum_{k=1}^{N_l}\int_{\Omega_k}\int_{A^k}\delta\mathbf{u}_\tau\left\{\left[(\mathbf{F}_\tau)^T(\boldsymbol{I}_p)^T\right]\times\right.$$

$$\left[\tilde{\mathbf{C}}^k_{pp}\mathbf{D}_p(\mathbf{F}_\tau)+\tilde{\mathbf{C}}^k_{pp}\mathbf{A}_p(\mathbf{F}_\tau)+\tilde{\mathbf{C}}^k_{pn}\mathbf{D}_{np}(\mathbf{F}_\tau)+\right.$$

$$\left.\delta_D\tilde{\mathbf{C}}^k_{pn}\mathbf{A}_n(\mathbf{F}_\tau)+\tilde{\mathbf{C}}^k_{pn}\mathbf{D}_{nz}(\mathbf{F}_\tau)\right]+$$

$$\left[(\mathbf{F}_\tau)^T(\boldsymbol{I}_{np})^T\right]\times$$

$$\left[\tilde{\mathbf{C}}^k_{np}\mathbf{D}_p(\mathbf{F}_\tau)+\tilde{\mathbf{C}}^k_{np}\mathbf{A}_p(\mathbf{F}_\tau)+\tilde{\mathbf{C}}^k_{nn}\mathbf{D}_{np}(\mathbf{F}_\tau)+\right.$$

$$\left.\left.\delta_D\tilde{\mathbf{C}}^k_{nn}\mathbf{A}_n(\mathbf{F}_\tau)+\tilde{\mathbf{C}}^k_{nn}\mathbf{D}_{nz}(\mathbf{F}_\tau)\right]\right\}\mathbf{u}_s\,d\Omega_k dz=$$

$$-\sum_{k=1}^{N_l}\int_{\Omega_k}\int_{A^k}\delta\mathbf{u}_\tau\left[\rho^k(\mathbf{F}_\tau)^T(\mathbf{F}_s)\right]\ddot{\mathbf{u}}_s\,d\Omega_k dz \tag{7.84}$$

and in compact form for the k-layer can be written as

$$\begin{array}{lll}\delta\mathbf{u}_\tau: & \mathbf{K}^{k\tau s}_{uu}\mathbf{u}_s+\mathbf{M}^{k\tau s}_{uu}\ddot{\mathbf{u}}_s=0 & \\ \Gamma^m_k: & \boldsymbol{\Pi}^{k\tau s}_{uu}\mathbf{u}_s=\boldsymbol{\Pi}^{k\tau s}_{uu}\bar{\mathbf{u}}_s & \Gamma^g_k:\mathbf{u}_s=\bar{\mathbf{u}}_s\end{array} \tag{7.85}$$

where

$$\mathbf{K}^{k\tau s}_{uu}=\int_{A^k}\left\{-\left[(\mathbf{F}_\tau)^T(\mathbf{D}_p)^T+(\mathbf{F}_\tau)^T(\mathbf{A}_p)^T\right]\times\right.$$

$$\left[\tilde{\mathbf{C}}^k_{pp}\mathbf{D}_p(\mathbf{F}_\tau)+\tilde{\mathbf{C}}^k_{pp}\mathbf{A}_p(\mathbf{F}_\tau)+\tilde{\mathbf{C}}^k_{pn}\mathbf{D}_{np}(\mathbf{F}_\tau)+\right.$$

$$\left.\delta_D\tilde{\mathbf{C}}^k_{pn}\tilde{\mathbf{A}}^k_n\mathbf{D}_{np}(\mathbf{F}_\tau)+\tilde{\mathbf{C}}^k_{pn}\mathbf{D}_{nz}(\mathbf{F}_\tau)\right]-$$

$$\left[(\mathbf{F}_\tau)^T(\mathbf{D}_{np})^T+(\mathbf{F}_\tau)^T(\delta_D\mathbf{A}_n)^T+(\mathbf{F}_\tau)^T(\mathbf{D}_{nz})^T\right]\times$$

$$\left[\tilde{\mathbf{C}}^k_{np}\mathbf{D}_p(\mathbf{F}_\tau)+\tilde{\mathbf{C}}^k_{np}\mathbf{A}_p(\mathbf{F}_\tau)+\tilde{\mathbf{C}}^k_{nn}\mathbf{D}_{np}(\mathbf{F}_\tau)+\right.$$

$$\left.\left.\delta_D\tilde{\mathbf{C}}^k_{nn}\tilde{\mathbf{A}}^k_n\mathbf{D}_{np}(\mathbf{F}_\tau)+\tilde{\mathbf{C}}^k_{nn}\mathbf{D}_{nz}(\mathbf{F}_\tau)\right]\right\}H^k_\alpha H^k_\beta\mathrm{d}z$$

$$\boldsymbol{\Pi}^{k\tau s}_{uu}=\int_{A^k}\left\{\left[(\mathbf{F}_\tau)^T(\boldsymbol{I}_p)^T\right]\times\right.$$

$$\left[\tilde{\mathbf{C}}^k_{pp}\mathbf{D}_p(\mathbf{F}_\tau)+\tilde{\mathbf{C}}^k_{pp}\mathbf{A}_p(\mathbf{F}_\tau)+\tilde{\mathbf{C}}^k_{pn}\mathbf{D}_{np}(\mathbf{F}_\tau)+\right.$$

$$\left.\delta_D\tilde{\mathbf{C}}^k_{pn}\mathbf{A}_n(\mathbf{F}_\tau)+\tilde{\mathbf{C}}^k_{pn}\mathbf{D}_{nz}(\mathbf{F}_\tau)\right]+$$

$$\left[(\mathbf{F}_\tau)^T\left(\boldsymbol{I}_{np}\right)^T\right]\times$$
$$\left[\tilde{\mathbf{C}}^k_{np}\mathbf{D}_p(\mathbf{F}_\tau)+\tilde{\mathbf{C}}^k_{np}\mathbf{A}_p(\mathbf{F}_\tau)+\tilde{\mathbf{C}}^k_{nn}\mathbf{D}_{np}(\mathbf{F}_\tau)+\right.$$
$$\left.\left.\delta_D\tilde{\mathbf{C}}^k_{nn}\mathbf{A}_n(\mathbf{F}_\tau)+\tilde{\mathbf{C}}^k_{nn}\mathbf{D}_{nz}(\mathbf{F}_\tau)\right]\right\}H^k_\alpha H^k_\beta\,\mathrm{d}z$$
$$\mathbf{M}^{k\tau s}_{u\,u}=\int_{A^k}\left[\rho^k(\mathbf{F}_\tau)^T(\mathbf{F}_s)\right]H^k_\alpha H^k_\beta\,\mathrm{d}z$$
$$(7.86)$$

For the sake of completeness the nine components of the fundamental primary differential nucleus $\mathbf{K}_{u\,u}$ are following reported

$$\begin{aligned}
K^{k\,\tau_{u\alpha}\,s_{u\alpha}}_{u\,u}&=\tilde{C}^k_{55}\left[J^{k\tau_{u\alpha_z}s_{u\alpha_z}}_{\alpha\,\beta}-\frac{\delta_D}{R_{\alpha_k}}J^{k\tau_{u\alpha_z}s_{u\alpha}}_{\beta}-\frac{\delta_D}{R_{\alpha_k}}J^{k\tau_{u\alpha}s_{u\alpha_z}}_{\beta}+\frac{\delta_D}{\left(R_{\alpha_k}\right)^2}J^{k\tau_{u\alpha}\,s_{u\alpha}}_{\frac{\beta}{\alpha}}\right]\\
&-\tilde{C}^k_{11}J^{k\tau_{u\alpha}\,s_{u\alpha}}_{\frac{\beta}{\alpha}}\left(\frac{\partial}{\partial\alpha}\right)_{s_{u\alpha}}\left(\frac{\partial}{\partial\alpha}\right)_{\tau_{u\alpha}}-\tilde{C}^k_{16}J^{k\tau_{u\alpha}\,s_{u\alpha}}\left(\frac{\partial}{\partial\alpha}\right)_{\tau_{u\alpha}}\left(\frac{\partial}{\partial\beta}\right)_{s_{u\alpha}}\\
&-\tilde{C}^k_{16}J^{k\tau_{u\alpha}\,s_{u\alpha}}\left(\frac{\partial}{\partial\alpha}\right)_{s_{u\alpha}}\left(\frac{\partial}{\partial\beta}\right)_{\tau_{u\alpha}}-\tilde{C}^k_{66}J^{k\tau_{u\alpha}\,s_{u\alpha}}_{\frac{\alpha}{\beta}}\left(\frac{\partial}{\partial\beta}\right)_{s_{u\alpha}}\left(\frac{\partial}{\partial\beta}\right)_{\tau_{u\alpha}}\\
K^{k\,\tau_{u\alpha}\,s_{u\beta}}_{u\,u}&=\tilde{C}^k_{45}\left[J^{k\tau_{u\alpha_z}s_{u\beta_z}}_{\alpha\,\beta}-\frac{\delta_D}{R_{\beta_k}}J^{k\tau_{u\alpha_z}s_{u\beta}}_{\alpha}-\frac{\delta_D}{R_{\alpha_k}}J^{k\tau_{u\alpha}s_{u\beta_z}}_{\beta}+\frac{\delta_D}{R_{\alpha_k}R_{\beta_k}}J^{k\tau_{u\alpha}\,s_{u\beta}}\right]\\
&-\tilde{C}^k_{12}J^{k\tau_{u\alpha}\,s_{u\beta}}\left(\frac{\partial}{\partial\alpha}\right)_{\tau_{u\alpha}}\left(\frac{\partial}{\partial\beta}\right)_{s_{u\beta}}-\tilde{C}^k_{16}J^{k\tau_{u\alpha}\,s_{u\beta}}_{\frac{\beta}{\alpha}}\left(\frac{\partial}{\partial\alpha}\right)_{s_{u\alpha}}\left(\frac{\partial}{\partial\alpha}\right)_{\tau_{u\beta}}\\
&-\tilde{C}^k_{26}J^{k\tau_{u\alpha}\,s_{u\beta}}_{\frac{\alpha}{\beta}}\left(\frac{\partial}{\partial\beta}\right)_{s_{u\beta}}\left(\frac{\partial}{\partial\beta}\right)_{\tau_{u\alpha}}-\tilde{C}^k_{66}J^{k\tau_{u\alpha}\,s_{u\beta}}\left(\frac{\partial}{\partial\alpha}\right)_{s_{u\beta}}\left(\frac{\partial}{\partial\beta}\right)_{\tau_{u\alpha}}\\
K^{k\,\tau_{u\alpha}\,s_{uz}}_{u\,u}&=\tilde{C}^k_{45}\left[J^{k\,\tau_{u\alpha_z}\,s_{uz}}_{\alpha}\left(\frac{\partial}{\partial\beta}\right)_{s_{uz}}-\frac{1}{R_{\alpha_k}}J^{k\tau_{u\alpha}\,s_{uz}}\left(\frac{\partial}{\partial\beta}\right)_{s_{uz}}\right]\\
&+\tilde{C}^k_{55}\left[J^{k\,\tau_{u\alpha_z}\,s_{uz}}_{\beta}\left(\frac{\partial}{\partial\alpha}\right)_{s_{uz}}-\frac{1}{R_{\alpha_k}}J^{k\tau_{u\alpha}\,s_{uz}}_{\frac{\beta}{\alpha}}\left(\frac{\partial}{\partial\alpha}\right)_{s_{uz}}\right]\\
&-\tilde{C}^k_{11}\frac{1}{R_{\alpha_k}}J^{k\tau_{u\alpha}\,s_{uz}}_{\frac{\beta}{\alpha}}\left(\frac{\partial}{\partial\alpha}\right)_{\tau_{u\alpha}}-\tilde{C}^k_{12}\frac{1}{R_{\beta_k}}J^{k\tau_{u\alpha}\,s_{uz}}\left(\frac{\partial}{\partial\alpha}\right)_{\tau_{u\alpha}}\\
&-\tilde{C}^k_{13}J^{k\tau_{u\alpha}\,s_{uz_z}}_{\beta}\left(\frac{\partial}{\partial\alpha}\right)_{\tau_{u\alpha}}-\tilde{C}^k_{16}\frac{1}{R_{\alpha_k}}J^{k\tau_{u\alpha}\,s_{uz}}\left(\frac{\partial}{\partial\beta}\right)_{\tau_{u\alpha}}\\
&-\tilde{C}^k_{26}\frac{1}{R_{\beta_k}}J^{k\tau_{u\alpha}\,s_{uz}}_{\frac{\alpha}{\beta}}\left(\frac{\partial}{\partial\beta}\right)_{\tau_{u\alpha}}-\tilde{C}^k_{36}J^{k\tau_{u\alpha}\,s_{uz}}_{\alpha}\left(\frac{\partial}{\partial\beta}\right)_{\tau_{u\alpha}}\\
K^{k\,\tau_{u\beta}\,s_{u\alpha}}_{u\,u}&=\tilde{C}^k_{45}\left[J^{k\tau_{u\beta_z}s_{u\alpha_z}}_{\alpha\,\beta}-\frac{\delta_D}{R_{\beta_k}}J^{k\tau_{u\beta}\,s_{u\alpha_z}}_{\alpha}-\frac{\delta_D}{R_{\alpha_k}}J^{k\tau_{u\beta_z}s_{u\alpha}}_{\beta}+\frac{\delta_D}{R_{\alpha_k}R_{\beta_k}}J^{k\tau_{u\beta}\,s_{u\alpha}}\right]
\end{aligned}$$

$$
\begin{aligned}
&- \tilde{C}^k_{12} J^{k\tau_{u_\beta} s_{u_\alpha}} \left(\frac{\partial}{\partial\alpha}\right)_{s_{u_\alpha}} \left(\frac{\partial}{\partial\beta}\right)_{\tau_{u_\beta}} - \tilde{C}^k_{16} J^{k\tau_{u_\beta} s_{u_\alpha}}_{\frac{\beta}{\alpha}} \left(\frac{\partial}{\partial\alpha}\right)_{s_{u_\alpha}} \left(\frac{\partial}{\partial\alpha}\right)_{\tau_{u_\beta}} \\
&- \tilde{C}^k_{26} J^{k\tau_{u_\beta} s_{u_\alpha}}_{\frac{\alpha}{\beta}} \left(\frac{\partial}{\partial\beta}\right)_{s_{u_\alpha}} \left(\frac{\partial}{\partial\beta}\right)_{\tau_{u_\beta}} - \tilde{C}^k_{66} J^{k\tau_{u_\beta} s_{u_\alpha}} \left(\frac{\partial}{\partial\alpha}\right)_{\tau_{u_\beta}} \left(\frac{\partial}{\partial\beta}\right)_{s_{u_\alpha}}
\end{aligned}
$$

$$
\begin{aligned}
K^{k\,\tau_{u_\beta}\,s_{u_\beta}}_{u\,u} &= \tilde{C}^k_{44} \left[J^{k\tau_{u_{\beta_z}} s_{u_{\beta_z}}}_{\alpha\,\beta} - \frac{\delta_D}{R_{\beta_k}} J^{k\tau_{u_{\beta_z}} s_{u_\beta}}_{\alpha} - \frac{\delta_D}{R_{\beta_k}} J^{k\tau_{u_\beta} s_{u_{\beta_z}}}_{\alpha} + \frac{\delta_D}{(R_{\beta_k})^2} J^{k\tau_{u_\beta} s_{u_\beta}}_{\frac{\alpha}{\beta}} \right] \\
&- \tilde{C}^k_{22} J^{k\tau_{u_\beta} s_{u_\beta}}_{\frac{\alpha}{\beta}} \left(\frac{\partial}{\partial\alpha}\right)_{s_{u_\beta}} \left(\frac{\partial}{\partial\alpha}\right)_{\tau_{u_\beta}} - \tilde{C}^k_{26} J^{k\tau_{u_\beta} s_{u_\beta}} \left(\frac{\partial}{\partial\alpha}\right)_{s_{u_\beta}} \left(\frac{\partial}{\partial\beta}\right)_{\tau_{u_\beta}} \\
&- \tilde{C}^k_{26} J^{k\tau_{u_\beta} s_{u_\beta}} \left(\frac{\partial}{\partial\alpha}\right)_{\tau_{u_\beta}} \left(\frac{\partial}{\partial\beta}\right)_{s_{u_\beta}} - \tilde{C}^k_{66} J^{k\tau_{u_\beta} s_{u_\beta}}_{\frac{\beta}{\alpha}} \left(\frac{\partial}{\partial\alpha}\right)_{s_{u_\alpha}} \left(\frac{\partial}{\partial\beta}\right)_{\tau_{u_\beta}}
\end{aligned}
$$

$$
\begin{aligned}
K^{k\,\tau_{u_\beta}\,s_{u_z}}_{u\,u} &= \tilde{C}^k_{44} \left[J^{k\,\tau_{u_{\beta_z}} s_{u_z}}_{\alpha} \left(\frac{\partial}{\partial\beta}\right)_{s_{u_z}} - \frac{1}{R_{\beta_k}} J^{k\tau_{u_\beta} s_{u_z}}_{\frac{\alpha}{\beta}} \left(\frac{\partial}{\partial\beta}\right)_{s_{u_z}} \right] \\
&+ \tilde{C}^k_{45} \left[J^{k\,\tau_{u_{\beta_z}} s_{u_z}}_{\beta} \left(\frac{\partial}{\partial\alpha}\right)_{s_{u_z}} - \frac{1}{R_{\beta_k}} J^{k\tau_{u_\beta} s_{u_z}} \left(\frac{\partial}{\partial\alpha}\right)_{s_{u_z}} \right] \\
&- \tilde{C}^k_{12} \frac{1}{R_{\alpha_k}} J^{k\tau_{u_\beta} s_{u_z}} \left(\frac{\partial}{\partial\beta}\right)_{\tau_{u_\beta}} - \tilde{C}^k_{22} \frac{1}{R_{\beta_k}} J^{k\tau_{u_\beta} s_{u_z}}_{\frac{\alpha}{\beta}} \left(\frac{\partial}{\partial\beta}\right)_{\tau_{u_\beta}} \\
&- \tilde{C}^k_{23} J^{k\tau_{u_\beta} s_{u_{zz}}}_{\alpha} \left(\frac{\partial}{\partial\beta}\right)_{\tau_{u_\beta}} - \tilde{C}^k_{16} \frac{1}{R_{\alpha_k}} J^{k\tau_{u_\beta} s_{u_z}}_{\frac{\beta}{\alpha}} \left(\frac{\partial}{\partial\alpha}\right)_{\tau_{u_\beta}} \\
&- \tilde{C}^k_{26} \frac{1}{R_{\beta_k}} J^{k\tau_{u_\beta} s_{u_z}} \left(\frac{\partial}{\partial\alpha}\right)_{\tau_{u_\beta}} - \tilde{C}^k_{36} J^{k\tau_{u_\alpha} s_{u_z}}_{\beta} \left(\frac{\partial}{\partial\alpha}\right)_{\tau_{u_\alpha}}
\end{aligned}
$$

$$
\begin{aligned}
K^{k\,\tau_{u_z}\,s_{u_\alpha}}_{u\,u} &= \tilde{C}^k_{11} \frac{1}{R_{\alpha_k}} J^{k\tau_{u_z} s_{u_\alpha}}_{\frac{\beta}{\alpha}} \left(\frac{\partial}{\partial\alpha}\right)_{s_{u_\alpha}} + \tilde{C}^k_{12} \frac{1}{R_{\beta_k}} J^{k\tau_{u_z} s_{u_\alpha}} \left(\frac{\partial}{\partial\alpha}\right)_{s_{u_\alpha}} \\
&+ \tilde{C}^k_{13} J^{k\tau_{u_{zz}} s_{u_\alpha}}_{\beta} \left(\frac{\partial}{\partial\alpha}\right)_{s_{u_\alpha}} + \tilde{C}^k_{16} \frac{1}{R_{\alpha_k}} J^{k\tau_{u_z} s_{\alpha z}} \left(\frac{\partial}{\partial\beta}\right)_{s_{u_\alpha}} \\
&+ \tilde{C}^k_{26} \frac{1}{R_{\beta_k}} J^{k\tau_{u_z} s_{\alpha}}_{\frac{\alpha}{\beta}} \left(\frac{\partial}{\partial\beta}\right)_{s_{u_\alpha}} + \tilde{C}^k_{36} J^{k\tau_{u_{zz}} s_{\alpha z}}_{\alpha} \left(\frac{\partial}{\partial\beta}\right)_{s_{u_\alpha}} \\
&- \tilde{C}^k_{45} \left[J^{k\,\tau_{u_z} s_{\alpha z}}_{\alpha} \left(\frac{\partial}{\partial\beta}\right)_{\tau_{u_z}} - \frac{1}{R_{\alpha_k}} J^{k\tau_{u_z} s_{u_\alpha}} \left(\frac{\partial}{\partial\beta}\right)_{\tau_{u_z}} \right] \\
&- \tilde{C}^k_{55} \left[J^{k\,\tau_{u_z} s_{u_{\alpha z}}}_{\beta} \left(\frac{\partial}{\partial\alpha}\right)_{\tau_{u_z}} - \frac{1}{R_{\alpha_k}} J^{k\tau_{u_z} s_{\alpha z}}_{\frac{\beta}{\alpha}} \left(\frac{\partial}{\partial\alpha}\right)_{\tau_{u_z}} \right]
\end{aligned}
$$

$$
K^{k\,\tau_{u_z}\,s_{u_\beta}}_{u\,u} = \tilde{C}^k_{12} \frac{1}{R_{\alpha_k}} J^{k\tau_{u_z} s_{u_\beta}} \left(\frac{\partial}{\partial\beta}\right)_{s_{u_\beta}} + \tilde{C}^k_{22} \frac{1}{R_{\beta_k}} J^{k\tau_{u_z} s_{u_\beta}}_{\frac{\alpha}{\beta}} \left(\frac{\partial}{\partial\beta}\right)_{s_{u_\beta}}
$$

$$
\begin{aligned}
&+ \tilde{C}^k_{23} J^{k\tau_{u_{zz}} s_{u_\beta}}_{\alpha} \left(\frac{\partial}{\partial \beta}\right)_{s_{u_\beta}} + \tilde{C}^k_{16} \frac{1}{R_{\alpha_k}} J^{k\tau_{u_z} s_{u_\beta}}_{\frac{\beta}{\alpha}} \left(\frac{\partial}{\partial \alpha}\right)_{s_{u_\beta}} \\
&+ \tilde{C}^k_{26} \frac{1}{R_{\beta_k}} J^{k\tau_{u_z} s_{u_\beta}} \left(\frac{\partial}{\partial \alpha}\right)_{s_{u_\beta}} + \tilde{C}^k_{36} J^{k\tau_{u_{zz}} s_{u_\beta}}_{\beta} \left(\frac{\partial}{\partial \alpha}\right)_{s_{u_\beta}} \\
&- \tilde{C}^k_{45} \left[J^{k\,\tau_{u_z} s_{u_{\beta z}}}_{\beta} \left(\frac{\partial}{\partial \alpha}\right)_{\tau_{u_z}} - \frac{1}{R_{\beta_k}} J^{k\tau_{u_z} s_{u_\beta}} \left(\frac{\partial}{\partial \alpha}\right)_{\tau_{u_z}} \right] \\
&- \tilde{C}^k_{44} \left[J^{k\,\tau_{u_z} s_{u_{\beta z}}}_{\alpha} \left(\frac{\partial}{\partial \beta}\right)_{\tau_{u_z}} - \frac{1}{R_{\beta_k}} J^{k\tau_{u_z} s_{u_\beta}}_{\frac{\alpha}{\beta}} \left(\frac{\partial}{\partial \beta}\right)_{\tau_{u_z}} \right] \\
K^{k\,\tau_{u_z} s_{u_z}}_{u\,u} &= \tilde{C}^k_{11} \frac{1}{\left(R_{\alpha_k}\right)^2} J^{k\tau_{u_z} s_{u_z}}_{\frac{\beta}{\alpha}} + \tilde{C}^k_{22} \frac{1}{\left(R_{\beta_k}\right)^2} J^{k\tau_{u_z} s_{u_z}}_{\frac{\alpha}{\beta}} + \tilde{C}^k_{33} J^{k\tau_{u_{zz}} s_{u_{zz}}}_{\alpha\,\beta} \\
&+ \tilde{C}^k_{13} \frac{1}{R_{\alpha_k}} \left(J^{k\tau_{u_{zz}} s_{u_z}}_{\beta} + J^{k\tau_{u_z} s_{u_{zz}}}_{\beta} \right) + \tilde{C}^k_{23} \frac{1}{R_{\beta_k}} \left(J^{k\tau_{u_{zz}} s_{u_z}}_{\alpha} + J^{k\tau_{u_z} s_{u_{zz}}}_{\alpha} \right) \\
&- \tilde{C}^k_{44} J^{k\tau_{u_z} s_{u_z}}_{\frac{\alpha}{\beta}} \left(\frac{\partial}{\partial \beta}\right)_{s_{u_z}} \left(\frac{\partial}{\partial \beta}\right)_{\tau_{u_z}} - \tilde{C}^k_{55} J^{k\tau_{u_z} s_{u_z}}_{\frac{\beta}{\alpha}} \left(\frac{\partial}{\partial \alpha}\right)_{s_{u_z}} \left(\frac{\partial}{\partial \alpha}\right)_{\tau_{u_z}} \\
&- \tilde{C}^k_{45} J^{k\tau_{u_z} s_{u_z}} \left(\frac{\partial}{\partial \alpha}\right)_{s_{u_z}} \left(\frac{\partial}{\partial \beta}\right)_{\tau_{u_z}} - \tilde{C}^k_{45} J^{k\tau_{u_z} s_{u_z}} \left(\frac{\partial}{\partial \alpha}\right)_{\tau_{u_z}} \left(\frac{\partial}{\partial \beta}\right)_{s_{u_z}} \\
&+ 2\,\tilde{C}^k_{12} \frac{1}{R_{\alpha_k} R_{\beta_k}} J^{k\tau_{u_z} s_{u_z}}
\end{aligned}
\tag{7.87}
$$

and those of the fundamental primary boundary nucleus $\mathbf{\Pi}_{u\,u}$ are

$$
\begin{aligned}
\Pi^{k\,\tau_{u_\alpha} s_{u_\alpha}}_{u\,u} &= n_\alpha \tilde{C}^k_{11} J^{k\tau_{u_\alpha} s_{u_\alpha}}_{\frac{\beta}{\alpha}} \left(\frac{\partial}{\partial \alpha}\right)_{s_{u_\alpha}} + n_\beta \tilde{C}^k_{66} J^{k\tau_{u_\alpha} s_{u_\alpha}}_{\frac{\alpha}{\beta}} \left(\frac{\partial}{\partial \beta}\right)_{s_{u_\alpha}} \\
&+ n_\beta \tilde{C}^k_{16} J^{k\tau_{u_\alpha} s_{u_\alpha}} \left(\frac{\partial}{\partial \alpha}\right)_{s_{u_\alpha}} + n_\alpha \tilde{C}^k_{16} J^{k\tau_{u_\alpha} s_{u_\alpha}} \left(\frac{\partial}{\partial \beta}\right)_{s_{u_\alpha}} \\
\Pi^{k\,\tau_{u_\alpha} s_{u_\beta}}_{u\,u} &= n_\alpha \tilde{C}^k_{16} J^{k\tau_{u_\alpha} s_{u_\beta}}_{\frac{\beta}{\alpha}} \left(\frac{\partial}{\partial \alpha}\right)_{s_{u_\beta}} + n_\beta \tilde{C}^k_{26} J^{k\tau_{u_\alpha} s_{u_\beta}}_{\frac{\alpha}{\beta}} \left(\frac{\partial}{\partial \beta}\right)_{s_{u_\beta}} \\
&+ n_\beta \tilde{C}^k_{66} J^{k\tau_{u_\alpha} s_{u_\beta}} \left(\frac{\partial}{\partial \alpha}\right)_{s_{u_\beta}} + n_\alpha \tilde{C}^k_{12} J^{k\tau_{u_\alpha} s_{u_\beta}} \left(\frac{\partial}{\partial \beta}\right)_{s_{u_\beta}} \\
\Pi^{k\,\tau_{u_\alpha} s_{u_z}}_{u\,u} &= n_\alpha \frac{1}{R_{\alpha_k}} \tilde{C}^k_{11} J^{k\tau_{u_\alpha} s_{u_z}}_{\frac{\beta}{\alpha}} + n_\alpha \frac{1}{R_{\beta_k}} \tilde{C}^k_{12} J^{k\tau_{u_\alpha} s_{u_z}} + n_\alpha \tilde{C}^k_{13} J^{k\tau_{u_\alpha} s_{u_{zz}}}_{\beta} \\
&+ n_\beta \frac{1}{R_{\alpha_k}} \tilde{C}^k_{16} J^{k\tau_{u_\alpha} s_{u_z}}_{\frac{\alpha}{\beta}} + n_\beta \frac{1}{R_{\beta_k}} \tilde{C}^k_{26} J^{k\tau_{u_\alpha} s_{u_z}}_{\frac{\alpha}{\beta}} + n_\beta \tilde{C}^k_{36} J^{k\tau_{u_\alpha} s_{u_{zz}}}_{\alpha} \\
\Pi^{k\,\tau_{u_\beta} s_{u_\alpha}}_{u\,u} &= n_\alpha \tilde{C}^k_{16} J^{k\tau_{u_\beta} s_{u_\alpha}}_{\frac{\beta}{\alpha}} \left(\frac{\partial}{\partial \alpha}\right)_{s_{u_\alpha}} + n_\beta \tilde{C}^k_{26} J^{k\tau_{u_\beta} s_{u_\alpha}}_{\frac{\alpha}{\beta}} \left(\frac{\partial}{\partial \beta}\right)_{s_{u_\alpha}} \\
&+ n_\beta \tilde{C}^k_{12} J^{k\tau_{u_\beta} s_{u_\alpha}} \left(\frac{\partial}{\partial \alpha}\right)_{s_{u_\alpha}} + n_\alpha \tilde{C}^k_{66} J^{k\tau_{u_\beta} s_{u_\alpha}} \left(\frac{\partial}{\partial \beta}\right)_{s_{u_\alpha}}
\end{aligned}
$$

$$
\begin{aligned}
\Pi_{u\,u}^{k\,\tau_{u_\beta}\,s_{u_\beta}} &= n_\alpha\,\tilde{C}_{66}^k J_{\frac{\beta}{\alpha}}^{k\tau_{u_\beta} s_{u_\beta}}\left(\frac{\partial}{\partial\alpha}\right)_{s_{u_\beta}} + n_\beta\,\tilde{C}_{22}^k J_{\frac{\alpha}{\beta}}^{k\tau_{u_\beta} s_{u_\beta}}\left(\frac{\partial}{\partial\beta}\right)_{s_{u_\beta}}\\
&\quad + n_\beta\,\tilde{C}_{26}^k J^{k\tau_{u_\beta} s_{u_\beta}}\left(\frac{\partial}{\partial\alpha}\right)_{s_{u_\beta}} + n_\alpha\,\tilde{C}_{26}^k J^{k\tau_{u_\beta} s_{u_\beta}}\left(\frac{\partial}{\partial\beta}\right)_{s_{u_\beta}}\\
\Pi_{u\,u}^{k\,\tau_{u_\beta}\,s_{u_\alpha}} &= n_\alpha\,\frac{1}{R_{\alpha_k}}\,\tilde{C}_{16}^k J_{\frac{\beta}{\alpha}}^{k\tau_{u_\beta} s_{u_z}} + n_\alpha\,\frac{1}{R_{\beta_k}}\,\tilde{C}_{26}^k J^{k\tau_{u_\beta} s_{u_z}} + n_\alpha\,\tilde{C}_{36}^k J_{\beta}^{k\tau_{u_\beta} s_{u_{z_z}}}\\
&\quad + n_\beta\,\frac{1}{R_{\alpha_k}}\,\tilde{C}_{12}^k J_{\frac{\alpha}{\beta}}^{k\tau_{u_\beta} s_{u_z}} + n_\beta\,\frac{1}{R_{\beta_k}}\,\tilde{C}_{22}^k J_{\frac{\alpha}{\beta}}^{k\tau_{u_\beta} s_{u_z}} + n_\beta\,\tilde{C}_{23}^k J_{\alpha}^{k\tau_{u_\beta} s_{u_{z_z}}}\\
\Pi_{u\,u}^{k\,\tau_{u_z}\,s_{u_\alpha}} &= -n_\alpha\,\frac{1}{R_{\alpha_k}}\,\tilde{C}_{55}^k J_{\frac{\beta}{\alpha}}^{k\tau_{u_z} s_{u_\alpha}} + n_\alpha\,\tilde{C}_{55}^k J_{\beta}^{k\tau_{u_\alpha} s_{u_{\alpha_z}}} - n_\beta\,\frac{1}{R_{\alpha_k}}\,\tilde{C}_{45}^k J^{k\tau_{u_z} s_{u_\alpha}}\\
&\quad + n_\beta\,\tilde{C}_{45}^k J_{\alpha}^{k\tau_{u_z} s_{u_{\alpha_z}}}\\
\Pi_{u\,u}^{k\,\tau_{u_z}\,s_{u_\beta}} &= -n_\alpha\,\frac{1}{R_{\beta_k}}\,\tilde{C}_{45}^k J^{k\tau_{u_z} s_{u_\beta}} + n_\alpha\,\tilde{C}_{45}^k J_{\beta}^{k\tau_{u_\alpha} s_{u_{\beta_z}}} - n_\beta\,\frac{1}{R_{\beta_k}}\,\tilde{C}_{44}^k J_{\frac{\alpha}{\beta}}^{k\tau_{u_z} s_{\beta}}\\
&\quad + n_\beta\,\tilde{C}_{44}^k J_{\alpha}^{k\tau_{u_z} s_{u_\beta}}\\
\Pi_{u\,u}^{k\,\tau_{u_z}\,s_{u_z}} &= n_\alpha\,\tilde{C}_{55}^k J_{\frac{\beta}{\alpha}}^{k\tau_{u_z} s_{u_z}}\left(\frac{\partial}{\partial\alpha}\right)_{s_{u_z}} + n_\beta\,\tilde{C}_{44}^k J_{\frac{\alpha}{\beta}}^{k\tau_{u_z} s_{u_z}}\left(\frac{\partial}{\partial\beta}\right)_{s_{u_z}}\\
&\quad + n_\beta\,\tilde{C}_{45}^k J^{k\tau_{u_z} s_{u_z}}\left(\frac{\partial}{\partial\alpha}\right)_{s_{u_z}} + n_\alpha\,\tilde{C}_{45}^k J^{k\tau_{u_z} s_{u_z}}\left(\frac{\partial}{\partial\beta}\right)_{s_{u_z}}
\end{aligned}
\tag{7.88}
$$

Finally the fundamental primary mass nucleus $\mathbf{M}_{u\,u}$ components can be written as:

$$
\begin{array}{lll}
M_{u\,u}^{k\,\tau_{u_\alpha}\,s_{u_\alpha}} = \rho^k J^{k\,\tau_{u_\alpha}\,s_{u_\alpha}} & M_{u\,u}^{k\,\tau_{u_\alpha}\,s_{u_\beta}} = 0 & M_{u\,u}^{k\,\tau_{u_\alpha}\,s_{u_z}} = 0\\
M_{u\,u}^{k\,\tau_{u_\beta}\,s_{u_\alpha}} = 0 & M_{u\,u}^{k\,\tau_{u_\beta}\,s_{u_\beta}} = \rho^k J^{k\,\tau_{u_\beta}\,s_{u_\beta}} & M_{u\,u}^{k\,\tau_{u_\beta}\,s_{u_z}} = 0\\
M_{u\,u}^{k\,\tau_{u_z}\,s_{u_\alpha}} = 0 & M_{u\,u}^{k\,\tau_{u_z}\,s_{u_\beta}} = 0 & M_{u\,u}^{k\,\tau_{u_z}\,s_{u_z}} = \rho^k J^{k\,\tau_{u_z}\,s_{u_z}}
\end{array}
\tag{7.89}
$$

7.5.3 RMVT-based governing differential equations of doubly-curved shells

By applying the Gauss theorem in Eq. (7.74) at RMVT the governing differential equations and natural boundary conditions for the mixed case are derived and given in the following

$$
\sum_{k=1}^{N_l}\int_{\Omega_k}\int_{A^k}\delta\mathbf{u}_\tau\Big\{\Big[-\left(\mathbf{F}_{\tau_u}\right)^T\left(\mathbf{D}_p\right)^T + \left(\mathbf{F}_{\tau_u}\right)^T\left(\mathbf{A}_p\right)^T\Big]
\times\Big[\tilde{\mathbf{C}}_{pp}^k\mathbf{D}_p\left(\mathbf{F}_{s_u}\right) + \tilde{\mathbf{C}}_{pp}^k\mathbf{A}_p\left(\mathbf{F}_{s_u}\right)\Big]\Big\}\mathbf{u}_s +
$$

$$
\begin{aligned}
&\delta \mathbf{u}_\tau \Big\{ \Big[-(\mathbf{F}_{\tau_u})^T (\mathbf{D}_p)^T + (\mathbf{F}_{\tau_u})^T (\mathbf{A}_p)^T \Big] \Big[\tilde{\mathbf{C}}^k_{pn} (\mathbf{F}_{s_\sigma}) \Big] + \\
&\quad \Big[-(\mathbf{F}_{\tau_u})^T (\mathbf{D}_{np})^T + (\mathbf{F}_{\tau_u})^T (\delta_D \mathbf{A}_n)^T + (\mathbf{F}_{\tau_u})^T (\mathbf{D}_{nz})^T \Big] \times \\
&\quad (\mathbf{F}_{s_\sigma}) \Big\} \boldsymbol{\sigma}^k_{n_s} + \\
&\delta \boldsymbol{\sigma}^k_{n_\tau} \Big\{ \Big(\mathbf{F}_{\tau_\sigma}\Big)^T \big[-(\mathbf{F}_{s_u}) (\mathbf{D}_{np}) + (\mathbf{F}_{s_u}) (\delta_D \mathbf{A}_n) + (\mathbf{F}_{s_u}) (\mathbf{D}_{nz}) \big] + \\
&\quad \Big[-(\mathbf{F}_{\tau_\sigma})^T (\mathbf{D}_p)^T + (\mathbf{F}_{s_u})^T (\mathbf{A}_p)^T \Big] \Big[\tilde{\mathbf{C}}^k_{pn} (\mathbf{F}_{s_u}) \Big] \Big\} \mathbf{u}_s + \\
&\delta \boldsymbol{\sigma}^k_{n_\tau} \Big[\tilde{\mathbf{C}}^k_{nn} (\mathbf{F}_{\tau_\sigma})^T (\mathbf{F}_{s_\sigma}) \Big] \boldsymbol{\sigma}^k_{n_s} \, d\Omega_k dz + \\
\sum_{k=1}^{N_l} \int_{\Gamma_k} \int_{A^k} &\delta \mathbf{u}_\tau \Big\{ \Big[(\mathbf{F}_{\tau_u})^T (\boldsymbol{I}_p)^T \Big] \Big[\tilde{\mathbf{C}}^k_{pp} \mathbf{D}_p (\mathbf{F}_{s_u}) + \tilde{\mathbf{C}}^k_{pp} \mathbf{A}_p (\mathbf{F}_{s_u}) \Big] \Big\} \mathbf{u}_s + \\
&\delta \mathbf{u}_\tau \Big\{ \Big[(\mathbf{F}_{\tau_u})^T (\boldsymbol{I}_{np})^T \Big] \times \\
&\quad (\mathbf{F}_{s_\sigma}) + \Big[(\mathbf{F}_\tau)^T (\boldsymbol{I}_p)^T \Big] \tilde{\mathbf{C}}^k_{pn} (\mathbf{F}_{s_\sigma}) \Big\} \boldsymbol{\sigma}^k_{n_s} \, d\Omega_k dz = - \\
\sum_{k=1}^{N_l} \int_{\Omega_k} \int_{A^k} &\delta \mathbf{u}_\tau \Big[\rho^k (\mathbf{F}_{\tau_u})^T (\mathbf{F}_{s_\sigma}) \Big] \ddot{\mathbf{u}}_s \, d\Omega_k dz
\end{aligned}
\tag{7.90}
$$

and in compact form:

$$
\begin{aligned}
\delta \mathbf{u}_\tau : &\quad \mathbf{K}^{k\tau s}_{u\,u} \mathbf{u}_s + \mathbf{K}^{k\tau s}_{u\,\sigma} \boldsymbol{\sigma}_{n_s} = -\mathbf{M}^{k\tau s}_{u\,u} \ddot{\mathbf{u}}_s \\
\delta \boldsymbol{\sigma}_{n_\tau} : &\quad \mathbf{K}^{k\tau s}_{\sigma\,u} \mathbf{u}_s + \mathbf{K}^{k\tau s}_{\sigma\,\sigma} \boldsymbol{\sigma}_{n_s} = 0 \\
\Gamma^m_k : &\quad \boldsymbol{\Pi}^{k\tau s}_u \mathbf{u}_s + \boldsymbol{\Pi}^{k\tau s}_\sigma \boldsymbol{\sigma}_{n_s} = \boldsymbol{\Pi}^{k\tau s}_u \bar{\mathbf{u}}_s + \boldsymbol{\Pi}^{k\tau s}_\sigma \bar{\boldsymbol{\sigma}}_{n_s} \qquad \Gamma^g_k : \mathbf{u}_s = \bar{\mathbf{u}}_s
\end{aligned}
\tag{7.91}
$$

where

$$
\begin{aligned}
\mathbf{K}^{k\tau s}_{u\,u} = \int_{A^k} &\Big\{ \Big[-(\mathbf{F}_{\tau_u})^T (\mathbf{D}_p)^T + (\mathbf{F}_{\tau_u})^T (\mathbf{A}_p)^T \Big] \times \\
&\Big[\tilde{\mathbf{C}}^k_{pp} \mathbf{D}_p (\mathbf{F}_{s_u}) + \tilde{\mathbf{C}}^k_{pp} \mathbf{A}_p (\mathbf{F}_{s_u}) \Big] \Big\} H^k_\alpha H^k_\beta \, \mathrm{d}z \\
\mathbf{K}^{k\tau s}_{u\,\sigma} = \int_{A^k} &\Big\{ \Big[-(\mathbf{F}_{\tau_u})^T (\mathbf{D}_p)^T + (\mathbf{F}_{\tau_u})^T (\mathbf{A}_p)^T \Big] \times \\
&\Big[\tilde{\mathbf{C}}^k_{pn} (\mathbf{F}_{s_\sigma}) \Big] + \\
&\Big[-(\mathbf{F}_{\tau_u})^T (\mathbf{D}_{np})^T + (\mathbf{F}_{\tau_u})^T (\delta_D \mathbf{A}_n)^T + (\mathbf{F}_{\tau_u})^T (\mathbf{D}_{nz})^T \Big] \times \\
&\Big(\mathbf{F}_{s_\sigma}\Big) \Big\} H^k_\alpha H^k_\beta \, \mathrm{d}z \\
\mathbf{K}^{k\tau s}_{\sigma\,u} = \int_{A^k} &\Big\{ \Big(\mathbf{F}_{\tau_\sigma}\Big)^T \Big[-(\mathbf{F}_{s_u}) (\mathbf{D}_{np}) + (\mathbf{F}_{s_u}) (\delta_D \mathbf{A}_n) + (\mathbf{F}_{s_u}) (\mathbf{D}_{nz}) \Big] + \\
&\Big[-(\mathbf{F}_{\tau_\sigma})^T (\mathbf{D}_p)^T + (\mathbf{F}_{\tau_\sigma})^T (\mathbf{A}_p)^T \Big] \times
\end{aligned}
$$

$$
\begin{aligned}
&\left[\tilde{\mathbf{C}}^k_{pn}\left(\mathbf{F}_{s_u}\right)\right]\Big\} H^k_\alpha H^k_\beta \,\mathrm{d}z \\
\mathbf{K}^{k\tau s}_{\sigma\sigma} &= \int_{A^k}\Big\{\left[\tilde{\mathbf{C}}^k_{nn}\left(\mathbf{F}_{\tau_\sigma}\right)^T\left(\mathbf{F}_{s_\sigma}\right)\right]\Big\} H^k_\alpha H^k_\beta \,\mathrm{d}z \\
\mathbf{\Pi}^{k\tau s}_{u} &= \int_{A^k}\Big\{\left[\left(\mathbf{F}_{\tau_u}\right)^T\left(\boldsymbol{I}_p\right)^T\right]\times \\
&\left[\tilde{\mathbf{C}}^k_{pp}\mathbf{D}_p\left(\mathbf{F}_{\tau_u}\right)+\tilde{\mathbf{C}}^k_{pp}\mathbf{A}_p\left(\mathbf{F}_{\tau_u}\right)\right]\Big\} H^k_\alpha H^k_\beta \,\mathrm{d}z \\
\mathbf{\Pi}^{k\tau s}_{\sigma} &= \int_{A^k}\Big\{\left[\left(\mathbf{F}_{\tau_\sigma}\right)^T\left(\boldsymbol{I}_{np}\right)^T\right]\left(\mathbf{F}_{s_\sigma}\right)+ \\
&\left[\left(\mathbf{F}_{\tau_\sigma}\right)^T\left(\boldsymbol{I}_p\right)^T\right]\tilde{\mathbf{C}}^k_{pn}\left(\mathbf{F}_{s_\sigma}\right)\Big\} H^k_\alpha H^k_\beta \,\mathrm{d}z
\end{aligned}
\tag{7.92}
$$

Similarly to the PVD case once performed the matrix calculus in Eq. (7.92) the governing differential equations are derived in their explicit form.

7.6 COUPLED AND UNCOUPLED THERMOELASTIC EQUATIONS

Both coupled and uncoupled thermoelastic equations are derived by using the variational statements provided in Chapter 3 [10–14]. In particular, for the sake of conciseness the governing equations are derived in their weak-form, the strong form can be easily obtained by using the Gauss theorem given in Eq. (7.74) or taken from the papers available in literature.

7.6.1 Coupled thermoelastic variational formulation

A fully coupled thermoelastic variational statement is obtained by virtue of an extension of the principle of the virtual displacement (PVD), which takes into account the internal thermal virtual work. By considering a laminate of N_l layers and the volume V_k for each layer k as an integral on the in-plane surface Ω_k and an integral in the thickness direction domain A_k, the variational statement can be written as

$$
\begin{aligned}
&\sum_{k=1}^{N_l}\int_{\Omega_k}\int_{A_k}\left(\delta\boldsymbol{\epsilon}^{k^T}_{pG}\boldsymbol{\sigma}^k_{pC}+\delta\boldsymbol{\epsilon}^{k^T}_{nG}\boldsymbol{\sigma}^k_{nC}-\delta\theta^k\eta^k_C-\delta\boldsymbol{\vartheta}^{k^T}_{pG}\boldsymbol{h}^k_{pC}-\delta\boldsymbol{\vartheta}^{k^T}_{nG}\boldsymbol{h}^k_{nC}\right)\mathrm{d}\Omega_k\,\mathrm{d}z \\
&=\sum_{k=1}^{N_l}\delta L^k_{ext}-\sum_{k=1}^{N_l}\delta L^k_{F_{ine}}
\end{aligned}
\tag{7.93}
$$

where the δL^k_{ext} and $\delta L^k_{F_{ine}}$ are the external and inertial virtual works at the k-layer level, respectively. In the case of free vibration analysis, in the variational statement in Eq. (7.93) the terms $\left(\delta \boldsymbol{\vartheta}_{pG}^{k^T} \boldsymbol{h}_{pC}^k\right)$ and $\left(\delta \boldsymbol{\vartheta}_{nG}^{k^T} \boldsymbol{h}_{nC}^k\right)$ are neglected because a gradient of temperature variation does not exist. Furthermore, the external virtual work δL_e^k is not accounted for because there are not external forces applied. Under these hypotheses the coupled thermoelastic variational statement becomes:

$$\sum_{k=1}^{N_l} \int_{\Omega_k} \int_{A_k} \left(\delta \boldsymbol{\varepsilon}_{pG}^{k^T} \boldsymbol{\sigma}_{pC}^k + \delta \boldsymbol{\varepsilon}_{nG}^{k^T} \boldsymbol{\sigma}_{nC}^k - \delta \theta^k \eta_C^k\right) \mathrm{d}\Omega_k \, \mathrm{d}z = -\sum_{k=1}^{N_l} \delta L_{F_{ine}}^k \quad (7.94)$$

In Eqs. (7.93) and (7.94) the first variation represents a minimization condition with respect to the primary variables, namely, displacements and temperature unknowns. Displacements can have a multilayer or layer level kinematics description whereas the temperature is always defined using a layer-wise approach. Therefore, when ESL models are used a mixed assembly procedure must be employed. On the contrary, when LW models are undertaken the usual assembly procedure for the layer-wise case can be easily and straightforwardly applied. For the sake of completeness, both procedures are carefully described in Ref. [14]. The weak form of the governing equations can be derived by substituting the thermoelastic constitutive equations given in chapter 6. Substituting the latter in Eq. (7.94) the explicit expressions of the internal and inertial virtual works in terms of unknown coefficients and trial functions are obtained as follows

$$\begin{aligned}
\delta L_{int}^k = & \int_{\Omega^k} \int_{A^k} \delta \mathbf{U}_{\tau i}^{k^T} \left[\mathbf{D}_p (\mathbf{F}_\tau \boldsymbol{\Psi}_i)\right]^T \tilde{\mathbf{C}}_{pp}^k \mathbf{D}_p (\mathbf{F}_s \boldsymbol{\Psi}_j) \mathbf{U}_{sj}^k \, d\Omega^k \, dz \\
& + \int_{\Omega^k} \int_{A^k} \delta \mathbf{U}_{\tau i}^{k^T} \left[\mathbf{D}_p (\mathbf{F}_\tau \boldsymbol{\Psi}_i)\right]^T \tilde{\mathbf{C}}_{pn}^k \mathbf{D}_{np} (\mathbf{F}_s \boldsymbol{\Psi}_j) \mathbf{U}_{sj}^k \, d\Omega^k \, dz \\
& + \int_{\Omega^k} \int_{A^k} \delta \mathbf{U}_{\tau i}^{k^T} \left[\mathbf{D}_p (\mathbf{F}_\tau \boldsymbol{\Psi}_i)\right]^T \tilde{\mathbf{C}}_{pn}^k \mathbf{D}_{nz} (\mathbf{F}_s \boldsymbol{\Psi}_j) \mathbf{U}_{sj}^k \, d\Omega^k \, dz \\
& + \int_{\Omega^k} \int_{A^k} \delta \mathbf{U}_{\tau i}^{k^T} \left[\mathbf{D}_{np} (\mathbf{F}_\tau \boldsymbol{\Psi}_i)\right]^T \tilde{\mathbf{C}}_{np}^k \mathbf{D}_p (\mathbf{F}_s \boldsymbol{\Psi}_j) \mathbf{U}_{sj}^k \, d\Omega^k \, dz \\
& + \int_{\Omega^k} \int_{A^k} \delta \mathbf{U}_{\tau i}^{k^T} \left[\mathbf{D}_{np} (\mathbf{F}_\tau \boldsymbol{\Psi}_i)\right]^T \tilde{\mathbf{C}}_{nn}^k \mathbf{D}_{np} (\mathbf{F}_s \boldsymbol{\Psi}_j) \mathbf{U}_{sj}^k \, d\Omega^k \, dz \\
& + \int_{\Omega^k} \int_{A^k} \delta \mathbf{U}_{\tau i}^{k^T} \left[\mathbf{D}_{np} (\mathbf{F}_\tau \boldsymbol{\Psi}_i)\right]^T \tilde{\mathbf{C}}_{nn}^k \mathbf{D}_{nz} (\mathbf{F}_s \boldsymbol{\Psi}_j) \mathbf{U}_{sj}^k \, d\Omega^k \, dz \\
& + \int_{\Omega^k} \int_{A^k} \delta \mathbf{U}_{\tau i}^{k^T} \left[\mathbf{D}_{nz} (\mathbf{F}_\tau \boldsymbol{\Psi}_i)\right]^T \tilde{\mathbf{C}}_{np}^k \mathbf{D}_p (\mathbf{F}_s \boldsymbol{\Psi}_j) \mathbf{U}_{sj}^k \, d\Omega^k \, dz \\
& + \int_{\Omega^k} \int_{A^k} \delta \mathbf{U}_{\tau i}^{k^T} \left[\mathbf{D}_{nz} (\mathbf{F}_\tau \boldsymbol{\Psi}_i)\right]^T \tilde{\mathbf{C}}_{nn}^k \mathbf{D}_{np} (\mathbf{F}_s \boldsymbol{\Psi}_j) \mathbf{U}_{sj}^k \, d\Omega^k \, dz \\
& + \int_{\Omega^k} \int_{A^k} \delta \mathbf{U}_{\tau i}^{k^T} \left[\mathbf{D}_{nz} (\mathbf{F}_\tau \boldsymbol{\Psi}_i)\right]^T \tilde{\mathbf{C}}_{nn}^k \mathbf{D}_{nz} (\mathbf{F}_s \boldsymbol{\Psi}_j) \mathbf{U}_{sj}^k \, d\Omega^k \, dz
\end{aligned}$$

$$
\begin{aligned}
&+\int_{\Omega^k}\int_{A^k} \delta\mathbf{U}_{\tau i}^{k^T}\left[\mathbf{D}_p\left(\mathbf{F}_\tau\,\boldsymbol{\Psi}_i\right)\right]^T\tilde{\boldsymbol{\lambda}}_p^k\left(F_{s_{u_z}}\,\phi_{T_j}\right)\Theta_{sj}^k\,d\Omega^k\,dz\\
&+\int_{\Omega^k}\int_{A^k} \delta\mathbf{U}_{\tau i}^{k^T}\left[\mathbf{D}_{np}\left(\mathbf{F}_\tau\,\boldsymbol{\Psi}_i\right)\right]^T\tilde{\boldsymbol{\lambda}}_n^k\left(F_{s_{u_z}}\,\phi_{T_j}\right)\Theta_{sj}^k\,d\Omega^k\,dz\\
&+\int_{\Omega^k}\int_{A^k} \delta\mathbf{U}_{\tau i}^{k^T}\left[\mathbf{D}_{nz}\left(\mathbf{F}_\tau\,\boldsymbol{\Psi}_i\right)\right]^T\tilde{\boldsymbol{\lambda}}_n^k\left(F_{s_{u_z}}\,\phi_{T_j}\right)\Theta_{sj}^k\,d\Omega^k\,dz\\
&+\int_{\Omega^k}\int_{A^k} \delta\Theta_{\tau j}^k\left(F_{\tau_{u_z}}\,\phi_{T_i}\right)\left[\left(\boldsymbol{\lambda}_p^k\right)^T\mathbf{D}_p\left(\mathbf{F}_s\,\boldsymbol{\Psi}_j\right)\right]\mathbf{U}_{sj}^k\,d\Omega^k\,dz\\
&+\int_{\Omega^k}\int_{A^k} \delta\Theta_{\tau j}^k\left(F_{\tau_{u_z}}\,\phi_{T_i}\right)\left[\left(\boldsymbol{\lambda}_n^k\right)^T\mathbf{D}_{np}\left(\mathbf{F}_s\,\boldsymbol{\Psi}_j\right)\right]\mathbf{U}_{sj}^k\,d\Omega^k\,dz\\
&+\int_{\Omega^k}\int_{A^k} \delta\Theta_{\tau j}^k\left(F_{\tau_{u_z}}\,\phi_{T_i}\right)\left[\left(\boldsymbol{\lambda}_n^k\right)^T\mathbf{D}_{nz}\left(\mathbf{F}_s\,\boldsymbol{\Psi}_j\right)\right]\mathbf{U}_{sj}^k\,d\Omega^k\,dz\\
&+\int_{\Omega^k}\int_{A^k} \delta\Theta_{\tau j}^k\left[\left(F_{\tau_{u_z}}\,\phi_{T_i}\right)\chi\,\left(F_{s_{u_z}}\,\phi_{T_j}\right)\right]\Theta_{sj}^k\,d\Omega^k\,dz\\
\delta L_{F_{in}}^k &= \int_{\Omega^k}\int_{A^k} \delta\mathbf{U}_{\tau i}^{k^T}\rho^k\left[\left(\mathbf{F}_\tau\boldsymbol{\Psi}_i\right)^T\left(\mathbf{F}_s\boldsymbol{\Psi}_j\right)\right]\ddot{\mathbf{U}}_{sj}\,d\Omega^k\,dz
\end{aligned}
\tag{7.95}
$$

The general expression of the internal virtual work is given as

$$
\begin{aligned}
\delta L_{int}^k = \delta\mathbf{U}_{\tau i}^{k^T}\boldsymbol{K}_{uu}^{k\tau sij}\mathbf{U}_{sj}^k + \delta\mathbf{U}_{\tau i}^{k^T}\boldsymbol{K}_{u\theta}^{k\tau sij}\Theta_{sj}^k +\\
\delta\Theta_{\tau i}^{k^T}\boldsymbol{K}_{\theta u}^{k\tau sij}\mathbf{U}_{sj}^k + \delta\Theta_{\tau i}^{k^T}\boldsymbol{K}_{\theta\theta}^{k\tau sij}\Theta_{sj}^k
\end{aligned}
\tag{7.96}
$$

and for the inertial virtual work as

$$
\delta L_{F_{in}}^k = \delta\mathbf{U}_{\tau i}^{k^T}\mathbf{M}_{uu}^{k\tau sij}\ddot{\mathbf{U}}_{sj}^k \tag{7.97}
$$

Comparing Eq. (7.96) with Eq. (7.95) the final form of the fundamental primary nuclei is obtained as follows

$$
\begin{aligned}
\mathbf{K}_{uu}^{k\tau sij} = \int_{\Omega^k}\int_{A^k}\Bigg\{&\left[\mathbf{D}_p\left(\mathbf{F}_\tau\,\boldsymbol{\Psi}_i\right)\right]^T\times\\
&\left[\tilde{\mathbf{C}}_{pp}^k\mathbf{D}_p\left(\mathbf{F}_s\boldsymbol{\Psi}_j\right)+\tilde{\mathbf{C}}_{pn}^k\mathbf{D}_{np}\left(\mathbf{F}_s\boldsymbol{\Psi}_j\right)+\tilde{\mathbf{C}}_{pn}^k\mathbf{D}_{nz}\left(\mathbf{F}_s\boldsymbol{\Psi}_j\right)\right]\\
&+\left[\mathbf{D}_{np}\left(\mathbf{F}_\tau\,\boldsymbol{\Psi}_i\right)\right]^T\times\\
&\left[\tilde{\mathbf{C}}_{np}^k\mathbf{D}_p\left(\mathbf{F}_s\boldsymbol{\Psi}_j\right)+\tilde{\mathbf{C}}_{nn}^k\mathbf{D}_{np}\left(\mathbf{F}_s\boldsymbol{\Psi}_j\right)+\tilde{\mathbf{C}}_{nn}^k\mathbf{D}_{nz}\left(\mathbf{F}_s\boldsymbol{\Psi}_j\right)\right]\\
&+\left[\mathbf{D}_{nz}\left(\mathbf{F}_\tau\,\boldsymbol{\Psi}_i\right)\right]^T\times\\
&\Big[\tilde{\mathbf{C}}_{np}^k\mathbf{D}_p\left(\mathbf{F}_s\boldsymbol{\Psi}_j\right)+\tilde{\mathbf{C}}_{nn}^k\mathbf{D}_{np}\left(\mathbf{F}_s\boldsymbol{\Psi}_j\right)+\\
&\tilde{\mathbf{C}}_{nn}^k\mathbf{D}_{nz}\left(\mathbf{F}_s\boldsymbol{\Psi}_j\right)\Big]\Bigg\}\,\mathrm{d}\Omega^k\,\mathrm{d}z
\end{aligned}
$$

$$
\begin{aligned}
\mathbf{K}_{u\,\theta}^{k\tau sij} = \int_{\Omega^k}\int_{A^k} \Big\{ & \left[\mathbf{D}_p\left(\mathbf{F}_\tau\,\boldsymbol{\Psi}_i\right)\right]^T \tilde{\boldsymbol{\lambda}}_p^k \left(F_{s_{u_z}}\,\phi_{T_j}\right) \\
& + \left[\mathbf{D}_{np}\left(\mathbf{F}_\tau\,\boldsymbol{\Psi}_i\right)\right]^T \tilde{\boldsymbol{\lambda}}_n^k \left(F_{s_{u_z}}\,\phi_{T_j}\right) \\
& + \left[\mathbf{D}_{nz}\left(\mathbf{F}_\tau\,\boldsymbol{\Psi}_i\right)\right]^T \tilde{\boldsymbol{\lambda}}_n^k \left(F_{s_{u_z}}\,\phi_{T_j}\right) \Big\}\, \mathrm{d}\Omega^k\,\mathrm{d}z \\
\mathbf{K}_{\theta\,u}^{k\tau sij} = \int_{\Omega^k}\int_{A^k} \Big\{ & \left(F_{\tau_{u_z}}\,\phi_{T_i}\right)\left[\left(\boldsymbol{\lambda}_p^k\right)^T \mathbf{D}_p\left(\mathbf{F}_s\,\boldsymbol{\Psi}_j\right)\right] \\
& + \left(F_{\tau_{u_z}}\,\phi_{T_i}\right)\left[\left(\boldsymbol{\lambda}_n^k\right)^T \mathbf{D}_{np}\left(\mathbf{F}_s\,\boldsymbol{\Psi}_j\right)\right] \\
& + \left(F_{\tau_{u_z}}\,\phi_{T_i}\right)\left[\left(\boldsymbol{\lambda}_n^k\right)^T \mathbf{D}_{nz}\left(\mathbf{F}_s\,\boldsymbol{\Psi}_j\right)\right] \Big\}\, \mathrm{d}\Omega^k\,\mathrm{d}z \\
\mathbf{K}_{\theta\,\theta}^{k\tau sij} = \int_{\Omega^k}\int_{A^k} \Big\{ & \left[\left(F_{\tau_{u_z}}\,\phi_{T_i}\right)\,\chi\,\left(F_{s_{u_z}}\,\phi_{T_j}\right)\right] \Big\}\, \mathrm{d}\Omega^k\,\mathrm{d}z \\
\mathbf{M}_{u\,u}^{k\tau sij} = \int_{\Omega^k}\int_{A^k} \Big\{ & \rho^k \left[\left(\mathbf{F}_\tau\,\boldsymbol{\Psi}_i\right)^T \left(\mathbf{F}_s\,\boldsymbol{\Psi}_j\right)\right] \Big\}\, \mathrm{d}\Omega^k\,\mathrm{d}z
\end{aligned}
\tag{7.98}
$$

After performing the matrix product in Eq. (7.98) the fundamental secondary nuclei are derived. Each fundamental secondary nucleus of the UF has to be expanded individually according to the expansion order chosen for the displacement components and the number of Ritz terms introduced in the expansion. When the expansions have been performed then the fundamental secondary nuclei are arranged to generate the fundamental primary nuclei related at the particular theory employed. The discrete form of the governing equations is finally obtained in terms of fundamental primary nuclei as

$$
\begin{aligned}
\delta\mathbf{U}_{\tau i}^{k^T}: &\quad \mathbf{K}_{u\,u}^{k\tau sij}\,\mathbf{U}_{sj}^k + \mathbf{K}_{u\,\theta}^{k\tau sij}\,\boldsymbol{\Theta}_{sj}^k = -\mathbf{M}_{u\,u}^{k\tau sij}\,\ddot{\mathbf{U}}_{sj}^k \\
\delta\boldsymbol{\Theta}_{\tau i}^{k^T}: &\quad \mathbf{K}_{\theta\,u}^{k\tau sij}\,\mathbf{U}_{sj}^k + \mathbf{K}_{\theta\,\theta}^{k\tau sij}\,\boldsymbol{\Theta}_{sj}^k = 0
\end{aligned}
\tag{7.99}
$$

The strong-form of Eqs. (7.99) can be derived by using the Gauss theorem given in Eq. (7.74).

7.6.2 Uncoupled thermoelastic variational formulation

In the case of uncoupled formulation the temperature is given as an external load and it is no more a primary variable in the analysis. An example of a typical case of uncoupled thermoelastic formulation is the thermoelastic stability analysis. In the latter case the variational statement can be given as

$$\sum_{k=1}^{N_l} \int_{\Omega^k} \int_{A^k} \left(\delta \boldsymbol{\varepsilon}_{pG}^{k\ T} \boldsymbol{\sigma}_{pC}^{k} + \delta \boldsymbol{\varepsilon}_{nG}^{k\ T} \boldsymbol{\sigma}_{nC}^{k} \right) \mathrm{d}\Omega^k \, \mathrm{d}z = \sum_{k=1}^{N_l} \delta L_{ext}^{k} \tag{7.100}$$

More specifically, the external virtual work leads to the initial stress matrix which contains, on the leading diagonal, the thermal loadings due to the temperature rises through-the-thickness. The latter is arbitrarily assigned in the formulation. The three non-zero terms of the initial stress are given as follows

$$\begin{aligned} K_{\sigma_{u_x u_x}}^{\tau_{u_x} s_{u_x}} &= \sigma_{xx0}^{\vartheta} J^{k\tau_{u_x} s_{u_x}}\, {}^{u_x}_{u_x} I_{mp}^{11}\, {}^{u_x}_{u_x} I_{nq}^{00} + \sigma_{yy0}^{\vartheta} J^{k\tau_{u_x} s_{u_x}}\, {}^{u_x}_{u_x} I_{mp}^{00}\, {}^{u_x}_{u_x} I_{nq}^{11} \\ K_{\sigma_{u_y u_y}}^{\tau_{u_y} s_{u_y}} &= \sigma_{xx0}^{\vartheta} J^{k\tau_{u_y} s_{u_y}}\, {}^{u_y}_{u_y} I_{mp}^{11}\, {}^{u_y}_{u_y} I_{nq}^{00} + \sigma_{yy0}^{\vartheta} J^{k\tau_{u_y} s_{u_y}}\, {}^{u_y}_{u_y} I_{mp}^{00}\, {}^{u_y}_{u_y} I_{nq}^{11} \\ K_{\sigma_{u_z u_z}}^{\tau_{u_z} s_{u_z}} &= \sigma_{xx0}^{\vartheta} J^{k\tau_{u_z} s_{u_z}}\, {}^{u_z}_{u_z} I_{mp}^{11}\, {}^{u_z}_{u_z} I_{nq}^{00} + \sigma_{yy0}^{\vartheta} J^{k\tau_{u_z} s_{u_z}}\, {}^{u_z}_{u_z} I_{mp}^{00}\, {}^{u_z}_{u_z} I_{nq}^{11} \end{aligned} \tag{7.101}$$

Results in the next chapters are obtained by using both uniform and non-uniform temperature distributions. In the non-linear case, the temperature rise is given as: i) power law, encompassing therefore uniform and linear distributions, ii) the solution of the one-dimensional Fourier equation of heat conduction and iii) sinusoidal. Each case is accurately described in the following subsections.

Uniform temperature rise

The plate initial temperature is assumed to be T_i. The temperature is uniformly raised to a final value T_f in which the plate buckles. The temperature change is given by:

$$\Delta T = T_f - T_i \tag{7.102}$$

Linear temperature rise

The temperature of the top surface is T_t and it is considered to vary linearly from T_t to the bottom surface temperature T_b. Therefore, the temperature rise through-the-thickness is given by:

$$T(z) = \Delta T \left(\frac{z}{h} + \frac{1}{2} \right) + T_t \tag{7.103}$$

where $\Delta T = T_b - T_t$.

Non-linear temperature rise

In this case, the temperature distribution through-the-thickness has been given according to the following three approaches:

1. In the first case, the temperature of the top surface is T_t and it is considered to vary from T_t to the bottom surface temperature T_b in which the plate buckles, according to a power law variation through-the-thickness. Therefore, the temperature rise through-the-thickness is given as

$$T(z) = \Delta T\left(\frac{z}{h} + \frac{1}{2}\right)^{\chi} + T_t \tag{7.104}$$

where χ is the temperature index $0 < \chi < \infty$. The linear temperature rise is obtained as a particular case by setting $\chi = 1$.

2. In the second case, the one-dimensional Fourier's heat conduction equation,

$$\begin{cases} \dfrac{\mathrm{d}}{\mathrm{d}z}\left[K(z)\dfrac{\mathrm{d}T}{\mathrm{d}z}\right] = 0 & -h/2 < z < h/2 \\ T = T_c & z = h/2 \\ T = T_m & z = -h/2 \end{cases} \tag{7.105}$$

is solved. $K(z)$ is the coefficient of thermal conduction, T_c and T_m denote the temperature changes at the ceramic side and the metal side, respectively. Similar to the coefficients of elastic moduli and thermal expansion, the coefficient of heat conduction is also assumed as a power form of coordinate variable z as

$$K(z) = (K_c - K_m)\, V_c^k + K_m \tag{7.106}$$

Equation (7.105) can be solved by using a polynomial power series expansion given as

$$T(z) = T_m + \frac{(T_c - T_m)}{C}\left(\frac{z}{h} + \frac{1}{2}\right)\sum_{i=0}^{N_T}\left[(-1)^i \frac{\left(\frac{z}{h} + \frac{1}{2}\right)^{ip}(K_c - K_m)^i}{(ip+1)\,K_m}\right] \tag{7.107}$$

where N_T is the number of series' terms, which for the case of non-uniform temperature rise is obtained from a convergence study. C is defined as follows

$$C = \sum_{i=0}^{N_T}\left[(-1)^i \frac{(K_c - K_m)^i}{(ip+1)\,K_m}\right] \tag{7.108}$$

3. In the third case, the temperature distribution across the thickness direction follows a sinusoidal law as

$$T(z) = \left\{1 - \cos\left[\frac{\pi}{2}\left(\frac{z}{h} + \frac{1}{2}\right)\right]\right\} + T_t \tag{7.109}$$

REFERENCES

1. Ritz W. Über eine neue Methode zur Lösung gewisser Variationsprobleme der mathematischen Physik [About a new method for the solution of certain variational problems of mathematical physics]. J Reine Angew Math 1909;135:1–61.
2. Reddy JN. Energy principles and variational methods in applied mechanics. 2nd edition. New Jersey: John Wiley & Sons; 2002.
3. Reddy JN. Variational methods in theoretical mechanics. 2nd edition. Berlin: Springer-Verlag; 1982.
4. Fazzolari FA. Quasi-3D beam models for the computation of eigenfrequencies of functionally graded beams with arbitrary boundary conditions. Compos Struct 2016;154:239–55.
5. Fazzolari FA, Carrera E. Advanced variable kinematics Ritz and Galerkin formulations for accurate buckling and vibration analysis of anisotropic laminated composite plates. Compos Struct 2011;94(1):50–67.
6. Fazzolari FA, Carrera E. Accurate free vibration analysis of thermo-mechanically pre/post-buckled anisotropic multilayered plates based on a refined hierarchical trigonometric Ritz formulation. Compos Struct 2012;95:381–402.
7. Fazzolari FA, Carrera E. Advances in the Ritz formulation for free vibration response of doubly-curved anisotropic laminated composite shallow and deep shells. Compos Struct 2013;101:111–28.
8. Fazzolari FA. Reissner's mixed variational theorem and variable kinematics in the modelling of laminated composite and FGM doubly-curved shells. Composites, Part B, Eng 2016;89:408–23.
9. Fazzolari FA, Banerjee JR. Axiomatic/asymptotic PVD/RMVT-based shell theories for free vibrations of anisotropic shells using an advanced Ritz formulation and accurate curvature descriptions. Compos Struct 2014;108:91–110.
10. Fazzolari FA, Carrera E. Free vibration analysis of sandwich plates with anisotropic face sheets in thermal environment by using the hierarchical trigonometric Ritz formulation. Composites, Part B, Eng 2013;50:67–81.
11. Fazzolari FA. Natural frequencies and critical temperatures of functionally graded sandwich plates subjected to uniform and non-uniform temperature distributions. Compos Struct 2015;121:197–210.
12. Fazzolari FA, Carrera E. Stability analysis of FGM sandwich plates by using variable-kinematics Ritz models. Mech Adv Mat Struct 2016;23(9):1–27.
13. Fazzolari FA. Modal characteristics of P- and S-FGM plates with temperature-dependent materials in thermal environment. J Therm Stresses 2016;39(7):854–73.
14. Fazzolari FA, Carrera E. Coupled thermoelastic effect in free vibration analysis of anisotropic multilayered plates and FGM plates by using a variable-kinematics Ritz formulation. Eur J Mech A, Solids 2014;44:157–74.

PART 3

Thermal Stress Analysis: Results and Applications

CHAPTER 8

Through-the-thickness thermal fields in one-layer and multilayered structures

8.1 INTRODUCTION

Stress fields related to temperature variations often represent a contributing factor, and, in some cases are the main cause of the failure of structures. Thin walled members of reactor vessels, turbines as well as the structures of future supersonic and hypersonic vehicles, such as that of high-speed civil transport and advanced tactical fighters, are particularly susceptible to failure resulting from excessive stress levels induced by thermal or combined thermo-mechanical loadings. Furthermore, thermal deformations play a fundamental role in multilayered thin-film regions, comprising optical mirrors. Due to their detrimental implications, the effects of both high temperature and mechanical loadings have to be considered in the earliest stages of the design process of such structures [1]. Nevertheless, large portions of the above mentioned structures are made up of multilayered plate and shell constructions such as traditional sandwich panels, panels made of anisotropic composite materials or layered isotropic structures used as thermal protection.

An accurate description of local stress fields in the layers becomes mandatory to prevent thermally loaded structures failure mechanisms. Early [2,3], and recent [4–6] exact three dimensional solutions have shown that appropriate structural modellings are required to describe what was summarized in Ref. [7] with the acronym C_z^0-Requirements: the so called *zig-zag* form of displacement fields and the *interlaminar continuity* for transverse stress fields are C^0-continuous functions in the plate thickness direction z.

Several reviews are available on this topic. Among these mention can be made of the excellent surveys by Tauchert [8], Noor and Burton [9], Argyris and Tenek [10] and the recent book by Reddy [11]. Interested readers can refer to these for a complete overview and literature. A few contributions that have been considered useful by the author and which

Thermal Stress Analysis of Composite Beams, Plates and Shells.
DOI: http://dx.doi.org/10.1016/B978-0-12-849892-7.00011-X

are related to bending of anisotropic plates are discussed in the following text.

The first reported study on thermal bending of anisotropic thin plates was reported by Pell [12]. Thereafter Stavsky [13] developed a thermoelastic theory for heterogeneous anisotropic plates. Wu and Tauchert [14,15] considered simply supported orthotropic plates. Thick plate theory which accounts for transverse shear deformation effects, was then considered by Ambartsumian [16], Rath and Das [17], Tholkachev and Shpektorov [18]; Whitney's [19] isothermal theory was extended to thermal problems by Reddy et al. [20,21]. Attempts to partially introduce the C_z^0-requirements have been made both in the field of Equivalent Single Layer Model (ESLM) and Layer-Wise Models (LWM). According to Reddy [11], in comparison with LWM analyses, ESLM ones preserve the independence of the number of the unknowns from the numbers of the layers, permitting advantageous extension to computational mechanics. Refined models that partially account for C_z^0-requirements have been considered by Khoroshun [22], Pankratova et al. [23], Cho Striz and Bert [24], Kheider and Reddy [25] and by Murakami [4]. In particular, this last paper concluded that "in order to predict rapid variation of transverse normal strains, a plate theory with cubic variation of in-plane displacement in each layer, rather than over the entire plate thickness, should be adopted. Otherwise full three dimensional analyses are recommended." In other words, due to the intrinsic through-the-thickness variation of thermal loadings, a layer-wise description is required to accurately describe the local response of layered plates.

The author [7], [26–34] has recently proposed and evaluated mixed LW and ESL plate and shell models based on Reissner's Mixed Variational Theorem RMVT [35,36] that are able to completely fulfil C_z^0-requirements. This means that, with respect to the classical formulation based on the Principle of Virtual Displacement (PVD), the proposed theory a priori fulfils the interlaminar equilibrium of the transverse shear stresses and permits their evaluation without requiring any post-processing process such as those used in most of the available analyses or in the predictor-corrector procedures of Noor and Burton [37].

8.2 DESCRIPTION OF THE FOUR SAMPLE PROBLEMS

A large investigation has been conducted in order to trace the temperature profiles $T(z)$ and their influence on the response of layered plates. Stresses

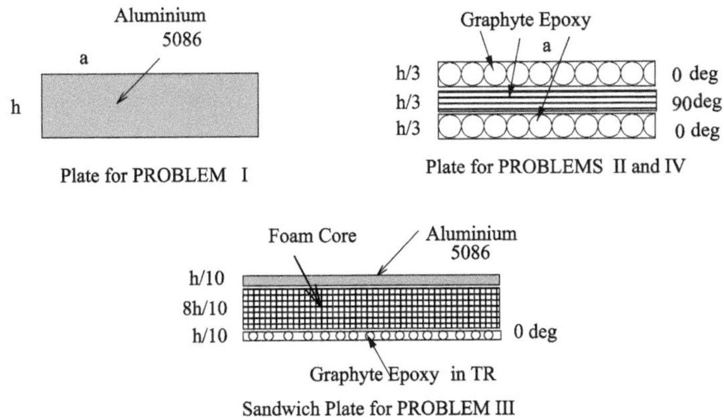

Figure 8.1 Lay-out of the considered plate problems I–IV.

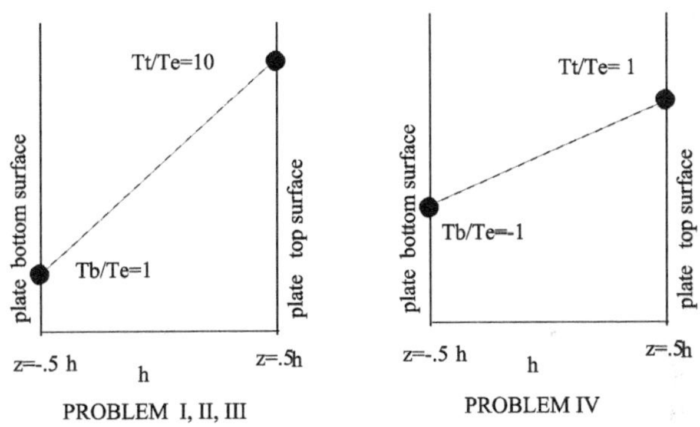

Figure 8.2 Temperature at the top/bottom plate surfaces for the considered Problems I-IV.

and displacement fields have been computed by accounting several temperature fields as well as multilayered lay-outs and geometries. The most significant results have been selected and presented in the next sections. These are related to four sample problems which data are depicted and described in Figs. 8.1, 8.2 and Table 8.1. Usual notations have been used as far as mechanical properties are concerned, see [38]. The four sample problems are:

- Problem I: *One layered Aluminum 5086 plate*
  It consists of a classical one layer isotropic plate made by aluminum alloy. The temperature conditions on the plate top/bottom surfaces are given in Fig. 8.2. Such a test problem has been chosen to investigate

Table 8.1 Thermomechanical properties of the considered materials.

Aluminum 5086
$E = 70300\ [N/mm^2]$, $\nu = .33\ [-]$, $\alpha = 24.E-6\ [°C^1]$, $K = 130\ [W/(m\,°C]$
Graphite Epoxy in [5]
$E_L = 172.72E6\ [N/mm^2]$, $E_T = E36.909E6\ [N/mm^2]$,
$G_{LT} = 3.45E6\ [N/mm^2]$, $G_{TT} = 1.38E6\ [N/mm^2]$, $\nu_{LT} = \nu_{TT} = .25\ [-]$
$\alpha_L = .57E{-}6\ [°C^{-1}]$, $\alpha_T = 35.6E{-}6\ [°C^{-1}]$
$K_L = 36.42\ [W/(m\,°C)]$, $K_T = .96\ [W/(m\,°C)]$
Graphite Epoxy in [6] (given in not dimensioned form)
$E_L/E_T = 25$, $G_{LT}/E_T = .5$, $G_{TT}/E_T = .3$
$\nu_{LT} = \nu_{TT} = 0.25$, $\alpha_T/\alpha_L = 1125$,
the values of K_L, K_T are as those in [5].
Sandwich Core
$(E, G) = .001 \times (E^\star, G^\star)$, $(\alpha, K) = 1000 \times (\alpha^\star, K^\star)$
$\star$ values of Aluminum 5086

the role that the made analyses play in the case of traditional isotropic one layered plates.

- Problem II: *Three-layered Graphite-Epoxy plate by Tungikar and Rao*
This is a three layered plate that was addressed in Ref. [5]. It consists in symmetrically laminated cross-ply plate. The temperature conditions of the plate top and bottom surfaces are those depicted in Fig. 8.2.
- Problem III: *Three-layered Sandwich plate*
It consists in a sandwich plate with unsymmetric faces. The top face is made by aluminum alloys while the bottom face consists of unidirectional graphite epoxy fibers. The properties of the core material have been obtained by those of aluminum alloys as stated in Table 8.1. Two cases of temperature conditions according to Fig. 8.2 have been investigated. These have been denoted as Problem IIIA and IIIB, respectively.
- Problem IV: *Three-layered Graphite-Epoxy plate by Bhaskar, Varadan and Ali*
This problem has the same geometries and mechanics of Problem II, while the temperature conditions (which are depicted in Fig. 8.2) coincide with those introduced by Bhaskar, Varadan and Ali [6]. A three-dimensional 3D solution to this problem in the case of linear profile $T_a(z)$ was also given in Ref. [6]. According to these authors the mechanical data are given in non-dimensional form. Thermal properties are those given in Ref. [5].

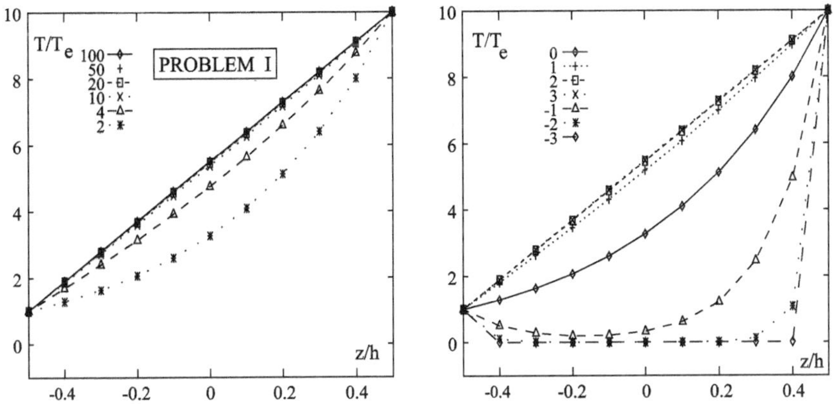

Figure 8.3 PROBLEM I. Temperature profiles $T(z)$ for different plate thickness ratio a/h (left) and thermal conductivity K_z (right, case $a/h = 2$, labels denote the exponent p of the used K_z according to the formula $K_z = K_z \times 10^p$).

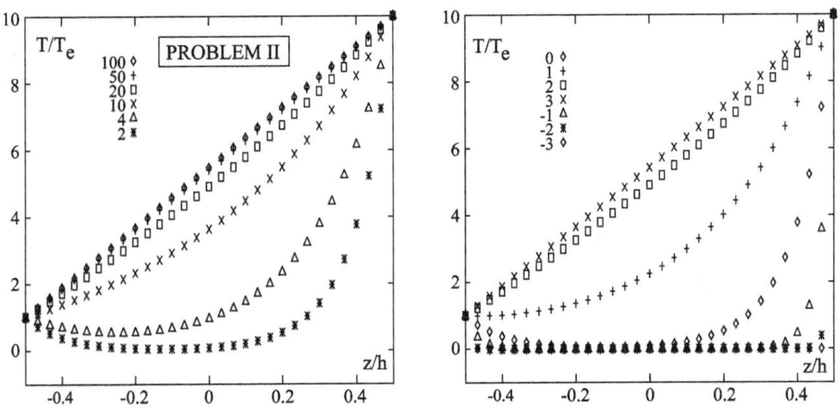

Figure 8.4 PROBLEM II. Details already provided in Fig. 8.3.

8.3 NUMERICAL ILLUSTRATIONS FOR TEMPERATURE PROFILES

The temperature has been always referred to as the environmental temperature T_e. According to the solution technique described in section 8.2, the heat conduction problem has been solved in several cases and results are documented in Figs. 8.3–8.5.

Figure 8.3 shows the effects of plate/thickness ratio a/h and thermal conductivity in the thickness direction on $T(z)$. Problem I has been addressed. It is concluded that very thin and/or high value of K_z are required

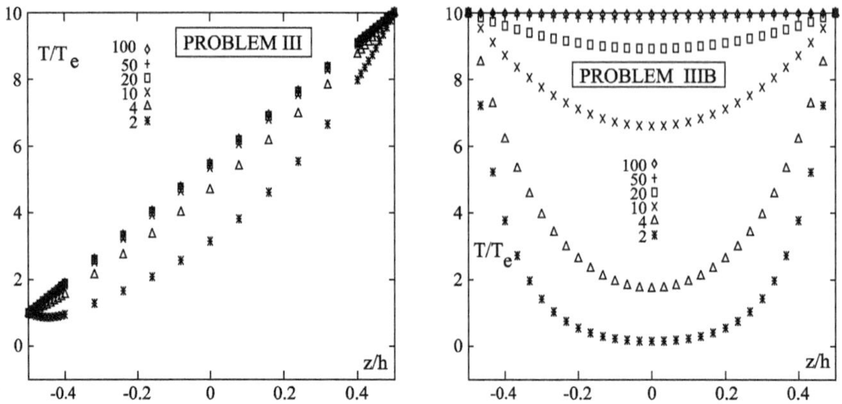

Figure 8.5 Temperature profiles $T(z)$ of a sandwich for different plate thickness ratios a/h. Problem IIIA (left), Problem IIIB (right).

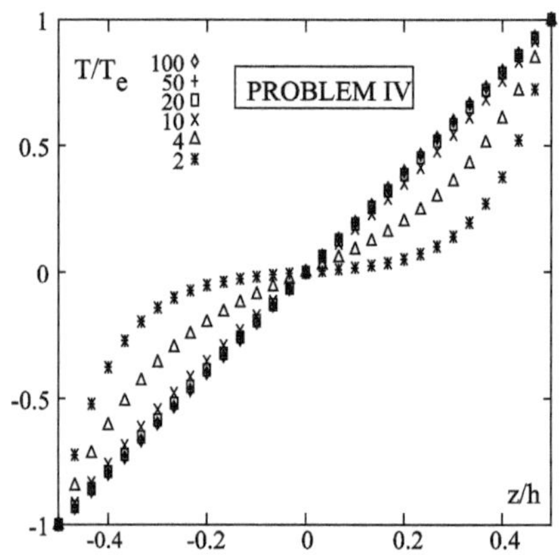

Figure 8.6 Temperature profiles for different plate thickness ratios a/h. Bhaskar, Varadan and Ali problem, [6] herein Problem IV.

to obtain a linear form of $T(z)$. In particular, thick and moderately thick plates exhibit a $T(z)$ form which differ very from the linear one even though a traditional one-layered, isotropic plate has been considered. Low value of the thermal conductivity leads to constant profile $T(z)$.

The same investigation has been conducted in Fig. 8.4 where a layered plate has been considered. To notice that the differences between linear and nonlinear form of the temperature profiles found in Fig. 8.3 are more evident in the case of layered plate. This is confirmed in both problem IIIA, IIIB and IV documented in Figs. 8.5 and 8.6.

Table 8.2 Influence of temperature profile on the response of thick and thin plates. Problem I and II results.

a/h	Problem I		Problem II	
	T_a	T_c	T_a	T_c
	Transverse displacement u_z at $z = h/2$			
2	.4300E−4	.4090E−4	.4164E−4	.2130E−4
4	.2182E−4	.2148E−4	.1381E−4	.9355E−5
10	.1572E−4	.1567E−4	.4087E−5	.3766E−5
20	.1485E−4	.1483E−4	.2454E−5	.2419E−5
50	.1460E−4	.1460E−4	.1977E−5	.1974E−5
100	.1457E−4	.1457E−4	.1908E−5	.1907E−5
	In-plane stress σ_{xx} at $z = \mp h/2$			
2	1.944	−1.225	.1865E+4	−.3255E+3
	−4.492	−8.810	.4532E+4	.1314E+4
4	−.0730	−1.149	.4023E+3	−.6730E+3
	−7.361	−8.741	.2646E+4	.1085E+4
10	−.7174	−.9094	−.2652E+3	−.5705E+3
	−8.260	−8.502	.1658E+4	.1252E+3
20	−.8120	−.8608	−.3658E+3	−.4561E+3
	−8.392	−8.453	.1479E+4	.1360E+4
50	−.8385	−.8464	−.3934E+3	−.4088E+3
	−8.429	−8.439	.1422E+4	.1403E+4
100	−.8423	−.8443	−.3974E+3	−.4013E+3
	−8.434	−8.437	.1414E+4	.1409E+3
	Transverse shear stress σ_{xz}			
	$z = \mp .3h$		$\mp h/6$	
2	−.6199E−0	−.1225E+1	−.2391E+3	.2686E+2
	.6199E−0	.1225E+1	.3602E+3	.3680E+2
4	−.9504E−1	−.2124E−0	−.7482E+2	.3672E+2
	.9504E−1	.2124E−0	.2366E+3	.1004E+3
10	−.6929E−2	−.1506E−1	−.4522E+1	.1359E+2
	.6929E−2	.1506E−1	.1201E+3	.9857E+2
20	−.8099E−3	−.1911E−2	−.1322E+1	.4208E+1
	.8099E−3	.1911E−2	.6359E+2	.6020E+2
50	−.5195E−4	−.1228E−3	.9933E+0	.1193E+1
	.5195E−4	.1228E−3	.2589E+2	.2566E+2
100	−.6496E−5	−.1536E−4	.5311E+0	.5564E+0
	.6496E−5	.1536E−4	.1298E+2	.1295E+2

Table 8.3 Influence of temperature profile on the response of thick and thin plates. Problem IIIA and IIIB results

a/h	Problem III-A		Problem III-B	
	T_a	T_c	$T_a - c$	$T_c - c$
	Transverse displacement u_z at $z = h/2$			
2	.1369E−1	.1144E−1	.1946E−1	.1466E−1
10	.1927E−2	.1906E−2	.3427E−2	.3368E−2
50	.9682E−4	.9693E−4	.1506E−3	.1505E−3
	In-plane stress σ_{xx} at $z = \mp h/2$			
2	.2599E+4	.2065E+4	.5859E+4	.4170E+4
	.7380E+3	.6143E+3	.1051E+4	.7875E+3
10	.7795E+4	.7653E+4	.1726E+5	.1639E+5
	.1303E+3	.1291E+3	.2242E+3	.2201E+3
	.6364E+2	.6370E+2	.1099E+3	1094E+4
50	−.1228E+4	−.1240E+4	−.1063E+4	−.1063E+4
	.3434E+2	.3448E+2	.6195E+2	.6189E+2
	.2093E+2	.2105E+2	.3924E+2	.3923E+2
	Transverse shear stress σ_{xz} at $z = \mp 9/10h$			
2	−.1135E+2	−.7631E+1	−.3520E+2	−.2149E+2
	.8277E+1	.8177E+1	.8176E+0	.4016E+1
10	−.1479E+1	−.1404E+1	−.5327E+1	−.5194E+1
	−.7141E+0	.2136E+1	.1228E+1	.1246E+0
	.9302E+0	.9496E+0	.4088E+0	.4117E+0
50	−.4774E+0	−.4694E+0	−.1370E+1	−.1368E+1
	.2168E+0	.2245E+0	−.1067E+0	−.1063E+0
	.5763E−1	.6131E−1	−.1421E+0	.1420E+0

A general conclusion is that thick and moderately thick plates show a temperature profile $T_c(z)$ which can be very much different by the linear case $T_a(z)$.

8.4 RESULTS ON PLATE RESPONSE

Detailed numerical analyses of the thermomechanical response for Problems I-IV have been reported in Tables 8.2–8.4 and Figs. 8.7–8.11. The following in-plane and out-of-plane displacements and stresses have been quoted in diagrams and tables:

$$u_x = U_x,\ u_z = U_z/(a/h)^2,$$
$$\sigma_{xx} = S_{xx},\ \sigma_{xz} = S_{xz},\ \sigma_{zz} = S_{zz}$$

Table 8.4 Influence of temperature profile on the response Bhaskar Varadan and Ali [6] plate problem

a/h	Problem IV		
	T_a*	T_a	T_c
		u_z at $z = h/2$	
2	96.79	96.79	75.60
10	17.39	17.39	16.92
50	10.50	10.50	10.48
		σ_{xx} at $z = \mp h/2$	
2	1390.	1390.	939.4
	1390.	1390.	939.4
10	1026.	1026.	982.0
	1026.	1026.	982.0
50	967.5	967.5	965.7
	967.5	967.5	965.7
		σ_{xz} at $z = \mp h/6$	
2	63.92	63.92	32.19
	63.92	63.92	32.19
10	60.54	60.64	58.65
	60.54	60.64	58.65
50	14.07	14.07	14.06
	14.07	14.07	14.06

* 3D analysis by Bhaskar, Varadan and Ali.

where U_x, U_z, S_{xx}, S_{xz}, S_{zz} are the maximum values (amplitude of their harmonic distribution, see Part I) of related displacements and stresses in the plate. The case in which the temperature profile has been assumed linear, T_a, has been always compared to the T_c case. Results are given for very thick $a/h = 2$, thick $a/h = 4$, moderately thick $a/h = 10$ and thin plates $a/h = 100$.

As far as problem I is concerned (Table 8.2 and Fig. 8.7) the following main comments can be drawn.

- T_a and T_c results merge in thin plate cases.
- The discrepancy between T_a and T_c results for thick plates is subordinate to the variables considered. Mayor discrepancy has to be registered in the case of stresses with respect to displacement evaluations.

As far as problem II is concerned (Table 8.2 and Fig. 8.8) the following further comments can be made.

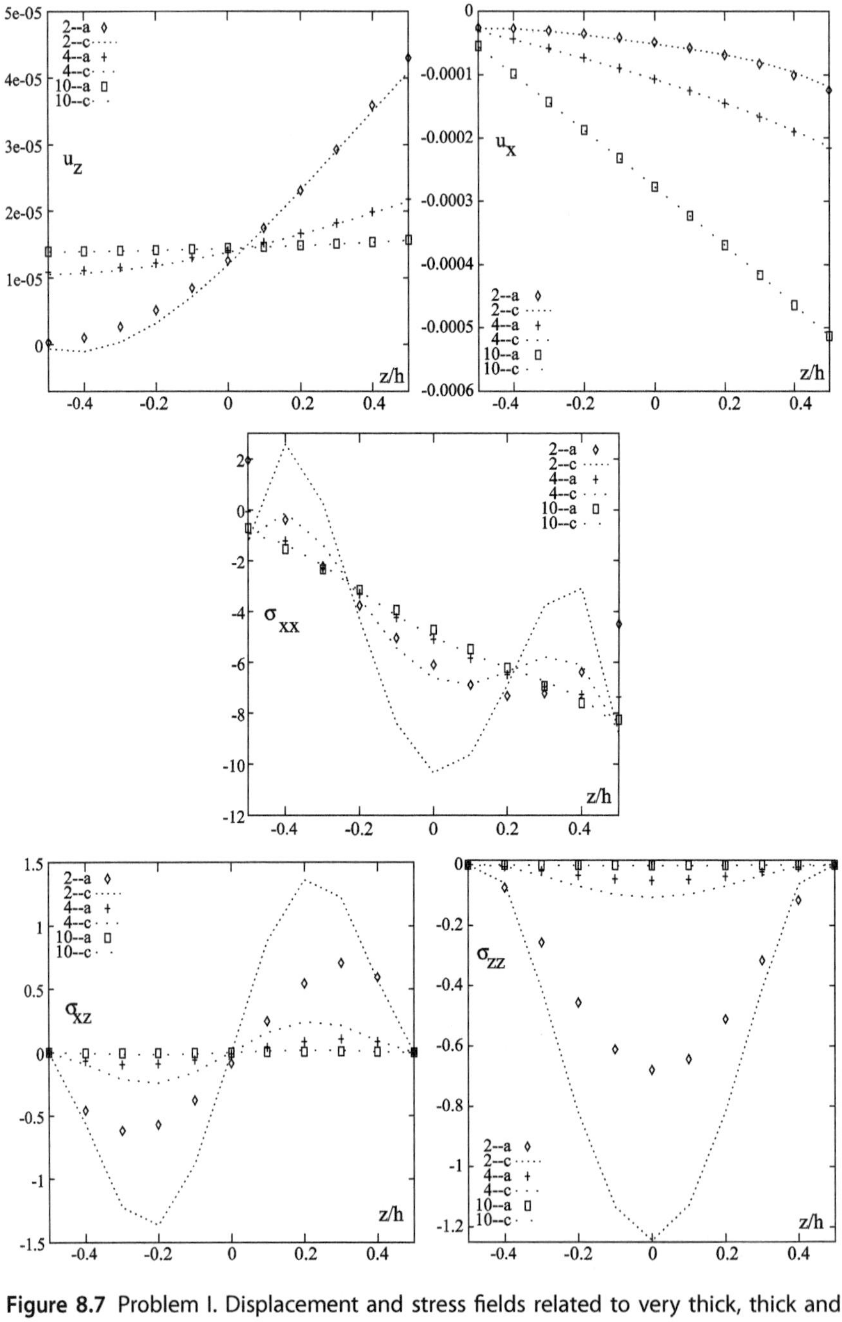

Figure 8.7 Problem I. Displacement and stress fields related to very thick, thick and moderately thick plates.

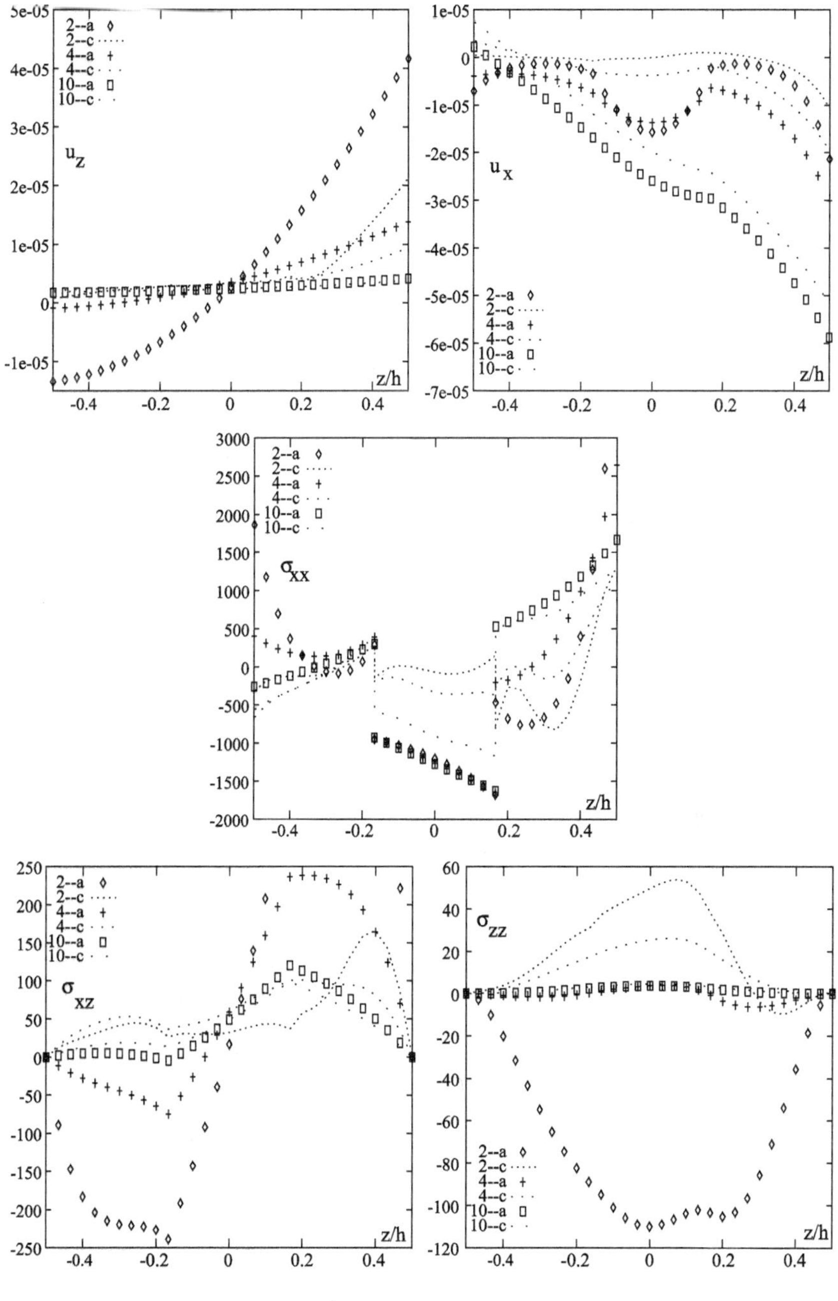

Figure 8.8 PROBLEM II. Details already provided in Fig. 8.7.

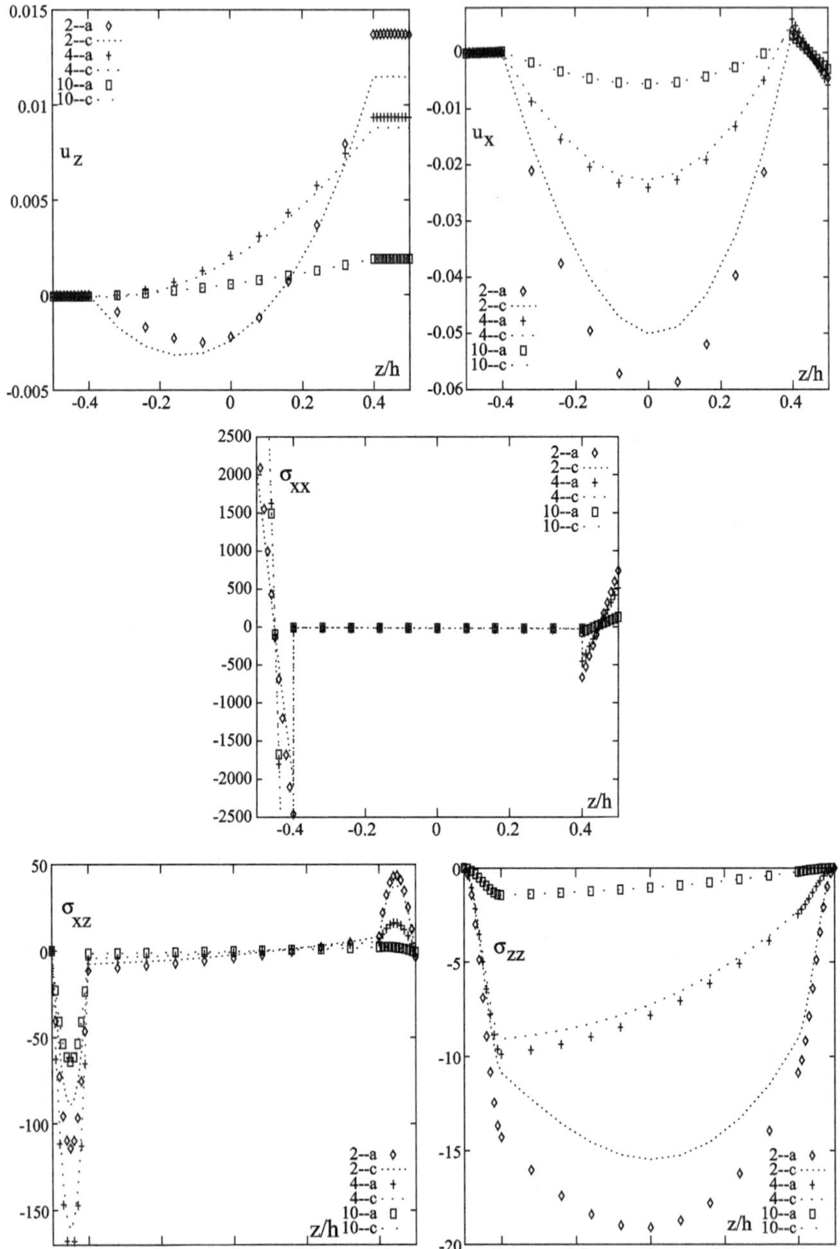

Figure 8.9 PROBLEM III-A. Details already provided in Fig. 8.7.

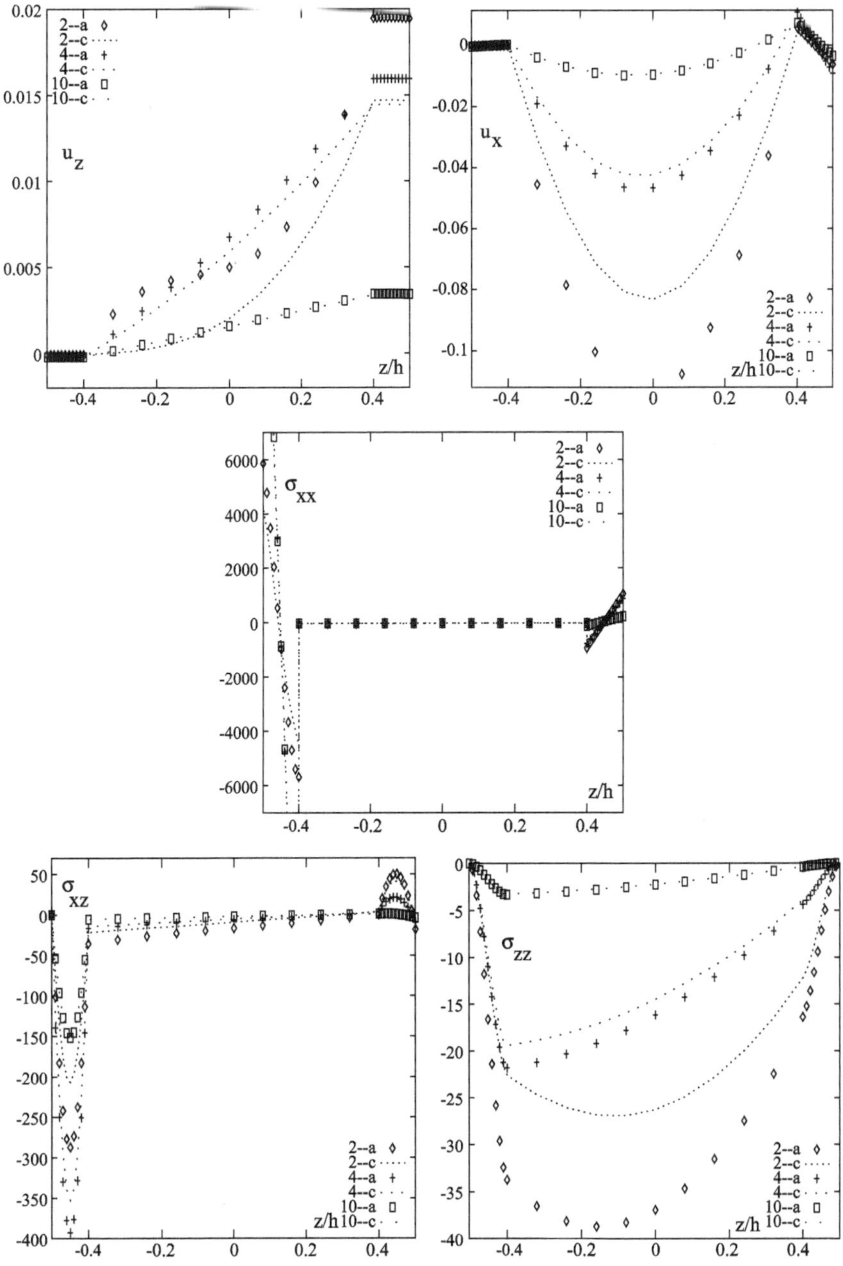

Figure 8.10 PROBLEM III-B. Details already provided in Fig. 8.7.

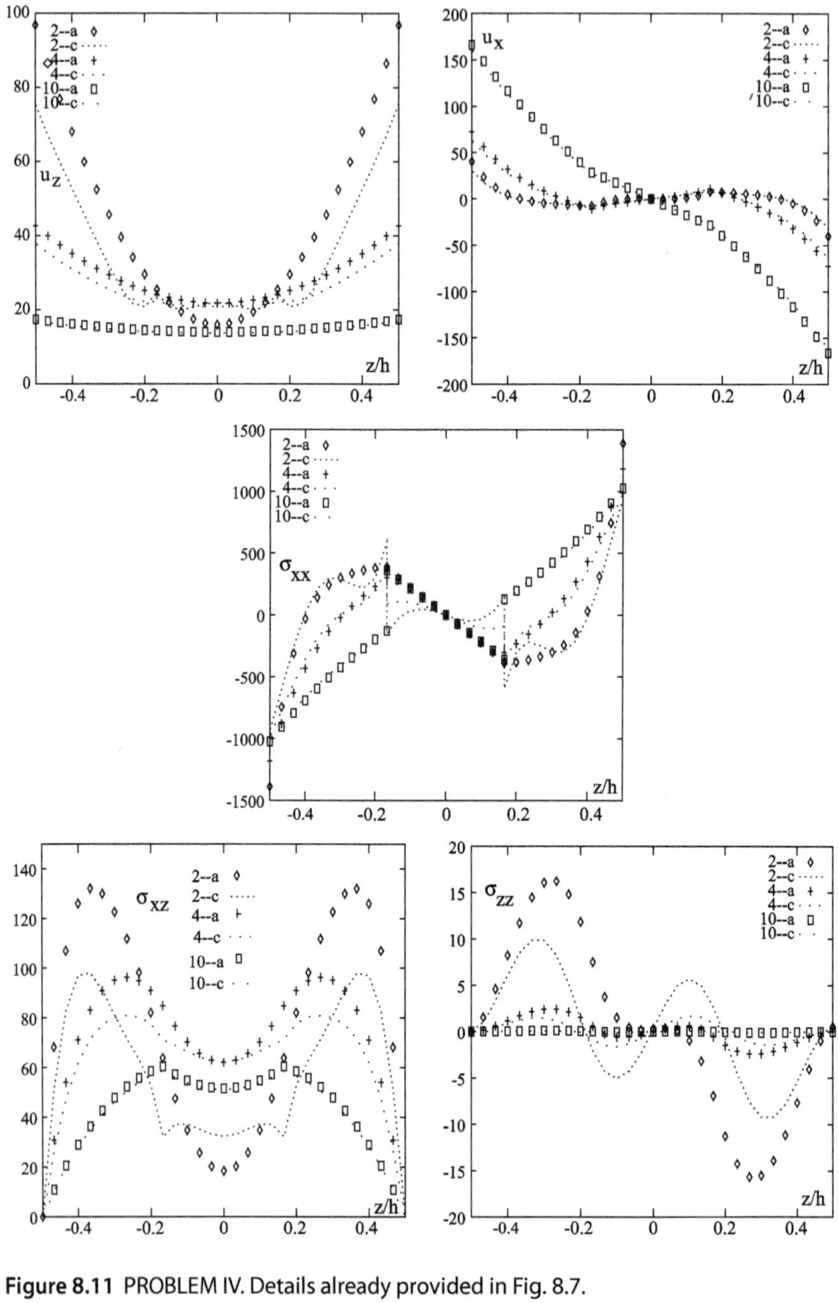

Figure 8.11 PROBLEM IV. Details already provided in Fig. 8.7.

- The discrepancy between T_a and T_c results for thick layered plates is much more evident with respect to that found in the isotropic one layer case of Problem I.
- The layer-wise form of stress and displacement fields is confirmed.

As far as sandwich plate Problem IIIA, IIIB is concerned (Table 8.3 and Figs. 8.9 and 8.10) the following can be remarked.

- Problem II comments are confirmed.
- The discrepancy between T_a and T_c analyses exhibited by Problem IIIA is confirmed by Problem IIIB results. That is, such a discrepancy is barely influenced by top/bottom temperature values of the plates.
- The largest discrepancy between T_a and T_c analyses has been found with correspondence to the two sandwich faces.

The Bhaskar, Varadan and Ali thermomechanical problem has been investigated in Table 8.4 and Fig. 8.11. The comments made above have been confirmed. In particular the effectives of the conducted investigation have been proved by the fact that the comparison between present analyses and 3D results in Ref. [6] shows that the adopted LM4 plate theories furnish a 3D description of stress and displacement field in a layer plate subjected to thermal loadings.

REFERENCES

1. Thornton EA. Thermal structures for aerospace applications. Reston (VA): American Institute of Aeronautics and Astronautics Journal Educational Series; 1996.
2. Srinivas S, Rao AK. A note on flexure of thick rectangular plates and laminates with variation of temperature across the thickness. Bull Acad Pol Sci Ser Sci Tech 1972;20:229–34.
3. Bapu Rao MN. 3D analysis of thermally loaded thick plates. Nucl Eng Des 1979;55:353–61.
4. Murakami H. Assessment of plate theories for treating the thermomechanical response of layered plates. Compos Eng 1993;3(2):137–49.
5. Tungikar VB, Rao KM. Three dimensional exact solution of thermal stresses in rectangular composite laminates. Compos Struct 1994;27:419–27.
6. Bhaskar K, Varadan TK, Ali JSM. Thermoelastic solution for orthotropic and anisotropic composite laminates. Composites, Part B 1996;27:415–20.
7. Carrera E. A class of two dimensional theories for multilayered plates analysis. Atti Accad Sci Torino, Mem Sci Fis 1995;19–20:49–87.
8. Tauchert TR. Thermally induced flexure, buckling and vibration of plates. Appl Mech Rev 1991;44(8):347–60.
9. Noor AK, Burton WS. Computational models for high-temperature multilayered composite plates and shells. Appl Mech Rev 1992;45(10):419–46.
10. Argyris J, Tenek L. Recent advances in computational thermostructural analysis of composite plates and shells with strong nonlinearities. Appl Mech Rev 1997;50(5):285–306.

11. Reddy JN. Mechanics of laminated composite plates, theory and analysis. CRC Press; 1997.
12. Pell WH. Thermal deflection of anisotropic thin plates. 1946.
13. Stavsky Y. Thermoelasticity of heterogeneous oleotropic plates. J Eng Mech Div 1963;89:89–105.
14. Wu CH, Tauchert TR. Thermoelastic analysis of laminated plates 1. Symmetric specially orthotropic laminates. J Therm Stresses 1980;3:247–59.
15. Wu CH, Tauchert TR. Thermoelastic analysis of laminated plates 2. Antisymmetric angle-ply and cross-ply laminates. J Therm Stresses 1980;3:365–78.
16. Ambartsumian SA. Theory of anisotropic plates. Stamford (CT): Technomic; 1970 [translation by Russian, edited by T. Cheron, Ashton J.E.].
17. Das YC, Rath BK. Thermal bending of moderately thick rectangular plates. AIAA J 1972;10(10):1349–51.
18. Tolkachev VM, Shpektorov VM. Modification of Reissner plate theory for contact problems with temperature loads. Sov Appl Mech (Engl Transl) 1980;16:56–9.
19. Whitney JM. The effects of transverse shear deformation on the bending of laminated plates. J Compos Mater 1969;3:534–47.
20. Reddy JN, Bert CW, Hsu YS, Reddy VS. Thermal bending of rectangular plates of bi-modulus composite materials. J Mech Eng Sci 1980;22:297–304.
21. Reddy JN, Hsu YS. Effects of shear deformation and anisotropy on the thermal bending of layered composite plates. J Therm Stresses 1980;3:475–93.
22. Khoroshun LP. Method of constructing equations of the shear theory of thermoelasticity of laminate plates and shells. Sov Appl Mech (Engl Transl) 1981;16(10):851–8.
23. Pankratova ND, Rasskazov AO, Bondar AG, Bondarskii AG. Thermostress state of shear-pliable multilayer orthotropic shells and plates. Sov Appl Mech (Engl Transl) 1988;23(7):658–63.
24. Cho KN, Bert CW, Striz AG. Thermal stress analysis of laminate using higher order individual-layer theory. J Therm Stresses 1989;12:321–32.
25. Kheider AA, Reddy JN. Thermal stress and deflections of cross-ply laminated plates using refined theories. J Therm Stresses 1991;14(4):419–38.
26. Carrera E. C_z^0-requirements: models for the two-dimensional analysis of multilayered structures. Compos Struct 1997;37:373–84.
27. Carrera E. Evaluation of layer-wise mixed theories for laminated plates analysis. AIAA J 1998;36:830–9.
28. Carrera E. Mixed layer-wise theories for multilayered plates analysis. Compos Struct 1998;43:57–70.
29. Carrera E. Layer-wise mixed models for accurate vibration analysis of multilayered plates. J Appl Mech 1998;65:820–9.
30. Carrera E. A Reissner's mixed variational theorem applied to vibrational analysis of multilayered shell. J Appl Mech 1998;66(1):69–78.
31. Carrera E. Multilayered shell theories that account for a layer-wise mixed description. Part I. Governing equations. AIAA J 1999;37(9):1107–16.
32. Carrera E. Multilayered shell theories that account for a layer-wise mixed description. Part II. Numerical evaluations. AIAA J 1999;37(9):1117–24.
33. Carrera E. A study of transverse normal stress effects on vibration of multilayered plates and shells. J Sound Vib 1999;225(5):803–29.
34. Carrera E. Single-layer vs multi-layers plate modelings on the basis of Reissner's mixed theorem. AIAA J 2000;38(2):342–52.

35. Reissner E. On a certain mixed variational theory and a proposed applications. Int J Numer Methods Eng 1984;20:1366–8.
36. Reissner E. On a mixed variational theorem and on a shear deformable plate theory. Int J Numer Methods Eng 1986;23:193–8.
37. Noor AK, Burton WS. Stress and free vibration analyses of multilayered composite plates. Compos Struct 1989;11:183–204.
38. Jones RM. Mechanics of composite materials. New York: McGraw-Hill; 1975.

CHAPTER 9

Static response of uncoupled thermoelastic problems

9.1 INTRODUCTION

A satisfactory thermal stress analysis is only possible, if advanced and refined structural models are developed in such a way that correctly evaluates the stiffness matrix of the structures under investigation, and if the correct thermal load is detected. The computation of the latter is a crucial issue and, indeed, its accuracy may affect significantly the results, even more than the proper modelling of other aspects of the problem.

Studies involving the thermo-elastic behaviour using classical or first-order theories are described by Kant and Khare [1] and Khdeir and Reddy [2]. In recent years, several higher-order two-dimensional models have been developed for such problems, which consider only an assumed temperature profile through the thickness. Among these, of particular interest is the higher-order model by Whu and Chen [3]. The same temperature profile is used by Khare et al. [4] to obtain a closed-form solution for the thermomechanical analysis of laminated and sandwich shells. Khdeir [5] and Khdeir et al. [6] assume a linear or constant temperature profile through the thickness. Barut et al. [7] analyze the non-linear thermoelastic behaviour of shells by means of the Finite Element Method (FEM), but the assigned temperature profile is linear. In the framework of the arbitrary distribution of temperature through the thickness, Miller et al. [8] and Dumir et al. [9] are worth mentioning, in the first a classical shell theory for composite shells is given, the second remarks the importance of the zig-zag form of displacements in the thermal analysis of composite shells. In the case of shells, further investigations were made by Hsu et al. [10] for both closed form and FEM analysis, and by Ding [11], who developed a weak formulation for the case of state equations including the boundary conditions.

In the last few years many contributions have been proposed to investigate the thermal effects in composite structures. In Ref. [12] a study on the influence of the through-the-thickness temperature profile on the thermo-mechanical response of multilayered anisotropic thick and thin plates has been addressed. The partially coupled stress problem was considered by solving Fourier's heat conduction equation. The importance of

Thermal Stress Analysis of Composite Beams, Plates and Shells.
DOI: http://dx.doi.org/10.1016/B978-0-12-849892-7.00012-1

mixed theories for a correct prediction of transverse shear/normal stresses due to thermal loadings has been remarked in Refs. [13,14]. A fully coupled thermo-mechanical analysis applied to plate structure is employed in Ref. [15]. Different types of thermal loadings such as uniform, triangular, bi-triangular (tentlike) and localized in-plane distribution were considered in Ref. [16]. A further comprehensive thermal stress analysis of laminated composite structures was provided in Ref. [17]. Extension to Functionally Graded Materials (FGMs) has been carried out in Ref. [18]. A thermal stability analysis of functionally graded material, isotropic and sandwich plates was proposed in Ref. [19], the Ritz method was employed and uniform, linear, and non-linear temperature profiles were taken into account. A complete thermal stress investigation for composite shells has been presented in Ref. [20]. Fourier's heat conduction equation was employed for laminated composite plate and shell structures in Ref. [21].

9.2 THERMAL STRESS ANALYSIS OF LAMINATED COMPOSITES BY A VARIABLE KINEMATIC MITC9 SHELL ELEMENT

This chapter is composed of two parts. The first one is devoted to the assessment of the shell element based on the Unified Formulation by the static analysis of simply supported plates, cylindrical shells and spherical shells. All of them are evaluated applying a thermal load with a bi-sinusoidal in-plane distribution. Before the assessment results of the static analysis a brief discussion about the evaluation of the temperature profile is given. Using the theory that provides the most accurate results, the second part presents some benchmark solutions relative to plates, cylindrical shells and spherical shells with particular lamination and boundary conditions.

9.2.1 Temperature profile evaluation

The temperature profile along the thickness direction is given in Fig. 9.1 for the plate structure and the cylindrical shell panel, for both the assumed linear profile and the calculated one.

For the three layered composite plate structure (see Fig. 9.1) the calculated profile is plotted for different thickness ratios a/h. It is evident that for thin plates the temperature profile can be assumed almost linear, conversely for thick plates the temperature behaviour is very far from

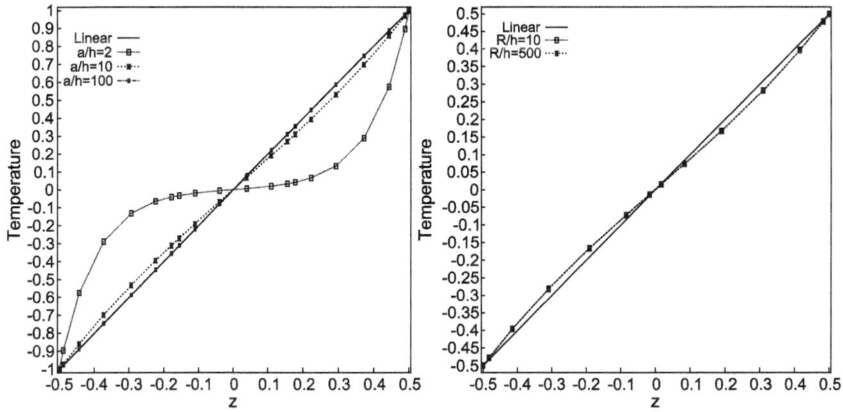

Figure 9.1 Temperature profiles for different values of length-to-thickness ratio ($\frac{a}{h}$) (Composite plate) and of radius-to-thickness ratio ($\frac{R}{h}$) (Composite cylinder).

the linear one, and large errors can be committed if the temperature profile is assumed as linear. For the two layered composite cylindrical shell panel, see Fig. 9.1, the calculated profile is plotted for different thickness to radius ratios R/h and thickness ratio $a/h = 10$. One can note that the effect of the curvature on the distribution of the temperature profile is negligible and the difference with the linear profile is only due to the thickness ratio $a/h = 10$. The proposed evaluations of temperature profile clarify the importance of a calculated temperature profile for thick plates and shells in order to avoid large errors in the approximation of thermal load.

9.2.2 Assessment

To assess the efficiency of the developed MITC9 shell element (see Fig. 9.4), three reference problems are considered: the first one is a cross-ply square multilayered plate with lamination (0°/90°/0°) and simply-supported boundary conditions and is compared with the 3D elasticity solution given by Bhaskar et al. [22]. The second problem is a square cylindrical panel, analytically analyzed, with lamination (0°/90°) and simply-supported boundary conditions. The last one is a square spherical panel, analytically analyzed, with lamination (0°/90°). The boundary conditions are simply-supported. All of them are analyzed by applying a thermal load

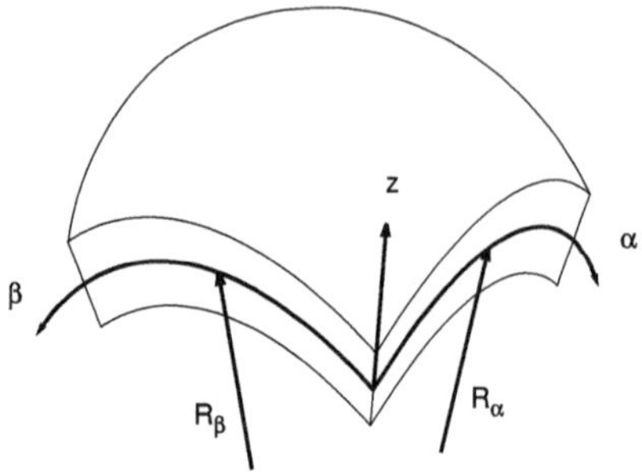

Figure 9.2 Reference system of the double curvature shell.

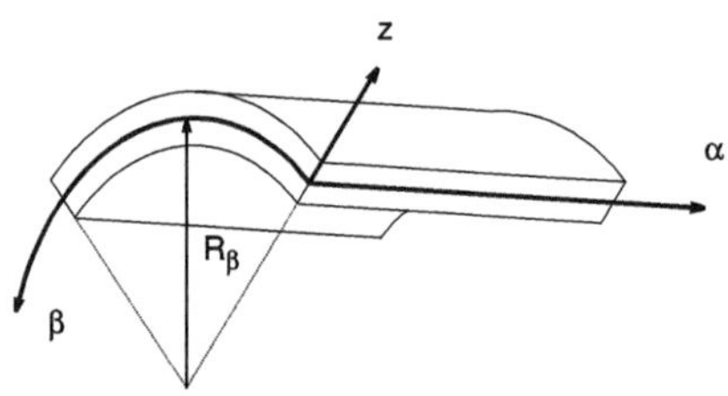

Figure 9.3 Reference system of the cylindrical shell.

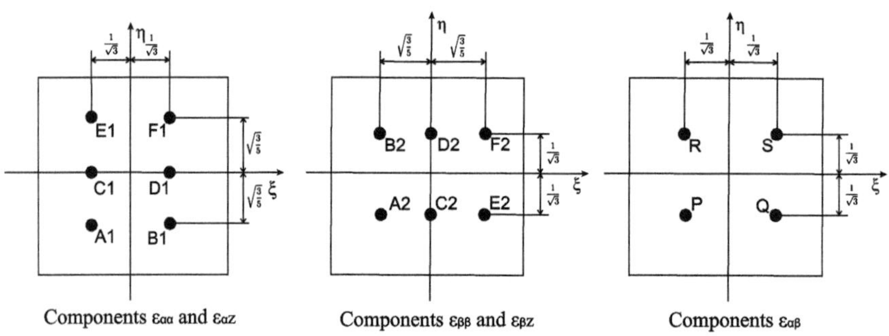

Figure 9.4 Tying points for the MITC9 shell finite element.

with a bi-sinusoidal in-plane distribution:

$$\theta(\alpha, \beta, z) = \hat{\theta}(z)\, sin\left(\frac{m\pi\alpha}{a}\right) sin\left(\frac{n\pi\beta}{b}\right) \tag{9.1}$$

where $m = n = 1$. These three problems are briefly described in the following sections.

Table 9.1 Physical data for multilayered plate, cylindrical and spherical shell

Material	Composite	Carbon
$\frac{E_L}{E_T}$	25.0	25.0
$\frac{G_{LT}}{G_{TT}}$	$\frac{0.5}{0.2}$	$\frac{0.5}{0.2}$
ν	0.25	0.25
$\frac{\alpha_T}{\alpha_L}$	1125.0	3.0
$\frac{K_L}{K_T}$	$\frac{36.42}{0.96}$	$\frac{36.42}{0.96}$

Table 9.2 Convergence study. Plate with thickness ratio $a/h = 100$, cylindrical panel and spherical panel with radius to thickness ratio $R/h = 500$. All the cases are computed for the calculated temperature profile T_c and with a *LW4* theory

	Mesh	4 × 4	6 × 6	8 × 8	10 × 10	Analytical
Plate	w	10.27	10.26	10.25	10.25	10.25
	σ_{xz}	7.466	7.213	7.102	7.084	7.069
Cylindrical	w	1.0966	1.0955	1.0953	1.0952	1.0953
	$\sigma_{\alpha z}$	−1.7090	−1.7131	−1.7093	−1.6983	−1.6957
Spherical	w	1.0958	1.0948	1.0946	1.0945	1.0945
	$\sigma_{\alpha z}$	−2.2848	−2.2065	−2.1562	−2.1461	−2.1403

Multilayered plate

The structure analyzed by Bhaskar et al. [22] is a composite multilayered square plate with lamination (0°/90°/0°). The physical properties of the material of the plate, composite, are given in Table 9.1. The geometrical dimensions are: $a = b = 1\,m$. The temperature boundary conditions are: $\hat{\theta}_{top} = 1\,K$, $\hat{\theta}_{bottom} = -1\,K$. The results are presented for different thickness ratios $a/h = 2, 10, 50, 100$. A mesh grid of 10 × 10 elements is taken to ensure the convergence of the solution (see Table 9.2). The values of the transversal displacement w and the transverse shear stress $\sigma_{\alpha z}$ are listed in Table 9.3 for the temperature profile calculated solving the Fourier heat conduction equation and compared with the assumed linear temperature profile. Other results in terms of transverse shear stress and transversal displacement are shown in Figs. 9.5–9.8.

All the FEM solutions, with an assumed linear temperature profile, lead to accurate results with respect to the 3D solution [22], that makes use of the same linear profile assumption. The results agree also with the analytical solutions provided for all the thickness ratios, except in the case

Table 9.3 Plate with lamination (0°/90°/0°). Transverse displacement $w = w(a/2, b/2) * h_{tot}$, evaluated at $z = \pm h/2$. Transverse shear stress $\sigma_{xz} = \sigma_{xz}(a, 0)$, evaluated at $z = +h/6$

	a/h		2	10	50	100
w	3D [22]		96.79	17.39	10.50	10.26
σ_{xz}	3D [22]		63.92	60.54	14.07	7.073
w	$LW4_a$	T_a	96.78	17.39	10.50	10.26
		T_c	49.09	16.39	10.47	10.25
	$LW4$	T_a	96.77	17.39	10.50	10.26
		T_c	48.85	16.39	10.47	10.25
	$LW1$	T_a	89.23	17.62	11.14	10.91
		T_c	44.17	16.69	11.11	10.91
	$ESLZ3$	T_a	94.85	17.37	10.50	10.26
		T_c	50.08	16.41	10.47	10.25
	$ESL4_a$	T_a	98.21	16.90	10.47	10.25
		T_c	49.55	15.93	10.44	10.25
	$ESL4$	T_a	98.20	16.90	10.47	10.25
		T_c	49.29	15.93	10.44	10.25
	$ESL2$	T_a	83.45	14.96	10.38	10.23
		T_c	40.87	14.09	10.35	10.22
σ_{xz}	$LW4_a$	T_a	63.82	60.54	14.07	7.073
		T_c	30.11	57.07	14.04	7.069
	$LW4$	T_a	63.93	60.66	14.10	7.088
		T_c	30.00	57.18	14.07	7.084
	$LW1$	T_a	42.54	58.78	13.69	6.883
		T_c	31.69	56.35	13.21	6.879
	$ESLZ3$	T_a	27.42	52.61	12.45	6.263
		T_c	23.42	50.43	12.43	6.260
	$ESL4_a$	T_a	37.25	36.33	8.251	4.143
		T_c	24.04	34.47	8.232	4.140
	$ESL4$	T_a	37.30	36.41	8.268	4.152
		T_c	23.96	34.55	8.250	4.149
	$ESL2$	T_a	11.58	16.21	3.624	1.819
		T_c	6.065	15.31	3.616	1.818

of FSDT model. Indeed, plate elements that present a constant transverse normal strain such as *FSDT* lead to inaccurate results for both thick and thin plates. It is confirmed what was found in Ref. [12]: at least a parabolic expansion for the displacements (u, v, w) is required to capture the linear thermal strains that are related to a linear through-the-thickness temper-

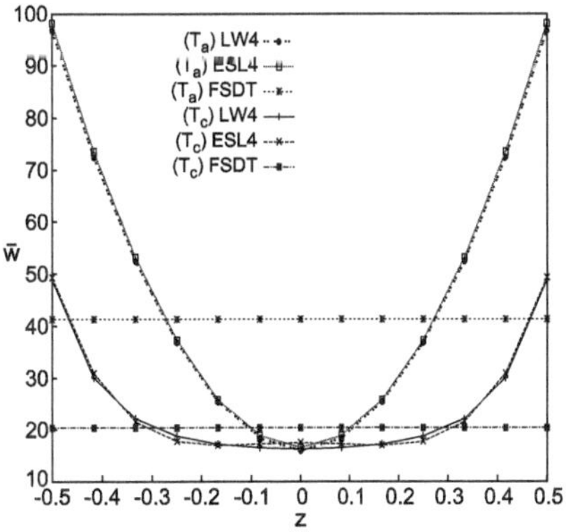

Figure 9.5 Transverse displacement w of a composite plate with $(a/h) = 2$.

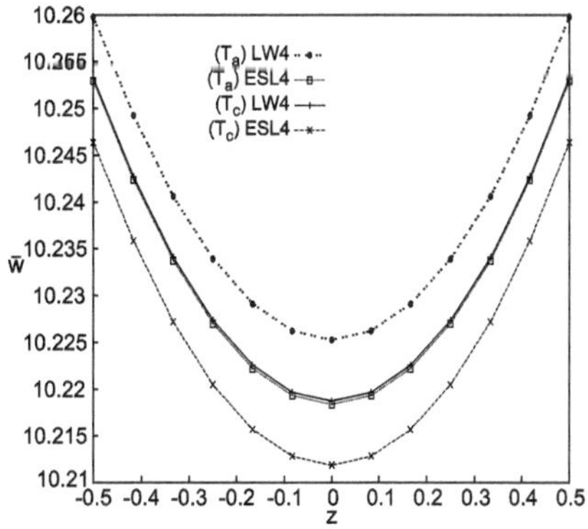

Figure 9.6 Transverse displacement w of a composite plate with $(a/h) = 100$.

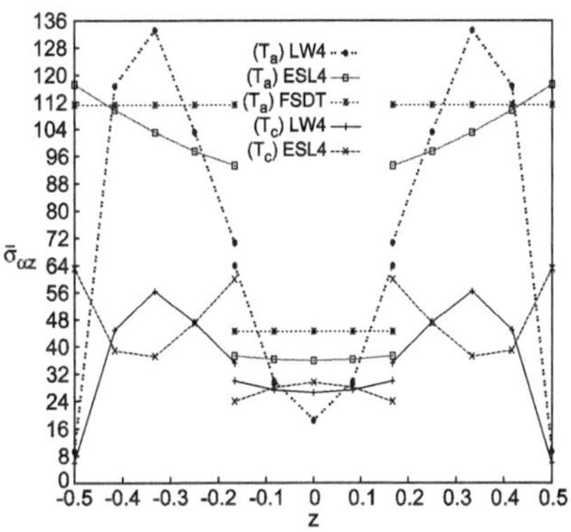

Figure 9.7 Transverse shear stress $\sigma_{\alpha z}$ of a composite plate with $(a/h) = 2$.

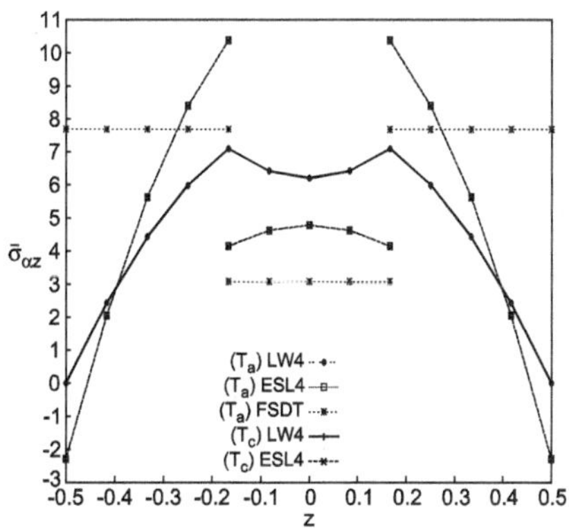

Figure 9.8 Transverse shear stress $\sigma_{\alpha z}$ of a composite plate with $(a/h) = 100$.

ature distribution. The results obtained with the calculated temperature profile are close to them of the assumed linear profile for plates with thickness ratios $a/h = 50, 100$, while for plates with thickness ratios $a/h = 2, 10$ the thermal profile is clearly non-linear and results are different from the linear cases even if the displacement field is approximated by refined models. In general, LW theories perform better than ESL ones and generally a lower-order expansion of the displacements is sufficient. Equivalent single layer analyses are quite satisfactory only for the transverse displacement if applied to thin plates $a/h = 100$, but not for the solution of the transverse shear stresses, as shown in Figs. 9.5–9.8. On the other hand, higher-order LW theories lead to better results but computationally more expensive.

Multilayered cylindrical panel

In this section, a cylindrical composite panel with lamination (0°/90°) is analyzed (see Fig. 9.3). The lamination angle is 0° for the bottom layer and 90° for the top layer. The geometrical dimensions are: $a = 1\,m$ and $b = 1\,m$, global thickness $h_{tot} = 0.1\,m$, curvature radius $R_\alpha = \infty$. The physical properties of the carbon are given in Table 9.1. The temperature boundary conditions are: $\hat{\theta}_{top} = 0.5\,K$, $\hat{\theta}_{bottom} = -0.5\,K$ for all the cases. The results are compared with the corresponding closed form solutions obtained with the Navier method and are presented for different radius to thickness ratios $R_\beta/h_{tot} = 10\,, 50\,, 100\,, 500$ with the corresponding curvature radii $R_\beta = 1\,, 5\,, 10\,, 50$. A mesh grid of 10 × 10 elements is taken to ensure the convergence of the solution (see Table 9.2). The values of the transversal displacement w and the transverse shear stress $\sigma_{\alpha z}$ are listed in Table 9.4 for the temperature profile calculated solving the Fourier's heat conduction equation and compared with the assumed linear temperature profile. Other results in terms of transverse shear stress and transversal displacement are shown in Figs. 9.9–9.12. All the FEs, in both calculated and assumed linear cases, lead to accurate results with respect to the analytical solutions for all the thickness ratios, except for FSDT elements. The difference between the calculated temperature profile and the assumed linear one is a constant and it is not affected by the curvature of the cylinder R_β; this difference is due to the thickness ratio which is $a/h = 10$. In general, LW theories perform better than ESL ones and often also with a lower-order expansion of the unknowns. Equivalent single layer analyses are quite satisfactory for the transverse displacement, even when lower radii to thickness ratios are considered ($R/h = 10$), but not for the transverse shear stress, as shown in Figs. 9.9–9.12.

Multilayered spherical panel

In this section, a square, spherical panel is analyzed (see Fig. 9.2). The temperature boundary conditions are: $\hat{\theta}_{top} = 0.5\,K$, $\hat{\theta}_{bottom} = -0.5\,K$ for all the cases. The results are compared with the analytical solutions obtained with the Navier method. A mesh grid of 10 × 10 elements is taken to ensure the convergence of the solution (see Table 9.2). The geometrical dimensions are: $a = 1\,m$ and $b = 1\,m$, global thickness $h_{tot} = 0.1\,m$ and curvature radii $R_\alpha = R_\beta = R$. The physical properties of the carbon are given in Table 9.1. The results are presented for different radius to thickness ratios $R/h_{tot} = 10\,, 50\,, 100\,, 500$ with the corresponding curvature radius

Table 9.4 Cylindrical panel with lamination (0°/90°). Transverse displacement $w = w(a/2, b/2)$, transverse shear stress $\sigma_{\alpha z} = \sigma_{\alpha z}(a, 0) * 10^2$, evaluated at $z = 0$

	R_β/h		10	50	100	500
w	$LW4_a$	T_a	0.7450	1.1192	1.1359	1.1412
		T_c	0.7188	1.0748	1.0904	1.0953
	$LW4$	T_a	0.7450	1.1192	1.1359	1.1412
		T_c	0.7158	1.0743	1.0902	1.0952
	$LW1$	T_a	0.7712	1.1538	1.1706	1.1759
		T_c	0.7412	1.1082	1.1243	1.1293
	$ESLZ3$	T_a	0.7454	1.1177	1.1342	1.1396
		T_c	0.7147	1.0717	1.0875	1.0926
	$ESL4_a$	T_a	0.7461	1.1194	1.1360	1.1413
		T_c	0.7199	1.0751	1.0907	1.0955
	$ESL4$	T_a	0.7461	1.1194	1.1360	1.1413
		T_c	0.7170	1.0746	1.0904	1.0955
	$ESL2$	T_a	0.7455	1.1152	1.1316	1.1369
		T_c	0.7150	1.0695	1.0852	1.0902
	$FSDT$	T_a	0.8745	1.2781	1.2941	1.2979
		T_c	0.8367	1.2229	1.2382	1.2419
$\sigma_{\alpha z}$	$LW4_a$	T_a	−10.901	−3.7541	−2.8789	−2.2428
		T_c	−10.051	−3.1516	−2.3086	−1.6957
	$LW4$	T_a	−10.923	−3.7615	−2.8845	−2.2471
		T_c	−10.011	−3.1485	−2.3086	−1.6983
	$LW1$	T_a	−8.3115	−4.0011	−3.5188	−3.1781
		T_c	−7.7544	−3.6140	−3.1507	−2.8235
	$ESLZ3$	T_a	−10.522	−3.5686	−2.7832	−2.2277
		T_c	−9.7816	−3.1176	−2.3651	−1.8329
	$ESL4_a$	T_a	−6.8978	−1.7276	−1.2097	−0.8599
		T_c	−6.3568	−1.3735	−0.8747	−0.5374
	$ESL4$	T_a	−6.9120	−1.7309	−1.2120	−0.8614
		T_c	−6.3345	−1.3701	−0.8732	−0.5377
	$ESL2$	T_a	−5.6195	−1.7814	−1.4294	−1.2006
		T_c	−5.3090	−1.6296	−1.2921	−1.0727
	$FSDT$	T_a	−4.8037	−0.6032	−0.2571	−0.0436
		T_c	−4.5916	−0.5767	−0.2457	−0.0418

$R = 1.0, 5.0, 10.0, 50.0$. The lamination angle is 0° for the bottom layer and 90° for the top layer. The values of the transversal displacement w and the transverse shear stress $\sigma_{\alpha z}$ are listed in Table 9.5 for the temperature profile calculated solving the Fourier heat conduction equation and

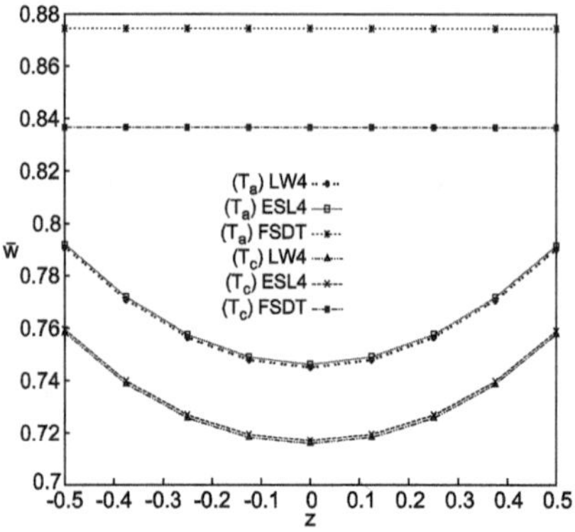

Figure 9.9 Transverse displacement w along the thickness, with radius to thickness ratio $(R/h) = 10$. Composite cylindrical panel.

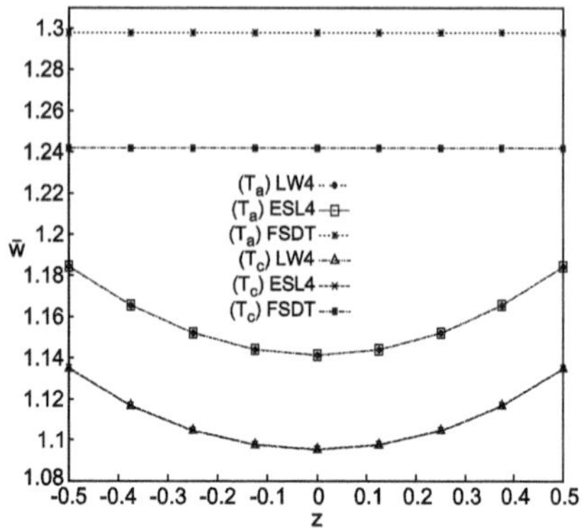

Figure 9.10 Transverse displacement w along the thickness, with radius to thickness ratio $(R/h) = 500$. Composite cylindrical panel.

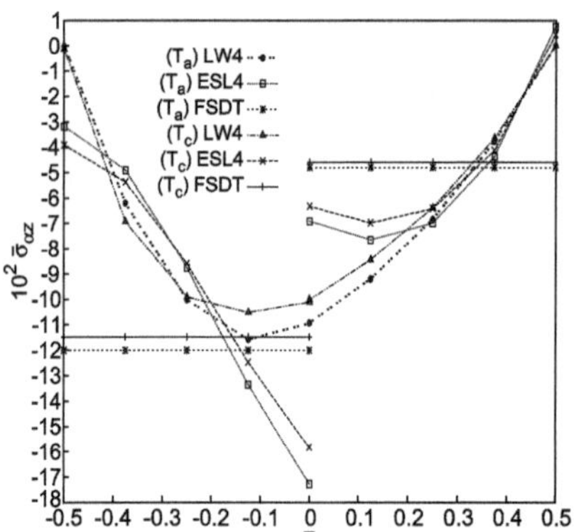

Figure 9.11 Transverse shear stress $\sigma_{\alpha z}$ along the thickness, with radius to thickness ratio $(R/h) = 10$. Composite cylindrical panel.

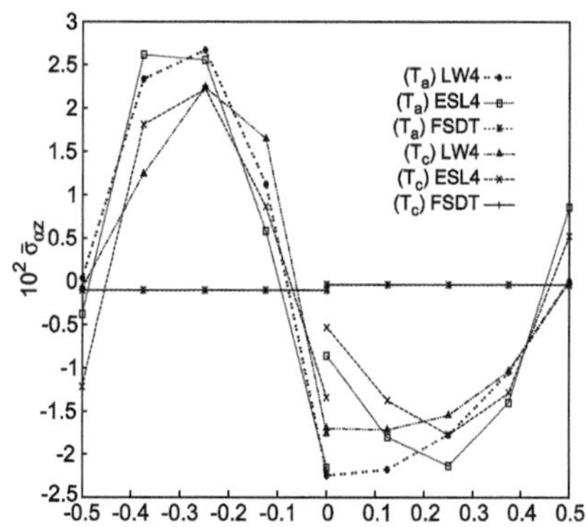

Figure 9.12 Transverse shear stress $\sigma_{\alpha z}$ along the thickness, with radius to thickness ratio $(R/h) = 500$. Composite cylindrical panel.

compared with the assumed linear temperature profile. Other results in terms of transverse shear stress and transversal displacement are shown in Figs. 9.13–9.16. Considerations similar to the previous problem can be made. All the FEs, in both calculated and assumed linear cases, lead to accurate results with respect to the analytical solutions for all the thickness ratios, except for FSDT elements. The difference between the calculated temperature profile solutions and the assumed linear one is constant and it

Table 9.5 Spherical panel with lamination (0°/90°). Transverse displacement $w = w(a/2, b/2)$, evaluated at $z = 0$, transverse shear stress $\sigma_{\alpha z} = \sigma_{\alpha z}(a, 0) * 10^2$ evaluated at $z = -h/4$

	R/h		10	50	100	500
w	$LW4_a$	T_a	0.3299	1.0507	1.1174	1.1404
		T_c	0.3203	1.0087	1.0725	1.0945
	$LW4$	T_a	0.3299	1.0507	1.1174	1.1405
		T_c	0.3240	1.0091	1.0726	1.0945
	$LW1$	T_a	0.3386	1.0836	1.1516	1.1751
		T_c	0.3325	1.0414	1.1062	1.1285
	$ESLZ3$	T_a	0.3306	1.0496	1.1159	1.1388
		T_c	0.3235	1.0071	1.0701	1.0918
	$ESL4_a$	T_a	0.3309	1.0511	1.1176	1.1406
		T_c	0.3213	1.0093	1.0728	1.0947
	$ESL4$	T_a	0.3309	1.0511	1.1176	1.1406
		T_c	0.3250	1.0096	1.0729	1.0947
	$ESL2$	T_a	0.3315	1.0477	1.1134	1.1361
		T_c	0.3248	1.0054	1.0679	1.0895
	$FSDT$	T_a	0.3927	1.1967	1.2709	1.2965
		T_c	0.3837	1.1459	1.2163	1.2406
$\sigma_{\alpha z}$	$LW4_a$	T_a	24.096	1.1199	−1.3854	−2.5674
		T_c	23.379	1.3972	−1.0041	−2.1403
	$LW4$	T_a	24.131	1.1212	−1.3877	−2.5714
		T_c	23.289	1.3831	−1.0122	−2.1461
	$LW1$	T_a	19.500	1.5598	−0.5309	−1.5931
		T_c	18.818	1.6718	−0.3315	−1.3536
	$ESLZ3$	T_a	21.061	2.4154	0.1275	−1.0771
		T_c	20.328	2.5096	0.3181	−0.8407
	$ESL4_a$	T_a	21.281	0.8662	−1.3842	−2.4603
		T_c	20.593	1.0635	−1.0932	−2.1270
	$ESL4$	T_a	21.312	0.8673	−1.3865	−2.4641
		T_c	20.521	1.0521	−1.1000	−2.1325
	$ESL2$	T_a	20.284	3.0723	0.9925	−0.0983
		T_c	19.598	3.1406	1.1473	0.0970
	$FSDT$	T_a	21.845	3.9515	1.5581	0.2307
		T_c	21.011	3.8024	1.5047	0.2240

is not affected by the curvature R_β. In general, LW theories perform better than ESL ones and generally a lower-order expansion of the unknowns is sufficient. Equivalent single layer analyses are quite satisfactory only for the

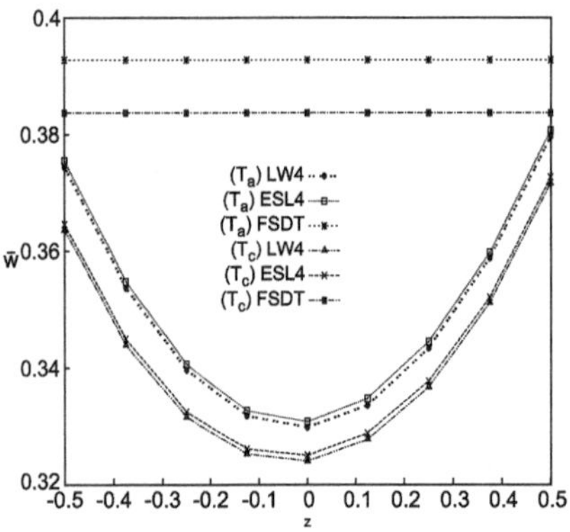

Figure 9.13 Transverse displacement w along the thickness, with radius to thickness ratio $(R/h) = 10$. Composite spherical panel.

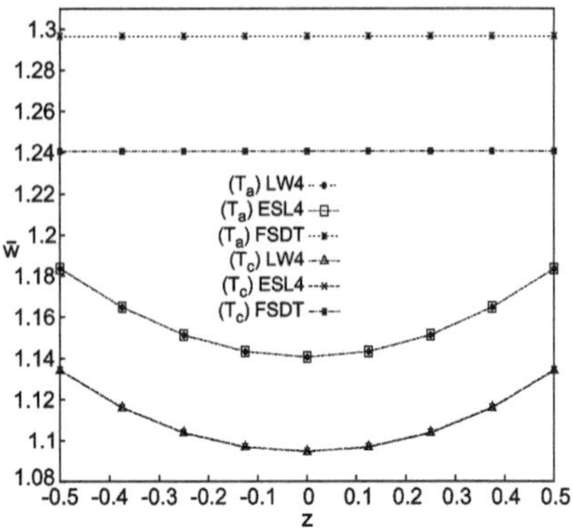

Figure 9.14 Transverse displacement w along the thickness, with radius to thickness ratio $(R/h) = 500$. Composite spherical panel.

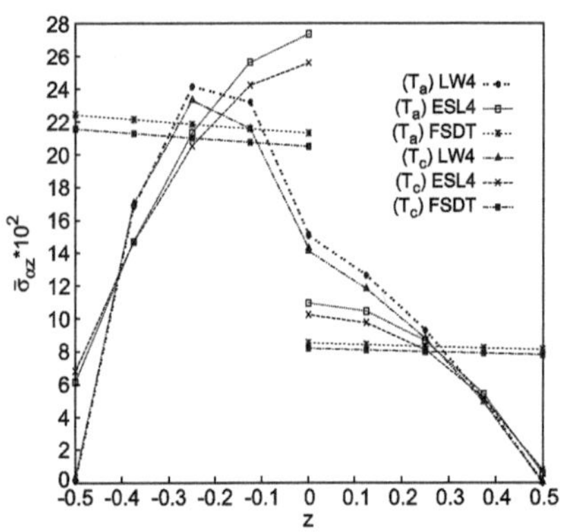

Figure 9.15 Transverse shear stress $\sigma_{\alpha z}$ along the thickness, with radius to thickness ratio $(R/h) = 10$. Composite spherical panel.

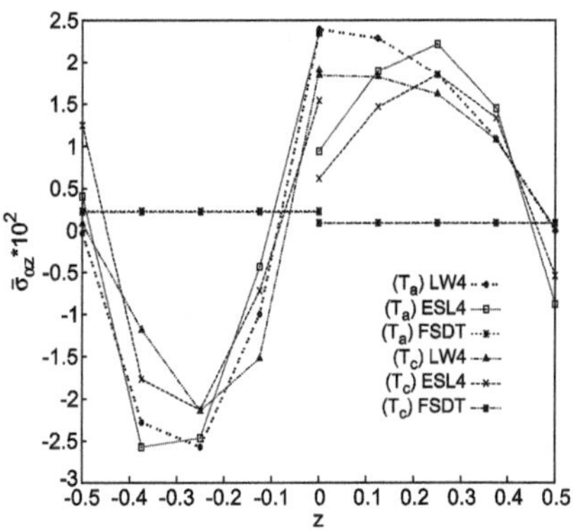

Figure 9.16 Transverse shear stress $\sigma_{\alpha z}$ along the thickness, with radius to thickness ratio $(R/h) = 500$. Composite spherical panel.

transverse displacement, even for lower radii to thickness ratios $(R/h = 10)$. Figures 9.13–9.16 show a different behaviour for the transverse shear stress, where higher-order LW models are required to get accurate results.

9.2.3 FEM benchmark solutions

Similar plates, cylindrical shells and spherical shells are analyzed, considering two new problems that have not reference analytical solutions:

1. Structures with lamination ±45° under bi-sinusoidal load and simply-supported boundary conditions.
2. Structures with clamped-free boundary conditions: edges parallel to β-direction clamped and those parallel to α-direction free. The lamination is equal to the assessment cases.

Anti-symmetric lamination ±45°

The first structure analyzed is a composite multilayered square plate with lamination (−45°/45°/−45°). The physical properties of the material, the geometrical data and the temperature boundary conditions are the same of the assessment cases. The structure is simply supported. The results are presented for different thickness ratios $a/h = 10, 100$. The same mesh grid of 10×10 elements of the assessment cases is taken to ensure the convergence of the solution. The values of the transversal displacement w and the transverse shear stress $\sigma_{\alpha z}$ are listed in Table 9.6. The second structure analyzed is a composite square cylindrical panel with lamination (−45°/45°). The lamination angle is −45° for the bottom layer and 45° for the top layer. The physical properties of the material, the geometrical data and the temperature boundary conditions are the same of the assessment cases. The structure is simply supported. The results are presented for different radius to thickness ratios $R/h = 10, 100$. The same mesh grid of 10×10 elements of the assessment cases is taken to ensure the convergence of the solution. The values of the transversal displacement w and the transverse shear stress $\sigma_{\alpha z}$ are listed in Table 9.6. The last structure analyzed is a composite square spherical panel with lamination (−45°/45°). The lamination angle is −45° for the bottom layer and 45° for the top layer. The physical properties of the material, the geometrical data and the temperature boundary conditions are the same of the assessment cases. The structure is simply supported. The results are presented for different radius to thickness ratios $R/h = 10, 100$. The same mesh grid of 10×10 elements of the assessment cases is taken to ensure the convergence of the solution. The values of the transversal displacement w and the transverse shear stress $\sigma_{\alpha z}$ are listed in Table 9.6.

Clamped-free boundary conditions

In this part, the structures are considered with clamped-free boundary conditions: edges parallel to β-direction clamped and those parallel to α-direction free.

Table 9.6 Benchmark problems. Case 1 with anti-symmetric lamination ±45°. Plate, cylindrical and spherical panel, transverse displacement $w = 10\,w(a/2, b/2, +h/2)$ and transverse shear stress $\sigma_{\alpha z} = 10^2\,\sigma_{\alpha z}(a, b/2, 0)$

			Plate (a/h)		Cylindrical (R/h)		Spherical (R/h)	
			10	100	10	100	10	100
w	$LW4$	T_a	1375.4	4607.8	4.8796	7.4058	2.6295	7.4971
		T_c	1293.5	4604.7	4.5741	7.0165	2.5533	7.1029
	$LW1$	T_a	1274.3	4895.2	4.7358	7.5077	2.4930	7.6005
		T_c	1203.2	4892.2	4.4477	7.1343	2.4305	7.2202
	$ESLZ3$	T_a	1307.7	4586.4	4.7120	7.3351	2.5257	7.4200
		T_c	1232.7	4583.4	4.4154	6.9528	2.4542	7.0325
	$ESL4$	T_a	1250.8	4550.1	4.7274	7.3978	2.5388	7.4776
		T_c	1177.8	4547.1	4.4239	7.0045	2.4615	7.0793
	$ESL2$	T_a	1164.1	4501.7	4.7187	7.2171	2.5756	7.3229
		T_c	1096.3	4498.8	4.4192	6.8330	2.5029	6.9316
	$FSDT$	T_a	1294.9	7222.5	5.0729	7.5539	2.6941	7.6548
		T_c	1219.4	7217.9	4.8079	7.2151	2.6483	7.3099
$\sigma_{\alpha z}$	$LW4$	T_a	−85.932	−56.181	15.624	2.4665	24.592	3.7828
		T_c	−81.708	−56.148	14.483	1.9577	22.964	3.2169
	$LW1$	T_a	−98.335	−56.066	17.223	5.5156	23.450	6.7916
		T_c	−93.436	−56.033	16.299	5.1621	22.161	6.3780
	$ESLZ3$	T_a	−97.289	−55.800	17.350	2.6837	26.823	4.0857
		T_c	−92.420	−55.767	16.512	2.5503	25.369	3.8853
	$ESL4$	T_a	−80.357	−52.295	16.651	0.3544	27.067	1.8573
		T_c	−76.298	−52.264	15.711	0.2248	25.347	1.6474
	$ESL2$	T_a	−61.532	−37.094	15.936	2.8891	23.900	4.2028
		T_c	−58.089	−37.072	15.343	2.9668	22.689	4.2054
	$FSDT$	T_a	−92.857	−45.626	12.861	2.2035	20.014	3.2885
		T_c	−87.595	−45.598	12.219	2.0645	18.994	3.0974

The first structure analyzed is a composite multilayered square plate. The physical properties of the material, the lamination angle, the geometrical data and the temperature boundary conditions are the same of the assessment cases. The results are presented for different thickness ratios $a/h = 10, 100$. The same mesh grid of 10×10 elements of the assessment cases is taken to ensure the convergence of the solution. The values of the transversal displacement w and the transverse shear stress $\sigma_{\alpha z}$ are listed in Table 9.7. The second structure analyzed is a composite square cylindrical panel. The physical properties of the material, the lamination angle, the geometrical data and the temperature boundary conditions are the same of the assessment cases. The results are presented for different radius to thick-

Table 9.7 Benchmark problems. Case 2 with clamped-free boundary condition. Plate, cylindrical and spherical panel, transverse displacement $w = 10\,w(a/2, b/2, +h/2)$ and transverse shear stress $\sigma_{\alpha z} = 10^2\,\sigma_{\alpha z}(a, b/2, 0)$

			Plate (a/h)		**Cylindrical** (R/h)		**Spherical** (R/h)	
			10	100	10	100	10	100
w	$LW4$	T_a	1226.8	3096.9	4.5476	5.3307	−0.3479	3.7373
		T_c	1157.7	3094.9	4.3447	5.1018	−0.1750	3.6265
	$LW1$	T_a	1210.9	3274.4	4.6164	5.3874	−0.3642	3.8575
		T_c	1145.2	3272.3	4.4100	5.1571	−0.1923	3.7404
	$ESLZ3$	T_a	1222.0	3096.8	4.5239	5.2945	−0.3407	3.7334
		T_c	1155.2	3094.9	4.3237	5.0705	−0.1681	3.6250
	$ESL4$	T_a	1183.3	3083.9	4.5479	5.3215	−0.3423	3.7477
		T_c	1116.2	3081.9	4.3460	5.0941	−0.1696	3.6368
	$ESL2$	T_a	986.40	3048.6	4.5152	5.2473	−0.3369	3.7263
		T_c	929.51	3046.7	4.3093	5.0195	−0.1652	3.6119
	$FSDT$	T_a	1039.7	4356.4	4.7747	5.5564	−0.5836	3.9916
		T_c	979.72	4353.6	4.5553	5.3129	−0.4107	3.8624
$\sigma_{\alpha z}$	$LW4$	T_a	−2619.8	−522.26	−3.9875	−5.1179	15.086	0.2373
		T_c	−2519.9	−522.01	−3.9154	−5.0207	13.754	−0.0535
	$LW1$	T_a	−2655.5	−579.97	−1.5318	−2.3709	14.765	1.8361
		T_c	−2520.6	−579.67	−1.5150	−2.3397	13.594	1.5630
	$ESLZ3$	T_a	−2436.3	−537.18	−2.9825	−3.9052	14.614	0.8821
		T_c	−2303.2	−536.90	−2.8354	−3.7432	13.481	0.6988
	$ESL4$	T_a	−2695.3	−465.54	−3.3266	−4.2840	14.134	0.4614
		T_c	−2582.7	−465.30	−3.2708	−4.2082	12.905	0.1929
	$ESL2$	T_a	−2287.1	−240.08	−2.8190	−3.6334	13.936	0.7940
		T_c	−2159.2	−239.94	−2.7079	−3.5120	12.809	0.5937
	$FSDT$	T_a	−4195.5	−446.19	−6.0953	−6.9657	13.570	−2.0926
		T_c	−3955.6	−445.91	−5.7896	−6.6468	12.506	−2.1219

ness ratios $R/h = 10, 100$. The same mesh grid of 10×10 elements of the assessment cases is taken to ensure the convergence of the solution. The values of the transversal displacement w and the transverse shear stress $\sigma_{\alpha z}$ are listed in Table 9.7. The last structure analyzed is a composite square spherical panel. The physical properties of the material, the lamination angle, the geometrical data and the temperature boundary conditions are the same of the assessment cases. The results are presented for different radius to thickness ratios $R/h = 10, 100$. The same mesh grid of 10×10 elements of the assessment cases is taken to ensure the convergence of the solution. The values of the transversal displacement w and the transverse shear stress $\sigma_{\alpha z}$ are listed in Table 9.7.

REFERENCES

1. Kant T, Khare RK. Finite element thermal stress analysis of composite laminates using a higher-order theory. J Therm Stresses 1994;17(2):229–55.
2. Khdeir AA, Reddy JN. Thermal stresses and deflections of cross-ply laminated plates using refined plate theories. J Therm Stresses 1991;14(4):419–38.
3. Zhen W, Wanji C. A global-local higher order theory for multilayered shells and the analysis of laminated cylindrical shell panels. Compos Struct 2008;84(4):350–61.
4. Khare RK, Kant T, Garg AK. Closed-form thermo-mechanical solutions of higher-order theories of cross-ply laminated shallow shells. Compos Struct 2003;59(3):313–40.
5. Khdeir AA. Thermoelastic analysis of cross-ply laminated circular cylindrical shells. Int J Solids Struct 1996;33(27):4007–17.
6. Khdeir AA, Rajab MD, Reddy JN. Thermal effects on the response of cross-ply laminated shallow shells. Int J Solids Struct 1992;29(5):653–67.
7. Barut A, Madenci E, Tessler A. Nonlinear thermoelastic analysis of composite panels under non-uniform temperature distribution. Int J Solids Struct 2000;37(27):3681–713.
8. Miller CJ, Millavec WA, Richer TP. Thermal stress analysis of layered cylindrical shells. AIAA J 1981;19(4):523–30.
9. Dumir PC, Nath JK, Kumari P, Kapuria S. Improved efficient zigzag and third order theories for circular cylindrical shells under thermal loading. J Therm Stresses 2008;31(4):343–67.
10. Hsu YS, Reddy JN, Bert CW. Thermoelasticity of circular cylindrical shells laminated of bimodulus composite materials. J Therm Stresses 1981;4(2):155–77.
11. Ding K. Thermal stresses of weak formulation study for thick open laminated shell. J Therm Stresses 2008;31(4):389–400.
12. Carrera E. Temperature profile influence on layered plates response considering classical and advanced theories. AIAA J 2002;40(9):1885–96.
13. Carrera E. An assessment of mixed and classical theories for the thermal stress analysis of orthotropic multilayered plates. J Therm Stresses 2000;23(9):797–831.
14. Robaldo A, Carrera E. Mixed finite elements for thermoelastic analysis of multilayered anisotropic plates. J Therm Stresses 2007;30:165–94.
15. Nali P, Carrera E, Calvi A. Advanced fully coupled thermo-mechanical plate elements for multilayered structures subjected to mechanical and thermal loading. Int J Numer Methods Eng 2011;85:869–919.
16. Carrera E, Ciuffreda A. Closed-form solutions to assess multilayered-plate theories for various thermal stress problems. J Therm Stresses 2004;27:1001–31.
17. Carrera E, Cinefra M, Fazzolari FA. Some results on thermal stress of layered plates and shells by using unified formulation. J Therm Stresses 2013;36:589–625.
18. Brischetto S, Leetsch R, Carrera E, Wallmersperger T, Kröplin B. Thermo-mechanical bending of functionally graded plates. J Therm Stresses 2008;31(3):286–308.
19. Fazzolari FA, Carrera E. Thermal stability of FGM sandwich plates under various through-the-thickness temperature distributions. J Therm Stresses 2014;37:1449–81.
20. Brischetto S, Carrera E. Thermal stress analysis by refined multilayered composite shell theories. J Therm Stresses 2009;32(1–2):165–86.
21. Brischetto S, Carrera E. Heat conduction and thermal analysis in multilayered plates and shells. Mech Res Commun 2011;38:449–55.
22. Bhaskar K, Varadan TK, Ali JSM. Thermoelastic solutions for orthotropic and anisotropic composite laminates. Composites, Part B 1996;27:415–20.

CHAPTER 10

Free vibration response of uncoupled thermoelastic problems

10.1 INTRODUCTION

The free vibration behaviour of thin-walled structures under thermal and mechanical loadings is one of the most researched fields in aerospace applications due to the extremely harsh thermal environmental conditions at which aircraft structure components are subjected. It becomes very important to predict in an accurate manner the dynamics characteristics of laminated composite and sandwich plates in thermal environment being the most used thin-walled structures in aerospace applications. To this aim, some three dimensional (3D) solutions of mathematical theory of elasticity have been provided, nevertheless 3D elasticity solutions exist in a very few cases, namely, simple boundary conditions and geometries. In order to deal with more complex structures, 2D plate models have been hugely developed during the last decades. The first accurate treatment of plates can be attributed to Lagrange [1] and Germain [2] 9th century. A good historical review of the development can be referred to in the books of Soedel [3] and Timoshenko [4]. This theory is now referred to as the classical plate theory (CPT). It uses the pure bending concept of plates in the development of the equations, where normals to the middle surface remain straight and normal. It is valid for small deformation of thin plates. The inclusion of shear deformation in the fundamental equations of plates is due to Reissner [5] and Mindlin [6]. Theories that account for shear deformation are now referred to as thick plate theories or shear deformation plate theories. During the last two decades, a variable kinematics 2D models approach, for composite laminated plates and shells, has been widely used [7]. Both Navier-type closed form and finite element solutions based on the same formulation were given in many other works. Recently, Fazzolari et al. [8] proposed two advanced Ritz and Galerkin formulations expanding the scope of work of the UF. The formulations were successfully employed in a wide range of applications [9–11]. The

Thermal Stress Analysis of Composite Beams, Plates and Shells.
DOI: http://dx.doi.org/10.1016/B978-0-12-849892-7.00013-3

same authors provided a further generalization in [12] where the capability to retain or discard a particular degree or freedom in the displacement model with respect to its effectiveness and independently from its position, was added, leading to the axiomatic/asymptotic-like plate models. All the well known plate theories such as CLPT (classical lamination plate theory) based on Chauchy [13], Poisson [14] or Kircchoff [15] assumptions type, FSDT (First order shear deformation theory) [5,6], TSDT (Third order Vlasov-Reddy theory) [16] and others HSDT (High order shear deformation theories) can be obtained as particular case. With regards to sandwich plates subjected to thermal loadings a 3D elasticity solution has been provided by Noor and Burton [17]. The same subject has been widely researched employing analytical and approximate solution methods. Tenneti and Chandrashekhara [18] presented non-linear thermal dynamic analysis of graphite/aluminum composite plates subjected to rapid heating using a higher order shear deformation theory. Numerical results were presented for the nonlinear dynamic response of metal matrix based laminated plates using a full mesh of nine-noded isoparametric elements. Locke [19] evaluated the effect of heating on the vibration frequencies and mode shapes of a single-layer boron/epoxy laminate with free edge conditions. Librescu et al. [20] presented a study of the vibration response of transversely isotropic flat and curved panels subjected to temperature fields and to mechanical loads into the post-buckling load range. Matsunaga studied the free vibration of angle-ply laminated composite and sandwich plates under thermal loading [21]. The aim of the present chapter is to investigate accurately the global-local free vibration response of sandwich plates with anisotropic face sheets in thermal environment. Two different sandwich plate configurations have been accounted for and in particular the effects of several parameters such as environmental temperature, length-to-thickness ratio, face-to-thickness ratio and stacking sequence have been discussed. All the analyses have been computed by choosing a set of trigonometric trial functions in the Ritz expansion and the accuracy of the proposed formulation has been proved comparing the results both with those present in literature including the 3D elasticity solution and with the 3D FEM solution.

10.2 SANDWICH PLATE WITH CROSS-PLY FACE SHEETS

A preliminary validation and assessment of the presented formulation is carried out in Tables 10.3, 10.4 and 10.5, where the equivalent single layer, zig-zag and layer-wise hierarchical plate models have been employed

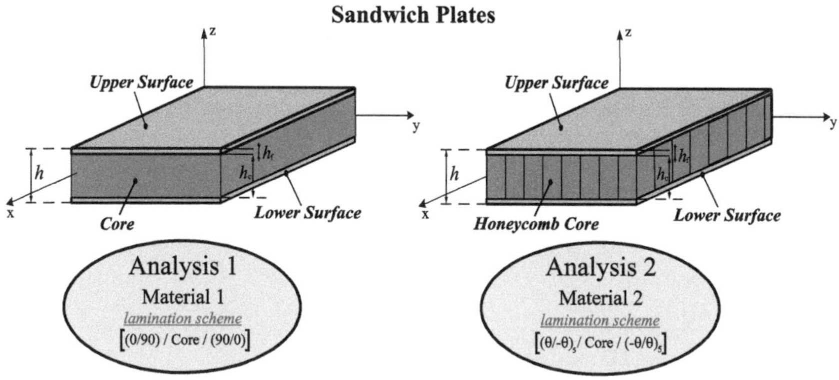

Figure 10.1 Sandwich plates geometry.

Table 10.1 List of acronyms used in tables to denote three-dimensional and two-dimensional plate theories

Acronym	Description
3D	3D exact elasticity solution by Kulkarni and Kapuria, [22]
HSDT L-G	HSDT local-global by Shariyat, [27]
ZZA	Zig-Zag Analytical by Kulkarni and Kapuria, [22]
ZZF	Zig-Zag FEM by Chakrabarti and Sheink, [28]
RTO	Refined Third Order by Kulkarni and Kapuria, [22]

Table 10.2 Sandwich plate with laminated cross-ply face sheets

Material 1						
Face Sheets						
E_1 [GPa]	E_2, E_3 [GPa]	G_{12}, G_{13} [GPa]	G_{23} [GPa]	ν_{12}, ν_{13}	ν_{23}	ρ [Kg/m^3]
276	6.9	6.9	6.9	0.25	0.30	681.8
Core						
E_1 [GPa]	E_2, E_3 [GPa]	G_{12}, G_{13} [GPa]	G_{23} [GPa]	ν_{12}, ν_{13}	ν_{23}	ρ [Kg/m^3]
0.5776	0.5776	0.1079	0.22215	0.0025	0.0025	1000

respectively, and compared to the 3D elasticity solution [22], 3D FEM solution obtained by using the 20 nodes SOLID191 element in ANSYS environment [23] and other analytical or FEM solutions available in literature (See Table 10.1). The sandwich plate taken into account is shown in *Analysis 1* in Fig. 10.1 and the material properties are given in Table 10.2. The first six dimensionless circular frequency parameters have been computed. In this particular case due to the chosen trigonometric trial functions and the cross-ply lamination of the face sheets, the Ritz analysis leads to

Table 10.3 First six dimensionless circular frequency parameters $\hat{\omega} = 100\,\omega\, a \sqrt{\frac{\rho^c}{E_1^f}}$, of simply supported square sandwich plates, lamination scheme $[0°/90°/Core/90°/0°]$, $h_f/h = 0.10$, using ED plate models and varying the length-to-thickness ratio

a/h	Theory	Circular frequency parameters						Average Err. (%)
		$\hat{\omega}_1$	$\hat{\omega}_2$	$\hat{\omega}_3$	$\hat{\omega}_4$	$\hat{\omega}_5$	$\hat{\omega}_6$	
10	3D	9.8281	15.5057	18.0752	21.6965	22.2022	26.9150	
	3D FEM†	10.114	14.8689	19.6125	20.7679	22.4418	26.7424	(0.579)
	HSDT L-G	9.8227	15.4959	18.0677	21.6912	22.2074	26.941	(−0.011)
	ZZA	9.8300	15.5100	18.0800	21.7030	22.2180	26.9300	(0.038)
	ZZF	10.052	14.410	18.963	19.422	21.250	24.489	(−3.944)
	RTO	12.088	20.615	22.152	27.675	30.143	35.329	(28.85)
		Present ED models						
	ED_{111}	18.4370	39.6209	40.1640	53.5657	64.9594	65.0662	(141.1)
	ED_{222}	18.3294	39.3267	39.8725	53.0178	64.2674	64.3820	(139.0)
	ED_{333}	11.1069	18.3475	20.3391	25.0300	26.6689	30.3581	(15.36)
	ED_{444}	11.1064	18.3475	20.3380	25.0297	26.6678	30.3561	(15.35)
	ED_{555}	10.6327	17.0994	19.7191	23.8309	24.6962	29.7656	(9.871)
	ED_{666}	10.6326	17.0991	19.7191	23.8308	24.6942	29.5406	(9.728)
	ED_{777}	10.3020	16.5101	18.9504	22.9256	23.7939	28.3214	(5.700)
	ED_{888}	10.3020	16.5100	18.9504	22.9256	23.7930	28.3214	(5.699)
20	3D	7.6882	13.8455	15.9204	19.6563	20.6760	24.9485	
	3D FEM	7.8250	13.0503	17.3013	19.0819	20.2218	24.5643	(−0.325)
	HSDT L-G	7.6876	13.8342	15.9188	19.6407	20.6683	24.9375	(−0.043)
	ZZA	7.6890	13.8475	15.9240	19.6600	20.6805	24.9540	(0.018)
	ZZF	7.929	13.045	17.320	18.838	20.097	24.143	(−0.675)
	RTO	8.721	17.705	18.530	24.105	27.714	32.136	(23.86)
		Present ED models						
	ED_{111}	10.4918	25.9842	27.3879	36.8741	48.7763	50.7054	(87.15)
	ED_{222}	10.4421	25.8635	27.2730	36.6588	48.5298	50.4674	(86.24)
	ED_{333}	8.3105	16.0414	17.4256	22.2138	24.6027	27.8205	(12.82)
	ED_{444}	8.3097	16.0410	17.4242	22.2128	24.6026	27.8186	(12.81)
	ED_{555}	8.0756	15.0449	17.0319	21.2653	22.7817	27.1542	(7.982)
	ED_{666}	8.0754	15.0449	17.0317	21.2653	22.7816	27.1540	(7.981)
	ED_{777}	7.9264	14.6234	16.5178	20.6041	22.0459	26.1559	(4.792)
	ED_{888}	7.9263	14.6234	16.5177	20.6040	22.0458	26.1557	(4.792)

† ANSYS FEM mesh 40 × 40 × 9 elements.

equal results obtained by means of the Navier-type closed-form solution, this phenomenon is known as one-term Ritz solution ($M = N = 1$). In Table 10.3, ESL models are investigated, by increasing the number of terms in the displacement field the solution quickly converges towards the exact one although even for the highest expansion order, namely ED_{888} model, both the average and the maximum errors, with respect to the 3D elasticity solution are still higher than the 5% to both moderately thick and thin sandwich plates. In Table 10.4 at the equivalent single layer displacement models has been added the MZZF, $M(z) = (-1)^k \zeta_k$, whose properties can

Table 10.4 First six dimensionless circular frequency parameters $\hat{\omega} = 100\,\omega\, a\sqrt{\frac{\rho^c}{E_1^f}}$, of simply supported square sandwich plates, lamination scheme $[0°/90°/\mathrm{Core}/90°/0°]$, $h_f/h = 0.10$, using EDZ plate models and varying the length-to-thickness ratio

a/h	Theory	Circular frequency parameters						Average Err. (%)
		$\hat{\omega}_1$	$\hat{\omega}_2$	$\hat{\omega}_3$	$\hat{\omega}_4$	$\hat{\omega}_5$	$\hat{\omega}_6$	
10	3D	9.8281	15.5057	18.0752	21.6965	22.2022	26.9150	
	3D FEM†	10.114	14.8689	19.6125	20.7679	22.4418	26.7424	(0.579)
	HSDT L-G	9.8227	15.4959	18.0677	21.6912	22.2074	26.941	(−0.011)
	ZZA	9.8300	15.5100	18.0800	21.7030	22.2180	26.9300	(0.038)
	ZZF	10.052	14.410	18.963	19.422	21.250	24.489	(−3.944)
	RTO	12.088	20.615	22.152	27.675	30.143	35.329	(28.85)
		Present EDZ models						
	EDZ_{111}	18.4369	39.6201	40.1639	53.5650	64.9591	65.0643	(141.1)
	EDZ_{222}	18.3294	39.3263	39.8725	53.0176	64.2674	64.3813	(139.0)
	EDZ_{333}	10.4366	16.7567	19.2424	23.2820	24.1448	28.7912	(7.290)
	EDZ_{444}	10.4352	16.7552	19.2381	23.2775	24.1434	28.7816	(7.272)
	EDZ_{555}	10.4158	16.7035	19.2202	23.2334	24.0743	28.7591	(7.067)
	EDZ_{666}	10.4158	16.7032	19.2202	23.2331	24.0720	28.7590	(7.065)
	EDZ_{777}	10.1914	16.2719	18.7445	22.6345	23.4052	28.0184	(4.363)
	EDZ_{888}	10.1914	16.2719	18.7444	22.6345	23.4047	28.0183	(4.363)
20	3D	7.6882	13.8455	15.9204	19.6563	20.6760	24.9485	
	3D FEM	7.8250	13.0503	17.3013	19.0819	20.2218	24.5643	(−0.325)
	HSDT L-G	7.6876	13.8342	15.9188	19.6407	20.6683	24.9375	(−0.043)
	ZZA	7.6890	13.8475	15.9240	19.6600	20.6805	24.9540	(0.018)
	ZZF	7.929	13.045	17.320	18.838	20.097	24.143	(−0.675)
	RTO	8.721	17.705	18.530	24.105	27.714	32.136	(23.86)
		Present EDZ models						
	EDZ_{111}	10.4918	25.9840	27.3878	36.8739	48.7756	50.7053	(87.15)
	EDZ_{222}	10.4421	25.8633	27.2730	36.6587	48.5293	50.4673	(86.24)
	EDZ_{333}	7.9904	14.8144	16.7112	20.8732	22.3734	26.5258	(6.103)
	EDZ_{444}	7.9893	14.8132	16.7089	20.8705	22.3720	26.5213	(6.090)
	EDZ_{555}	7.9784	14.7654	16.6982	20.8316	22.2890	26.5067	(5.888)
	EDZ_{666}	7.9782	14.7654	16.6981	20.8316	22.2888	26.5066	(5.887)
	EDZ_{777}	7.8724	14.4444	16.3773	20.3828	21.7262	25.8983	(3.695)
	EDZ_{888}	7.8722	14.4443	16.3772	20.3827	21.7262	25.8981	(3.694)

† FEM mesh 40 × 40 × 9 elements.

be found in Refs. [12,24,25]. Its introduction affects the results accuracy indeed the average and the maximum errors decrease more than the 1% reaching, for the EDZ_{888} model and moderately thick sandwich plate, the 4.363% and the 5.416% respectively. In the case of thin sandwich plate the errors percentage become 3.694% and 5.079%. In Table 10.5 layer-wise models are assessed, for all the tested expansion orders, both the average and the maximum errors are approximately zero, becoming zero for higher order such as LD_{333}, LD_{522}, LD_{444} and LD_{555}. The LD_{333} is the LW model with the less number of degrees of freedom (48 DOFs) able to

Table 10.5 First six dimensionless circular frequency parameters $\hat{\omega} = 100\,\omega\, a\sqrt{\frac{\rho^c}{E_1^f}}$, of simply supported square sandwich plates, lamination scheme $[0°/90°/\text{Core}/90°/0°]$, $h_f/h = 0.10$, using LD plate models and varying the length-to-thickness ratio

a/h	Theory	Circular frequency parameters $\hat{\omega}_1$	$\hat{\omega}_2$	$\hat{\omega}_3$	$\hat{\omega}_4$	$\hat{\omega}_5$	$\hat{\omega}_6$	Average Err. (%)
10	3D	9.8281	15.5057	18.0752	21.6965	22.2022	26.9150	
	3D FEM†	10.114	14.8689	19.6125	20.7679	22.4418	26.7424	(0.579)
	HSDT L-G	9.8227	15.4959	18.0677	21.6912	22.2074	26.941	(−0.011)
	ZZA	9.8300	15.5100	18.0800	21.7030	22.2180	26.9300	(0.038)
	ZZF	10.052	14.410	18.963	19.422	21.250	24.489	(−3.944)
	RTO	12.088	20.615	22.152	27.675	30.143	35.329	(28.85)
		Present LW models						
	LD_{111}	9.8292	15.5080	18.0785	21.7007	22.2086	26.9225	(0.020)
	LD_{222}	9.8281	15.5059	18.0752	21.6967	22.2039	26.9164	(0.003)
	LD_{225}	9.8281	15.5059	18.0752	21.6967	22.2039	26.9164	(0.003)
	LD_{252}	9.8281	15.5058	18.0752	21.6966	22.2039	26.9163	(0.002)
	LD_{522}	9.8281	15.5057	18.0752	21.6966	22.2022	26.9151	(0.000)
	LD_{333}	9.8281	15.5057	18.0752	21.6965	22.2022	26.9150	(0.000)
	LD_{444}	9.8281	15.5057	18.0752	21.6965	22.2022	26.9150	(0.000)
	LD_{555}	9.8281	15.5057	18.0752	21.6965	22.2022	26.9150	(0.000)
20	3D	7.6882	13.8455	15.9204	19.6563	20.6760	24.9485	
	3D FEM	7.8250	13.0503	17.3013	19.0819	20.2218	24.5643	(−0.325)
	HSDT L-G	7.6876	13.8342	15.9188	19.6407	20.6683	24.9375	(−0.043)
	ZZA	7.6890	13.8475	15.9240	19.6600	20.6805	24.9540	(0.018)
	ZZF	7.929	13.045	17.320	18.838	20.097	24.143	(−0.675)
	RTO	8.721	17.705	18.530	24.105	27.714	32.136	(23.86)
		Present LW models						
	LD_{111}	7.6887	13.8466	15.9223	19.6585	20.6783	24.9515	(0.010)
	LD_{222}	7.6882	13.8455	15.9205	19.6563	20.6761	24.9486	(0.000)
	LD_{225}	7.6882	13.8455	15.9205	19.6563	20.6761	24.9486	(0.000)
	LD_{252}	7.6882	13.8455	15.9205	19.6563	20.6761	24.9486	(0.000)
	LD_{522}	7.6882	13.8455	15.9205	19.6563	20.6761	24.9486	(0.000)
	LD_{333}	7.6882	13.8455	15.9204	19.6563	20.6760	24.9485	(0.000)
	LD_{444}	7.6882	13.8455	15.9204	19.6563	20.6760	24.9485	(0.000)
	LD_{555}	7.6882	13.8455	15.9204	19.6563	20.6760	24.9485	(0.000)

† FEM mesh 40 × 40 × 9 elements.

match perfectly the 3D elasticity solution leading at zero average and maximum errors. Furthermore, it is important to notice how other models such as ZZA and RTO [22] not only lead to less accurate results but introduce 7623 DOFs and 2023 DOFs respectively, increasing considerably the CPU time. This result proves again the efficiency of the layer-wise models embedded in the HTRF and how its application is mandatory when studying the free vibration behaviour of thick sandwich plates with a higher face-to-thickness ratio. In Figs. 10.2 and 10.3 the first six mode shapes along with

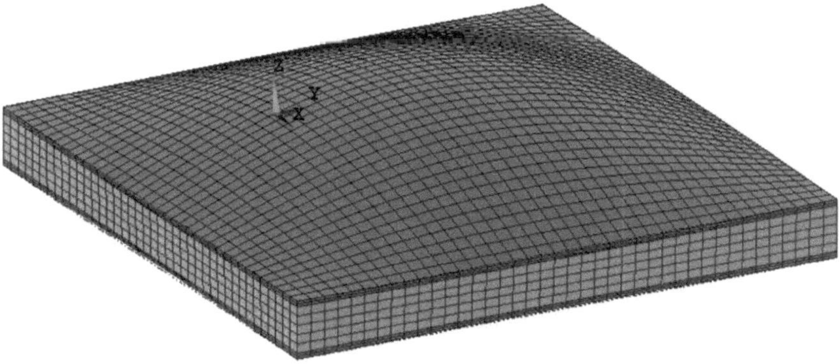

(a) 1st Mode $\hat{\omega}_1 = 10.114$

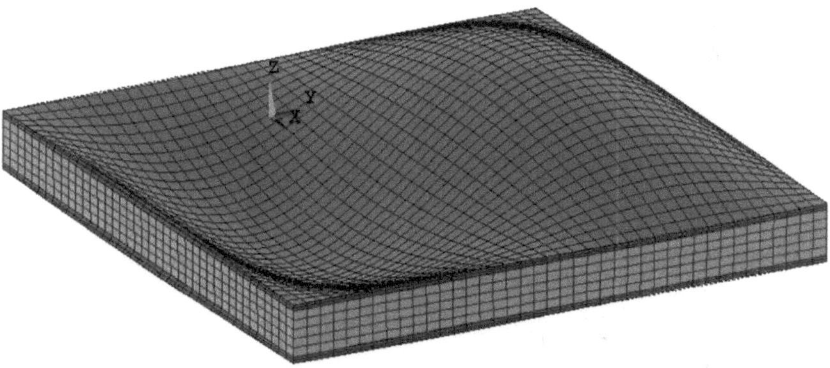

(b) 2st Mode $\hat{\omega}_2 = 14.8689$

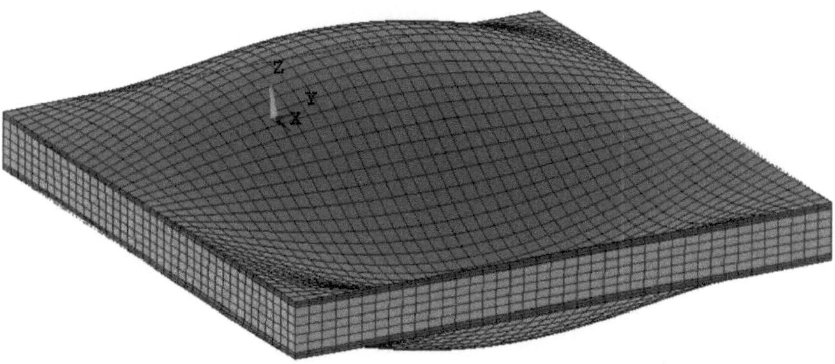

(c) 3st Mode $\hat{\omega}_3 = 19.6165$

Figure 10.2 First, second and third flexural modes of the sandwich plate in *Analysis 1* and dimensionless circular frequency parameters $\hat{\omega} = 100\,\omega\,a\sqrt{\frac{\rho^c}{E_1^f}}$.

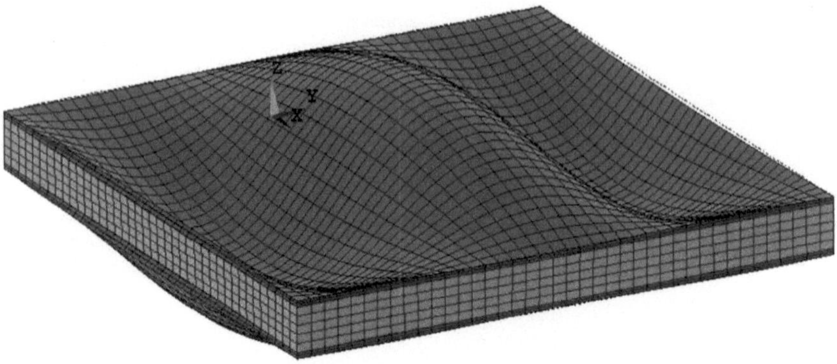

(a) 4st Mode $\hat{\omega}_4 = 20.7679$

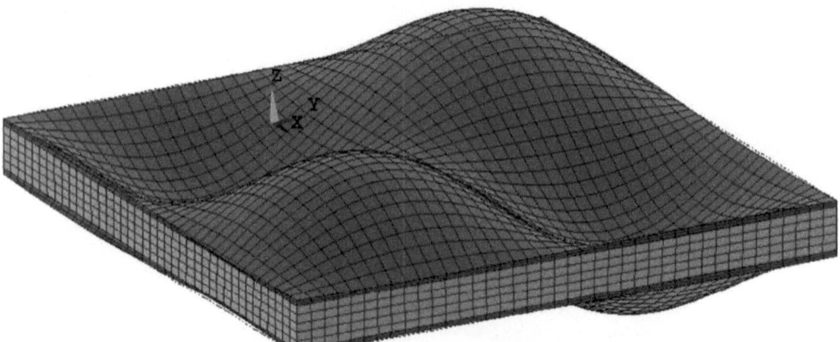

(b) 5st Mode $\hat{\omega}_5 = 22.4418$

(c) 6st Mode $\hat{\omega}_6 = 26.7424$

Figure 10.3 Fourth, fifth and sixth flexural modes of the sandwich plate in *Analysis 1* and dimensionless circular frequency parameters $\hat{\omega} = 100\,\omega\,a\sqrt{\frac{\rho^c}{E_1^f}}$.

Table 10.6 Sandwich plate with laminated angle-ply face sheets

Material 2			
Face Sheets			
E_1/E_2	G_{12}/E_2	G_{23}/E_2	ν_{12}
19.0	0.52	0.338	0.32
ν_{13}	ν_{23}	α_1/α_0	α_2/α_0
0.32	0.49	0.001	1.0
Core			
E_1/E_2^f	E_2/E_2^f	E_3/E_2^f	G_{12}/E_2^f
3.2×10^{-5}	2.9×10^{-5}	0.4	2.4×10^{-3}
G_{13}/E_2^f	G_{23}/E_2^f	ν_{12}	ν_{13}
7.9×10^{-2}	6.6×10^{-2}	0.99	3.0×10^{-5}
ν_{23}	α_1/α_0	α_2/α_0	$\rho_c/\rho^{f(1)}$
3.0×10^{-5}	1.36	1.36	0.07

the dimensionless circular frequency parameters of the moderately thick sandwich plate ($a/h=10$) are shown.

10.3 SANDWICH PLATE WITH ANGLE-PLY FACE SHEETS

The effect of the lamination angle of the face sheets on the dimensionless circular frequency parameters is studied in Tables 10.7, 10.8 and 10.9, respectively. The analysis carried out is referred to as *Analysis 2* in Fig. 10.1. The material properties of the honeycomb core and face sheets are given in Table 10.6. The general stacking sequence is $\left[(\theta°/-\theta°)_5\,/\text{core}/\,(-\theta°/\theta°)_5\right]$, where $\theta=15°,\,30°,\,45°$. The results are compared with those proposed by Matsunaga [21] and those obtained via ANSYS using the 4 nodes SHELL181 element. A good agreement is found from thick to thin plates above all employing the LD_{222} model. Moreover, compared to ANSYS results the LD_{222} model gives a significant refinement with respect to those provided by Matsunaga [21]. The used half-wave number is $M=N=1$ and a further increase does not affect the results considerably. The convergence of the presented formulation has been widely proved in Refs. [8, 9]. For all the analyzed stacking sequences the introduction of the MZZF in the ESL model does not affect the accuracy remarkably this can be related to the higher number of the face sheets employed with respect to

Table 10.7 Fundamental dimensionless circular frequency parameter $\hat{\omega} = \omega h \sqrt{\frac{\rho^f}{E_2}}$, of simply supported square sandwich plates, lamination scheme $[(15^\circ/-15^\circ)_5/\text{Core}/(-15^\circ/15^\circ)_5]$ and varying the face-to-thickness and the length-to-thickness ratios

a/h		h_f/h			
		0.05		0.1	
2	ANSYS†	0.1341^{+1}	$\Delta_{FEM}\%$	0.1120^{+1}	$\Delta_{FEM}\%$
	Matsunaga [21]	0.1357^{+1}	(1.19)	0.1155^{+1}	(3.13)
Present models	LD_{222}	0.1341^{+1}	(0.00)	0.1142^{+1}	(1.96)
	EDZ_{888}	0.1360^{+1}	(1.42)	0.1164^{+1}	(3.93)
	ED_{999}	0.1345^{+1}	(0.30)	0.1162^{+1}	(3.75)
	ED_{444}	0.1395^{+1}	(4.06)	0.1180^{+1}	(5.36)
10	ANSYS†	0.1544^{+0}	$\Delta_{FEM}\%$	0.1474^{+0}	$\Delta_{FEM}\%$
	Matsunaga [21]	0.1558^{+0}	(0.91)	0.1488^{+0}	(0.95)
Present models	LD_{222}	0.1546^{+0}	(0.13)	0.1480^{+0}	(0.41)
	EDZ_{888}	0.1559^{+0}	(0.97)	0.1497^{+0}	(1.56)
	ED_{999}	0.1551^{+0}	(0.45)	0.1495^{+0}	(1.42)
	ED_{444}	0.1576^{+0}	(2.07)	0.1512^{+0}	(2.58)
100	ANSYS†	0.1912^{-2}	$\Delta_{FEM}\%$	0.2042^{-2}	$\Delta_{FEM}\%$
	Matsunaga [21]	0.1916^{-2}	(0.21)	0.2033^{-2}	(−0.44)
Present models	LD_{222}	0.1911^{-2}	(−0.05)	0.2042^{-2}	(0.00)
	EDZ_{888}	0.1917^{-2}	(0.26)	0.2045^{-2}	(0.15)
	ED_{999}	0.1917^{-2}	(0.26)	0.2045^{-2}	(0.15)
	ED_{444}	0.1921^{-2}	(0.47)	0.2047^{-2}	(0.24)

† FEM mesh 50 × 50 elements.

$\Delta_{FEM}\% = \frac{\hat{\omega} - \hat{\omega}_{FEM}}{\hat{\omega}_{FEM}} \times 100$.

the previous case (*Analysis 1*). Thereby, as can be observed from the results in Tables 10.7, 10.8 and 10.9 increasing the ESL expansion of one order is more efficient than introducing the MZZF. For moderately thick ($a/h = 10$) and thin ($a/h = 100$) sandwich plates the fundamental circular frequency parameter increases when increasing the lamination angle in the considered rage ($\theta = 15^\circ$, 30°, 45°), a different behaviour is shown by the thick sandwich plate ($a/h = 2$). The difference percentage between the presented models and the FEM results is not affected prominently from the lamination angle but increases when increasing the face-to-thickness ratio h_f/h.

Table 10.8 Fundamental dimensionless circular frequency parameter $\hat{\omega} = \omega h \sqrt{\frac{\rho^f}{E_2}}$, of simply supported square sandwich plates, lamination scheme $[(30°/-30°)_5/\text{Core}/(-30°/30°)_5]$ and varying the face-to-thickness and the length-to-thickness ratios

a/h		h_f/h			
		0.05		0.1	
2	ANSYS†	0.1411^{+1}	$\Delta_{FEM}\%$	0.1154^{+1}	$\Delta_{FEM}\%$
	Matsunaga [21]	0.1426^{+1}	(1.06)	0.1195^{+1}	(3.55)
Present models	LD_{222}	0.1411^{+1}	(0.00)	0.1183^{+1}	(2.51)
	EDZ_{888}	0.1431^{+1}	(1.42)	0.1205^{+1}	(4.42)
	ED_{999}	0.1414^{+1}	(0.21)	0.1204^{+1}	(4.33)
	ED_{444}	0.1469^{+1}	(4.11)	0.1221^{+1}	(5.81)
10	ANSYS†	0.1743^{+0}	$\Delta_{FEM}\%$	0.1647^{+0}	$\Delta_{FEM}\%$
	Matsunaga [21]	0.1755^{+0}	(0.69)	0.1662^{+0}	(0.91)
Present models	LD_{222}	0.1748^{+0}	(0.29)	0.1657^{+0}	(0.60)
	EDZ_{888}	0.1760^{+0}	(0.98)	0.1676^{+0}	(1.76)
	ED_{999}	0.1752^{+0}	(0.52)	0.1674^{+0}	(1.64)
	ED_{444}	0.1780^{+0}	(2.12)	0.1692^{+0}	(2.73)
100	ANSYS†	0.2195^{-2}	$\Delta_{FEM}\%$	0.2345^{-2}	$\Delta_{FEM}\%$
	Matsunaga [21]	0.2195^{-2}	(0.00)	0.2328^{-2}	(−0.72)
Present models	LD_{222}	0.2195^{-2}	(0.00)	0.2345^{-2}	(0.00)
	EDZ_{888}	0.2200^{-2}	(0.23)	0.2348^{-2}	(0.13)
	ED_{999}	0.2200^{-2}	(0.23)	0.2348^{-2}	(0.13)
	ED_{444}	0.2204^{-2}	(0.41)	0.2350^{-2}	(0.21)

† FEM mesh 50 × 50 elements.

$\Delta_{FEM}\% = \frac{\hat{\omega}-\hat{\omega}_{FEM}}{\hat{\omega}_{FEM}} \times 100.$

10.4 EFFECT OF THE THERMAL ENVIRONMENT ON THE FREE VIBRATION RESPONSE

In Figs. 10.4 and 10.5 the effect of the temperature on the modal displacements u_x, u_y and u_z is evaluated. The modal displacements are dimensionalized with respect to the maximum displacement value u_i^{max} with $i = x, y, z$, respectively. The study has been focused on the sandwich plate given in *Analysis 2* (see Fig. 10.1) and using a lamination angle $\theta = 45°$. As can be observed from both figures, all the modal displacements are noticeably affected by the change of the length-to-thickness ratio a/h. On the contrary, the influence of the temperature variation is relevant only

Table 10.9 Fundamental dimensionless circular frequency parameter $\hat{\omega} = \omega h\sqrt{\frac{\rho^f}{E_2}}$, of simply supported square sandwich plates, lamination scheme $[(45°/-45°)_5/\text{Core}/(-45°/45°)_5]$ and varying the face-to-thickness and the length-to-thickness ratios

a/h		h_f/h			
		0.05		0.1	
2	ANSYS†	0.1256^{+1}	$\Delta_{FEM}\%$	0.1169^{+1}	$\Delta_{FEM}\%$
	Matsunaga [21]	0.1230^{+1}	(−2.07)	0.1204^{+1}	(2.99)
Present models	LD_{222}	0.1228^{+1}	(−2.23)	0.1192^{+1}	(1.97)
	EDZ_{888}	0.1230^{+1}	(−2.07)	0.1214^{+1}	(3.85)
	ED_{999}	0.1230^{+1}	(−2.07)	0.1213^{+1}	(3.76)
	ED_{444}	0.1232^{+1}	(−1.91)	0.1230^{+1}	(5.22)
10	ANSYS†	0.1822^{+0}	$\Delta_{FEM}\%$	0.1716^{+0}	$\Delta_{FEM}\%$
	Matsunaga [21]	0.1826^{+0}	(0.22)	0.1718^{+0}	(0.12)
Present models	LD_{222}	0.1818^{+0}	(−0.22)	0.1712^{+0}	(−0.23)
	EDZ_{888}	0.1832^{+0}	(0.55)	0.1732^{+0}	(0.93)
	ED_{999}	0.1823^{+0}	(0.05)	0.1730^{+0}	(0.82)
	ED_{444}	0.1854^{+0}	(1.76)	0.1750^{+0}	(1.98)
100	ANSYS†	0.2324^{-2}	$\Delta_{FEM}\%$	0.2482^{-2}	$\Delta_{FEM}\%$
	Matsunaga [21]	0.2322^{-2}	(−0.09)	0.2462^{-2}	(−0.81)
Present models	LD_{222}	0.2324^{-2}	(0.00)	0.2482^{-2}	(0.00)
	EDZ_{888}	0.2328^{-2}	(0.17)	0.2485^{-2}	(0.12)
	ED_{999}	0.2328^{-2}	(0.17)	0.2485^{-2}	(0.12)
	ED_{444}	0.2332^{-2}	(0.34)	0.2487^{-2}	(0.20)

† FEM mesh 50 × 50 elements.
$\Delta_{FEM}\% = \frac{\hat{\omega}-\hat{\omega}_{FEM}}{\hat{\omega}_{FEM}} \times 100$.

in the case of thick sandwich plate with a higher face-to-thickness ratio and for the dimensionless modal transverse displacement u_z/u_z^{max}. In Fig. 10.6, the dynamic behaviour of the same sandwich plate, in a thermal environment, has been investigated for two different lamination angles $\theta = 45°$ and $\theta = 90°$ varying the length-to-thickness ratio. More specifically, the first three modes of the sandwich plates have been monitored increasing the environment temperature. In all the analyzed cases the first mode decreases monotonically approaching to zero as soon as the environment temperature reaches the critical one. The second and the third frequencies decrease as well and what interesting to note is that the rate of decline increases when decreasing the length-to-thickness ratio. The mode shifting

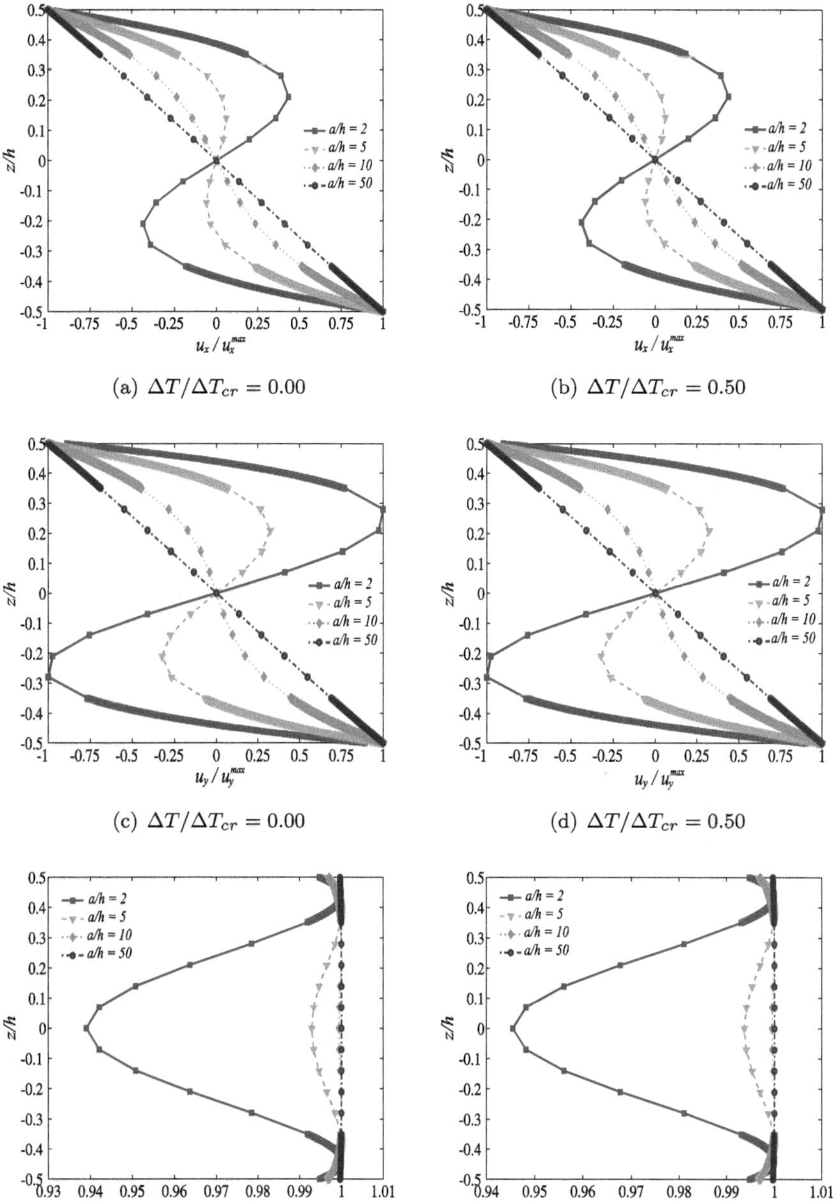

(a) $\Delta T/\Delta T_{cr} = 0.00$

(b) $\Delta T/\Delta T_{cr} = 0.50$

(c) $\Delta T/\Delta T_{cr} = 0.00$

(d) $\Delta T/\Delta T_{cr} = 0.50$

(e) $\Delta T/\Delta T_{cr} = 0.00$

(f) $\Delta T/\Delta T_{cr} = 0.50$

Figure 10.4 Dimensionless displacement modes $u_x(0, b/2)/u_x^{max}$, $u_y(a/2, 0)/u_y^{max}$ and $u_z(a/2, b/2)/u_z^{max}$ of simply supported square sandwich plates with stacking sequence $\left[(45^\circ/-45^\circ)_5/\text{Core}/(-45^\circ/45^\circ)_5\right]$, $h_f/h = 0.15$ and using a ED_{444} model.

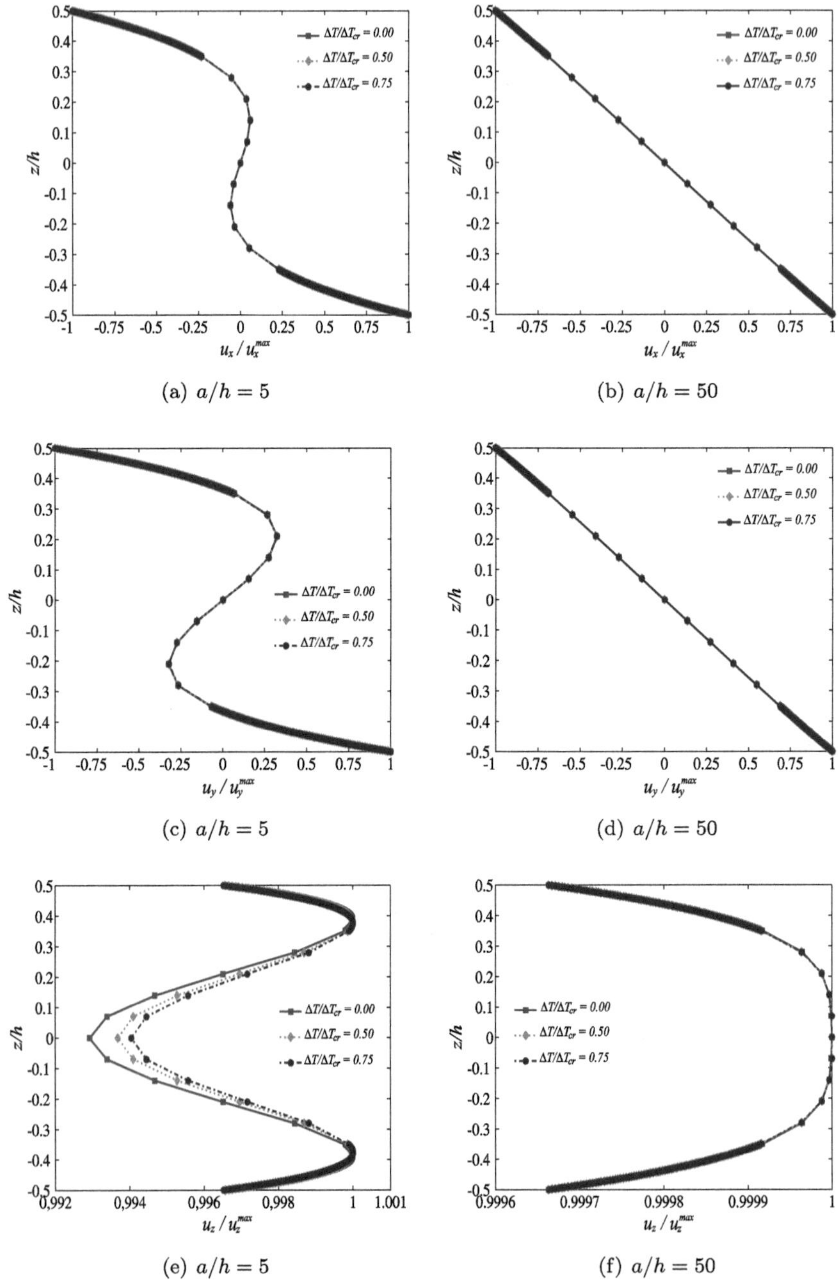

Figure 10.5 Dimensionless displacement modes $u_x(0, b/2)/u_x^{max}$, $u_y(a/2, 0)/u_y^{max}$ and $u_z(a/2, b/2)/u_z^{max}$ of simply supported square sandwich plates with stacking sequence $\left[(45°/-45°)_5\,/\text{Core}/\,(-45°/45°)_5\right]$, $h_f/h = 0.15$ and using a ED_{444} model.

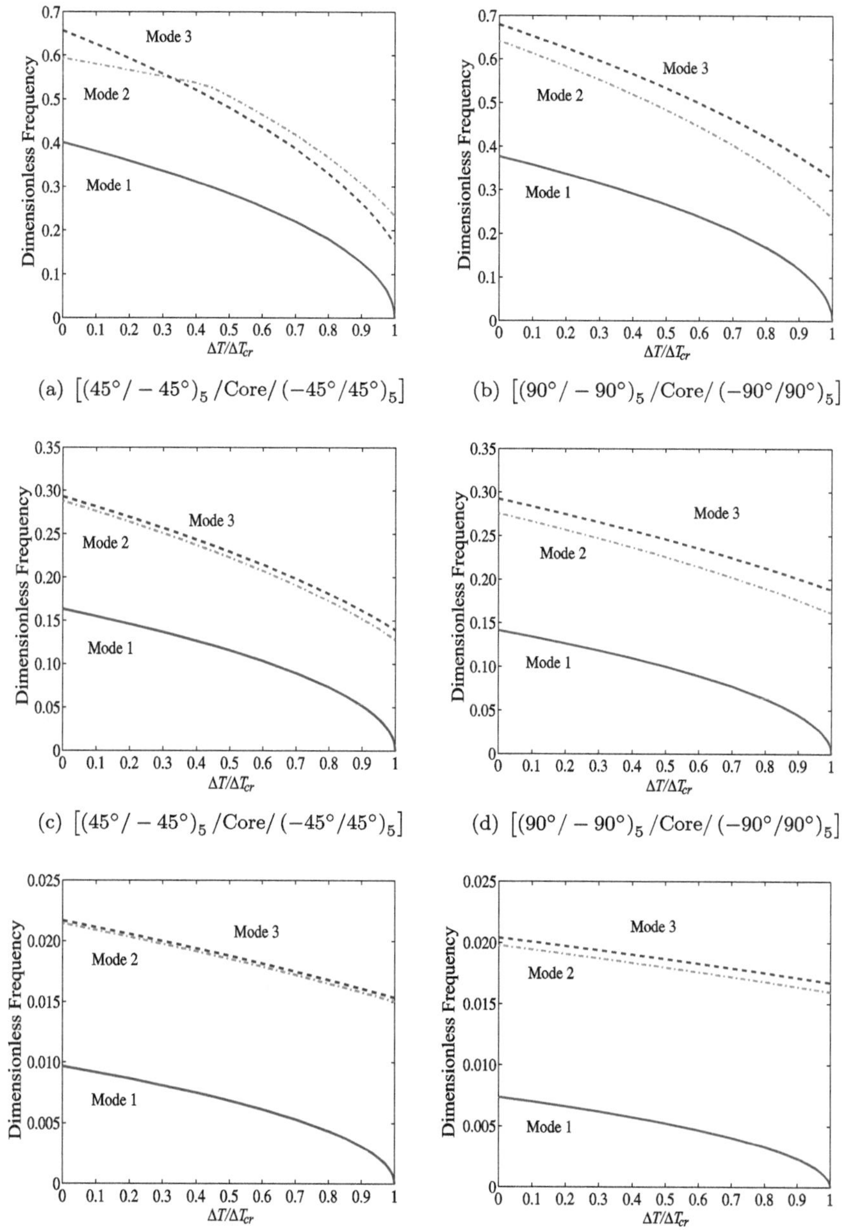

(a) $[(45°/-45°)_5\,/\text{Core}/\,(-45°/45°)_5]$ (b) $[(90°/-90°)_5\,/\text{Core}/\,(-90°/90°)_5]$

(c) $[(45°/-45°)_5\,/\text{Core}/\,(-45°/45°)_5]$ (d) $[(90°/-90°)_5\,/\text{Core}/\,(-90°/90°)_5]$

(e) $[(45°/-45°)_5\,/\text{Core}/\,(-45°/45°)_5]$ (f) $[(90°/-90°)_5\,/\text{Core}/\,(-90°/90°)_5]$

Figure 10.6 Effect of the environment temperature change on the first three dimensionless circular frequency parameters $\hat{\omega} = \omega h \sqrt{\frac{\rho^f}{E_2}}$ of simply supported square sandwich plates varying the length-to-thickness ratio, $a/h = 5$ ((a) and (b)), $a/h = 10$ ((c) and (d)) and $a/h = 50$ ((e) and (f)), $h_f/h = 0.15$ and using a ED_{444} model.

(or modal interchange) phenomenon, namely, the modes change in order of appearance [9,26], is observed only in the case of thick sandwich plate $a/h = 5$ and face sheets with lamination angle $\theta = 45°$.

REFERENCES

1. Lagrange JL. Note communiquee aux commissaires pour le prix de la surface elastique (December 1811) [Note communicated to the commissioners for the price of the surface elastic]. Ann Chim 1828;39:149–207.
2. Germain S. Researches sur la theorie des surfaces elastiques [Researches on the theory of elastic surfaces]. Paris, 1821.
3. Soedel W. Vibrations of shells and plates, vol. 39. New York: Marcel Dekker; 2004. p. 149–207.
4. Timoshenko SP. History of strength of materials, vol. 39. New York: Dover Publication; 2004. p. 149–207.
5. Reissner E. The effect of transverse shear deformation on the bending of elastic plates. J Appl Mech 1945;67:A67–77.
6. Mindlin R. Influence of rotary inertia and shear on flexural motions of isotropic elastic plates. J Appl Mech 1951;18:31–8.
7. Carrera E. Theories and finite elements for multilayered plates and shells: a unified compact formulation with numerical assessment and benchmarking. Arch Comput Methods Eng 2003;10(3):216–96.
8. Fazzolari FA, Carrera E. Advanced variable kinematics Ritz and Galerkin formulation for accurate buckling and vibration analysis of laminated composite plates. Compos Struct 2011;94:50–67.
9. Fazzolari FA, Carrera E. Accurate free vibration analysis of thermo-mechanically pre/post-buckled anisotropic multilayered plates based on a refined hierarchical trigonometric Ritz formulation. Compos Struct 2013;95:381–402.
10. Fazzolari FA, Carrera E. Thermo-mechanical buckling analysis of anisotropic multilayered composite and sandwich plates by using refined variable-kinematics theories. J Therm Stress 2013;36:321–50.
11. Fazzolari FA, Carrera E. Coupled thermoelastic effect in free vibration analysis of anisotropic multilayered plates by using an advanced variable-kinematics Ritz formulation. Eur J Mech Solid/A 2014;44:157–74.
12. Fazzolari FA, Banerjee JR. Axiomatic/asymptotic PVD/RMVT-based shell theories for free vibrations of anisotropic shells using an advanced Ritz formulation and accurate curvature descriptions. Comp Struc 2014;108:91–110.
13. Chauchy AL. Sur l'equilibre et le mouvement des corps elastique. Excer Math 1828;3:328–55.
14. Poisson SD. Memoire sur l'equilibre et le mouvement des corps elastique [Memorandum on the equilibrium and motion of elastic bodies]. Ann Chim 1828;3:337–55.
15. Kirchhoff GR. Uber das gleichgewicht und diw bewegung einer elasticien [About the equilibrium and motion of elastic bodies]. J Angew Math 1850;40:51–88.
16. Reddy JN. A simple high order theory for laminated composite plates. J Appl Mech 1984;95:745–52.
17. Noor AK, Burton WS. Three-dimensional solutions for thermal buckling multilayered anisotropic plates. J Eng Mech 1992;118:638–701.

18. Tenneti R, Chandrashekhara K. Nonlinear thermal dynamic analysis of graphite/aluminum composite plates. AIAA J 1994;32:1931–3.
19. Locke J. Vibration analysis of heated anisotropic plates with free edge conditions. J Aircraft 1994;31:696–702.
20. Librescu L, Lin W, Nemeth MP, Starnes JHJ. Vibration of geometrically imperfect panels subjected to thermal and mechanical loads. J Spacecraft Rockets 1996;33:285–91.
21. Matsunaga H. Free vibration and stability of angle-ply laminated composite and sandwich plates under thermal loading. Compos Struct 2007;77:249–62.
22. Kulkarni SD, Kapuria S. Free vibration analysis of composite and sandwich plates using an improved discrete Kirchhoff quadrilateral element based on third order zigzag theory. Comput Mech 2008;42(6):803–24.
23. ANSYS v10.0 theory manual. Southpointe (PA): ANSYS Inc.; 2006.
24. Demasi L. Refined multilayered plate elements based on Murakami zig-zag functions. Compos Struct 2005;70:308–16.
25. Ferreira AJM, Roque CMC, Carrera E, Cinefra M, Polit O. Radial basis functions collocation and a unified formulation for bending, vibration and buckling analysis of laminated plates, according to a variation of Murakami's zig-zag theory. Eur J Mech A, Solids 2011;3:559–70.
26. Giunta G, Biscani F, Belouettar S, Ferreira AJM, Carrera E. Free vibration analysis of composite beam via refined theories. Composites, Part B 2013;44(1):540–52.
27. Shariyat M. A generalized global–local high-order theory for bending and vibration analyses of sandwich plates subjected to thermo-mechanical loads. Int J Mech Sci 2010;52(3):495–514.
28. Chakrabarti A, Sheikh AH. Vibration of laminate-faced sandwich plate by a new refined element. Proc Inst Mech Eng, G J Aerosp Eng 2004;17:123–34.

CHAPTER 11

Static and dynamic responses of coupled thermoelastic problems

11.1 INTRODUCTION

Thermo-mechanical coupling effects play a crucial role in the operational life of aerospace structures. They often represent the main cause of failure and as a consequence they should be considered in the earlier stages of the design process. Thin-walled parts of reactor vessels, turbines and the structures of future supersonic and hypersonic vehicles (such as high-speed civil transport and advanced tactical fighters) are particularly susceptible to failure resulting from excessive stress level induced by thermal or combined thermo-mechanical loadings [1]. Thermal deformation also plays a fundamental role in multilayered thin-film zones, comprising optical mirrors [2], as well as space reflector antennas, which require stringent geometric tolerance compared to traditional structures [3].

Thus, there is the need to fully understand the complex behaviour of thermal structures. In this respect, the research community has focused its efforts towards the development of advanced thermoelastic formulations. Various authors (see Refs. [4–7]) have shown that refined structural models are required when dealing with thermal problems. More specifically, advanced models require the fulfilment of the C_z^0-requirements summarized in Ref. [8]. The latter account for zig-zag effects and inter-laminar continuity condition. Both displacements and transverse and normal stresses must be C_z^0-continuous functions in the plate/shell-thickness direction. On the importance of using refined theories, it is worth mentioning the conclusion drawn by Murakami [2]: "in order to predict rapid variation of transverse normal strain, a plate theory with cubic variation of in-plane displacement in each layer, rather than over the entire plate thickness should be adopted. Otherwise full three-dimensional analyses are recommended." In other words, due to the intrinsic through-the-thickness variation of thermal loadings, a Layer-Wise (LW) kinematic description is required to calculate accurately the local response of layered structures. Further insights on this topic were given by Bhaskar and Varadan [9]. They provided various thermoelastic solutions for laminated composite structures and underlined the

Thermal Stress Analysis of Composite Beams, Plates and Shells.
DOI: http://dx.doi.org/10.1016/B978-0-12-849892-7.00014-5

importance of using refined theories to cope with thermal structures problems.

Thermo-mechanical problems are usually solved by employing 3D Finite Elements (FEs) available in commercial FE software. These codes are based on *partially coupled* thermoelastic formulations. Thus, they can provide the mechanical deformation due to a thermal loading, but do not allow the computation of the temperature change caused by a mechanical loading. The latter, as already highlighted in this book, is not always a negligible effect. Moreover, solid elements are particularly inconvenient when multilayered structures have to be analyzed. In fact, at least three solid elements should be placed through the thickness of each layer in order to have reasonable results and this leads to heavy computational costs. It is implicit that the use of plate elements instead of solid elements would decrease the number of DOFs involved in the analysis.

Refined thermo-mechanical FEs based on the formulation proposed in Ref. [10] are employed in the following analyses. Only inplane stress results are listed in the present chapter since the exact solution for transverse stresses is not available in the reference paper. However, the full three dimensional state of stress, including all transverse stress components, is required in case of failure analysis of composite multilayered structures. Additional information on this topic can be found in Refs. [11–13]. The proposed FEs are *fully-coupled*: if a mechanical loading is applied to the structure, the temperature variation caused by mechanical deformations can be calculated as an instantaneous effect. This last point is crucial when thermographic experiments are conducted in order to recover the three-dimensional stress field starting from a temperature distribution [14,15]. There are some other rare cases when coupled interaction from mechanical to thermal field can be significant. This is the case of heat dissipation processes or large thermo-mechanical excitations that can be found in forming processes.

Analytical solutions for the stress analysis of multilayered plates subjected to thermal loading were proposed in Ref. [16], with particular emphasis on stress field calculation referring to the Principle of Virtual Displacement (PVD) and Reissner's Mixed Variational Theorem (RMVT) variational statements. The corresponding plate FEs were proposed in Refs. [17,18], respectively for PVD and RMVT applications. The effect of the thermo-mechanical coupling was briefly presented in Ref. [19]. Fourier's law was applied to calculate the plate-thickness steady-state temperature profile in previous articles.

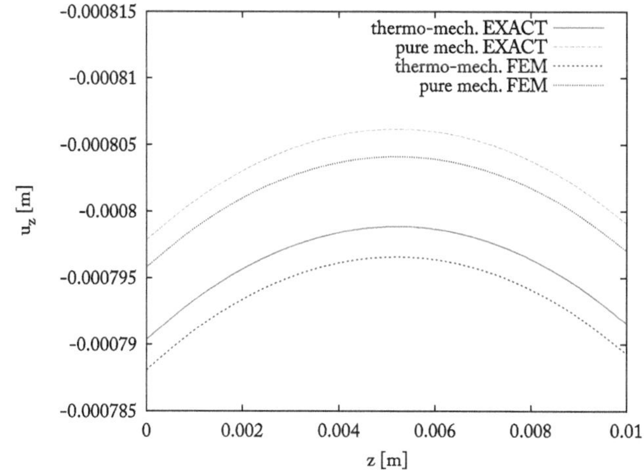

Figure 11.1 Through-the-thickness displacement at the plate position $(\frac{a}{2}, \frac{b}{2})$ – a regular 23 × 23 mesh of Q4 LD4 FEs is employed.

11.2 MECHANICAL LOADING: STATIC INSTANTANEOUS THERMO-MECHANICAL ANALYSIS

A simply supported square aluminum plate is considered to investigate the full coupling between temperature and mechanical fields. The analysis of a simple isotropic plate is proposed. Results are shown in dimensional form so that numbers can be judged by engineering sense. The mechanical properties are: $E = 73$ $[GPa]$, $\nu = 0.3$, $\alpha = 25E{-}6$ $[K^{-1}]$, $C = 897$ $[J/(kg \cdot K)]$, $\rho = 2800$ $[kg/m^3]$. The plate dimensions are: length and width $a = b = 10$ $[m]$, thickness $h = 1$ $[m]$. The plate is loaded at the top face by the bi-sinusoidal pressure of peak value $p_m = 2E7$ $[Pa]$. Such load leads to a stress condition very close to the material elastic limit: $\sigma_{xx} = \sigma_{yy} \simeq 404E6$ $[Pa]$. The present case study provides an example of the maximum thermal variation caused by a mechanical loading, within the frame of a linear static analysis. A 23 × 23 mesh of LD4 Q4 elements is employed. In general there is not a substantial difference in terms of displacements between those two analyses for the displacements u_x and u_y. A slight difference can be seen in Fig. 11.1, where the transverse displacement u_z is plotted along the z-axis, which has the origin at the bottom of the plate. The difference between this two analyses is very small and it is due to the fact that for a fully coupled thermo-mechanical analysis, a small part of the work is employed to modify the temperature of the plate. The u_z displacement for this kind of analysis is smaller than the one obtained by a simple mechanical analysis,

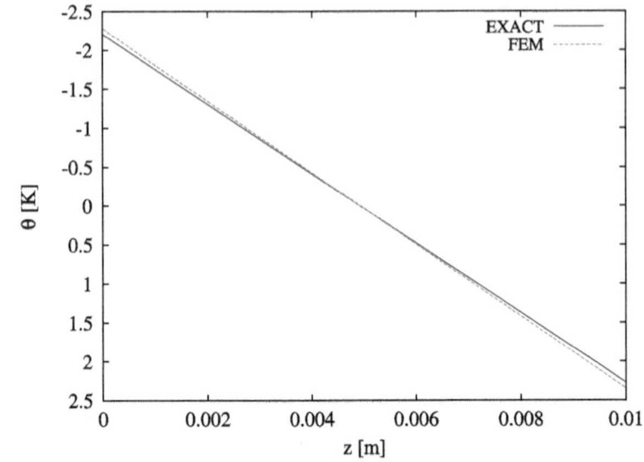

Figure 11.2 Through-the-thickness temperature profile at the plate position $(\frac{a}{2}, \frac{b}{2})$ – a regular 23 × 23 mesh of Q4 LD4 FEs is employed.

Figure 11.3 View of the plate under static deformation; the colors at the top and bottom faces indicate the instantaneous temperature variation – unit of measurement $[K]$. For interpretation of the references to color in this figure legend, the reader is referred to the web version of this chapter.

because a part of the work is used to develop the temperature variation. The same amount of work is subtracted from the work used to deform the plate. It can be seen that only a small part of work has been used to modify the temperature. In fact, Fig. 11.2 shows that the maximum thermal variation is lower than 2.5 $[K]$ (with the stresses close to the elastic limit). Figure 11.3 gives a three-dimensional representation of the thermal field calculated at the top/bottom face of the plate (the thickness axis is scaled unequally in order to make the temperature field more visible). It is implicit that the calculated temperature profile concerns the first moment in which the *static* load is applied to the structure. In fact, after the transient heat change in the plate itself, the temperature variation would be equal to zero throughout the plate and the displacement field will become the one calculated in a pure mechanical analysis.

To be remarked that, in accordance with the proposed approach, the inclusion of thermo-mechanical full coupling leads to the addition of one DOF at each row and column of the stiffness fundamental nucleus, with consequent increase of the computational time. The author experienced that solving a generic fully coupled static analysis by using direct methods requires almost twice than the corresponding pure mechanical problem. Additional efforts should be devoted in order to optimize the solver referring to iterative methods for sparse systems.

11.3 THERMAL LOADING: HIGHER-ORDER EFFECTS ON DISPLACEMENTS AND STRESS RESULTS

In this section several Q9 FEM results of thermo-mechanical static analysis of multilayered anisotropic plates obtained compared to the 3D exact solutions found by Bhaskar and Varadan [9]. Numerical results are presented for the bending of a (0°/90°/0°) square ($a = b$) laminate due to the temperature field given by

$$\theta = \overline{\theta}(2z/h)sin(\frac{\pi x}{a})sin(\frac{\pi y}{b}), \tag{11.1}$$

where h is the total thickness of the laminate and the origin of the z-axis is at the centre of the plate. Different thickness ratio are considered, ranging from very thick ($a/h = 2$) to very thin ($a/h = 100$) plates. The following material properties, typical of high-modulus graphite/epoxy, are assumed:

$$\begin{aligned} &E_L/E_T = 25;\\ &G_{LT}/E_T = 0.5;\\ &G_{TT}/E_T = 0.2;\\ &\nu_{LT}/\nu_{TT} = 0.25;\\ &\alpha_T/\alpha_L = 1125, \end{aligned}$$

where L and T refer to directions parallel and perpendicular, respectively, to the fibres. In the FEM analysis, the dimensional values given for material properties are though coherently with the above given dimensionless ratios.

The distribution of temperature along the thickness, for given thermal boundary conditions at the top and at the bottom surfaces, can be obtained by solution of the heat conduction equation (see, for example, the work of

Table 11.1 z values are given in parentheses; (x, y) values are: $(a/2, a/2)$ for $\widetilde{u}_z$, $\widetilde{\sigma}_{xx}$ and $\widetilde{\sigma}_{yy}$; $(0, a/2)$ for $\widetilde{u}_x$; $(a/2, 0)$ for $\widetilde{u}_y$ and $(0, 0)$ for $\widetilde{\sigma}_{xy}$ – a regular 6×6 mesh of Q9 LD1 FEs is employed

$\frac{a}{h}$	$sol.(\mp\frac{h}{2})$	$\widetilde{u}_x$	$\widetilde{u}_y$	$\widetilde{u}_z$	$\widetilde{\sigma}_{xx}$	$\widetilde{\sigma}_{yy}$	$\widetilde{\sigma}_{xy}$
2	3D [9]	±20.04	±151.4	96.79	±1390	±635.4	±269.3
2	FEM	±12.14	±155.0	86.71	±643.4	±741.6	±268.6
4	3D [9]	±18.11	±81.83	42.69	±1183	±856.1	±157.0
4	An [16]	–	–	41.24	–	–	–
4	FEM	±15.95	±84.05	40.95	±895.2	±965.3	±160.7
10	3D [9]	±16.61	±31.95	17.39	±1026	±1014	±76.29
10	FEM	±17.01	±33.02	17.61	±935.6	±1132	±80.46
20	3D [9]	±16.17	±20.34	12.12	±982.0	±1051	±57.35
20	FEM	±17.06	±21.32	12.66	±929.0	±1170	±61.75
50	3D [9]	±16.02	±16.71	10.50	±967.5	±1063	±51.41
50	FEM	±17.05	±17.70	11.14	±925.1	±1182	±55.87
100	3D [9]	±16.00	±16.17	10.26	±965.4	±1065	±50.53
100	An [16]	–	–	10.92	–	–	–
100	FEM	±17.04	±17.19	10.91	±923.6	±1184	±54.97
CLT	EX. [9]	±15.99	±15.99	10.18	±964.6	±1065	±50.24

Tungikar and Rao [6]). However, only simple linear antisymmetric (with respect to z) thermal variation is considered as this would lead to bending (without stretching) of a symmetric laminate and would be adequate to bring out the importance of non-classical influences such as shear deformation and thickness stretch.

The deflections and stresses are presented in terms of the following dimensionless parameters (sign ~):

$$\widetilde{u}_z = \frac{u_z}{h\alpha_L\overline{\theta}(\frac{a}{h})^2}, \qquad (\widetilde{u}_x, \widetilde{u}_y) = \frac{(u_x, u_y)}{h\alpha_L\overline{\theta}(a/h)}, \qquad \widetilde{\sigma}_{ij} = \frac{\sigma_{ij}}{E_T\alpha_L\overline{\theta}}.$$

In Tables 11.1–11.4 there is a comparison between the 3D exact solution [9] and the FEM results obtained with the proposed thermo-mechanical FEs, with the through-the-thickness expansion ranging from one to four. Even if the employed mesh is not particularly refined, FEM results obtained with higher-order thickness expansion (e.g. LD3/LD4) are in very good agreement with the exact solution, both for displacement and for in-plane stresses, wherever is the plate thickness ratio. In the other side, results obtained by LD1 FEs are in general accurate only for thin multilayered plates. See the graphic view of Tables 11.1–11.4 in Figs. 11.4(a)–11.4(f).

Table 11.2 z values are given in parentheses; (x, y) values are: $(a/2, a/2)$ for $\tilde{u}_z$, $\tilde{\sigma}_{xx}$ and $\tilde{\sigma}_{yy}$; $(0, a/2)$ for $\tilde{u}_x$; $(a/2, 0)$ for $\tilde{u}_y$ and $(0, 0)$ for $\tilde{\sigma}_{xy}$ – a regular 6×6 mesh of Q9 LD2 FEs is employed

$\frac{a}{h}$	$sol.(\mp\frac{h}{2})$	$\tilde{u}_x$	$\tilde{u}_y$	$\tilde{u}_z$	$\tilde{\sigma}_{xx}$	$\tilde{\sigma}_{yy}$	$\tilde{\sigma}_{xy}$
2	3D [9]	±20.04	±151.4	96.79	±1390	±635.4	±269.3
2	FEM	±16.77	±140.3	88.26	±1141	±675.0	±252.0
4	3D [9]	±18.11	±81.83	42.69	±1183	±856.1	±157.0
4	An [16]	–	–	42.25	–	–	–
4	FEM	±17.47	±79.81	41.50	±1159	±860.3	±156.1
10	3D [9]	±16.61	±31.95	17.39	±1026	±1014	±76.29
10	FEM	±16.58	±31.77	17.32	±1052	±1013	±77.70
20	3D [9]	±16.17	±20.34	12.12	±982.0	±1051	±57.35
20	FEM	±16.19	±20.27	12.11	±1012	±1050	±58.62
50	3D [9]	±16.02	±16.71	10.50	±967.5	±1063	±51.41
50	FEM	±16.04	±16.67	10.50	±997.0	±1062	±52.57
100	3D [9]	±16.00	±16.17	10.26	±965.4	±1065	±50.53
100	An [16]	–	–	10.26	–	–	–
100	FEM	±16.01	±16.16	10.26	±994.0	±1063	±51.65
CLT	EX. [9]	±15.99	±15.99	10.18	±964.6	±1065	±50.24

Table 11.3 z values are given in parentheses; (x, y) values are: $(a/2, a/2)$ for $\tilde{u}_z$, $\tilde{\sigma}_{xx}$ and $\tilde{\sigma}_{yy}$; $(0, a/2)$ for $\tilde{u}_x$; $(a/2, 0)$ for $\tilde{u}_y$ and $(0, 0)$ for $\tilde{\sigma}_{xy}$ – a regular 6×6 mesh of Q9 LD3 FEs is employed

$\frac{a}{h}$	$sol.(\mp\frac{h}{2})$	$\tilde{u}_x$	$\tilde{u}_y$	$\tilde{u}_z$	$\tilde{\sigma}_{xx}$	$\tilde{\sigma}_{yy}$	$\tilde{\sigma}_{xy}$
2	3D [9]	±20.04	±151.4	96.79	±1390	±635.4	±269.3
2	FEM	±18.55	±142.3	90.33	±1300	±656.7	±257.8
4	3D [9]	±18.11	±81.83	42.69	±1183	±856.1	±157.0
4	An [16]	–	–	42.68	–	–	–
4	FEM	±17.82	±80.42	41.92	±1191	±855.0	±157.8
10	3D [9]	±16.61	±31.95	17.39	±1026	±1014	±76.29
10	FEM	±16.61	±31.81	17.34	±1055	±1012	±77.82
20	3D [9]	±16.17	±20.34	12.12	±982.0	±1051	±57.35
20	FEM	±16.19	±20.28	12.11	±1012	±1050	±58.64
50	3D [9]	±16.02	±16.71	10.50	±967.5	±1063	±51.41
50	FEM	±16.04	±16.67	10.50	±997.1	±1062	±52.57
100	3D [9]	±16.00	±16.17	10.26	±965.4	±1065	±50.53
100	An [16]	–	–	10.26	–	–	–
100	FEM	±16.01	±16.16	10.26	±994.1	±1063	±51.65
CLT	EX. [9]	±15.99	±15.99	10.18	±964.6	±1065	±50.24

Table 11.4 z values are given in parentheses; (x, y) values are: $(a/2, a/2)$ for $\tilde{u}_z$, $\tilde{\sigma}_{xx}$ and $\tilde{\sigma}_{yy}$; $(0, a/2)$ for $\tilde{u}_x$; $(a/2, 0)$ for $\tilde{u}_y$ and $(0, 0)$ for $\tilde{\sigma}_{xy}$ – a regular 6 × 6 mesh of Q9 LD4 FEs is employed

$\frac{a}{h}$	$sol.(\mp\frac{h}{2})$	$\tilde{u}_x$	$\tilde{u}_y$	$\tilde{u}_z$	$\tilde{\sigma}_{xx}$	$\tilde{\sigma}_{yy}$	$\tilde{\sigma}_{xy}$
2	3D [9]	±20.04	±151.4	96.79	±1390	±635.4	±269.3
2	FEM	±18.65	±142.4	90.40	±1309	±654.7	±258.4
4	3D [9]	±18.11	±81.83	42.69	±1183	±856.1	±157.0
4	An [16]	–	–	42.69	–	–	–
4	FEM	±17.82	±80.44	41.92	±1191	±854.9	±157.8
10	3D [9]	±16.61	±31.95	17.39	±1026	±1014	±76.29
10	FEM	±16.61	±31.81	17.34	±1055	±1012	±77.82
20	3D [9]	±16.17	±20.34	12.12	±982.0	±1051	±57.35
20	FEM	±16.19	±20.28	12.11	±1012	±1050	±58.64
50	3D [9]	±16.02	±16.71	10.50	±967.5	±1063	±51.41
50	FEM	±16.04	±16.67	10.50	±997.1	±1062	±52.57
100	3D [9]	±16.00	±16.17	10.26	±965.4	±1065	±50.53
100	An [16]	–	–	10.26	–	–	–
100	FEM	±16.01	±16.16	10.26	±994.1	±1063	±51.65
CLT	EX. [9]	±15.99	±15.99	10.18	±964.6	±1065	±50.24

From these results it is deduced that, when the plate is very thick, LD1 and LD2 modelling do not fully describe higher-order/three-dimensional effects. In other words, low-order kinematic descriptions are not appropriate to analyze extremely thick plates. In order to better investigate this point, it can be noted that the available analytical solution for the LD1 (Table 11.1, $a/h = 4$) is different from the 3D result and so the correspondent FE is not expected to converge to the 3D solution. The same reasoning can be applied for stress results (with amplified discrepancies). Figures 11.4(a)–11.4(f) illustrate that rising the order of the kinematic description permits to approach 3D results also in case of very thick plates. A slight error in the convergence of FEM results could be due to the fact that some quantities are calculated on the plate edges, where boundary conditions are also applied.

Figures 11.5(a), 11.5(b) show the convergence study respectively for $u_z(\frac{a}{2}, \frac{a}{2}, \frac{h}{2})$ and $\sigma_{zz}(\frac{a}{2}, \frac{a}{2}, \frac{h}{2})$, with $a/h = 20$. It can be concluded that FEM results converge to the 3D exact values [9] and that accurate results for displacements and in-plane stresses are obtained with Q9 FEs, even with a poor mesh. Moreover, it is confirmed that the convergence is much faster for displacements than for stresses. Further details as

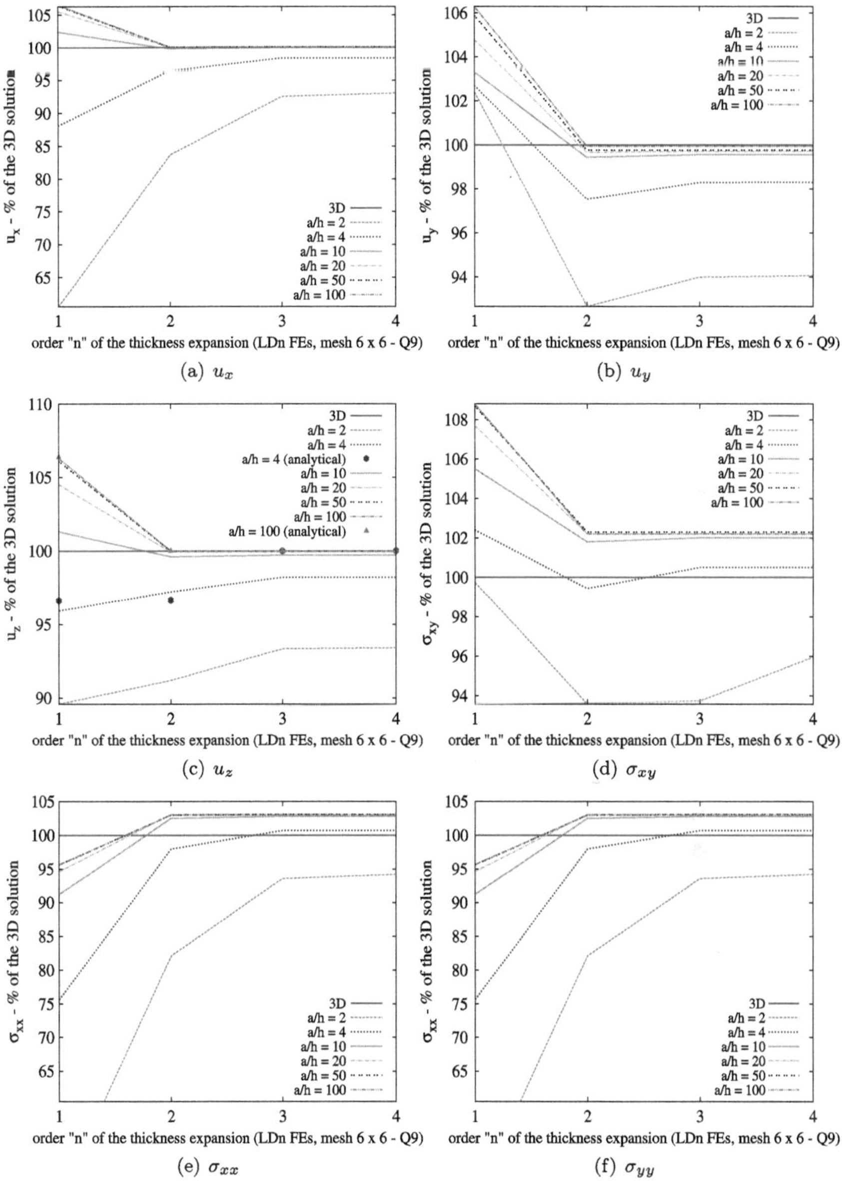

Figure 11.4 FEM and 3D solutions for displacement and in-plane stress components.

well as results on the thermomechanical coupling effect can be found in Ref. [19].

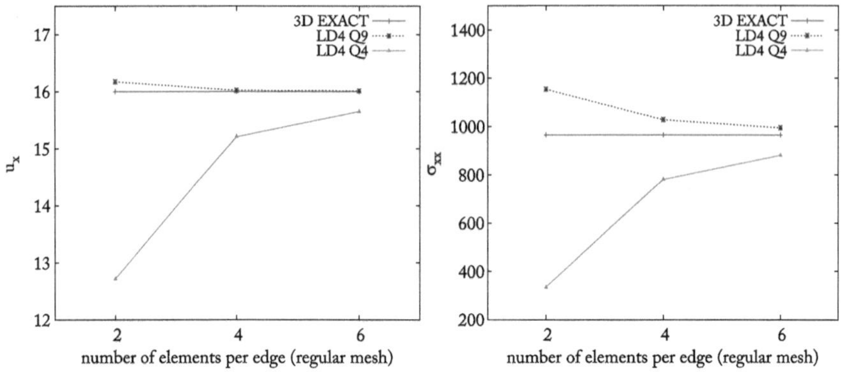

(a) u_x displacement component.

(b) σ_{xx} stress component.

Figure 11.5 Convergence study.

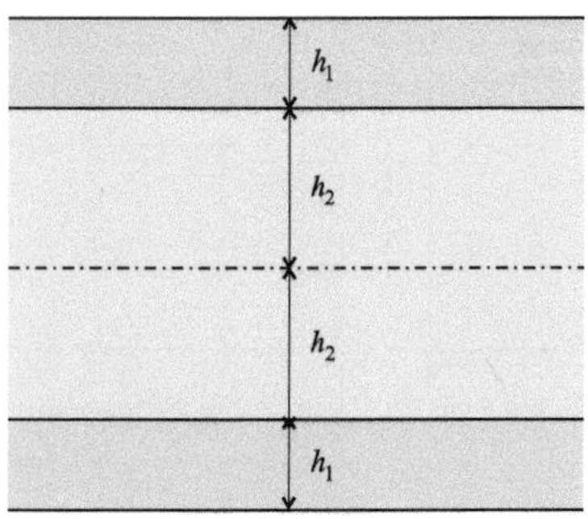

Figure 11.6 Plate-thickness geometry description.

11.4 THERMAL LOADING: ASSESSMENT OF TEMPERATURE PROFILE, STEADY-STATE SOLUTION

A simply supported three-layered square plate of edge $a = 0.1$ $[m]$ is considered. The top and bottom layer are made of aluminum, having equal thickness h_1. The middle layer is made of steel and has thickness $2 \times h_2$ (Fig. 11.6). Aluminum material properties are $E = 73.0E9$ $[Pa]$, $G = 27.239E9$ $[Pa]$, $\nu = 0.34$, $\alpha = 25.D - 6$ $[K^{-1}]$, $\kappa = 180$ $[W/(m \cdot K)]$, $\rho = 2800$ $[kg/m^3]$ and $C = 897\, J/(K \cdot kg)$, where α is the coefficient of thermal expansion. Steel coefficients are $E = 210.0E9$ $[Pa]$, $G = 80.77E9$ $[Pa]$, $\nu = 0.3$, $\alpha = 11.1E - 6$ $[K^{-1}]$, $\kappa = 13$ $[W/(m \cdot K)]$, $\rho = 7860$ $[kg/m^3]$, $C = 450\, J/(K \cdot kg)$. The T_{ref} is set to 298.15 $[K]$. A temperature variation respect to T_{ref} is imposed at the top and at the bottom face of the panel: $+10$ $[K]$ and -10 $[K]$ respectively. A coupled thermo-mechanical

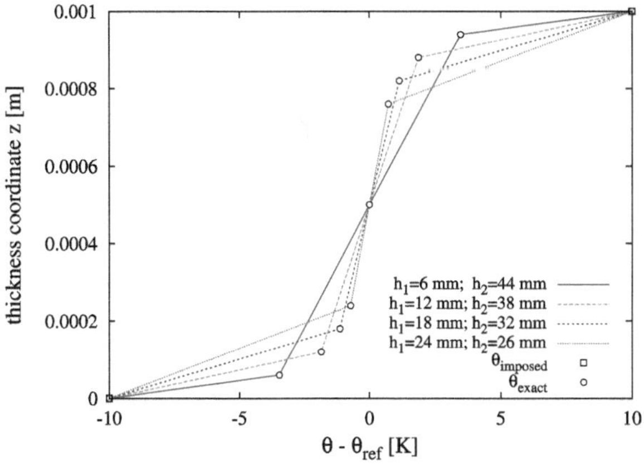

Figure 11.7 Variation of the plate-thickness temperature profile with the ratio h_1/h_2, total thickness constant – point $(\frac{a}{2}, \frac{a}{2})$.

static analysis was run with a regular mesh of 11 × 11 Q4 LD1 FEs to calculate the through-the-thickness temperature profile and the plate displacement caused by the imposed temperatures, at steady-state condition. The attention is restricted to the central point of the plate $(\frac{a}{2}, \frac{a}{2})$. Each curve in Fig. 11.7 shows a temperature profile along the thickness of the plate. Different geometrical configurations are considered by the variation of h_1 and h_2 mutual dimensions, keeping the same total thickness. The calculated temperatures profiles are in very good agreement with the exact solution obtained applying Fourier's law (interface points θ_{exact}). The middle plate displacement u_z is illustrated in Fig. 11.8 for the various choices of h_1/h_2 ratios. The configuration of minimum displacement is identified by the minimum of the curve. Figure 11.9 shows the variation of the temperature profile when κ_2 varies progressively from the value typical of steel to the value typical of aluminum. The agreement with the exact solution is reconfirmed. Moreover, if $\kappa_1 = \kappa_2$ the temperature profile is linear, as usual for a single-layered panel. Figure 11.10 shows the variation of u_z for different κ_2/κ_1 ratios, with κ_1 kept constant. It is shown that the ideal configuration of maximum displacement is for $\kappa_2 = 0$. Moreover, as κ_2 increases, the displacement becomes minor (until an asymptotic value not present in the figure for reasons of scale).

A three layered simply supported square plate with side $a = 0.1$ $[m]$ and thickness ratio $a/h = 100$ is considered in the following. The external layers are made of aluminum, with thickness $h/4$. The internal layer is made of

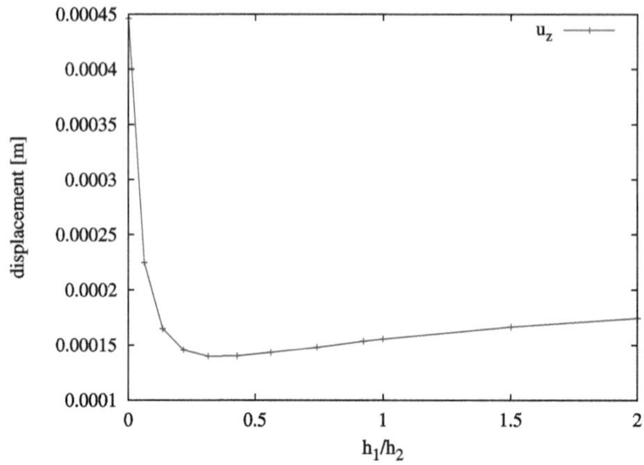

Figure 11.8 Variation of the displacement *uz* with the ratio h_1/h_2, total thickness constant – middle plate point.

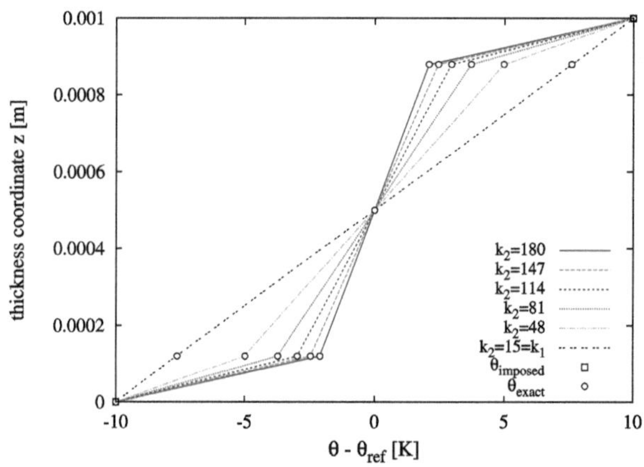

Figure 11.9 Variation of the plate-thickness temperature profile with the ratio κ_2/κ_1, κ_1 constant – point $(\frac{a}{2}, \frac{a}{2})$.

steel. The material properties are those listed for the above case study, with the only difference that $k = 80$ $[W/(m * K)]$ for the steel (difference introduced in order to make the figures more easy-to-read). The imposed θ is linear in x and y directions: concerning the top surface of the plate: $\theta - \theta_{ref}$ starts from 0 $[K]$, at point $(0, 0)$, and linearly increases along the bisector until 10 $[K]$. Correspondent opposite values are imposed to the plate bottom surface (Fig. 11.11). A coupled thermo-mechanical static analysis was run with a regular mesh of 11×11 Q4 LD1 FEs in order to calculate the

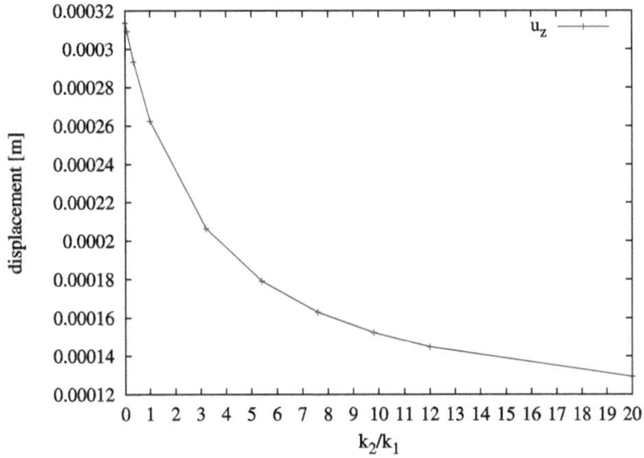

Figure 11.10 Variation of the displacement u_z with the ratio $\kappa_2/\kappa_1, \kappa_1$ constant – middle plate point.

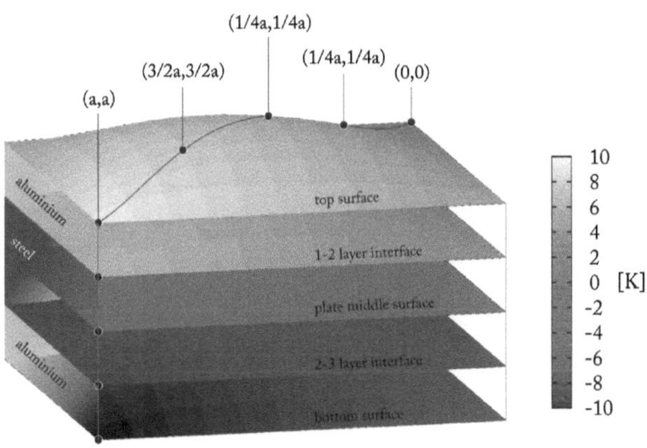

Figure 11.11 Temperature profile at the interfaces: $\theta - \theta_{ref}$. For interpretation of the references to color in this figure legend, the reader is referred to the web version of this chapter.

steady-state temperature distribution in the plate. The temperature field at the interfaces between layers is illustrated in Fig. 11.11. The temperature variation at the plate middle surface is also plotted. This temperature distribution is clearly equal to zero, in accordance with the problem symmetry.

Concluding, the cases study illustrated in this section show the usefulness of the formulated thermo-mechanical FEs, which permit to calculate in one single run the steady-state static deformation of a structure subjected to thermal loading. The separate application of the Fourier's law to obtain

Table 11.5 Undamped natural frequencies calculated for the pure mechanical case and for the thermo-mechanical coupled case [*Hz*] – a regular 20 × 20 mesh of Q4 ED1 FEs is employed

freq. n.	thickness ratio = 1/10			thickness ratio = 1/20		
	pure mech.	th.-mech.	difference	pure mech.	th.-mech.	difference
1	51740.9	52331.3	+1.14 %	27663.1	28018.1	+1.28%
2	99963.2	100893	+0.93 %	55888.7	56513.3	+1.12%
3	99963.2	101094	+1.13 %	55888.7	56649.8	+1.36%
4	140360	141719	+0.97 %	81005.2	81986.6	+1.21%
5	167999	168575	+0.34 %	99246.7	100378	+1.14%
6	168310	168860	+0.33 %	99864.7	101075	+1.21%
7	168310	169498	+0.71 %	121925	123248	+1.09%

the through-the-thickness temperature profile is not required. In fact, the temperature distribution through all the layers of the structure is automatically calculated by using the material thermal conductivity coefficients, which are considered together with the other constitutive coefficients in the fundamental nucleus.

11.5 THERMO-MECHANICAL DYNAMIC ANALYSIS OF ALUMINUM PLATE

A mechanically fully clamped square plate of aluminum (material properties in Sec. 11.4) is considered with a regular 20 × 20 mesh of Q4 ED1 FEs. Two plate thickness ratios are considered: 1/10 and 1/20, with thickness equal to 0.01 [*m*] and 0.005 [*m*] respectively. Pure mechanical and thermo-mechanical coupled cases are addressed to. No mechanical or thermal initial stresses are considered. Thermo-mechanical mode shapes give the shape of the structure's temperature distribution associated to corresponding natural frequencies, respecting the fact that compressed/expanded locations result hotter/colder respect to T_{ref}. In the time domain, the temperature variation of a fixed point would change sign alternatively and in accordance to the vibration frequency. In real cases, the change in temperature during vibration should be almost negligible. A more significant effect of coupling is described in Table 11.5, where the comparison between the undamped natural frequencies calculated in the two analysis is presented. It can be noted that natural frequencies are sensitive to thermo-mechanical coupling effects. Moreover, the percentage difference depends not only on thermal coefficients but also, in small quantity, on the thickness ratio (and consequently on the impact of boundary conditions). As conclusion, aluminum can be considered as one of those materials which requires the thermo-mechanical coupling in modelling, when very accurate results are needed for plate analysis.

REFERENCES

1. Thoronton EA. Thermal structures for aerospace applications. Reston (VA): AIAA Education Series; 1996.
2. Murakami H. Assessment of plate theories for treating the thermo-mechanical response of layered plates. Compos Eng 1993;3(2):137–49.
3. Creschik G, Palisoc A, Cassapakis C, Veal G, Mikulas MM. Sensitivity study of precision pressurized membrane reflector deformations. AIAA J 2000;38:308–14.
4. Srinivas S, Rao AK. A note on flexure of thick rectangular plates and laminates with variation of temperature across the thickness. Bull Acad Pol Sci Ser Sci Tech 1972;20:229–34.
5. Bapu Rao MN. 3D analysis of thermally loaded thick plates. Nucl Eng Des 1979;55:353–61.
6. Tungikar VB, Rao KM. Three dimensional exact solution of thermal stresses in rectangular composite laminate. Compos Struct 1994;27:419–27.
7. Bhaskar K, Varadan TK, Alii JSM. Thermoelastic solutions for orthotropic and anisotropic composite laminates. Composites Part B: Engineering 1996;27:415–20.
8. Carrera E. A class of two dimensional theories for multilayered plates analysis. Atti Accad Sci Torino, Mem Sci Fis 1995;19–20:49–87.
9. Bhaskar K, Varadan TK. A new theory for accurate thermal-mechanical flexural analysis of symmetric plates. Compos Struct 1999;45:227–32.
10. Carrera E, Brischetto S, Nali P. Variational statements and computational models for multifield problems and multilayered structures. Mech Adv Mat Struct 2008;15(3):182–98.
11. Rolfes R, Noor AK, Sparr H. Evaluation of transverse thermal stresses in composite plates based on first-order shear deformation theory. Comput Methods Appl Mech Eng 1998;167:355–68.
12. Rolfes R, Noack J, Taeschner M. High performance 3D-analysis of thermo-mechanically loaded composite structures. Compos Struct 1999;46:367–79.
13. Rohwer K, Rolfes R, Sparr H. Higher-order theories for thermal stresses in layered plates. Int J Solids Struct 2001;38:3673–87.
14. Spiessberger C, Gleiter A, Busse G. Aerospace applications of lockin-thermography with optical, ultrasonic, and inductive excitation. In: Int. Symposium on NDT in Aerospace. 2008.
15. Fantoni G, Merletti LG, Salerno A. Stato dell'arte della termografia Lock-In applicata a componenti di elicottero: analisi termoelastica e rilevazione di difetti. AIPnD 2008;27:30–4.
16. Carrera E. An assessment of mixed and classical theories for the thermal stress analysis of orthotropic multilayered plates. J Therm Stresses 2000;23:797–831.
17. Robaldo A, Carrera E, Benjeddou A. Unified formulation for finite element thermoelastic analysis of multilayered anisotropic composite plates. J Therm Stresses 2005;28:1031–65.

18. Robaldo A, Carrera E. Mixed finite elements for thermoelastic analysis of multilayered anisotropic plates. J Therm Stresses 2007;30(2):165–94.
19. Carrera E, Boscolo M, Robaldo A. Hierarchic multilayered plate elements for coupled multifield problems of piezoelectric adaptive structures: formulation and numerical assessment. Arch Comput Method E 2007;14(4):383–430.

CHAPTER 12

Thermal buckling

12.1 INTRODUCTION

A comprehensive literature review on the thermal-buckling analysis has been given by Thornton [1] and Tauchert [2], where thermal effects upon flexure, buckling and vibration of plates and shells were presented. Probably the first analyses of thermal buckling of shear deformable laminated plates are included in the work of Tauchert [3]. He used a first order shear deformation theory (FSDT) to analyze simply supported plates of antisymmetric angle-ply construction subjected to a uniform temperature rise. Yang and Sheih [4] employed the Galerkin method to investigate thermal buckling of initially stressed antisymmetric cross-ply plates. Chen et al. [5] considered both uniform and non-uniform temperature distributions using the finite element method. Noor and Burton [6] used predictor-corrector procedures for thermal buckling analysis. Prabhu and Dhanaraj [7] considered symmetrically laminated plates with different boundary conditions and used the finite element method in the analysis. A three dimensional solution for composite and sandwich plates has been provided by Noor [8, 9]. Kant and Babu [10], dealt with the same problem by employing shear deformable finite element models. Other contributions which are referred to pure mechanical or pure thermal loadings can be found in Refs. [11,12]. The first article on thermal buckling of composite laminates using a higher order shear deformation theory (HSDT) was provided by Chang [13]. The HSDT takes into account the thickness normal deformation effect, but the terms that represent the second-order variation of in-plane displacements through the thickness are not considered. Chang and Leu [14] used a theory that considers the transverse normal deformation and obtained analytical solutions for antisymmetric laminates using full stress-strain relations. The numerical results showed surprising discrepancies, almost independent of the slenderness, when compared with FSDT and Reddy's [15] third-order theory (TSDT). The stress resultants corresponding to the transverse normal stress are not included in the thermally induced fundamental state of stress, which has led to large differences. Later Rohwer [16] corrected this

Thermal Stress Analysis of Composite Beams, Plates and Shells.
DOI: http://dx.doi.org/10.1016/B978-0-12-849892-7.00015-7

by using a reduced stress-strain relationship. Shu and Sun [17] obtained a classical analytical solution for simply supported symmetric cross-ply laminates using a higher order formulation that accounts for the parabolic variation of transverse shear strains and shear stress continuity across each layer interface. All of the investigators in these studies, based on HSDT, obtained classical analytical solutions except in [17] where the finite element method was adopted. Although exhaustive literature has been produced concerning pure mechanical and pure thermal loadings, very little attention has been devoted to buckling caused by coexisting thermal and mechanical loads, which is the real case in the operation life of aerospace structures. Some results have been obtained by Shi [18] presented a finite element solution for the thermal buckling behaviour of laminated composite plates under combined mechanical and thermal loads. Elsami and Javaheri [19] investigated the buckling of composite cylindrical shells under mechanical and thermal loads. The governing equations were derived using Love-Kirchhoff hypothesis and Sander's non-linear strain-displacement relations. The present chapter provides a thorough assessment of both composite and sandwich plates subjected to single or combined mechanical and/or thermal loadings. A comprehensive set of results is obtained by using advanced and refined equivalent single layer (ESL), zig-zag (ZZ) and layer-wise (LW) plate theories with hierarchical capabilities. The latter are generate, in a systematic way, within the framework of the UF (more details can be found in Refs. [20–23]). The influence of many parameters such as lamination schemes, material properties, length-to-thickness ratio, face-to-plate thickness ratio, aspect ratio, orthotropic ratio, upon the critical loads and the critical temperatures is evaluated. The effect of transverse normal deformation is considered and its influence on the results is discussed. The influence of the boundary terms, when different from zero, is also taken into account.

12.2 THERMAL BUCKLING ANALYSIS OF LAMINATED COMPOSITE AND SANDWICH STRUCTURES

The results in terms of critical temperatures or mechanical critical buckling load have been computed in this section for both sandwich and laminated composite plates. The materials used in the analysis carried out are provided in Tables 12.1, 12.2 and 12.3.

Table 12.1 Sandwich plates

Material 1

Face Sheets			
E_1/E_2	G_{12}/E_2	G_{23}/E_2	ν_{12}
19.0	0.52	0.338	0.32
ν_{13}	ν_{23}	α_1/α_0	α_2/α_0
0.32	0.49	0.001	1.0
Core			
E_1/E_2^f	E_2/E_2^f	E_3/E_2^f	G_{12}/E_2^f
3.2×10^{-5}	2.9×10^{-5}	0.4	2.4×10^{-3}
G_{13}/E_2^f	G_{23}/E_2^f	ν_{12}	ν_{13}
7.9×10^{-2}	6.6×10^{-2}	0.99	3.0×10^{-5}
ν_{23}	α_1/α_0	α_2/α_0	$\rho_c/\rho^{f(1)}$
3.0×10^{-5}	1.36	1.36	0.07

Table 12.2 Composite plate

Material 2

E_1 [GPa]	$E_2 = E_3$ [GPa]	$G_{12} = G_{13}$ [GPa]	G_{23} [GPa]
127.6	11.6	6	1.8
$\nu_{12} = \nu_{13}$	ν_{23}	$\alpha_1 \frac{1}{°C}$	α_2
0.30	0.36	-0.9×10^{-6}	27×10^{-6}

Table 12.3 Composite plate

Material 3

E_1/E_2	G_{12}/E_2	G_{23}/E_2	ν_{12}
40.0	0.6	0.5	0.25
ν_{12}	ν_{12}	α_1/α_0	α_2/α_0
0.25	0.25	0.02	22.5

12.2.1 Analytical solution for thermal buckling analysis of sandwich plates

A validation of the results and an assessment of ESL, ZZ and LW models, is carried out. Results are compared to the 3D elasticity solution provided by Noor [8,9] for thermal buckling analysis of sandwich plates, with stacking sequence $\left[(0°/90°)_5/\text{Core}/(90°/0°)_5\right]$ and *Material 1* (see Table 12.1). The considered boundary condition is simply supported (SSSS), two different parameters, thickness ratio a/h and face-to-plate ratio h_f/h, are changed,

Table 12.4 Critical temperature parameter $\lambda_\vartheta = \alpha_0 T_{cr}$ for simply supported Sandwich plate, MAT1, ED models

a/h		h_f/h					
		0.025		0.1		0.15	
5	3D	0.8512	Error%	0.3820	Error%	0.2805	Error%
	ED_{111}	0.9579	(12.5)	0.5980	(56.5)	0.5169	(84.3)
	ED_{222}	0.9504	(11.7)	0.5873	(53.7)	0.5054	(80.2)
	ED_{333}	0.9072	(6.58)	0.4118	(7.80)	0.3129	(11.6)
	ED_{444}	0.8862	(4.11)	0.4054	(6.13)	0.3094	(10.3)
	ED_{555}	0.8702	(2.23)	0.4029	(5.47)	0.3062	(9.16)
	ED_{666}	0.8604	(1.08)	0.4029	(5.47)	0.3061	(9.13)
	ED_{777}	0.8523	(0.13)	0.3949	(3.37)	0.2989	(6.56)
	ED_{888}	0.8466	(−0.54)	0.3945	(3.27)	0.2988	(6.52)
20	3D	0.0929	Error%	0.0726	Error%	0.0623	Error%
	ED_{111}	0.1003	(7.97)	0.0835	(15.1)	0.0746	(19.7)
	ED_{222}	0.0978	(5.27)	0.0793	(9.23)	0.0704	(13.0)
	ED_{333}	0.0973	(4.74)	0.0748	(3.03)	0.0646	(3.69)
	ED_{444}	0.0950	(2.26)	0.0739	(1.79)	0.0642	(3.05)
	ED_{555}	0.0948	(2.05)	0.0738	(1.65)	0.0640	(2.73)
	ED_{666}	0.0936	(0.75)	0.0738	(1.65)	0.0639	(2.57)
	ED_{777}	0.0935	(0.65)	0.0735	(1.24)	0.0636	(2.09)
	ED_{888}	0.0929	(0.00)	0.0734	(1.10)	0.0635	(1.93)

and their influence on the results is critically discussed. In Table 12.4 the first assessment, of the *critical temperature parameter* λ_ϑ, involving ESL models is presented. Understandably, for all the theories, the error percentage grows up, increasing the face-to-plate thickness ratio h_f/h, indeed in this case the 3D effects become prominent. Vice-versa increasing the thickness ratio a/h, the error decreases steadily. The results are in good agreement with the 3D elasticity solution. When $h_f/h = 0.15$ and $a/h = 5$, 3D effects can not be neglected and the use of advanced high order ESL and ZZ theories accounting to the transverse normal deformation becomes mandatory in order to accurately determine the critical temperature parameters. A proof is the error difference between the ED_{111} and ED_{888} models: in the first case is about 84%, in the last one 6.52%. Similar observations can be drawn analyzing the results provided by using the EDZ models (see Table 12.5). The inclusion of the *Murakami's function* increases the rate of convergence, improving the results above all when low expansion orders are used. The results for the models ED_{111} and EDZ_{111} are higher than the others. This

Table 12.5 Critical temperature parameter $\lambda_{\vartheta} = \alpha_0 \, T_{cr}$ for simply supported Sandwich plate. MAT1, FDZ models

a/h		h_f/h					
		0.025		0.1		0.15	
5	3D	0.8512	Error%	0.3820	Error%	0.2805	Error%
	EDZ_{111}	0.9544	(12.1)	0.5891	(54.2)	0.5072	(80.8)
	EDZ_{222}	0.9470	(11.6)	0.5788	(51.5)	0.4963	(76.9)
	EDZ_{333}	0.9065	(6.50)	0.4112	(7.64)	0.3118	(11.2)
	EDZ_{444}	0.8855	(4.03)	0.4048	(5.97)	0.3083	(9.91)
	EDZ_{555}	0.8699	(2.20)	0.4022	(5.29)	0.3055	(8.91)
	EDZ_{666}	0.8602	(1.06)	0.4022	(5.29)	0.3055	(8.91)
	EDZ_{777}	0.8522	(0.12)	0.3946	(3.30)	0.2984	(6.38)
	EDZ_{888}	0.8464	(−0.56)	0.3942	(3.19)	0.2983	(6.35)
20	3D	0.0929	Error%	0.0726	Error%	0.0623	Error%
	EDZ_{111}	0.1002	(7.86)	0.0833	(14.7)	0.0744	(19.4)
	EDZ_{222}	0.0978	(5.27)	0.0791	(8.95)	0.0702	(12.7)
	EDZ_{333}	0.0973	(4.74)	0.0748	(3.03)	0.0645	(3.53)
	EDZ_{444}	0.0949	(2.15)	0.0739	(1.79)	0.0641	(2.89)
	EDZ_{555}	0.0947	(1.94)	0.0738	(1.65)	0.0640	(2.73)
	EDZ_{666}	0.0936	(0.75)	0.0738	(1.65)	0.0639	(2.57)
	EDZ_{777}	0.0935	(0.65)	0.0735	(1.24)	0.0635	(1.93)
	EDZ_{888}	0.0929	(0.00)	0.0734	(1.10)	0.0635	(1.93)

is a pathology that is peculiar of the linear models which retain the full 3D constitutive law, and that has been referred to as *Poisson Locking* or *Thickness Locking*. As can be observed in Table 12.6, concerning LW models, an excellent agreement with the 3D elasticity solution is shown by using low expansion orders. In this particular case employing high expansion orders seems don't affect the accuracy results and for low values of the ratio h_f/h the use of the high order theory underestimates the critical temperature parameter.

12.2.2 Convergence analysis for pure thermal and pure mechanical buckling loads, for symmetric angle-ply laminates

The convergence analysis is a crucial point when approximate solutions are required. From a theoretical point of view once proved that the trial functions are admissible in the original variational principle, the convergence

Table 12.6 Critical temperature parameter $\lambda_{\vartheta} = \alpha_0 T_{cr}$ for simply supported Sandwich plate, MAT1, LD models

a/h		h_f/h					
		0.025		0.1		0.15	
5	3D	0.8512	Error%	0.3820	Error%	0.2805	Error%
	LD_{111}	0.8447	(−0.76)	0.3839	(0.50)	0.2863	(2.07)
	LD_{222}	0.8415	(−1.14)	0.3819	(−0.02)	0.2850	(1.60)
	LD_{333}	0.8414	(−1.15)	0.3819	(−0.02)	0.2850	(1.60)
	LD_{444}	0.8414	(−1.15)	0.3819	(−0.02)	0.2850	(1.60)
	LD_{445}	0.8414	(−1.15)	0.3819	(−0.02)	0.2850	(1.60)
	LD_{454}	0.8414	(−1.15)	0.3819	(−0.02)	0.2850	(1.60)
	LD_{544}	0.8414	(−1.15)	0.3819	(−0.02)	0.2850	(1.60)
	LD_{555}	0.8414	(−1.15)	0.3819	(−0.02)	0.2850	(1.60)
20	3D	0.0929	Error%	0.0726	Error%	0.0623	Error%
	LD_{111}	0.0924	(−0.54)	0.0727	(0.14)	0.0627	(0.64)
	LD_{222}	0.0924	(−0.54)	0.0727	(0.14)	0.0627	(0.64)
	LD_{333}	0.0924	(−0.54)	0.0727	(0.14)	0.0627	(0.64)
	LD_{444}	0.0924	(−0.54)	0.0727	(0.14)	0.0627	(0.64)
	LD_{445}	0.0924	(−0.54)	0.0727	(0.14)	0.0627	(0.64)
	LD_{454}	0.0924	(−0.54)	0.0727	(0.14)	0.0627	(0.64)
	LD_{544}	0.0924	(−0.54)	0.0727	(0.14)	0.0627	(0.64)
	LD_{555}	0.0924	(−0.54)	0.0727	(0.14)	0.0627	(0.64)

to the true solution is guaranteed as the number of admissible function tends to infinity. In practical computations, the number of Ritz and/or Galerkin terms is limited by CPU time this truncation inevitably affects the results accuracy. Moreover, another related aspect to take into account is the stability of the solution, since numerical error may occur due to the ill conditioning when many admissible functions are required to obtain a good convergence. All these reasons justify the importance to carry out a convergence analysis whenever an approximate solution is applied. In Table 12.7 convergence analysis, in the case of mechanical buckling loads is performed by using a ED_{222} model and Ritz (R), Galerkin (G) and Generalized Galerkin (GG) methods. A four layers laminate $\left[\theta/-\theta\right]_s$ and *Material 2* (see Table 12.2) is studied. A good convergence, for all the employed methodologies, is reached by using $M = N = 12$ as half-waves number. Note that a square selection strategy is adopted, i.e., the same number of Ritz or Galerkin terms is used in the expansion along x and y directions. GM has a higher rate of convergence which leads to higher

Table 12.7 Convergence analysis of the buckling load σ_{xx} [MPa] for angle-ply laminate by using ED_{222} theory

m, n	RM	GM	GGM	RM	GM	GGM
	$[15^\circ/-15^\circ]_s$			$[30^\circ/-30^\circ]_s$		
2	61.247	62.464	61.247	74.464	77.251	74.488
4	60.348	61.974	60.380	71.576	75.129	71.733
6	60.033	61.849	60.097	70.526	74.542	70.881
8	59.873	61.801	59.957	69.971	74.293	70.480
10	59.775	61.777	59.870	69.620	74.160	70.241
12	59.707	61.763	59.808	69.374	74.078	70.079
	$[45^\circ/-45^\circ]_s$			$[75^\circ/-75^\circ]_s$		
2	81.715	85.041	81.742	47.555	46.664	47.549
4	77.767	81.487	77.900	44.844	44.648	45.132
6	76.363	80.554	76.789	44.459	44.411	44.833
8	75.601	80.150	76.276	44.289	44.322	44.718
10	75.108	79.929	75.977	44.190	44.276	44.656
12	74.758	79.792	75.777	44.124	44.248	44.616

buckling loads when compared to R and GGM. This difference, between GM and GGM, can be explained by taking into account the boundary terms which are different form zero in angle-ply laminates, and are retained in the GGM whilst are discarded in the GM. When isotropic plates or cross-ply laminates are studied then the three methodologies lead to the same results, as already proved in Ref. [22]. In Table 12.8 the same procedure has been computed to the evaluation of the critical temperature variations. In this case the influence of the boundary terms is strongly dependent both from the half-waves number and from the lamination angles.

12.2.3 Influence of lamination angle, thickness ratio, aspect ratio and orthotropic ratio on the critical temperature parameters, for symmetric angle-ply laminates

The critical temperature parameters of a five-layer laminated composite plates, with stacking layup $[\theta/-\theta/\theta/-\theta/\theta]$ and made up of the *Material 3* (see Table 12.3) have been computed. In particular, in Fig. 12.1 the influence of the lamination angle and the variable kinematics, on the critical temperature, are shown. As can be observed, the lamination angle $\theta = 45^\circ$ maximizes the critical temperature parameter $\lambda_\vartheta = \alpha_0 T_{cr}$. Moreover, the

Table 12.8 Convergence analysis of the critical temperature variation ΔT_{cr} [K] for angle-ply laminate by using ED_{222} theory

m, n	RM	GM	GGM	RM	GM	GGM
	$[15°/-15°]_s$			$[30°/-30°]_s$		
2	168.562	165.408	168.544	264.991	274.924	265.075
4	158.949	158.262	159.970	253.400	264.517	254.016
6	157.585	157.419	158.909	249.699	262.133	250.985
8	156.981	157.105	158.503	247.766	261.144	249.583
10	156.629	156.941	158.283	246.545	260.616	248.752
12	156.394	156.842	158.140	245.693	260.294	248.186
	$[45°/-45°]_s$			$[75°/-75°]_s$		
2	290.789	302.644	290.884	168.562	165.408	168.544
4	277.057	290.953	277.608	158.949	158.262	159.970
6	272.011	287.667	273.591	157.585	157.419	158.909
8	269.271	286.251	271.737	156.981	157.105	158.503
10	267.499	285.483	270.655	156.629	156.941	158.283
12	266.242	285.010	269.932	156.394	156.842	158.140

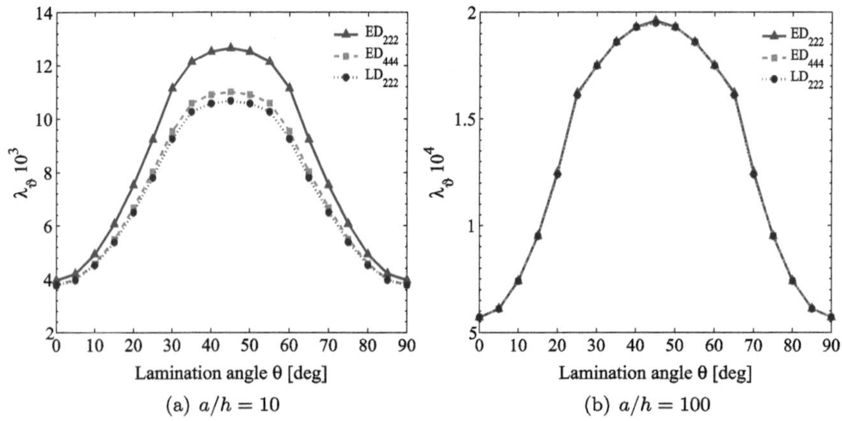

Figure 12.1 Critical temperature parameter $\lambda_{\vartheta} = \alpha_0 T_{cr}$, varying the lamination angle and the plate theory, $\frac{a}{b} = 1$, *Material 3*.

layerwise model LD_{222}, due to its high accuracy, when dealing with both thin and thick plates, reached at a reasonable computational cost, has been selected kinematics model to carry out further investigations. In Fig. 12.2 the critical temperature parameter is computed for different lamination angles, different length-to-thickness ratio, different aspect ratio and different

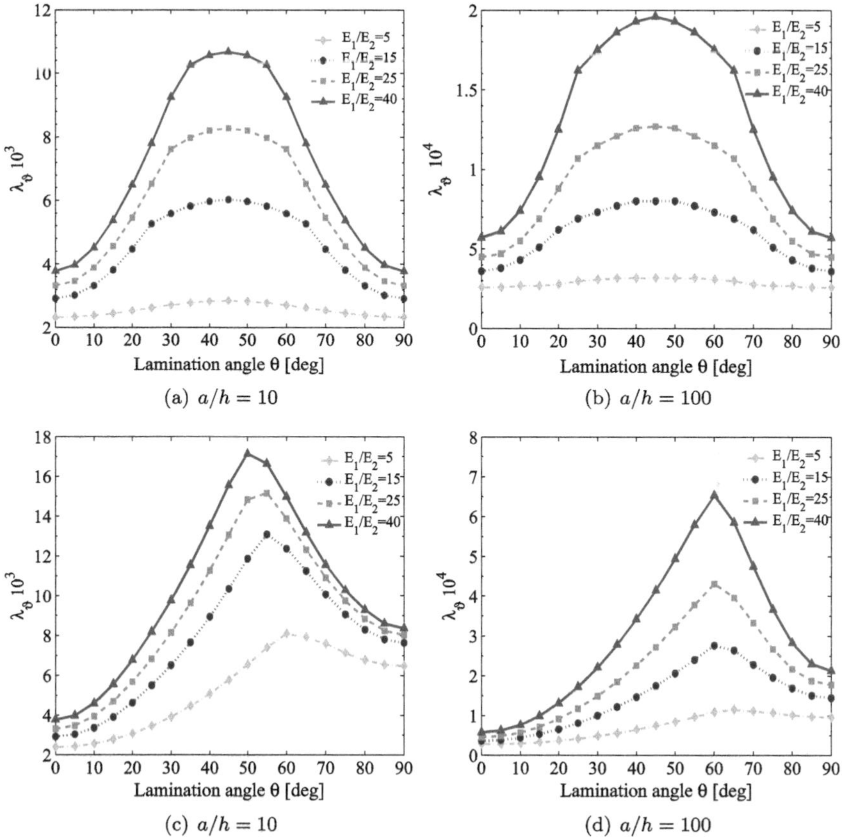

Figure 12.2 Critical temperature parameter $\lambda_\vartheta = \alpha_0 \, T_{cr}$, varying θ and E_1/E_2, and using *Material 3*. In Figs. (a) and (b) $\frac{a}{b} = 1$, in Figs. (c) and (d) $\frac{a}{b} = 2$.

orthotropic ratio. In particular, in Figs. 12.2(a) and 12.2(b), the square plate case $a/b = 1$ is studied, the distribution of the critical temperature parameter with respect to the lamination angle is symmetric and reached a maximum value at the angle $\theta = 45°$ for both thick (see Fig. 12.2(c)) and thin (see Fig. 12.2(d)) plates. Moreover, λ_ϑ increases when increasing the orthotropic ratio E_1/E_2. It's interesting to note that once fixed the aspect ratio $a/b = 2$, varying the lamination angle, the angle which maximizes λ_ϑ is significantly dependent to the length-to-thickness ratio b/h, and in particular for thin plate ($a/h = 100$) this angle is $\theta = 60°$, for thick plate ($a/h = 10$), instead is $\theta = 50°$.

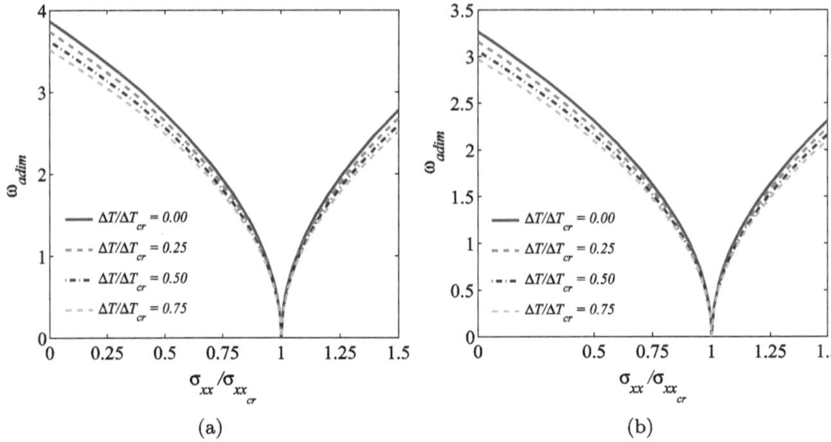

Figure 12.3 Influence of the thermo-mechanical loading upon the fundamental dimensionless circular frequency parameter $\omega_{adim} = \omega\sqrt{\frac{a^4\rho}{E_1h^2}}$. (a) $[45°/-45°]_s$; (b) $[0°/90°]_s$.

12.3 INFLUENCE OF THERMAL-MECHANICAL INTERACTION LOADINGS ON THE CIRCULAR FREQUENCY PARAMETERS

In Fig. 12.3 the trend of the fundamental circular frequency parameter of an in-plane axially stressed plate is described. The plate is further subjected to a change in the environmental temperature expressed as a percentage of the plate critical temperature. Understandably increasing the axial load, the fundamental circular frequency parameter decreases, approaching to zero when the applied in-plane load is close to the critical buckling load. Moreover, a further decrease of the frequency parameter is generated by an increase of the environmental temperature. If the in-plane axial load is increased behind its critical value then the frequency parameter starts to increase and keeps this trend in the post-critical field. In Fig. 12.4 the dynamic behaviour of a thermally loaded plated is represented. In particular, the trend of the first four circular frequency parameters, while increasing the thermally induced in-plane loadings, is depicted. Increasing the temperature variation the value of the thermal load increases as well and the circular frequency parameters decrease as expected. As can be seen in the same figure, the composite plate shows a similar overall dynamic behaviour when subjected to a pure thermal loading (Fig. 12.4(a)) and a thermo-

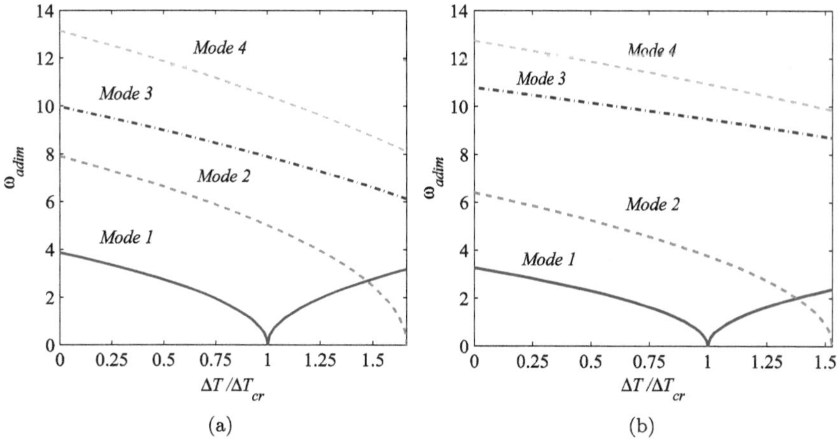

Figure 12.4 Influence of the temperature variation upon the first four dimensionless circular frequency parameters $\omega_{adim} = \omega\sqrt{\frac{a^4\rho}{E_1 h^2}}$. (a) $\sigma_{xx}/\sigma_{xx_{cr}} = 0.00$; (b) $\sigma_{xx}/\sigma_{xx_{cr}} = 0.75$.

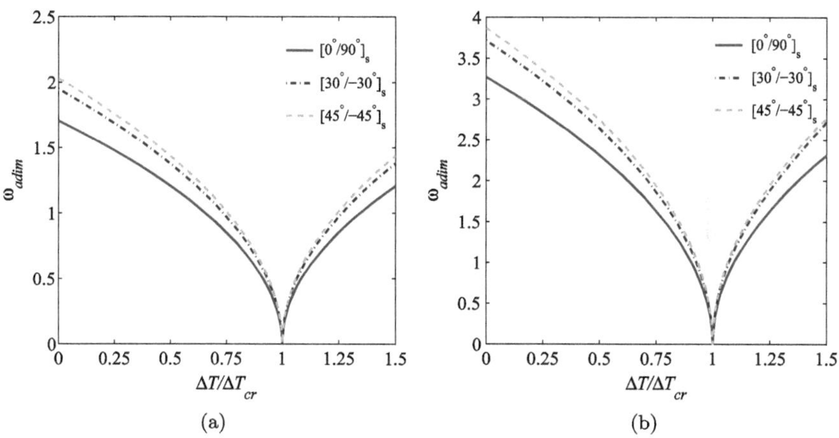

Figure 12.5 Influence of the lamination scheme upon the fundamental dimensionless circular frequency parameter $\omega_{adim} = \omega\sqrt{\frac{a^4\rho}{E_1 h^2}}$. (a) $\sigma_{xx}/\sigma_{xx_{cr}} = 0.00$; (b) $\sigma_{xx}/\sigma_{xx_{cr}} = 0.75$.

mechanical loading (Fig. 12.4(b)). In particular, in both figures, it is possible to observe in the post-critical field the well-known phenomenon of modal-interchange between the first two modes. In Fig. 12.5 the influence of the lamination scheme upon the fundamental circular frequency of a plate thermally and thermo-mechanically loaded is analyzed. In both cases the curves relative to a typical lamination scheme decrease monotonically approaching to zero increasing the applied load. The higher circular frequency parameter values are obtained adopting a lamination scheme $[45°/-45°]_s$ and the lower by using $[0°/90°]_s$, as expected the circular frequency parameter is lower when the plate is subjected to a thermo-mechanical loading.

REFERENCES

1. Thornton EA. Thermal buckling of plates and shells. Appl Mech Rev 1993;46(10):485–506.
2. Tauchert TR. Thermally induced flexure, buckling, and vibration of plates. Appl Mech Rev 1991;44(8):347–60.
3. Tauchert TR. Thermal buckling of thick antisymmetric angle-ply laminates. J Therm Stresses 1987;10:113–24.
4. Yang IH, Sheih JA. Generic thermal buckling of initially stressed antisymmetric cross-ply thick laminates. Int J Solids Struct 1988;24:1059–70.
5. Cheng WJ, Lin PD, Chen LW. Thermal buckling behaviour of thick composite laminated plates under non-uniform temperature distribution. Comput Struct 1991;41:637–45.
6. Noor AK, Jeanne WS. Predictor corrector procedures for thermal buckling analysis of multilayered composite plates. Comput Struct 1991;40:1071–84.
7. Prabhu MR, Dhanaraj RC. Thermal buckling of laminated composite plates. Comput Struct 1994;53(5):1193–204.
8. Noor AK, Burton WS. Three-dimensional solutions for thermal buckling multilayered anisotropic plates. J Eng Mech 1992;118:638–701.
9. Noor AK, Peters JM, Burton WS. Three-dimensional solutions for initially stressed structural sandwiches. J Eng Mech 1994;120:284–303.
10. Kant T, Babu CS. Refined high order finite element models for thermal buckling of laminated composite and sandwich plates. J Therm Stresses 2000;23:111–30.
11. Matsunaga H. Thermal buckling of angle-ply laminated composite and sandwich plates according to high-order deformation theory. Compos Struct 2006;72:177–92.
12. Noor AK, Burton WS. Three-dimensional solutions for thermal buckling and sensitivity derivatives of temperature-sensitive multilayered angle-ply plates. J Appl Mech 1992;59:848–56.
13. Chang JS. FEM analysis of buckling and thermal buckling of antisymmetric angle ply laminates according to transverse shear and normal deformable high order displacement theory. Comput Struct 1990;37(6):925–46.
14. Chang JS, Leu SY. Thermal buckling of antisymmetric angle-ply laminates in a uniform temperature field. Compos Sci Technol 1991;41:109–28.
15. Reddy JN. A simple high order theory for laminated composite plates. J Appl Mech 1984;51:745–52.
16. Rohwer K. Discussion on Cheng and Leu. Compos Sci Technol 1992;45:181–2.
17. Shu X, Sun XL. Thermomechanical buckling of laminated composite plates with higher-order transverse shear deformation. Comput Struct 1994;53:1–8.
18. Shi Y, Lee RY, Mei C. Coexisting thermal post-buckling of composite plates with initial imperfections using finite element modal method. J Therm Stresses 1999;22:595–614.

19. Elsami MR, Javaheri R. Buckling of composite cylindrical shells under mechanical and thermal loads. J Therm Stresses 1999;22:527–45.
20. Fazzolari FA, Carrera E. Thermo-mechanical buckling analysis of anisotropic multilayered composite and sandwich plates by using refined variable-kinematics theories. J Therm Stresses 2012;36(4):321–50.
21. Fazzolari FA. Natural frequencies and critical temperatures of functionally graded sandwich plates subjected to uniform and non-uniform temperature distributions. Compos Struct 2015;121:197–210.
22. Fazzolari FA, Carrera E. Advanced variable kinematics Ritz and Galerkin formulations for accurate buckling and vibration analysis of laminated composite plates. Compos Struct 2011;94(1):50–67.
23. Fazzolari FA, Carrera E. Accurate free vibration analysis of thermo-mechanically pre/post-buckled anisotropic multilayered plates based on a refined hierarchical trigonometric Ritz formulation. Compos Struct 2013;95:381–402.

CHAPTER 13

Thermal stresses in functionally graded materials

13.1 INTRODUCTION

FGMs represent a class of heterogeneous composite materials made up of a mixture of ceramics and metals that are characterized by the smooth and continuous variation in properties from the bottom to the top of the considered structural element. To this class also belong those materials in which the gradation is generated by varying the fibres volume fraction in laminated composites. In the general case, the material properties of FGMs are controlled by the variation of the volume fraction of the constituent materials. Being ultrahigh temperature-resistant materials, they are suitable for aerospace applications, such as aircraft, space vehicles, barrier coating and propulsion systems. Moreover, they have several advantages over other types of advanced materials like fibre-reinforced composites, indeed, problems like delamination, fibre failure, adverse hygroscopic effects due to moisture content etc. are effectively eliminated or non-existent.

With their potential applications, FGMs are steadfastly making headway in aerospace design. Thus, there is the need to fully analyze the free vibration and thermal stability characteristics of the FGMs. This necessity led the research community to investigate several configurations of isotropic and sandwich FGM plates by using a considerable amount of new structural theories.

In particular, El Meiche et al. [1] proposed a new hyperbolic shear deformation theory for buckling and vibration of functionally graded sandwich plates. Matsunaga [2] dealt with free vibration and stability of functionally graded plates according to a 2-D higher-order deformation theory. The same author [3] dealt with the thermal bucking of several FGM plate configurations using higher-order models accounting for the effects of transverse shear and normal deformations and rotatory inertia. The set of fundamental dynamic equations of a two-dimensional (2-D) higher-order theory for rectangular functionally graded (FG) plates were derived through Hamilton's principle and by using the method of power series expansion of displacement components. Abrate [4] coped with free

Thermal Stress Analysis of Composite Beams, Plates and Shells.
DOI: http://dx.doi.org/10.1016/B978-0-12-849892-7.00016-9

vibration, buckling, and static deflections of functionally graded plates. Dozio [5] studied the free vibration behaviour of sandwich plates with FGM core via variable-kinematic 2-D Ritz models using Chebyshev polynomials. The same author [6] recently proposed an exact free vibration analysis of Lévy FGM plates with higher-order shear and normal deformation theories.

Viola et al. [7–9] used an advanced generalized differential quadrature method (GDQ) for the analysis of FGM plate and shell structures with general shapes. Tornabene et al. [10–12] proposed new advances in the use of the GDQ for FGM structures. A comprehensive study on the stability of FGM plates has been provided by Bateni et al. [13]. The investigation was focused on the stability behaviour of thick FGM rectangular plates subjected to mechanical and thermal loads by using a four-variable refined plate theory. The buckling analysis of thick functionally graded plates under mechanical and thermal loadings was studied by Shariyat and Eslami [14] by using a third order shear deformation theory. Li et al. [15] applied the three-dimensional Ritz method based on the use of the Chebyshev polynomials for the free vibration analysis of functionally graded material sandwich plates. Javaheri and Elsami [16] provided a consistent assessment of the critical temperatures of functionally graded plates by using a higher-order theory. An edge-based smoothed finite element method has been developed by Nguyen-Xuan et al. [17] to analyze functionally graded plates. Zhu and Liew [18] computed the natural frequencies of moderately thick functionally graded plates by local Kriging meshless method. Zhao et al. [19] carried out a free vibration analysis of functionally graded plates using the element-free kp-Ritz method.

The present chapter provides a comprehensive thermal stability and free vibration analysis of both isotropic and FGM sandwich plates. Accuracy of the presented formulation has been examined by testing the developed hierarchical models [20–26] with results available in literature. Uniform, linear and non-linear temperature variations through the thickness layer plate direction have been taken into account. Several FGM sandwich plate configurations have been investigated. Results have been presented in terms of critical temperatures and natural frequencies. The effects of significant parameters such as volume fraction index, length-to-thickness ratio, aspect ratio and sandwich plate-type have been discussed.

Table 13.1 Materials' properties

	Al_2O_3	Aluminum	ZrO_2	(Ti 6Al 4V)$_1$	(Ti 6Al 4V)$_2$
E [GPa]	380	70	244.27	66.2	150.7
ν	0.3	0.3	0.3	0.3	0.298
$\rho\left[\frac{kg}{m^3}\right]$	3800	2707	3000	4429	4429
$\kappa\left[\frac{W}{mK}\right]$	10.4	204	1.7	1.0	1.0
$\alpha\left[\frac{1}{°C}\right]$	7.4×10^{-6}	23×10^{-6}	12.766×10^{-6}	10.3×10^{-6}	10.3×10^{-6}

Table 13.2 Comparison of the first ten natural frequencies (Hz) of simply-supported FGM isotropic plates Al_2O_3/Al with $a/h = 80$ ($a = b = 0.4$ m and $h = 0.005$ m)

p	Theory	Circular frequency parameters					
		ω_1	$\omega_{2,3}$	ω_4	$\omega_{5,6}$	$\omega_{7,8}$	$\omega_{9,10}$
0	He et al. [27]	144.66	360.53	569.89	720.57	919.74	1225.7
	Zhao et al. [19]	143.67	360.64	575.87	725.53	938.18	1238.8
	ED2 [5]	144.97	362.15	579.02	723.42	939.76	1229.3
	ED4 [5]	144.96	362.11	578.90	723.23	939.44	1228.7
	EDZ_{555}	144.96	362.11	578.90	723.23	939.44	1227.17
	EDZ_{222}	144.97	362.15	579.02	723.42	939.76	1227.71
	ED_{444}	144.96	362.11	578.90	723.23	939.44	1227.17
	ED_{225}	144.97	362.15	579.02	723.42	939.76	1227.71
	ED_{222}	144.97	362.15	579.02	723.42	939.76	1227.71
2000	He et al. [27]	268.92	669.40	1052.5	1338.5	1695.2	2280.9
	Zhao et al. [19]	268.60	674.38	1076.8	1356.9	1754.4	2316.9
	ED2	271.03	677.09	1082.6	1352.6	1757.1	2298.4
	ED4	271.01	677.00	1082.3	1352.2	1756.5	2297.4
	EDZ_{555}	271.01	676.98	1082.3	1352.2	1756.5	2294.5
	EDZ_{222}	271.02	677.07	1082.5	1352.5	1757.1	2295.5
	ED_{444}	271.01	676.98	1082.3	1352.2	1756.5	2294.5
	ED_{225}	271.02	677.07	1082.5	1352.5	1757.1	2295.5
	ED_{222}	271.02	677.07	1082.5	1352.5	1757.1	2295.5

13.2 NATURAL FREQUENCIES OF FGM ISOTROPIC AND SANDWICH PLATES

The ceramic and metallic materials involved in the analysis are given in Table 13.1. The first assessment is based on the computation of natural frequencies of FGM isotropic and sandwich plates. In particular, in Table 13.2,

Table 13.3 Comparison of the dimensionless fundamental frequency parameter $\hat{\omega} = \frac{\omega b^2}{h}\sqrt{\frac{\rho_c}{E_c}}$ of simply-supported FGM isotropic plates Al/Zr O_2 with $a/h = 10$

Theory	Mode	Volume fraction index				
		$p=0$	$p=1$	$p=5$	$p=10$	$p=10000$
FSDT [18]	1	5.7619	4.9437	4.6260	4.5018	4.1304
	2, 3	13.7980	11.8537	11.0456	10.7449	9.8904
EDZ444	1	5.7769	4.9572	4.6296	4.5065	4.1407
	2, 3	13.8050	11.8608	11.0095	10.7172	9.8950
EDZ222	1	5.7950	4.9707	4.6513	4.5273	4.1536
	2, 3	13.9017	11.9335	11.1241	10.8267	9.9642
ED444	1	5.7769	4.9572	4.6296	4.5065	4.1407
	2, 3	13.8050	11.8608	11.0095	10.7172	9.8950
ED225	1	5.7950	4.9707	4.6513	4.5273	4.1536
	2, 3	13.9017	11.9335	11.1241	10.8267	9.9642
ED222	1	5.7950	4.9707	4.6513	4.5273	4.1536
	2, 3	13.9017	11.9335	11.1241	10.8267	9.9642

Table 13.4 Comparison of the dimensionless frequency parameters $\hat{\omega} = \frac{\omega b^2}{h}\sqrt{\frac{\rho_c}{E_c}}$ of fully clamped FGM isotropic plates Al/Zr O_2 with $a/h = 10$

Theory	p	Circular frequency parameters					
		$\hat{\omega}_1$	$\hat{\omega}_2$	$\hat{\omega}_3$	$\hat{\omega}_4$	$\hat{\omega}_5$	$\hat{\omega}_6$
FSDT [18]	0	9.8710	18.8814	18.8814	26.4062	31.3034	31.6125
	0.5	8.9095	17.0768	17.0768	23.9094	28.3717	28.6494
	1	8.4916	16.2711	16.2711	22.7770	27.0244	27.2899
	2	8.1710	15.6252	15.6252	21.8475	25.8963	26.1540
ED222	0	10.1690	19.4720	19.4720	27.3497	32.3911	32.7070
	0.5	9.1741	17.5981	17.5981	24.7456	29.3349	29.6175
	1	8.7391	16.7551	16.7551	23.5531	27.9156	28.1855
	2	8.4055	16.0807	16.0807	22.5747	26.7276	26.9901

the first ten natural frequencies of simply-supported FGM isotropic plates made of $ZrO_2/(Ti\ 6Al\ 4V)_2$ obtained by using the HTRF are compared with those proposed by Zhao et al. [19] based on a FSDT element-free kp-Ritz, by He et al. [27] based on a CPT FEM and by Dozio [5] based on higher-order models and Chebyshev-Ritz formulation. It can be seen as for the first three modes the results difference with the CPT and FSDT is acceptable but for higher-order modes there is a significant discrepancy. It is interesting to note that despite the use of identical higher-order kinemat-

Table 13.5 Comparison of the dimensionless fundamental frequency parameter $\hat{\omega} = \frac{\omega b^2}{h}\sqrt{\frac{\rho_0}{E_0}}$ ($\rho_0 = 1\ \text{Kg/m}^3$ and $E_0 = 1$ GPa) of simply-supported sandwich plates with Al_2O_3/Al graded core

a/h	Theory	Fundamental frequency parameter				
		$p=0$	$p=0.5$	$p=1$	$p=2$	$p=5$
5	3D [17]	1.19580	1.25338	1.31569	1.39567	1.44540
	ED2 [6]	1.20511	1.25927	1.32039	1.40196	1.45402
	LD3 [6]	1.19580	1.25337	1.31569	1.39579	1.44551
	ED_{555}	1.19580	1.25339	1.31569	1.39568	1.44540
	ED_{333}	1.19601	1.25351	1.31583	1.39620	1.44610
	ED_{225}	1.20505	1.25922	1.32035	1.40195	1.45402
	ED_{222}	1.20510	1.25928	1.32039	1.40197	1.45403
10	3D	1.29751	1.34847	1.40828	1.49309	1.54980
	ED3	1.30054	1.35031	1.40970	1.49501	1.55248
	LD3	1.29750	1.34846	1.40828	1.49312	1.54983
	ED_{555}	1.29751	1.34847	1.40828	1.49309	1.54980
	ED_{333}	1.29757	1.34850	1.40831	1.49324	1.55001
	ED_{225}	1.30054	1.35031	1.40970	1.49501	1.55248
	ED_{222}	1.30054	1.35031	1.40970	1.49501	1.55248
100	3D	1.33931	1.38669	1.44491	1.53143	1.59105
	ED3	1.33934	1.38671	1.44493	1.53145	1.59108
	LD3	1.33931	1.38669	1.44491	1.53142	1.59105
	ED_{555}	1.33931	1.38670	1.44492	1.53143	1.59106
	ED_{333}	1.33931	1.38670	1.44492	1.53143	1.59106
	ED_{225}	1.33934	1.38672	1.44493	1.53145	1.59109
	ED_{222}	1.33934	1.38672	1.44493	1.53145	1.59109

ics models for higher modes the Chebyshev-Ritz formulation has a slight loss of accuracy with respect to the HTRF. In Table 13.3 the first three natural frequencies of simply-supported FGM isotropic moderately thick plates made of Al/ZrO_2, obtained by using several higher-order ESL and ZZ plate theories are compared with the local Kriging meshless method based on FSDT [18]. The results are in a good agreement for all of the given values of the volume fraction index p. In Table 13.4 the same FGM plate is analyzed for fully clamped boundary condition, once again the results match well. In Table 13.5 the fundamental frequency of a simply supported sandwich plates Al_2O_3/Al with FGM core ($h_c = 0.8\,h$) is com-

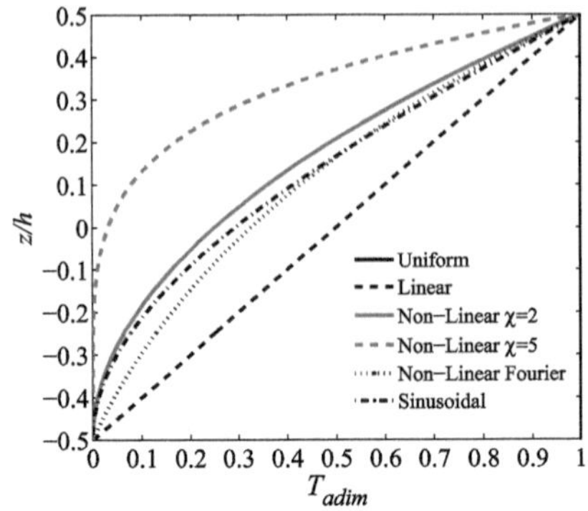

Figure 13.1 Dimensionless temperature distributions through-the-thickness direction.

Table 13.6 Comparison of the dimensionless frequency parameters $\hat{\omega} = \frac{\omega b^2}{h}\sqrt{\frac{\rho_0}{E_0}}$ ($\rho_0 = 1$ Kg/m^3 and $E_0 = 1$ GPa) simply-supported sandwich plates with Al_2O_3/Al graded core

Theory	**Circular frequency parameters**				
	$\hat{\omega}_1$	$\hat{\omega}_2$	$\hat{\omega}_3$	$\hat{\omega}_4$	$\hat{\omega}_5$
ED3 [6]	1.38669	3.46521	3.46521	5.54189	6.92533
LD3 [6]	1.38669	3.46520	3.46520	5.54189	6.92532
EDZ_{888}	1.38669	3.46521	3.46521	5.54189	6.92533
EDZ_{222}	1.38671	3.46533	3.46533	5.54221	6.92583
ED_{555}	1.38669	3.46521	3.46521	5.54189	6.92533
ED_{333}	1.38669	3.46521	3.46521	5.54189	6.92533
ED_{225}	1.38669	3.46521	3.46521	5.54189	6.92533
ED_{222}	1.38671	3.46533	3.46533	5.54221	6.92583

puted. Several advanced higher-order plate theories are compared with the 3D elasticity solution [15] and a layerwise theory [5]. The proposed theory perfectly matches the 3D elasticity solution, leading to a percentage error defined as $\Delta_{3D}\% = \frac{\hat{\omega}-\hat{\omega}_{3D}}{\hat{\omega}_{3D}} \times 100$ equal to zero for all the considered length-to-thickness ratios and volume fraction. In Table 13.6 for the same FGM sandwich plate configuration with $p = 0.5$ and $a/h = 100$ an excellent agreement of the first five natural frequencies is found with respect to a higher-order layerwise theory [5]. In Table 13.7 six different FGM sandwich plates with FGM face sheets and a ceramic core are investigated. The results are presented in terms of dimensionless fundamental circular frequency for several values of the volume fraction index and with $a/h = 10$.

Table 13.7 Comparison of the dimensionless fundamental frequency parameter $\hat{\omega} = \frac{\omega b^2}{h}\sqrt{\frac{\rho_0}{E_0}}$ ($\rho_0 = 1$ Kg/m^3 and $E_0 = 1$ GPa) simply-supported FGM isotropic plates Al_2O_3/Al

p	Theory	Fundamental frequency parameter					
		1 − 0 − 1	2 − 1 − 2	2 − 1 − 1	1 − 1 − 1	2 − 2 − 1	1 − 2 − 1
0.5	CPT [15]	1.47157	1.51242	1.54264	1.54903	1.58374	1.60722
	FSDT [15]	1.44168	1.48159	1.51035	1.51695	1.55001	1.57274
	TSDT [15]	1.44424	1.48408	1.51253	1.51922	1.55199	1.57451
	SSDT [15]	1.44436	1.48418	1.51258	1.51927	1.55202	1.57450
	3D-Ritz [15]	1.44614	1.48608	1.50841	1.52131	1.54926	1.57668
	EDZ_{888}	1.44621	1.48612	1.50846	1.52133	1.54929	1.57669
	ED_{999}	1.44620	1.48612	1.50846	1.52133	1.54929	1.57669
	ED_{444}	1.44624	1.48614	1.50847	1.52134	1.54931	1.57672
1	CPT	1.26238	1.32023	1.37150	1.37521	1.43247	1.46497
	FSDT	1.24031	1.29729	1.34637	1.35072	1.40555	1.43722
	TSDT	1.24320	1.30011	1.34888	1.35333	1.40789	1.43934
	SSDT	1.24335	1.30023	1.34894	1.35339	1.40792	1.43931
	3D-Ritz	1.24470	1.30181	1.33511	1.35523	1.39763	1.44137
	EDZ_{888}	1.24471	1.30183	1.33512	1.35522	1.39760	1.44135
	ED_{999}	1.24470	1.30183	1.33512	1.35521	1.39760	1.44135
	ED_{444}	1.24481	1.30186	1.33513	1.35524	1.39764	1.44142
5	CPT	0.95844	0.99190	1.08797	1.05565	1.16195	1.18867
	FSDT	0.94256	0.97870	1.07156	1.04183	1.14467	1.17159
	TSDT	0.94598	0.98184	1.07432	1.04466	1.14731	1.17397
	SSDT	0.94630	0.98207	1.07445	1.04481	1.14741	1.17399
	3D-Ritz	0.94476	0.98103	1.02942	1.04532	1.10983	1.17567
	EDZ_{888}	0.94433	0.98101	1.02933	1.04517	1.10940	1.17536
	ED_{999}	0.94433	0.98097	1.02931	1.04499	1.10940	1.17536
	ED_{444}	0.94630	0.98214	1.02993	1.04556	1.10994	1.17572
10	CPT	0.94321	0.95244	1.05185	1.00524	1.11883	1.13614
	FSDT	0.92508	0.93962	1.03580	0.99256	1.10261	1.12067
	TSDT	0.92839	0.94297	1.03862	0.99551	1.10533	1.12314
	SSDT	0.92875	0.94332	1.04558	0.99519	1.04154	1.13460
	3D-Ritz	0.92727	0.94078	0.98929	0.99523	1.06104	1.12466
	EDZ_{888}	0.92647	0.94044	0.98889	0.99519	1.06045	1.12422
	ED_{999}	0.92608	0.94044	0.98882	0.99484	1.06045	1.12416
	ED_{444}	0.92885	0.94284	0.99007	0.99596	1.06140	1.12468

Table 13.8 Critical temperatures ΔT_{cr} (°C) of simply supported (SSSS) square FGM isotropic plates

a/h	Theory	ΔT_{cr}				
		$p=0$	$p=0.5$	$p=1$	$p=2$	$p=5$
50	ES-FEM [17]	70.6998	39.4860	32.2723	28.5288	29.3283
	ED444	68.2055	38.6553	31.6979	28.0962	28.9625
100	ES-FEM [17]	17.7187	9.8946	8.0867	7.1492	7.3515
	ED444	17.0871	9.6821	7.9389	7.0379	7.2594

Table 13.9 Critical temperature ΔT_{cr} (°C) of fully clamped (CCCC) square FGM isotropic plates

a/h	Theory	ΔT_{cr}				
		$p=0$	$p=0.5$	$p=1$	$p=2$	$p=5$
50	ES-FEM [17]	188.2834	105.2699	86.0739	76.0781	78.0599
	ED222	185.8634	105.5901	86.6282	76.7304	78.8096
100	ES-FEM [17]	47.4967	26.5411	21.6980	19.1804	19.7017
	ED222	48.0005	27.2915	22.4017	19.8472	20.3664

The results are in excellent agreement with the 3D Ritz solution [15] and lead to a higher accuracy than other higher-order theories present in literature. It is interesting to note that for all of the considered values of p the lowest fundamental frequency is obtained by using the FGM sandwich plate configuration $1-0-1$ whilst the highest one with the configuration $1-2-1$. Moreover, it is observed in the same Table that when increasing the value of p the dimensionless fundamental frequency decreases.

13.3 CRITICAL TEMPERATURE OF FGM ISOTROPIC AND SANDWICH PLATES

In Tables 13.8 and 13.9, results, for simply supported and fully clamped boundary conditions, respectively, are compared against those proposed by Nguyen et al. [17] using the edge-smoothed finite element method. As can be seen the results in terms of critical temperature, for different values of the volume fraction index, are in excellent agreement, for both of the boundary conditions taken into account. Two different thickness-to-length ratios have been examined $a/h = 50$ and $a/h = 100$. As can be observed the critical temperature decreases when increasing both the volume fraction index (p) and the thickness-to-length ratio (a/h). As expected the fully

Table 13.10 Critical temperatures ΔT_{cr} (°C) of different FGM isotropic square plates with $p = 0$

Temperature Distribution	Theory	a/h				
		10	20	40	80	100
Uniform	CLPT [16]	1709.911	427.477	106.869	26.717	17.099
	HSDT [16]	1617.484	421.516	106.492	26.693	17.088
	EDZ_{888}	1599.294	420.146	106.404	26.688	17.087
	EDZ_{333}	1599.322	420.146	106.404	26.691	17.087
	ED_{999}	1599.293	420.146	106.404	26.688	17.087
	ED_{444}	1599.294	420.146	106.404	26.688	17.087
	ED_{222}	1609.305	420.844	106.449	26.691	17.088
Linear	CLPT	3409.821	844.955	203.738	43.434	24.198
	HSDT	3224.968	833.032	202.984	43.387	24.177
	EDZ_{888}	3188.250	830.287	202.808	43.377	24.172
	EDZ_{333}	3188.308	830.287	202.808	43.375	24.172
	ED_{999}	3188.250	830.286	202.808	43.376	24.174
	ED_{444}	3188.250	830.286	202.808	43.376	24.174
	ED_{222}	3208.314	831.684	202.898	43.382	24.177
Non-linear (Fourier)	CLPT	3409.821	844.955	203.738	43.434	24.198
	HSDT	3224.968	833.032	202.984	43.387	24.177
	EDZ_{888}	3188.250	830.286	202.808	43.371	24.172
	EDZ_{333}	3188.308	830.287	202.808	43.375	24.172
	ED_{999}	3188.250	830.286	202.808	43.376	24.174
	ED_{444}	3188.250	830.286	202.808	43.376	24.174
	ED_{222}	3208.314	831.684	202.898	43.382	24.177

clamped (CCCC) boundary condition leads to higher critical temperatures than those computed by using the simply supported (SSSS) one. In the case of CCCC boundary condition the convergence has been reached by using $M = N = 16$ as half-wave numbers. In Tables 13.10 and 13.11 an assessment of the developed ESL and ZZ plate theories is carried out comparing the results with the analytical solutions based on a CLPT and a HSDT, provided by Javaheri and Eslami [16]. The proposed results represent the critical temperature of Al_2O_3/Al FGM isotropic plates (see Table 13.1) with $T_m = 5\,°C$, $p = 0$ and $p = 5$, respectively, when subjected to uniform, linear and non-linear (Fourier) temperature rise through-the-thickness direction (see Fig. 13.1). Results match very well from moderately thick ($a/h = 10$) to thin ($a/h = 100$) FGM isotropic plates. The proposed higher order ESL

Table 13.11 Critical temperatures ΔT_{cr} (°C) of different FGM isotropic square plates with $p = 5$

Temperature Distribution	Theory	a/h				
		10	20	40	80	100
Uniform	CLPT [16]	726.571	181.642	45.410	11.352	7.265
	HSDT [16]	678.926	178.528	45.213	11.340	7.260
	EDZ_{888}	669.396	177.803	45.166	11.339	7.256
	EDZ_{333}	669.817	177.833	45.167	11.337	7.262
	ED_{999}	669.396	177.803	45.166	11.337	7.259
	ED_{444}	669.545	177.814	45.167	11.337	7.259
	ED_{222}	677.859	178.404	45.205	11.340	7.260
Linear	CLPT	1242.035	304.054	69.558	10.934	3.899
	HSDT	1160.024	298.693	69.219	10.913	3.891
	EDZ_{888}	1147.338	297.737	69.155	10.909	3.889
	EDZ_{333}	1148.089	297.789	69.158	10.908	3.892
	ED_{999}	1147.337	297.737	69.155	10.909	3.889
	ED_{444}	1147.595	297.756	69.156	10.909	3.889
	ED_{222}	1162.138	298.776	69.222	10.913	3.891
Non-linear (Fourier)	CLPT	1553.336	380.261	86.999	13.675	4.877
	HSDT	1450.769	373.557	86.568	13.648	4.866
	EDZ_{888}	1434.646	372.345	86.487	13.643	4.864
	EDZ_{333}	1435.591	372.411	86.491	13.643	4.871
	ED_{999}	1434.646	372.345	86.487	13.643	4.864
	ED_{444}	1434.969	372.369	86.488	13.643	4.864
	ED_{222}	1453.172	373.645	86.571	13.648	4.866

and ZZ theories making use of a higher number of degrees of freedom (DOFs) lead to a more refined results. It is interesting to note that the proposed ED_{222} plate model, especially for thin FGM isotropic plates, leads exactly to the same results of the HSDT given by Javaheri and Eslami [16]. The trends of the critical temperature, as already observed in Tables 13.8 and 13.9, decrease when increasing both the volume fraction index and the thickness-to-length ratio. As can be seen in Tables 13.10 and 13.11, the critical temperatures of a FGM isotropic plate, when subjected to a linear temperature distribution through-the-thickness, are higher than those evaluated considering a uniform temperature rise through-the-thickness for both $p = 0$ and $p = 5$. In sharp contrast, when compared to those computed when accounting for a non-linear temperature distribution, then are

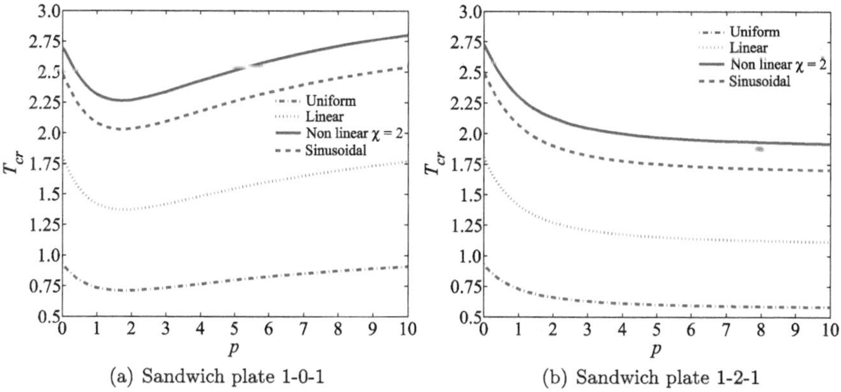

Figure 13.2 Critical temperature for different temperature rises through-the-thickness varying the volume fraction index p.

equal for $p = 0$ and lower for $p = 5$. Several FGM sandwich plates made up of $Zr0_2/(Ti\ 6Al\ 4V)_1$ (see Table 13.1), in Tables 13.12 ,13.13 and 13.14, are examined. In particular, different FGM sandwich plate configurations, which volume fractions are depicted in Fig. 10.2, are subjected to various temperature distributions through-the-thickness direction. The FGM sandwich plates are composed of FGM face sheets and a ceramic core. In the present investigation for the linear and non-linear temperature rises, $T_t = 25\,°C$. Results, in terms of critical buckling temperatures, are compared with those proposed by Zenkour and Sobhy [29]. In particular, the latter were obtained by using first-order shear deformation plate theory (FPT), higher-order shear deformation plate theory (HPT) and sinusoidal shear deformation plate theory (SDT). Most notably, the critical temperatures are computed for different values of the thickness-to-length ratio, $p = 2$ and with several temperature rise through-the-thickness, uniform, linear and non-linear ($\chi = 5$, see Fig. 13.1). As expected independently of the values assumed by the volume fraction index the critical temperature decreases when increasing the thickness-to-length ratio. The highest critical temperature is reached with the FGM sandwich plate configuration $1 - 0 - 1$. Moreover in Fig. 13.2 the critical temperature, computed for different temperature rises through-the-thickness is depicted with respect to the volume fraction index, for two FGM sandwich plate configurations, $1 - 0 - 1$ and $1 - 2 - 1$ respectively. As expected in both configurations, the non-linear temperature rises, namely, the one with $\chi = 2$ and the sinusoidal lead to the higher critical temperatures.

Table 13.12 Critical temperatures $T_{cr} = 10^{-3}\,\Delta T_{cr}$ (°C) of different sandwich square plates under uniform temperature rise and $p = 2$

Lamination scheme	**Theory**	a/h			
		5	10	25	50
1 − 0 − 1	SPT [28]	2.63459	0.71815	0.11789	0.02958
	HPT [28]	2.63018	0.71783	0.11788	0.02958
	FPT [28]	2.57355	0.71357	0.11776	0.02957
Present plate models	EDZ_{888}	2.53838	0.71029	0.11768	0.02957
	EDZ_{333}	2.54581	0.71088	0.11770	0.02960
	ED_{999}	2.53828	0.71028	0.11768	0.02957
	ED_{444}	2.54580	0.71088	0.11769	0.02957
	ED_{111}	3.04976	0.86525	0.14402	0.03621
2 − 1 − 2	SPT	2.39953	0.65098	0.10671	0.02677
	HPT	2.39637	0.65075	0.10670	0.02676
	FPT	2.34733	0.64710	0.10660	0.02676
Present plate models	EDZ_{888}	2.32049	0.64461	0.10653	0.02676
	EDZ_{333}	2.32393	0.64488	0.10654	0.02676
	ED_{999}	2.32049	0.64461	0.10653	0.02676
	ED_{444}	2.32393	0.64488	0.10654	0.02676
	ED_{111}	2.78709	0.78520	0.13038	0.03277
1 − 1 − 1	SPT	2.36195	0.64253	0.10541	0.02645
	HPT	2.35999	0.64238	0.10540	0.02645
	FPT	2.31737	0.63921	0.10532	0.02644
Present plate models	EDZ_{888}	2.29107	0.63675	0.10525	0.02644
	EDZ_{333}	2.29231	0.63684	0.10525	0.02644
	ED_{999}	2.29075	0.63672	0.10525	0.02644
	ED_{444}	2.29229	0.63684	0.10525	0.02644
	ED_{111}	2.75073	0.77555	0.12881	0.03237
1 − 2 − 1	SPT	2.42899	0.66689	0.10972	0.02754
	HPT	2.42873	0.66687	0.10972	0.02754
	FPT	2.39541	0.66436	0.10966	0.02754
Present plate models	EDZ_{888}	2.35962	0.66102	0.10956	0.02753
	EDZ_{333}	2.36129	0.66116	0.10956	0.02753
	ED_{999}	2.35962	0.66102	0.10956	0.02753
	ED_{444}	2.36123	0.66115	0.10956	0.02753
	ED_{111}	2.83790	0.80550	0.13409	0.03372

Table 13.13 Critical temperatures $T_{cr} = 10^{-3}\,\Delta T_{cr}$ (°C) of different sandwich square plates under linear temperature rise ($\chi = 1$) and $p = 2$

Lamination scheme	Theory	a/h			
		5	10	25	50
$1-0-1$	SPT [28]	5.21919	1.38631	0.18578	0.00916
	HPT [28]	5.21036	1.38566	0.18576	0.00917
	FPT [28]	5.09710	1.37714	0.18553	0.00915
Present plate models	EDZ_{888}	5.02496	1.37054	0.18535	0.00914
	EDZ_{333}	5.03986	1.37172	0.18538	0.00921
	ED_{999}	5.02475	1.37052	0.18535	0.00914
	ED_{444}	5.03982	1.37172	0.18538	0.00914
	ED_{111}	6.04724	1.68045	0.23803	0.02243
$2-1-2$	SPT	4.74906	1.25196	0.16341	0.00354
	HPT	4.74274	1.25150	0.16340	0.00354
	FPT	4.64467	1.24420	0.16320	0.00353
Present plate models	EDZ_{888}	4.58978	1.23919	0.16306	0.00352
	EDZ_{333}	4.59657	1.23972	0.16308	0.00352
	ED_{999}	4.58966	1.23919	0.16306	0.00352
	ED_{444}	4.59654	1.23973	0.16308	0.00352
	ED_{111}	5.52250	1.52037	0.21075	0.01554
$1-1-1$	SPT	4.67391	1.23506	0.16082	0.00289
	HPT	4.66999	1.23477	0.16081	0.00289
	FPT	4.58474	1.22842	0.16064	0.00288
Present plate models	EDZ_{888}	4.53086	1.22347	0.16050	0.00288
	EDZ_{333}	4.53336	1.22366	0.16051	0.00288
	ED_{999}	4.53023	1.22342	0.16050	0.00288
	ED_{444}	4.53331	1.22366	0.16051	0.00288
	ED_{111}	5.44983	1.50107	0.20761	0.01476
$1-2-1$	SPT	4.80799	1.28377	0.16944	0.00508
	HPT	4.80746	1.28375	0.16944	0.00508
	FPT	4.74083	1.27872	0.16931	0.00507
Present plate models	EDZ_{888}	4.66775	1.27202	0.16912	0.00506
	EDZ_{333}	4.67111	1.27228	0.16913	0.00506
	ED_{999}	4.66775	1.27202	0.16912	0.00506
	ED_{444}	4.67097	1.27228	0.16913	0.00507
	ED_{111}	5.62392	1.56096	0.21819	0.01744

Table 13.14 Critical temperatures $T_{cr} = 10^{-3}\,\Delta T_{cr}$ (°C) of different sandwich square plates under non-linear temperature rise ($\chi = 5$) and $p = 2$

Lamination scheme	Theory	a/h			
		5	10	25	50
1 − 0 − 1	SPT [28]	23.06830	6.12734	0.82115	0.04051
	HPT [28]	23.02926	6.12449	0.82107	0.04052
	FPT [28]	22.52869	6.08684	0.82005	0.04044
Present plate models	EDZ_{888}	21.70772	6.01748	0.81831	0.04040
	EDZ_{333}	21.76775	6.02255	0.81847	0.04028
	ED_{999}	21.70677	6.01741	0.81831	0.04038
	ED_{444}	21.76380	6.02248	0.81845	0.04039
	ED_{111}	25.71472	7.34202	1.05003	0.09908
2 − 1 − 2	SPT	22.38252	5.90053	0.77017	0.01668
	HPT	22.35275	5.89838	0.77011	0.01668
	FPT	21.89054	5.86398	0.76918	0.01662
Present plate models	EDZ_{888}	21.24283	5.82465	0.77048	0.01639
	EDZ_{333}	21.27146	5.82705	0.77052	0.01664
	ED_{999}	21.24229	5.82461	0.77045	0.01664
	ED_{444}	21.26902	5.82701	0.77052	0.01664
	ED_{111}	25.17843	7.11324	0.99501	0.07349
1 − 1 − 1	SPT	22.00152	5.81379	0.75703	0.01363
	HPT	21.98303	5.81247	0.75699	0.01363
	FPT	21.58175	5.78254	0.75619	0.01358
Present plate models	EDZ_{888}	21.22219	5.81520	0.76668	0.01375
	EDZ_{333}	21.23387	5.81632	0.76667	0.01375
	ED_{999}	21.21930	5.81520	0.76668	0.01375
	ED_{444}	21.23175	5.81629	0.76667	0.01375
	ED_{111}	25.17488	7.10418	0.99097	0.07054
1 − 2 − 1	SPT	21.54917	5.75380	0.75946	0.02279
	HPT	21.54679	5.75368	0.75946	0.02279
	FPT	21.24818	5.73116	0.75885	0.02275
Present plate models	EDZ_{888}	21.56980	5.95798	0.79567	0.02384
	EDZ_{333}	21.58801	5.95927	0.79570	0.02385
	ED_{999}	21.56978	5.95798	0.79567	0.02384
	ED_{444}	21.58557	5.95922	0.79570	0.02385
	ED_{111}	25.67606	7.28319	1.02583	0.08209

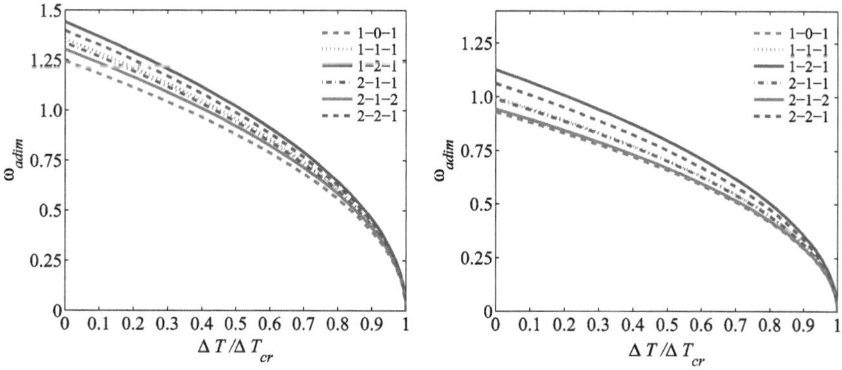

(a) Volume fraction index $p = 1$

(b) Volume fraction index $p = 10$

Figure 13.3 Effect of the temperature on the dimensionless fundamental circular frequency $\omega_{adim} = \frac{\omega b^2}{h}\sqrt{\frac{\rho_c}{E_c}}$ of different FGM sandwich plates.

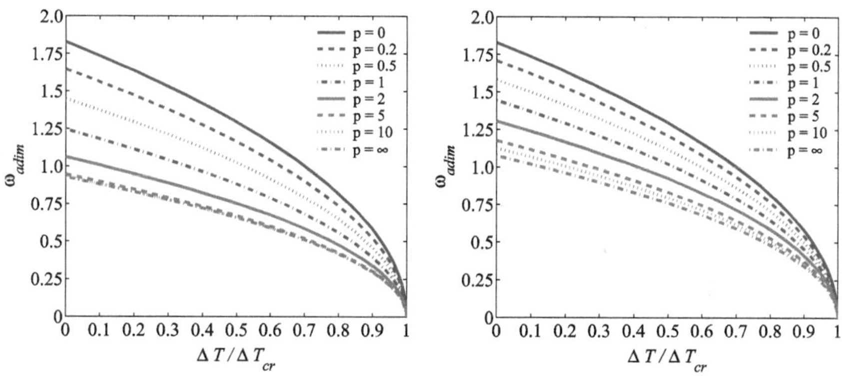

(a) Sandwich plate 1-0-1

(b) Sandwich plate 1-2-1

Figure 13.4 Effect of the temperature on the dimensionless fundamental frequency parameter $\omega_{adim} = \frac{\omega b^2}{h}\sqrt{\frac{\rho_c}{E_c}}$ varying the volume fraction index p.

13.4 FREE VIBRATION CHARACTERISTICS OF FGM SANDWICH PLATES IN THERMAL ENVIRONMENT

The effect of a uniform temperature distribution on the free vibration behaviour of FGM sandwich plates is taken into account in Figs. 13.3 and 13.4. In particular, in Fig. 13.3 the effect of the temperature on the dimensionless fundamental circular frequency of six different FGM sandwich plates is shown. As expected for all of the configurations the fundamental frequency decreases when increasing the temperature and approaching zero when the temperature is close to the critical one. In Fig. 13.4 for two fixed FGM sandwich plate configurations, $1-0-1$ and $1-2-1$, once again the dimensionless fundamental circular frequency is computed when increasing the environment temperature for different values of the volume fraction index. It is observed that increasing the value of p the fundamental frequency decreases and its trend when increasing the temperature as expected does not change significantly.

REFERENCES

1. El Meiche N, Tounsi A, Ziane N, Mechab I, Adda Bedia EA. A new hyperbolic shear deformation theory for buckling and vibration of functionally graded sandwich plate. Int J Mech Sci 2011;53(4):237–47.
2. Matsunaga H. Free vibration and stability of functionally graded plates according to a 2-D higher-order deformation theory. Compos Struct 2008;82(4):499–512.
3. Matsunaga H. Thermal buckling of functionally graded plates according to a 2D higher-order deformation theory. Compos Struct 2009;90(1):76–86.
4. Abrate S. Free vibration, buckling, and static deflections of functionally graded plates. Compos Sci Technol 2006;66(14):2383–94.
5. Dozio L. Natural frequencies of sandwich plates with FGM core via variable-kinematic 2-D Ritz models. Compos Struct 2013;96:561–8.
6. Dozio L. Exact free vibration analysis of Lévy FGM plates with higher-order shear and normal deformation theories. Compos Struct 2014;11:415–25.
7. Viola E, Rossetti L, Fantuzzi N. Numerical investigation of functionally graded cylindrical shells and panels using the generalized unconstrained third order theory coupled with the stress recovery. Compos Struct 2012;94:3736–58.
8. Viola E, Rossetti L, Fantuzzi N, Tornabene F. Static analysis of functionally graded conical shells and panels using the generalized unconstrained third order theory coupled with the stress recovery. Compos Struct 2014;112:44–65.
9. Viola E, Tornabene F. Free vibrations of three parameter functionally graded parabolic panels of revolution. Mech Res Commun 2009;36:587–94.
10. Tornabene F. Free vibration analysis of functionally graded conical, cylindrical shell and annular plate structures with a four-parameter power-law distribution. Comput Methods Appl Mech Eng 2009;198:2911–35.
11. Tornabene F. Free vibration analysis of functionally graded conical, cylindrical shell and annular plate structures with a four-parameter power-law distribution. Comput Method Appl M 2009;198:2911–35.
12. Tornabene F, Fantuzzi N, Bacciocchi M. Free vibrations of free-form doubly-curved shells made of functionally graded materials using higher-order equivalent single layer theories. Composites Part B: Engineering 2014;67:490–509.
13. Bateni M, Kiani Y, Eslami MR. A comprehensive study on stability of FGM plates. Int J Mech Sci 1976;75:134–44.
14. Shariyat BS, Eslami MR. Buckling of thick functionally graded plates under mechanical and thermal loads. Compos Struct 2007;78(3):433–9.
15. Li Q, Iu VP, Kou KP. Three-dimensional vibration analysis of functionally graded material sandwich plates. J Sound Vib 2008;311:498–515.

16. Javaheri R, Elsami MR. Thermal buckling of functionally graded plates based on higher order theory. J Therm Stresses 2002;25:603–25
17. Nguyen-Xuan H, Tran LV, Nguyen-Thoi T, Vu-Do HC. Analysis of functionally graded plates using an edge-based smoothed finite element method. Compos Struct 2011;93:3019–39.
18. Zhu P, Liew KM. Free vibration analysis of moderately thick functionally graded plates by local Kriging meshless method. Compos Struct 2011;93:2925–44.
19. Zhao X, Lee YY, Liew KM. Free vibration analysis of functionally graded plates using the element-free kp-Ritz method. J Sound Vib 2009;319:918–39.
20. Fazzolari FA. Natural frequencies and critical temperatures of functionally graded sandwich plates subjected to uniform and non-uniform temperature distributions. Compos Struct 2015;121:197–210.
21. Fazzolari FA. Modal characteristics of P- and S-FGM plates with temperature-dependent materials in thermal environment. J Therm Stresses 2016;39(7):854–73.
22. Fazzolari FA, Carrera E. Advanced variable kinematics Ritz and Galerkin formulations for accurate buckling and vibration analysis of laminated composite plates. Compos Struct 2011;94(1):50–67.
23. Fazzolari FA, Carrera E. Thermo-mechanical buckling analysis of anisotropic multi-layered composite and sandwich plates by using refined variable-kinematics theories. J Therm Stresses 2012;36(4):321–50.
24. Fazzolari FA, Carrera E. Accurate free vibration analysis of thermo-mechanically pre/post-buckled anisotropic multilayered plates based on a refined hierarchical trigonometric Ritz formulation. Compos Struct 2013;95:381–402.
25. Fazzolari FA, Carrera E. Coupled thermoelastic effect in free vibration analysis of anisotropic multilayered plates and FGM plates by using a variable-kinematics Ritz formulation. Eur J Mech A, Solids 2014;44:157–74.
26. Fazzolari FA, Carrera E. Free vibration analysis of sandwich plates with anisotropic face sheets in thermal environment by using the hierarchical trigonometric Ritz formulation. Composites, Part B, Eng 2013;50:67–81.
27. He XQ, Ng TY, Sivashanker S, Liew KM. Active control of FGM plates with integrated piezoelectric sensors and actuators. Int J Solids Struct 2001;38:1641–5.
28. Zenkour AM. A comprehensive analysis of functionally graded sandwich plates: part 2 – buckling and free vibration. Int J Solids Struct 2005;42(18–19):5243–58.
29. Zenkour AM, Sobhy M. Thermal buckling of various types of FGM sandwich plates. Compos Struct 2010;93(1):93–112.

CHAPTER 14

Thermal effect on flutter of panels

14.1 INTRODUCTION

During their operations life aerospace structures are subjected contemporaneously both to aerodynamics loads, which depend on aerodynamic pressure distributions and viscous forces, and aero-thermal effects which take into account surface heating-rate and inner temperature distributions. Aerospace structures are mainly composed by thin-walled structures like plates and shells, which are, therefore, when exposed to high speed airflow on the outer surface and at certain critical speed might become dynamically instable. Moreover, the rise of temperature due to aerodynamics heating, strongly decrease the structural load-bearing capacity affecting significantly the buckling phenomena. It becomes mandatory for design engineers to carry out dynamic and static stability analysis to prevent failures. The dynamic instability which affect thin-walled plate and shell structures is referred to as panel flutter. The first investigation of this phenomenon is traced back to Jordan [1], who presented this physical nature of the phenomenon. The subject was later on extensively researched, both from an experimental and analytical standpoint. A better understanding of the panel flutter, above all, from the mathematical modelling point of view, was given by Dowell [2] and Librescu [3]. Comprehensive reviews on the topic were proposed by Fung [4], Johns [5,6], Dowell [7] and Bismarck-Nasr [10].

During the last decades several methodologies such as, Finite Element Method (FEM), Galerkin, Ritz and other Meshless formulations have been extensively used to cope with the panel flutter of plate and shell structures. In particular, Bismarck-Nasr [8–11] proposed a FE formulation to deal with the aeroelasticity of plates and shells. Olson applied FEs to panel flutter [12,13]. Supersonic panel flutter was investigated by Sander et al. [14]. Chowdary et al. [15,16] employed a shear deformable FE to deal with supersonic flutter of laminated composite panels and skew composite plates. Ganaphati and Touratier [17] included thermal effects, taking into account several temperature distributions. Oh and Kim [18] examined supersonic flutter of cylindrical composite panels with large thermoelastic deflections. Supersonic flutter behaviour of composite skew flat panels under mechanical and thermal loadings was analyzed by Singha and Gana-

Thermal Stress Analysis of Composite Beams, Plates and Shells.
DOI: http://dx.doi.org/10.1016/B978-0-12-849892-7.00017-0

Table 14.1 Critical dimensionless dynamic pressure $\lambda^*_{D_{cr}} = \lambda_{D_{cr}} \frac{a^3}{D}$ and dimensionless flutter frequency $\omega^*_{cr} = \omega_{cr} \left(\frac{a^2}{\pi^2 h}\right) \sqrt{\frac{\rho}{D}}$ for simply supported isotropic plates

Theory	$\lambda^*_{D_{cr}}$	ω^*_{cr}
Bismark-Nasr [11]	512.65	1840.29
Sander et al. [14]	512.20	1844.00
Han and Yang [24]	512.33	1840.55
Lin et al. [25]	512.58	1847.50
Prakash and Ganapathi [26]	511.11	1840.29
Present ED_{222} plate model	512.61	1846.00

pathi [19]. Shiau and Liu [20] examined the nonlinear flutter behaviour of a two-dimensional simply supported composite laminated plate at high supersonic Mach number. The investigation was based on the Von Karman's large deflection plate theory and quasi-steady aerodynamic theory. Chandiramani et al. [21] studied the non-linear dynamic behaviour of a uniformly compressed, composite panel subjected to non-linear aerodynamic loading due to a high-supersonic co-planar flow is analyzed. Recently an interesting investigation on the panel flutter at low supersonic speeds was provided by Vedeneev [22]. The same author proposed [23] a new expression of the piston theory obtained starting from the un-steady gas pressure expression of the potential flow theory, in order to accurately describe the single-mode flutter phenomenon, which occurs at low supersonic Mach numbers.

14.2 FLUTTER BEHAVIOUR OF FLAT PANELS IN SUPERSONIC FLOW

A first validation is proposed in Table 14.1. In particular, the critical dimensionless dynamic pressure $\lambda^*_{D_{cr}}$ and the critical dimensionless frequency parameter ω^*_{cr} have been computed for an isotropic square plate ($a/h = 100$ and $a/b = 1$). The plate theory used in the analysis is the ED_{222} which leads to results which are in excellent agreement with those proposed in the literature. Moreover, the effect of the lamination angle θ on the dimensionless critical pressure $\lambda^*_{D_{cr}}$ and the dimensionless critical frequency parameter ω^*_{cr}, for a single orthotropic layer with material properties $E_1/E_1 = 5$, $E_2 = 2.7 \times 10^6$ Psi, $G_{12}/E_1 = 0.35$, $\nu_{12} = 0.3$ and $\rho = 0.192\,10^{-3}$ lb s^2/in^4, is shown in Fig. 14.1. In the same figure it is interesting to note the significant effect of the aspect ratio a/b on the flutter characteristics. In particular, for $a/b = 1$ dimensionless dynamic pressure $\lambda^*_{D_{cr}}$ decreases monotonically when

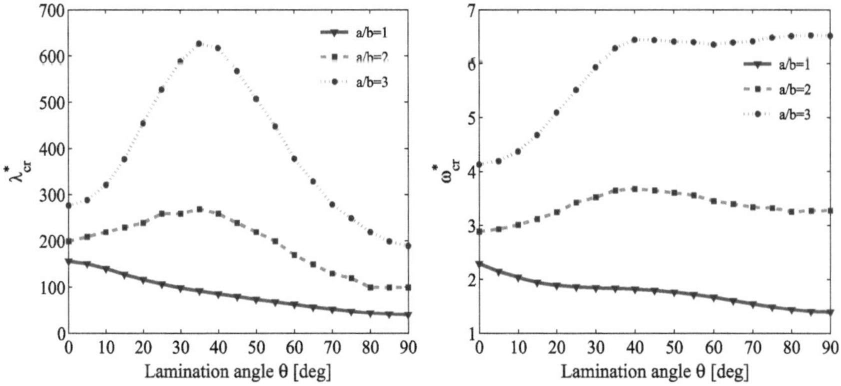

(a) (b)

Figure 14.1 Critical dynamic pressure $\lambda^*_{D_{cr}} = \lambda_{D_{cr}} \frac{a^3}{E_{22}\, h^3}$ and critical frequencies $\omega^*_{cr} = \omega_{cr}\left(\frac{a^2}{\pi^2 h}\right)\sqrt{\frac{\rho}{E_{22}}}$ of a single orthotropic layer varying the lamination angle θ.

Table 14.2 Mechanical properties of the materials considered

	Aluminum	Alumina	
E	70	380	Gpa
ν	0.3	0.3	–
ρ	2700	3940	Kg/m^3
α	23×10^{-6}	7.4×10^{-6}	1/K

increasing uniformly from 0° to 90° the lamination angle θ. For $a/b = 2$ and $a/b = 3$ the curves reach a maximum in the value of the dimensionless dynamic pressure $\lambda^*_{D_{cr}}$ at $\theta = 35°$.

14.3 AEROELASTIC INSTABILITIES OF FGM PANELS UNDER THERMO-MECHANICAL LOADS

This section shows some results obtained using the aero-thermo-elastic model introduced in the sections above. Different FGM material are considered but all of them are built using two materials, aluminum as metallic and alumina as ceramic, the mechanical and thermal properties are reported in Table 14.2. The volume fraction formulation used in the present model is the one given in chapter 6, section 6.2, in the case of classical isotropic FGM. In the same section it is possible to find the figures showing the behaviour of the parameter V_c for different values of p. The thermo-elastic and aero-elastic models have been assessed comparing the critical instability

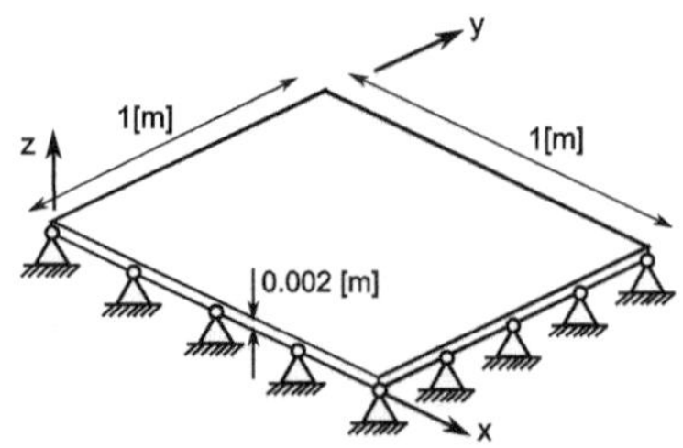

Figure 14.2 Geometry of the simply supported panel.

Table 14.3 ΔT_{cr} for different volume fraction index, p, evaluated using different structural models

	p				
	0	0.5	1	2	5
ESL3	68.22	38.72	31.75	28.14	29.02
ESL4	68.22	38.71	31.74	28.14	29.02
LW3	68.33	38.72	31.75	28.14	29.02
LW4	68.22	38.71	31.74	28.14	29.02
Refs. [27,28]	68.02	38.65	31.70	28.09	28.96
Ref. [29]	70.7	39.48	32.27	28.53	29.33

temperatures and the panel flutter speed with results from literature. Finally the aero-thermo-elastic analysis of a FGM plate is performed.

14.3.1 Thermo-elastic model assessment

A simply supported panel is considered to assess the thermo-elastic model. The geometry of the panel is reported in Fig. 14.2. The FGM mixtures law considered in this assessment are:

$$E(z) = (E_c - E_m)\, V_c(z) + E_m$$
$$\alpha(z) = (\alpha_c - \alpha_m)\, V(z) + \alpha_m$$
$$\nu(z) = \nu_0$$

where the subscript c stands for the ceramic material and m for the metallic. Table 14.3 shows the critical temperature of the panel for different values of volume fraction index evaluated using different structural models. The results are compared with those from literature. The present model provide a good accuracy in the evaluation of the critical temperature variation. When linear models are used, ESL2 and LW2, the results are non-accurate therefore, at the least a quadratic model is required to achieve the expected

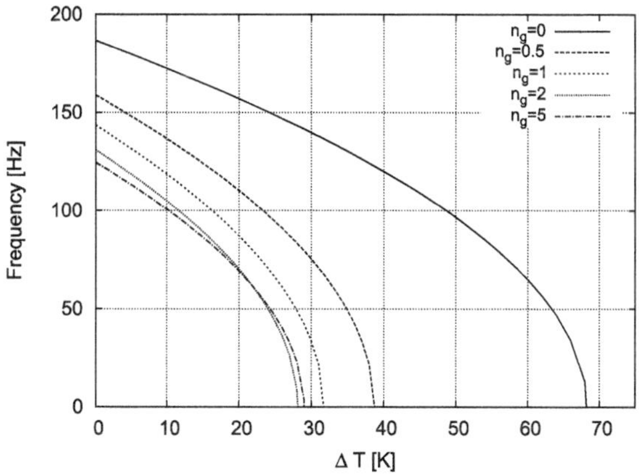

Figure 14.3 First natural frequency vs temperature.

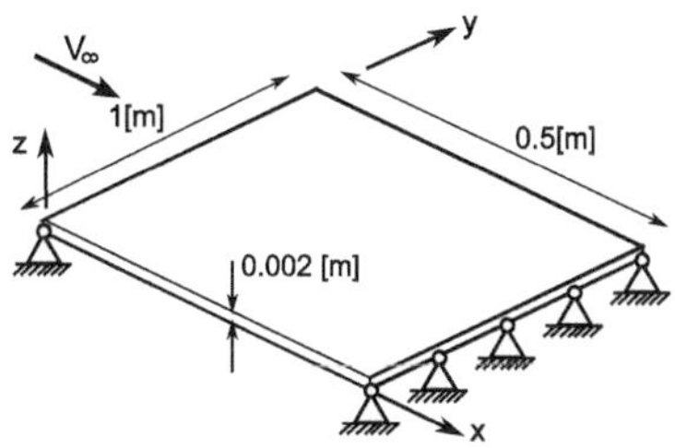

Figure 14.4 Geometry and reference system of the panel use in the aeroelastic assessment.

results. When a single layer is considered LW models do not introduce any advantages with respect to the ESL. When the volume fraction index increases, the percentage of aluminum in the FGM composition increases therefore a lower value of critical temperature is obtained. Figure 14.3 shows the evolution of the first natural frequency when the temperature is increased for different value of volume fraction index. This analysis shows that increasing the p the natural frequency decreases, also when the temperature increases the natural frequency decreases, when the temperature reach the critical value the frequency becomes zero.

14.3.2 Aero-elastic model assessment

The panel flutter speed of a simple isotropic panel, see Fig. 14.4 has been investigated in order to assess the aeroelastic model. The panel is built us-

Table 14.4 Panel flutter critical Mach and frequency for different structural models

Model	Ma_{cr}	f_{cr}
Krause [30]	4.5	66.03
ESL2	4.39	65.46
ESL3	4.36	65.31
ESL4	4.36	65.31
LW2	4.36	65.31
LW3	4.36	65.32
LW4	4.36	65.32

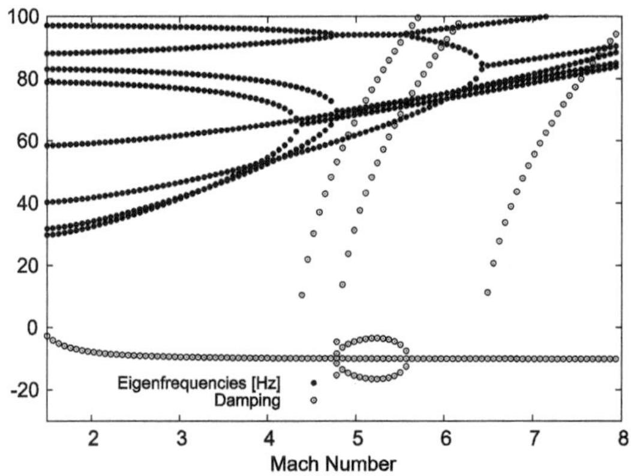

Figure 14.5 Evolution of the natural frequencies and damping of the panel at different Mach numbers.

ing an aluminum alloy. The results in terms of critical Mach number and frequency are reported in Table 14.4. Different structural models have been considered. The results show that both, the ESL and the LW models, provide a good agreement with the reference results. A linear formulation is enough to ensure a good accuracy. Figure 14.5 shows the evolution of the natural frequency and damping at different Mach numbers. The results show the classical aeroelastic instability, two frequencies merge together and at the same speed the damping becomes positive making the panel unstable.

14.3.3 Aero-thermo-elastic analysis of a FGM panel

In this section the aero-thermo-elastic model is considered. A square panel, see Fig. 14.6, is considered. Two sides are simply supported and two are free, therefore only the σ_{xx^θ} is considered. The FGM mixture equations are

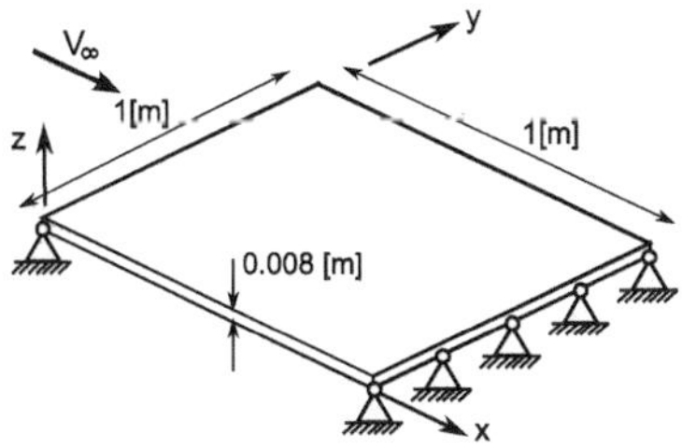

Figure 14.6 Geometry and reference system used in the aero-thermo-elastic analysis.

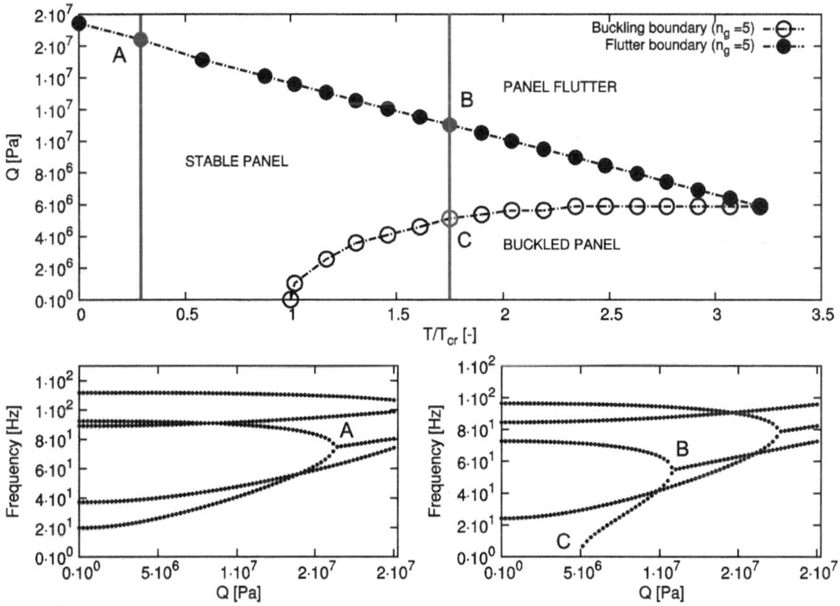

Figure 14.7 Stability boundaries for different temperatures and dynamic pressure.

reported below, where B denotes the bulk modulus that is strictly related with the elastic modulus. μ denotes the shear modulus.

$$
\begin{aligned}
\frac{B - B_m}{B_c - B_m} &= \frac{V_c}{1 + (1 - V_c)\frac{B_c - B_m}{B_m + \frac{4}{3}\mu_m}} \\
\frac{\mu - \mu_m}{\mu_c - \mu_m} &= \frac{V_c}{1 + (1 - V_c)\frac{\mu_c - \mu_m}{\mu_m + f_1}} \\
\frac{\alpha - \alpha_m}{\alpha_c - \alpha_m} &= \frac{\frac{1}{B} - \frac{1}{B_m}}{\frac{1}{B_c}\frac{1}{B_m}}
\end{aligned}
\tag{14.1}
$$

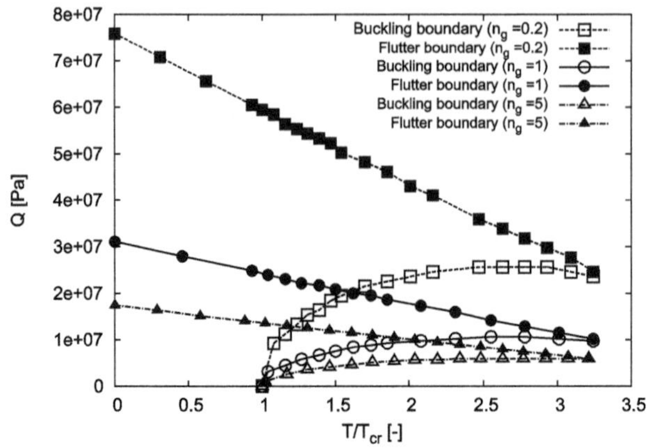

Figure 14.8 Stability boundaries for different *p* values.

Figure 14.7 shows the instability boundaries of the panel when *p* is equal to 5. In the lower part of the diagram there are reported the evolutions of the natural frequency before and after the critical temperature. If a non-critical temperature is considered, the only instability which is present is the panel flutter instability, point A. When the temperature is greater than the critical value, in the first part of the dynamic pressure domain, from zero up to point *C*, the panel is buckled. Between the point *C* and *B* the panel is stable while, after the point *C* a panel flutter phenomena arises. If all the temperature range is investigated a stability diagram can be drown, as the bigger one in Fig. 14.7. The upper limits are given by the panel flutter phenomena, while the lower boundary arises from the buckling phenomena. Figure 14.8 shows the stability boundary for different values of the volume fraction index. Higher is the alumina percentage, higher is the stability margin. This behaviour is due to the higher mechanical properties of the alumina with respect to the aluminum.

REFERENCES

1. Jordan PF. The physical nature of panel flutter. Aero Digest 1956:34–8.
2. Dowell EH. Aeroelasticity of plates and shells. 1st edition. Leyden (The Netherlands): Noordhoff International; 1975.
3. Librescu L. Elastostatics and kinetics of anisotropic and heterogeneous shell-type structures. 1st edition. Leyden (The Netherlands): Noordhoff International; 1975.
4. Fung YC. A summary of the theories and experiments on panel flutter. AGARD Manual on Aeroelasticity, Part III, Chapter 7. 1961.
5. Johns DJ. Survey of panel flutter. AGARD Advisory Report 1. 1965.

6. Johns DJ. A panel flutter review. AGARD Manual on Aeroelasticity, Part III, Chapter 7. 1961.
7. Dowell EH. Panel flutter: a review of the aeroelastic stability of plates and shells. AIAA J 1970;8(3):385–99.
8. Bismark-Nasr MN. Finite element analysis of aeroelasticity of plates and shells. Appl Mech Rev 1992;45(12):461–82.
9. Bismark-Nasr MN. Finite element method applied to the flutter of two parallel elastically coupled flat plates. Int J Numer Methods Eng 1977;11(7):1188–93.
10. Bismark-Nasr MN. On the sixteen degree of freedom rectangular plate element. Int J Comput Struct 1991;40(4):1059–60.
11. Bismark-Nasr MN. Structural dynamics in aeronautical engineering. 1st edition. AIAA Education Series; 1999.
12. Olson MD. Finite elements applied to panel flutter. AIAA J 1967;5(12):2267–70.
13. Olson MD. Some flutter solution using finite elements. AIAA J 1970;8(4):747–52.
14. Sander B, Bon C, Geradin M. Finite element analysis of supersonic panel flutter. Int J Numer Methods Eng 1973;7(3):379–94.
15. Chowdary TVR, Parthan S, Sinha PK. Finite element flutter analysis of laminated composite panels. Comput Struct 1994;53(2):245–51.
16. Chowdary TVR, Sinha PK, Parthan S. Finite element flutter analysis of composite skew panels. Comput Struct 1994;58(3):613–20.
17. Ganapathi M, Touratier M. Supersonic flutter analysis of thermally stressed laminated composite flat panels. Compos Struct 1996;4(2):241–8.
18. Oh IK, Kim DH. Vibration characteristics and supersonic flutter of cylindrical composite panels with large thermoelastic deflections. Compos Struct 2009;90(2):208–16.
19. Singha MK, Ganapathi M. A parametric study on supersonic flutter behavior of laminated composite skew flat panels. Compos Struct 2005;69(1):55–63.
20. Shiau LC, Lu LT. Nonlinear flutter of composite laminated plates. Math Comput Model 1990;14:983–8.
21. Chandiramani NK, Plaut RH, Librescu LI. Non-linear flutter of a buckled shear-deformable composite panel in a high-supersonic flow. Int J Non-Linear Mech 1995;30(2):149–67.
22. Vedeneev VV. Panel flutter at low supersonic speeds. J Fluids Struct 2012;29:79–96.
23. Vedeneev VV, Guvernyuk SV, Zubkov AF, Kolotnikov ME. Experimental observation of single mode panel flutter in supersonic gas flow. J Fluids Struct 2010;26(5):764–79.
24. Han A, Yang T. Nonlinear panel flutter using higher-order triangular finite elements. AIAA J 1983;10:1453–61.
25. Lin KJ, Lu PJ, Tarn JQ. Flutter analysis of anisotropic panels with patched cracks. J Aircr 1991;28:899–904.
26. Prakash T, Ganapathi M. Supersonic flutter characteristics of functionally graded flat panels including thermal effects. Compos Struct 2006;72:10–8.
27. Fazzolari FA. Natural frequencies and critical temperatures of functionally graded sandwich plates subjected to uniform and non-uniform temperature distributions. Compos Struct 2015;121:197–210.
28. Fazzolari FA. Thermal stability of FGM sandwich plate under various through-the-thickness temperature distributions. J Therm Stresses 2015;37(12):1449–81.
29. Nguyen-Xuan H, Tran LV, Nguyen-Thoi T, Vu-Do HC. Analysis of functionally graded plates using an edge-based smoothed finite element method. Compos Struct 2011;93:3019–33.
30. Krause H. Flattern flacher Schalen bei Ueberschallanstroemung. 1998.

INDEX

Thermal Stress Analysis of Composite Beams, Plates and Shells.
DOI: http://dx.doi.org/10.1016/B978-0-12-849892-7.00025-X

K

L

M

N

MIX
Papier | Fördert gute Waldnutzung
FSC® C083411

Kolibri 360 GmbH
Ferdinand-Jühlke-Straße 7
99095 Erfurt, Deutschland
produktsicherheit@kolibri360.de